Bridge and Structure Estimating

Other McGraw-Hill Books of Interest

BRANTLEY & BRANTLEY • *Building Materials Technology*
BREYER • *Design of Wood Structures*
BROCK ET AL. • *Field Inspection Handbook*
BROCKENBROUGH & MERRITT • *Structural Steel Designer's Handbook*
BROWN • *Foundation Behavior and Repair*
BROWN • *Practical Foundation Engineering Handbook*
CIVITELLO • *Construction Operations Manual of Policies and Procedures*
FAHERTY & WILLIAMSON • *Wood Engineering and Construction Handbook*
FOSTER ET AL. • *Construction Estimates from Take-Off to Bid*
GAYLORD & GAYLORD • *Structural Engineering Handbook*
LEVY • *Project Management in Construction*
MERRITT & RICKETTS • *Building Design and Construction Handbook*
MERRITT ET AL. • *Standard Handbook for Civil Engineers*
NEWMAN • *Design and Construction of Wood Framed Buildings*
NEWMAN • *Standard Handbook of Structural Details for Building Construction*
O'BRIEN • *Preconstruction Estimating: Budget Through Bid*
O'BRIEN • *CPM in Construction Management*
PALMER & COOMBS • *Construction Accounting and Financial Management*
PARMLEY • *Field Engineer's Manual*
RITZ • *Total Construction Project Management*
SHARP • *Behavior and Design of Aluminum Structures*
TONIAS • *Bridge Engineering*

Bridge and Structure Estimating

J. David Nardon

McGraw-Hill
New York San Francisco Washington, D.C. Auckland Bogotá
Caracas Lisbon London Madrid Mexico City Milan
Montreal New Delhi San Juan Singapore
Sydney Tokyo Toronto

Library of Congress Cataloging-in-Publication Data

Nardon, J. David
   Bridge and structure estimating / J. David Nardon.
      p.   cm.
   Includes index.
   ISBN 0-07-045669-0 (hardcover : alk. paper)
   1. Bridges—Design and construction—Estimates.   2. Structural engineering—Estimates.   I. Title.
TG153.N37   1995
624'.2'0299—dc20   95-23949
                                                                          CIP

McGraw-Hill

A Division of The **McGraw·Hill** Companies

Copyright © 1996 by The McGraw-Hill Companies, Inc. All rights reserved. Printed in the United States of America. Except as permitted under the United States Copyright Act of 1976, no part of this publication may be reproduced or distributed in any form or by any means, or stored in a data base or retrieval system, without the prior written permission of the publisher.

1 2 3 4 5 6 7 8 9 0   QM/QM   9 0 1 0 9 8 7 6

ISBN 0-07-045669-0

The sponsoring editor for this book was Larry Hager, the editing supervisor was Peggy Lamb, and the production supervisor was Pamela A. Pelton. It was set in Century Schoolbook by Estelita F. Green of McGraw-Hill's Professional Book Group composition unit.

Printed and bound by Quebecor Martinsburg.

This book is printed on acid-free paper.

McGraw-Hill books are available at special quantity discounts to use as premiums and sales promotions, or for use in corporate training programs. For more information, please write to the Director of Special Sales, McGraw-Hill, 11 West 19th Street, New York, NY 10011. Or contact your local bookstore.

   This entire publication is dedicated to the premise that the estimator's primary objectives are to achieve safety, quality, economy, and productivity. All examples and illustrations given are theoretical. The actual application of these means and methods must be given the attention to detail that the specific project, site condition, and operation warrants.
   No representations or responsibility is implied.

> Information contained in this work has been obtained by The McGraw-Hill Companies, Inc. ("McGraw-Hill"), from sources believed to be reliable. However, neither McGraw-Hill nor its authors guarantees the accuracy or completeness of any information published herein and neither McGraw-Hill nor its authors shall be responsible for any errors, omissions, or damages arising out of use of this information. This work is published with the understanding that McGraw-Hill and its authors are supplying information, but are not attempting to render engineering or other professional services. If such services are required, the assistance of an appropriate professional should be sought.

I-94 and M-39 Interchange, Allen Park, Mich. (*Courtesy of Michigan Department of Transportation.*)

Linn Cove Viaduct, Grand Father Mountain, N.C. Structure type, concrete segmental. Owner, National Park Service and Federal Highway Administration. (*Courtesy of Figg Engineering Group.*)

Throughout my career, there have been many people to whom I owe a great debt of thanks for my development of skills, dedication to commitment and quality, and integrity and honesty.

To my parents and grandparents for the guidance in establishing lifelong goals. To my mother for her endless support, and my father for the training, knowledge, and special work time together. To my father-in-law for the leadership morals and skills he taught me, and my mother-in-law for the hours of 1st edits of this book, and the constant reminder of the "midnight hour."

I would like to extend my appreciation to my employers for the trust and management advancement throughout my career.

Foremost, the two most special people in my life; my daughter Nichole and wife Paula. To Nichole, for the endless hours this book has consumed of our time together. Finally, to Paula; for always believing in me and being my sustenance, for keeping me focused, and for her dedication of countless hours in assisting with this project.

Walter B. Jones Bridge, Intracoastal Waterway, Hyde County, N.C. Structure type, conventional span, prestressed concrete girder and continuous structural steel. (*Courtesy of Federal Highway Administration.*)

Contents

Preface xv
Acknowledgments and Professional Credits xvii

Part 1 The Basics — 1

Chapter 1. Introduction to Estimating — 3
Types of Structures — 3
Company Structure — 4
Historical Data — 5

Chapter 2. Estimating Factors — 9
Equipment Ownership — 10
Labor Costs (Direct and Indirect) — 15
Labor Crew Determination — 18
Equipment Crew Determination — 21
Production Unit Factoring — 24

Part 2 Construction Systems, Methods, and Materials — 27

Chapter 3. Formwork and Falsework Applications and Systems — 29
Technical — 29
Applications for Conventional Formwork Systems — 84
Applications for Metal Panel Form Systems — 99

Chapter 4. Concrete Mixes, Technical Aspects — 105
Introduction — 105
Mix Design — 106
Placing and Finishing — 118
Curing Concrete — 125

Chapter 5. Equipment Requirements — 129
Primary and Major Equipment Units — 129
Secondary and Support Equipment Units — 143

Part 3 Fundamentals of Estimating — 145

Chapter 6. Scoping the Market — 147

Chapter 7. Contracts and Specifications — 149

Chapter 8. Contract Construction Drawings and Plans — 153

Chapter 9. Project Structuring and Format — 157

Chapter 10. Project Schedule — 159

Part 4 Takeoff and Cost Analysis Techniques — 161

Chapter 11. Foundation Preparation — 163

- Technical Section — 163
  - Introduction — 163
  - Definition — 163
  - Excavation — 163
  - Measurement and Payment — 167
- Takeoff Quantification — 168
  - Introduction — 168
  - Definition — 168
  - Takeoff — 168
  - Earthwork and Excavation Takeoff — 169
  - Hand Unit Time Cycle — 172
  - Excavator Time Cycle — 173
  - Production Factoring — 174
  - Earthwork Backfill — 174
  - Placement, Spreading, and Compaction — 178
  - Foundation Detailing — 179
  - Method 1. Undercut and Removal — 180
  - Method 2. Hand Chip and Dentil Work — 181
  - Method 3. Selected Removal — 183
  - Final Grading — 184
  - Dewatering and Subaquifer Preparation — 185
  - Questions of Estimate — 187
  - Cofferdam Cell Dewatering — 192
  - Foundation Preparation Summary — 196
- Component Takeoff — 196
  - Foundation Preparation — 196

Chapter 12. Foundation Piling — 199

- Technical Section — 199
  - Introduction — 199
  - Piling, Permanent Foundation, Specifications — 200
  - Types of Piling — 201
  - Sheet Piling, Permanent and Temporary — 231
- Takeoff Quantification — 238
  - Definition — 238
  - Introduction — 241
  - Takeoff — 250
  - Procedure — 251
  - Questions of Estimate — 252
  - Application — 253
  - Material Allocation — 263
- Component Takeoff — 268
  - Foundation Piling — 268

Chapter 13. Substructure Unit — 273

- Technical Section — 273
  - Introduction — 273
  - Definition — 273
  - Components — 274
- Takeoff Quantification — 281
  - Introduction — 281
  - Definition — 282
  - Takeoff — 282
  - Abutment Unit — 285

Pier Units	294
Retaining Wall Unit	303
Procedure	310
Questions of Estimate	311
Application	311
Operation Analysis, Abutments and Center Pier	313
Production Factoring, Time Duration	331
Material Allocation	332
Section Summary	333
Component Takeoff	333
Substructure Unit	333

Chapter 14. Beam Structure — 339

Technical Section	339
Introduction	339
Definition	339
Concrete Beams	340
Structural Steel, Girders	356
Takeoff Quantification	365
Introduction	365
Definition	366
Takeoff	366
Prestressed, Precast Concrete Beams	366
Structural Steel Girders	379
Procedure	394
Questions of Estimate	395
Application	396
Operation Analysis	398
Production Factoring, Time Duration	404
Material Allocation	406
Section Summary	407
Component Takeoff	408
Beam Structure Unit	408

Chapter 15. Superstructure — 411

Technical Section	411
Introduction: Conventional Superstructure	411
Definition	411
Introduction: Cast-in-Place Superstructure	427
Takeoff Quantification	439
Introduction	439
Definition	440
Takeoff Procedure	440
Superstructure Unit	441
Box Girder Unit	446
Superstructure Deck Section	449
Box Girder Section	483
Parapet and Median Rail	493
Procedure	496
Questions of Estimate	496
Application	497
Operation Analysis	499
Production Factoring, Time Duration	508
Material Allocation	509
Takeoff Quantification Section Summary	512
Component Takeoff	512
Superstructure Unit	512
Cast-in-Place Superstructure Unit	516

xii   Contents

Chapter 16. Approach Structure Unit, Concrete — 519

- Technical Section — 519
  - Introduction — 519
  - Concrete Approach Slab — 519
- Takeoff Quantification — 520
  - Introduction — 520
  - Definition — 521
  - Takeoff — 521
  - Parapet and Median Rail — 530
  - Curb and Gutter Section — 532
  - Procedure — 532
  - Questions of Estimate — 533
  - Application — 533
  - Operation Analysis — 534
  - Production Factoring, Time Duration — 543
  - Material Allocation — 543
  - Takeoff Quantification Section Summary — 545
- Component Takeoff — 546
  - Approach Structure Unit — 546

Chapter 17. Reinforcing Steel — 549

- Technical Section — 549
  - Introduction — 549
  - Reinforcing Steel and Wire Rope — 550
- Takeoff Quantification — 557
  - Introduction — 557
  - Definition — 558
  - Takeoff — 558
  - Procedure — 563
  - Questions of Estimate — 563
  - Application — 563
  - Operational Analysis — 570
  - Production Factoring, Time Duration — 578
  - Material Allocation — 579
  - Takeoff Quantification Section Summary — 580
- Component Takeoff — 580
  - Reinforcing Steel Unit — 580

Chapter 18. Structure Drainage — 583

- Technical Section — 583
  - Introduction — 583
  - Subsurface Drain Systems — 583
  - Superstructure Drain Systems — 584
- Takeoff Quantification — 585
  - Introduction — 585
  - Definition — 585
  - Takeoff — 585
  - Procedure — 589
  - Questions of Estimate — 589
  - Application — 590
  - Operation Analysis — 591
  - Production Factoring, Time Duration — 593
  - Material Allocation — 593
  - Takeoff Quantification Section Summary — 594
- Component Takeoff — 594

Contents xiii

Chapter 19. Slope Protection — 597

Technical Section — 597
  Introduction — 597
  Components — 597
Takeoff Quantification — 600
  Introduction — 600
  Definition — 600
  Takeoff — 600
  Procedure — 605
  Questions of Estimate — 605
  Application — 606
  Operation Analysis — 607
  Production Factoring, Time Duration — 612
  Material Allocation — 612
  Takeoff Quantification Section Summary — 614
Component Takeoff — 614

Chapter 20. Waterproofing and Joint Fillers — 617

Technical Section — 617
  Introduction — 617
  Waterproofing — 618
  Dampproofing — 620
  Waterstops — 620
Takeoff Quantification — 624
  Introduction — 624
  Definition — 625
  Takeoff — 625
  Procedure — 629
  Questions of Estimate — 629
  Application — 630
  Operation Analysis — 631
  Production Factoring, Time Analysis — 632
  Material Allocation — 633
  Takeoff Quantification Section Summary — 634
Component Takeoff — 634
  Structure Waterproofing — 634

Chapter 21. Exampled Component Cost Allocation — 635

Cost Estimate Analysis — 635
  Example 1. Pile Driving — 635
  Example 2. Abutment Wall Construction — 645

Part 5 Putting It Together — 667

Chapter 22. Material Specifications and Bid Preparation — 671

Material Specifications — 671
Bid Preparation and Closing — 671

Chapter 23. Low Bid and Award — 675

Low Bid — 675
Award — 676
Supporting Cause for Publishing — 677

Glossary   679
Index   687

Redwood Falls River Bridge, Minnesota (1935). Structure type, composite concrete arch. (*Courtesy of Federal Highway Administration.*)

Preface

Bridge and Structure Estimating guides the estimator with precise fundamentals and techniques of quantity takeoff and estimating procedures in the structure related field. This book was conceived with a direct, hands-on approach dealing with specific bridge and structure estimating. It is not a broad overview of general estimating nor does it lead to, or discuss in brief, other types of construction. It is fully intended to specialize in the field of bridges and structures. It was written as a detailed reference for the estimator to provide specific procedures, fundamentals, and techniques in estimating actual components of a bridge or concrete structure. It is a direct approach in means and method application, providing step-by-step procedures and information required to formulate a complete and accurate estimate. This book teaches a young upcoming estimator the basics needed or possibly suggests a few new schemes to the experienced estimator.

Bridge and Structure Estimating establishes the importance of hands-on field experience and the utilization of detailed facts needed to compile a competitive bid. The focus is on pulling together all the pieces needed to be a consistent estimator; i.e., selecting projects that fit your organization's operations, utilizing and/or acquiring the necessary equipment, appraising the manpower resources; and considering the current market arena. The following considerations are addressed in the succeeding chapters.

Can you do the project and be profitable?

Can you be precise and accurate on the preparation of the bid?

How can we utilize our expertise to gain an advantage?

Is the project a risk, and if so, can we calculate it?

What experience does the company bring to the project?

How will the project be financed?

What is the owners' payment schedule?

What are the insurance and bonding requirements?

Has consideration been given to new technology with efficiency and economics in mind?

The methods and procedures outlined are based on thoughts, ideas, and experiences. Certain facts will have to be examined based on the individuality of each company. The calculations contained are intended for estimating purposes and project budgeting. Individual judgment must determine the end result value due to variables particular to individual projects, company preferences, material characteristics, and performance efficiency. Specific fundamentals and techniques needed to acquire one's own bidding factors and strategies (not averaged historical data factors from hundreds of companies) are explained. The expertise and skills of a particular construction firm according to its ability to perform are demonstrated. A successful estimator cannot bid profitably using formulas acquired from averaging multiple company performances

Procedures required; from the initial concept of the project bidding process, through the hard dollar bid stage and closing, to the management aspects of the construction firm; are detailed. And lastly, the application of these fundamentals are applied toward a practical, a direct, and an effective operation within a specific project, presenting an understanding of the development of a component identification system.

On the technical side, emphasis is on structural components that will be useful with regard to practical applications. Professional engineers must: apply a budgetary cost to structures in order to view different aspects of design; have an understanding of the basic fundamentals of actual field construction techniques; realize equipment application, capacity, and availability; and understand the ever-changing technology in today's market. The designers must be fully aware of the procedures and components within their design to complement the construction field. Many times portions of, or components of, structures have a design that is theoretically improbable or not feasible to construct. Much time and money is often lost in the field during construction due to redesign.

Structure types focused on are: conventional precast and steel beam, flat slab, posttensioned segmental, cast-in-place, bascule, cable stayed, arch and truss designs, and specialty structures, all of which could be of land or marine use.

The methods and procedures outlined within are intended for estimating guidelines only. Discussions within the book associate directly with professionals of the bridge structure industry. It takes the reader step by step through the entire process of structure estimating and outlines specific examples and formulates a system of estimating techniques. Many variables will exist regarding specific performance on individual projects (including weather, site conditions, materials, equipment, experience factors, labor force, personal opinions and preferences, etc.). The formulas and activities presented are intended to give a framework along with a distinctive format. The intention is to develop individual interpretations and modifications necessary to adapt to the specific company at hand. The methods taught demonstrate the utilization of this system and techniques in any company structure, emphasizing fundamentals and a strong foundation.

J. David Nardon
February 15, 1996

Acknowledgments and Professional Credits

The authoring and completion of this book would not have been possible without the contribution of technical data, illustrations, charts, and pictures from the organizations and firms credited herewith. Much time, consideration, and coordination has been spent with this venture which has made this book a versatile and challenging project. My thanks and appreciation is extended to all who participated.

Supporting Firm	Subject and Contact
American Concrete Institute P.O. Box 19150 Detroit, MI 48219	Concrete Technical Support Mr. Robert Wiedyke
Associated Pile & Fitting Corp. Box 1048 Clifton, NJ 07014	Pile Points & Splices Mr. Frank Gillen
Barnes & Sweeny Enterprises P.O. Box 536 Novi, MI 48376	Reinforcing Materials Mr. George Barnes Mr. David "Scoop" Sherrill
Bethlehem Steel Corporation 1170 8th Avenue Bethlehem, PA 18016-7699	Steel Sheet Piling Mr. Henry Von Spreckelson
Bid-Well Corporation, Div. of CMI Box 97 Canton, SD 57013	Bridge Finish Machine Mr. Murray Rowe
D.S. Brown Company 300 East Cherry Street North Baltimore, OH 45872	Expansion Joint Devices Mr. Daniel Brown
Camlever Box 1249 Pomona, CA 91769	Concrete Buckets Mr. John Harris
Caterpillar Incorporated 100 NE Adams Street Peoria, IL 61629	Equipment Ownership Mr. Joe Wendland
Dayton Superior 721 Richard Street Miamisburg, OH 45342	Formwork Accessories Mr. Alan McClelland
Dywidag Systems International 320 Marmon Drive Bolingbrook, IL 60439	Posttensioning Systems Mr. Ron Bonomo

Acknowledgments and Professional Credits

Economy Forms Corporation
4301 NE 14th Street
Des Moines, IA 50316-0386

Concrete Formwork Systems
Ms. Jean Ann Rodibaugh
Mr. Keith Wilgef, Vice Pres.

Figg Engineering Group
424 North Calhoun Street
Tallahassee, FL 32301

Structure Illustrations
Ms. Linda McCallister

Grove Worldwide
Box 21
Shady Grove, PA 17256

Hydraulic Cranes
Mr. Doug Zoerb

Hartwig Manufacturing Corporation
5801 Packer Drive
Wausau, WI 54401

Structural Steel
Mr. Henry Wanserski

International Construction Equipment
301 Warehouse Drive
Matthews, NC 28105

Pile Driving Equipment
Mr. Bill Grier

Kiewit Construction Group Inc.
1000 Kiewit Plaza
Omaha, NE 68131-3374

Construction Firm
Mr. Brad Chapman V.P.

Linkbelt Construction Equipment
2651 Palumbo Drive
Lexington, KY 40683-3600

Lifting Cranes
Mr. Pete Wollison

Michigan Department of Transportation
P.O. Box 30050
Lansing, MI 48909

Standard Drawings
Mr. Steven Beck

Morgen Manufacturing Company
P.O. Box 160
Yankton, SD 57078

Concrete Conveyors
Mr. Drew Cope

National Construction Specialties
7742 Greenfield Road
Dearborn, MI 48126

Formwork and Concrete Accessories
Mr. William Robertson

Precast/Prestressed Concrete Institute
175 West Jackson Boulevard
Chicago, IL 60604

Precast Concrete Illus.
Mr. Sidney Freedman

Schwing America Incorporated
5900 Centerville Road
White Bear, MN 55127

Concrete Pumping Equipment
Mr. Brad Wucherpsennig

Shugart Manufacturing
P.O. Box 748
Chester, SC 29706

Construction Equipment
Mr. Frank Shugart

Topikal Incorporated
3390 Peachtree Road NE
Atlanta, GA 30326

Permanent Metal Formwork
Mr. Russ Woerheide

U.S. Department of Transportation
Federal Highway Administration
400 7th Street SW HFL-21
Washington, DC 20590

Vulcan Iron Works Incorporated
2909 Riverside Drive
Chattanooga, TN 37406-0402

Ralph Whitehead & Associates
1201 Greenwood Cliff
Charlotte, NC 28204

Technical Specifications & Photographs
Mr. G. A. Hay
Mr. Anthony F. Welch

Pile Driving Equipment
Mr. Warrington

Structure Photographs
Mr. Stuart Matthis

Percy V. Pennybacker Jr. Bridge, Lake Austin, Tex., 1984 First Place Federal Highway Administration's Excellence in Highway Design. Structure type, weathering structural steel arch. Owner, Texas Department of Highways and Public Transportation. (*Courtesy of Federal Highway Administration*)

Part 1

The Basics

West Cornwall Covered Bridge, West Cornwall, Conn. Structure type, wooden covered bridge. Owner, Connecticut Department of Transportation. (*Courtesy of Federal Highway Administration.*)

In this day and age of economic upturns and downswings in the construction market, an experienced estimator has to be able to cover all the bases. One must stay very aggressive but yet competitive in this highly volatile market and keep abreast of all the latest trends in equipment-oriented fields, laborsaving innovations, material applications, safety, and construction technology.

The estimator is the "bridge," linking the corporate arm to the field management and construction team. One must, of course, possess some knowledge of this type of construction and have the willingness to learn. One must be able to visualize the task, and have an understanding of equipment uses, capacities, labor functions, and production rates.

Chapter 1, "Introduction to Estimating," discusses types of structures, company structure, and historical data.

Chapter 2, "Estimating Factors," comprises five sections considered to be the core of estimating: equipment ownership, labor costs (direct and indirect), labor crew determination, equipment crew determination, and production unit factoring. It establishes costs and production factors of a specific company and relates them to the final hard dollar estimate.

Chapter 1

Introduction to Estimating

This book fully describes the objectives and fundamentals needed to become an estimator. It teaches the basics, the skills, and the requirements of good consistent quantity takeoffs, measurements, production estimates, pricing, and project skills. This introduction will start the flow of the chapters to follow. Each section begins with an overview of the estimating and takeoff techniques and outlines procedures needed to obtain the fundamentals for various types of concrete structure estimating.

The purpose of this scenario is not only to teach estimators to simply do quantity takeoffs of plans but also to fully educate them in the complete structure of a company, from corporate office to the accounting levels to the field construction team. This will prepare them for the "hard dollar estimates" and show the importance of open negotiations with the owners, to the finalization and the completion of the project. The book describes the importance of linking the corporate side to the field side.

Clearly outlined are the objectives needed to establish a reliable and practical system of fundamental estimating and quantity takeoff techniques.

Types of Structures

Throughout the book are illustrations of various types of existing bridge structures to enable the user to view these structures and comprehend their use, function, and design.

The illustrations within each section assist in the understanding of the various types of work discussed in the chapters, along with detailing various structure components. There is an illustrative view of all the various types of bridges and structures along with an explanation of the purpose, design, use, and economics of these structures.

Some examples of bridge structures are:

1. Conventional span, precast and steel
2. Arch design, concrete and steel
3. Truss design, steel and cantilever span
4. Flat slab
5. Cast-in-place concrete: cantilever, traveler, and falsework
6. Segmental, precast and posttension
7. Suspension bridges
8. Cable-stayed
9. Bascule

This creates an overview of the evolution of bridges in both construction technique and technology.

Company Structure

This section emphasizes the internal structure of a construction company. It teaches the estimator the importance of a well-structured office system, the limitations and growth potentials, along with experience and specialties associated with the firm, how the personnel of the organization ties the whole company together as one, and the key to structuring a good reputable firm. It shows how the financial status of the company plays a key role in estimating, from the asset value to the hard dollar financial status, in regard to bonding power and local, state, and federal prequalifications. This section also emphasizes the accounting systems of the company as to tracking costs and being of liquid status. It reflects how insurance markets affect construction firms and the estimators' responsibility of understanding liability and compensation insurance. It also shows affiliations with labor unions. Detail is given to the overhead system and its relationship to estimating. The understanding of plans and drawings is only a small part of becoming a good, experienced estimator. The section on company structure gives a full view of the corporate and field requirements.

The corporate side of the company structure comprises the owner(s), senior officers and corporate leadership, accounting personnel, and senior project management. This unit has the responsibilities of corporate decisions, accounting and finance procedures, business affair and operational duties, and final management of current and future projects.

The second tier of the company structure is middle management, which is a direct link to upper management. This unit is comprised of the direct division and regional project management, the estimating staff, general supervision, and plant maintenance supervisors. This unit has the responsibilities of future project procurement, project budget control, material and equipment acquisitions, direct project management, overall scheduling, and the function of the projects.

The third tier of the company structure is the field-level management and supervision, which is a direct link to the middle management. This unit consists of the day-to-day decision makers of each individual project, being the specific project managers, project engineers, and superintendents. The responsibilities include direct project scheduling, material deliveries and requirements, equipment procurement, and project performance and safety.

The fourth tier of the company structure is the general supervisors, field engineering, and individual crew supervisors. This unit answers directly to the project supervision and is responsible for the actual construction activities of the project. These personnel deal specifically with each structure component from start to finish.

The organization and structuring of a company, with qualified and responsible personnel and equipment, is essential to operating a profitable and prosperous enterprise. All levels of management, working as a team, are the nucleus of a well-orchestrated construction project.

Historical Data

This section highlights the important values of tracking and compiling past data of trends, completed projects, production factors, equipment changes, and various labor markets throughout a certain regional work area, and expands to further sections.

It touches on, and defines, the importance of the corporate requirements, and the field-level requirements, one of which is corporate and field data flow through management.

The corporate side handles the compiling of data in the areas of labor dollars in market regions and the labor and union affiliations, areas of profitability, geographical area situations, insurance fluctuations, and the general competition of an area, thus structuring overhead to fit. Tracking past projects with regard to the bidder's tabulations of prices, the similarities of projects in regard to the type of work bid, and marketing areas is discussed. Also, understanding the owner's estimated cost, how it was derived, and general market trends is addressed. It demonstrates the need of equipment tracking, from both the standpoint of the new cost and the maintenance and upkeep of your company's own fleet of equipment. This is essential for related ownership and operating costs. This, ultimately, has a great effect on estimating projects.

The field side deals solely with actual, direct project cost. This section emphasizes the importance of detail and accuracy in reporting field data from labor-hours and quantity production to material usage and equipment reporting. It emphasizes the need of reporting accurate and current data. The impact of weather conditions and the role they play on projects, as to the bid estimations versus the actual conditions, are also discussed.

This entire section, along with subsequent sections, structures estimators for the area of ever-changing formulas of cost and production with which they will be directly associated in estimating and bidding.

The method of record keeping for future reference is a critical component in the competitive field of construction. The historical data library of past performance, bid analysis, and activity procedures becomes a useful tool for a manager and estimator. The data library can be comprised of labor production and performance analysis, material characteristics and uses, specific equipment performance, component consumption such as fuel, hardware, and concrete overruns, and items specific to a certain region.

Performance data can be compiled from case histories on file and from actual projects currently being constructed. This information becomes the data for future reference. This will also prove to be an accurate method of compiling a project cost to budget analysis.

In establishing a system of tracking data for cost comparison and operation performance, a code of component accounts is defined. These reference the actual and specific component items for the project. The component account system identifies the operation components required for the quantity takeoff procedure, and follows through to the construction process for cost performance analysis, ultimately establishing the actual activity component craft-hour usage. The actual craft-hour usage is then ratioed to the actual field performed quantity for the specified component item, producing the craft-hour to unit production factor.

A code of accounts for historical data is defined from the structure unit initially, then branched to the component unit within the unit, then to the specific activity performance items within the component unit. For example, the structure unit for a bridge structure may be classified as the pier unit; the defined component unit is identified as a pier column. From this the specific required work operations are classified as form and strip columns, install embed items, place concrete columns, and rub and patch columns. A reporting and takeoff format is structured with the appropriate and defined unit of measurement:

Pier unit

    Operation: Formwork, install, 3′ round pier column, sf

    Operation: Embed item, install, 3′ round pier column, ea

    Operation: Place concrete, 3′ round pier column, cy

    Operation: Formwork, remove, 3′ round pier column, sf

    Operation: Rub and patch, 3′ round pier column, sf

This information can be established for specific regional analysis, certain project-type history, or labor performance market area. The recorded information must be tracked, recorded, and identified by the specific work component item and labor performance unit within the operation. This will build a database that is usable for any operation type, component item, and labor performance area of the contractor's demographic region.

A well-defined production performance accounting system will also identify actual site condition factors that may impact a future project's performance. The fact of specific component tracking and defining the actual conditions of the specific project must be considered. This may be item definitions of geographic characteristics, labor performances, certain material components distinctive to the area, and other specific features that are unique to the project.

The data must be accurate and concise to the identified component. From the time of the quantity takeoff, to the actual project performance, to the historical archives, to the calculation of component production factoring, back to the researching of similar component items for a bid analysis, this circle of tracking and establishing critical data must be followed through in a systematic and standard method. This must be defined and remain constant for the specific operation.

The historical data of past performance and cost allocation will be only as good as the information fed to it. Therefore, for the information to be a useful and reliable tool for future project analysis, the procedure set forth must be simple but accurate in a defined manner to easily track and record the required information.

Chapter 2

Estimating Factors

The first section, Equipment Ownership, contains the development of a specific company's equipment ownership and operating costs, which are derived from historical data and current cost development procedures. It covers types and sizes of equipment, new and used cost, purchasing policies, depreciations, insurance, and resale value. It teaches how to apply life cycles on major and minor equipment and how to classify them as such; how to apply direct operating costs such as fuel, parts, oil, and tires and tracks; and how to compile this to a daily cost. The other side of owning equipment is renting and/or leasing; thus consideration must be given to this in establishing costs and how to apply these methods to a company.

The second section, Labor Costs (Direct and Indirect), describes the costs associated with labor, and formulas to arrive at direct and indirect costs. Direct costs are actual wages paid to an employee, and the indirect costs are considered to be payroll taxes and insurances, fringe benefits, and any other required contributions of the employer. It also discusses how geographic areas and states differ in pay scales, understanding federal laws and work hours, the total labor cost involved in preparing an estimate, and the intensity of the labor force in bridge and structure work.

Both the third and fourth sections, Labor Crew Determination and Equipment Crew Determination, establish structuring average labor and equipment crews. They show how to set up work crews for bidding and how to formulate these into specific work items involved in this type of construction. They also demonstrate how to apply hard dollar costing to these crews once this "time saving" system is established. Also discussed is the ability to adjust crews, not only for regional or union demands but for special functions that may arise. They tie the specific equipment owned by a company to its costing, and not general informative off-the-shelf cost data that are not detailed specifically for individual use.

The fifth section, Production Unit Factoring, formulates production factors. This is the key element to competitive bidding. It shows how the data from past projects (historical) and task performance can be mapped into productive and competitive production factors. It identifies the applications of these in the use of moderate and average conditions relating to specific regions, and how to apply specific factors to a certain project. This theory is based on units produced versus craft-hours expended.

Within this section, the method of activity account identifying is discussed. A directory of codes for use of specific and like functions relating to the construction of bridges and concrete-related structures is established. It is the link to tracking similar information, and sets the fundamentals to be applied at any construction firm.

This entire section shows how to implement with specific historical data, acquire accurate bidding skills, and track future projects for the much-needed information.

This section is devoted to teaching an estimator how to adapt specific fundamentals to the unique features of any one company and its way of doing business. It shows the importance of understanding labor and equipment to better utilize them for what the intent is and its application.

Equipment Ownership

Many methods of calculating and allocating the ownership and operating costs of equipment exist. The ownership factors can be measured against allocation depreciations per year; hours used per year; or a straight-line charge-off over a time frame designated in years, months, days, and hours. The initial ownership cost of the equipment is defined as the cost of the base machine plus any attachments. The operating cost is defined as the necessary and required costs associated with fuel consumption, parts cost, lubricants and oils, tires and tracks, and any other defined costs required for the normal operation and performance of the equipment unit.

Regardless of the method of allocating owning and operating costs, the machine performance must ultimately be measured in unit cost per component, a measure that includes both production and associated costs. Purchasing reliable equipment, properly maintaining the equipment unit, and proper care and use are the key factors for long-term equipment use and service. The equipment unit is solely based on performance of cycle time and hourly production factors.

The efficiency of the equipment is also a consideration; a calculation factor of 85% productivity is recommended. An efficiency factor can be modified based on actual conditions and performance requirements. This will allow for the anticipated actual job efficiency, operator efficiency, and production characteristics.

The methods chosen for estimating equipment owning and operating costs may vary widely, depending on locality, industry practices, owner preferences, and intended usage and performance factors. The method discussed with this section, designed by Caterpillar Equipment Company and detailed in *The Performance Handbook* (by Caterpillar), when used in good judgment will provide reasonably accurate estimates for equipment ownership and operating costs. The listed formulas are intended as guidelines, basing the requirements on actual working conditions that determine the consumption of components for operating costs.

The identifiable conditions that must be acknowledged when estimating equipment performance are classified as "excellent," "good," "average," and "severe." These are factored on a basis of experience and comparison considering normal approximations of production.

Machinery users must balance productivity and cost to achieve optimum performance, in other words, to achieve the desired production and performance at the lowest possible cost. The approach most often used to measure equipment performance is a simple equation:

$$\frac{\text{Lowest possible hourly costs}}{\text{Highest possible hourly productivity}} = \text{top machine performance}$$

This section considers the cost aspect of performance. The hourly owning and operating costs for a given equipment unit can vary widely owing to the influence of many factors: the type of work required by the machine; the local usage component costs such as fuel, parts, tires and tracks, and lubricants; accessory and mobilization costs; shop and repair facility costs; and finance and interest rates. These are variable factors which need to be allocated with specific conditions.

The calculation of the hourly costs must be performed to the best reasonable degree of accuracy for which the machine will be expected to perform. This is governed by application and location. This procedure suggests the method of estimating hourly ownership and operating costs which will produce an accurate estimate. The following principles must apply to all calculations:

Equipment purchase prices must be obtained from local sources.

Calculations are based on the complete, equipped unit.

Multiplier factors are based on an equal comparison.

The designated applications and zones are by user's opinion.

The term hour is defined as performance time duration.

To first create a format to establish a systematic method of equipment cost procedures, an ownership form must be compiled. The equipment owner, in order to protect the investment and replace the unit, must recover over the machine's useful life an expected amount equal to the loss in resale value plus the associated costs of owning the unit including interest, insurance, and taxes. For accounting and estimating purposes, the estimated resale value must be calculated in during the ownership allocation procedure, which will establish the depreciation schedules in accordance with the equipment usage concern. This procedure is recommended to be performed by individuals who are appropriately educated in finance and taxes.

The equipment ownership and depreciation used in this procedure is based on simple straight-line write-off procedures based solely on the number of years or hours of expected and gainful unit use. It will be assumed that the consideration of depreciation, of owning and operating, is intended for and based on the useful life criteria, not tax write-off life. Factors other than operating conditions can influence the equipment unit's depreciation period, such as the owner's desire to accelerate the investment recovery, the purchase of the unit for a specific project duration, and local economic conditions.

Maintenance factors also play an important role in determining the economic life of an equipment unit. For example, operating conditions may suggest a 12,000-hour depreciation period, but poor maintenance may result in an economic decision of a 10,000-hour life. It must be understood that regular and qualified maintenance will enable the owner and user to retain the equipment unit to its fullest and economically expected life. Therefore, a knowledge of the intended use, operating conditions and maintenance practices, and special care concerns is essential in establishing the expected life duration for depreciation purposes.

The following format will provide a useful method of determining and calculating the ownership and operating costs associated with an equipment unit.

Ownership and operating cost format

Machine designation _____
Estimated ownership period (years) _____
Estimated usage (hours per year) _____
Ownership usage (total hours) _____

Ownership costs

1. Delivered price including attachments _____
2. Residual value at replacement _____% _____
3. a. Recoverable value (line 1 less line 2) _____
   b. Cost per hour (value divided by hours) _____
4. Interest costs: N = number of years

$$\frac{[(N+1)/2N] * \text{delivered price} * \text{simple interest \% rate}}{\text{Hours per year}} = \underline{\hspace{2cm}}$$

5. Insurance: N = number of years

$$\frac{[(N+1)/2N] * \text{delivered price} * \text{insurance \% rate}}{\text{Hours per year}} = \underline{\hspace{2cm}}$$

or $_____ per year/_____ hours/year = _____

6. Property tax: N = number of years

$$\frac{[(N+1)/2N] * \text{delivered price} * \text{tax \% rate}}{\text{Hours per year}} = \underline{\hspace{2cm}}$$

or $_____ per year/_____ hours/year = _____

7. Total hourly owning cost

(Lines 3b + 4 + 5 + 6) = _____

Operating costs

8. Fuel: Unit price * consumption =
9. Lubes, oils, grease, filters:

	Unit price * consumption		
Engine	_____	* _____	= _____
Transmission	_____	* _____	= _____
Final drive	_____	* _____	= _____
Hydraulics	_____	* _____	= _____
Grease	_____	* _____	= _____
Filters	_____	* _____	= _____
		Total =	_____

10. a. Tires: replacement cost/life in hours = _____
    b. Undercarriage:

    (Impact + abrasiveness + Z factor) * basic factor = _____

11. Repair reserve:

    (Extended use multiplier * basic repair factor) = _____

12. Special wear items: cost/life
13. Total operating costs

    (Lines 8 + 9 + 10a or b, + 11 + 12) = _____

14. Total owning and operating costs = _____
    Section 2 calculation, residual value at replacement
    Gross selling price:_____
        Less: a. Commission_____
              b. Make-ready cost_____
              c. Inflation_____
    Net residual value:_____
    Section 11 calculation, repair reserve conversion factors
        Labor rate ratio_____
        Parts cost ratio_____

1. **Delivered price:** The delivered price should include all costs associated with mobilization and setup at the initial delivery site, including taxes.

2. **Residual value at replacement:** All equipment will have some residual value at trade-in. Some methods of ownership may elect to depreciate the equipment value to zero, and others recognize a residual value at resale or trade-in. In many cases the potential value at resale or trade-in becomes a key factor in the purchasing decision. This may be a method of reducing the investment through recovery in depreciation charges.

   When resale or trade-in value is used in establishing owning and operating costs, local conditions and used equipment values must be considered. The key factor in resale or trade-in value is the number of accumulated hours on the unit, the type of life performance and work conditions, and the physical condition of the unit.

3. **Value recovery:** The delivered price less the estimated residual value results in the value required to be recovered through work. This divided by the estimated total usage hours produces the hourly cost required to equalize the asset value.

4. Interest: A method of recovering interest paid is disbursing it within the hourly charge of owning and operating cost. This is usually based on the average annual investment within the unit. The interest is considered to be the cost of using capital for the purchase of the equipment unit.

The interest is calculated over the number of years of use, realizing the average annual investment during the use period and applying the interest rate and the expected usage. If the specific insurance and tax costs are not known, the following formula is applied:

$$\frac{\text{Simple interest}}{N = \text{number of years}}$$

$$\frac{[(N+1)/2N] * \text{delivered price} * \text{simple interest rate \%}}{\text{Hours per year}}$$

5 and 6. Insurance and taxes: Insurance cost and property taxes can be calculated with one of two methods. If the specific annual cost is known, this figure should be multiplied by the estimated usage (hours/years) and used. If the specific insurance and tax costs are not known, the following formula is applied:

$$\frac{\text{Insurance}}{N = \text{number of years}}$$

$$\frac{[(N+1)/2N] * \text{delivered price} * \text{insurance rate \%}}{\text{Hours per year}}$$

$$\frac{\text{Property tax}}{N = \text{number of years}}$$

$$\frac{[(N+1)/2N] * \text{delivered price} * \text{tax rate \%}}{\text{Hours per year}}$$

8–11. Fuel and component consumption: Actual fuel consumption can be closely measured in the field and tracked for historical data. Initially, however, the fuel consumption must be estimated and predicted for the specific application. The application is determined by the engine load factor, which controls the engine fuel consumption. Assuming an engine is continuously producing full horsepower, it is operating at a load factor of 1.0.

Equipment units rarely operate at a load factor of 1.0 for extended periods. Idling, empty travel, and lesser load power usage rates will decrease and reduce the load factor. The fuel consumption is calculated by applying the estimated hourly consumption and multiplying the local fuel cost per measured unit, producing the hourly fuel cost.

The other operating costs are calculated and defined in a similar format and determination.

12. Special wear items: Costs for high-wear items such as hoist cable, boom welding, cutting edges, ripper units, and bucket teeth must be associated

separately. These costs vary widely depending on applications, materials, and operating techniques.

The previous listed examples can be formulated for individual equipment units or similar units within the fleet capturing an average unit cost.

A quick method of determining a straight-line ownership cost is determining the expected life duration in years. From this the expected monthly usage per year must be defined, and the expected daily usage per month. Within this period an estimated hourly usage determination must be identified, normally 1600 to 2000 hours per year.

An example would assume an expected life of 7 years. Within the yearly usage, it performs in a 10-month cycle, and a 20-day per month usage duration, utilizing an 8-hour-shift day. Therefore:

$$7 \text{ years} * 10 \text{ months per year} = 70\text{-month charge-off}$$

$$70 \text{ months} * 20\text{-day month} = 1400 \text{ usable days}$$

$$1400 \text{ days} * 8\text{-hour shift} = 11{,}200\text{-hour life}$$

Determining equipment ownership and operating costs are calculated and realized in many methods and formats. It must be understood that specific uses, depreciation methods, and long-term needs will be determined and utilized by each company's preference. The equipment fleet will be the company's largest single investment and asset.

Labor Costs (Direct and Indirect)

The labor cost associated with a project includes two forms of cost distribution, classified as direct labor and indirect labor, or payroll burden and fringe benefits. The direct labor is identified as the actual hourly wage paid to the employee, including overtime compensation. The indirect labor classification is defined as fringe benefits, payroll taxes, and payroll-driven insurance policies, payroll cost components that are paid by and are the responsibility of the company as an employer.

The direct labor costs are those paid to the employee in the form of hourly wages. These are formed from the base wage and overtime compensated wage adjustment. The overtime portion is calculated as a factor to the base wage. Dependent upon the region and local union agreements, the overtime compensation is calculated differently.

Depending on the region and owner or trade specifications, specific minimum wage determinations are made for skill classification and work performance. The carpenter classification may mandate a wage of $16 per hour in one region and $10 per hour in another region; and a cement finisher may be classified with a $17 wage for one area and a $9 wage for another.

Overtime compensation can be calculated as all hours in excess of 40 per workweek, or all hours in excess of 8 per work shift and 40 per workweek.

This calculation is normally factored at one and one-half times the base wage. In addition to this, some labor agreements specify that all hours in excess of 8 per work shift are calculated at one and one-half the base wage, and hours in excess of 10 per work shift are calculated at two times the base wage. Another wage compensation may be the requirement of show-up time, or guarantee time. These wage components, if specified and required, will be paid if the employee reports to work and the shift has been canceled, or if the employee starts work and the shift is cut short of a minimum specified hour guarantee.

The Federal Davis Bacon Wage Determination, International and Local Union Agreements, the U.S. Department of Labor Wage Decision, and other specified wage determinations clearly specify the minimum allowable wage for each project along with the overtime compensation requirements and special pay requirements.

The indirect labor components, considered fringe benefits and payroll burden, are calculated by various methods. Fringe benefits are components such as health and welfare compensation, monetary vacation allowance, pension and retirement funding, and in some geographic areas apprentice training and industry promotion funding. These benefits are normally paid by the employer. The vacation payment, since it is paid to the employee, is normally considered taxable income. The remaining benefits are considered as a contribution by the employer, for the employee, to other third-party management agencies and are not normally taxable. As with differing wage determinations and allocations, so may be the fringe benefits with respect to regional and trade specifications.

The fringe benefit calculation may or may not be formulated on a time and one-half basis. In other words, the benefits may be calculated on hours worked; for example, with a 50-hour workweek, the base wage is termed a 40 hour with 10 overtime calculation. The fringe benefits are based on a 50-hour calculation:

Base wage @ $16.00 hour
Overtime @ $1\frac{1}{2}$ factor over 40 hours (hours worked)
Fringe benefits:
   Health and welfare $2.00
   Pension $1.25
   Vacation $0.50
Calculation:
        Base wage: $16.00 * 40 = $640.00
        Overtime: ($16.00 * 1.5) * 10 = $240.00
                  Total wages = $880.00 (taxable)
        Vacation: $0.50 * 50 = $ 25.00 (taxable)
        Benefits: $3.25 * 50 = $162.50 (nontaxable)

The payroll tax portion of the employer contribution consists of the applicable components of social security tax, the employer share of state and federal income tax contributions, local tax contributions, workers' compensation insurance, and contractor public and property liability insurance. These forms of taxes and insurance are payroll-driven calculations and are based on a percentage of wage components without fringe benefits. Normally the base wage, the overtime compensation, and vacation wages are used in the formulation of the payroll tax burden calculation.

Workers' compensation is based on regional conditions, work type, and experience factor modifiers. The regional base is specific to the area of the work, the work type defines the insurance factor with respect to risk, and the experience modifier is solely based on the company's accident record and safety ability. This component is the most variable adjustment factor to a company's workers' compensation insurance rates.

The following is an example formula of payroll tax allocation:

Social security tax	@	7.65%
State unemployment tax	@	10.50% (with earning limits)
Federal unemployment tax	@	0.80% (with earning limits)
Workers' compensation	@	22.00%
Contractor public liability	@	3.00%
Contractor property liability	@	2.00%
Total		45.95%

Using the payroll calculation formula previously listed, the payroll tax burden is allocated as follows:

Calculation:	
Wages: $880.00 * 1.4595	= $1284.36
Vacation: $25.00 * 1.4595	= $ 36.49
Benefits:	= $ 162.50
Total employer wage concern	= $1483.35
Hours worked @ 50.0	
Average wage per hour: $1483.35/50	= $ 29.667

The "wage concern" to be identified by the employer for bidding purposes is the earned wages, benefits, and payroll taxes and insurance. The state and federal tax portions have earning limits that maximize the employer contribution amount. For bidding analysis and budget purposes, these limits are not considered; the full allocation is calculated against the wages.

The total wage concern is calculated for each employee as required and is divided by the hours worked within the appropriate pay period, providing an *average wage per hour* determination. This average wage is utilized for crew cost calculations within the crew formulation activity.

Within different regions, labor classifications, wage determinations, trade agreements, fringe benefit requirements, workers' compensation factors and modifiers, and state and local tax variances differ; the estimator must investigate this area of cost thoroughly and accurately in order to prepare an accurate and competitive cost analysis along with a realistic budget.

Labor Crew Determination

The formulation and sizing of labor work crews is based on the component operation for which they are required. The required component operation determines the number and skill type of personnel needed. Variation of these labor crew determinations is based on regional areas, trade agreements, production requirements, condition demands, and personal preference.

With certain trade agreements, there may be a requirement that the carpenter is permitted to perform only carpentry-related work, a cement finisher concrete-related work, etc., whereas in other regional work areas, a carpenter may be permitted to perform any type of function, a laborer can assist a carpenter within an area of expertise, and so on. As with wage determinations, these trade agreements, wage classification determinations, and skill requirements must be fully researched and understood prior to the bid, cost analysis, and crew determinations.

The labor production crews are compiled of the various crafts and trades required to adequately perform the defined activity. The various trade definitions include:

CS = crew supervisor

OP = equipment operators

CP = carpenters

CF = cement finishers

LB = laborers

TD = truck drivers

IW = ironworkers

RB = reinforcing steel placers (rod busters)

PD = pile drivers

Along with the primary classifications of the general skills, there is often a secondary classification of skill-level descriptions for each trade. The secondary classifications include skilled and unskilled, helpers and assistants, or class and group levels for the trades.

For the general project conditions, there are personnel classifications of

PM = project manager

PS = project superintendent

PE = project engineer

GS = general supervisor

PC = party chief

IP = instrument person (survey personnel)

RP = rod or chain person (survey personnel)

EM = equipment mechanic

OL = oilers

OS = office staff

Each labor crew must be properly sized and formed to provide the most productive and cost-efficient grouping possible. The crew supervisor selected for the specific labor crew should be classified by the controlling trade function of the crew: carpenter supervisor for a forming crew, cement supervisor for a concrete placement crew, etc. The production base for the operation is prorated and factored utilizing certain key trade personnel within the crew along with the required support people. The craft-hour factor utilization is solely based on the expended hours against the quantities performed. The labor crew must be determined for the actual trades required to perform the operation, an example being a concrete placement crew which has six personnel, performing a placement rate of 20 cubic yards of concrete per hour, for a cycle of an 8-hour shift, which will be formatted as follows:

Labor crew function, concrete placement crew, footer

Trade type	Number of personnel	Shift duration	Total hours
CS-Supervisor	1.0	8.5	8.5
OP	1.0	8.0	8.0
CF	2.0	8.0	16.0
LB	2.0	8.0	16.0
Total	6.0		48.5
Average hours per shift:		8.083	

The labor crew consists of one crew supervisor, skilled as a cement finisher, an operator for the placement crane, two cement finishers to finish the concrete, and two laborers to assist with the placement, vibration, and curing. This labor crew was formed by visualizing the required operation or structuring it based on past experience, taking into account the required craft.

The production assumes the crew will be productive for the full shift. The crew supervisor is allotted an additional ½ hour per shift for paperwork, tool gathering, site check and preparation, etc.

Once the labor crew is structured, the associated cost is applied to arrive at an average wage per craft-hour relative to the specific labor crew. This is calculated using the methods to determine the total wage concern, discussed in the previous section (Labor Costs).

The required cost allocations are the base wage for each craft, the overtime compensation method, the applicable fringe benefits, and the appropriate payroll tax and insurance factors.

The Basics

For this scenario, the following criteria are used:

The overtime compensation is based on 8 hours per shift and 40 hours per week. The work is performed in a region that does not have labor agreements or fringe benefit allocations. The payroll tax and insurance are as follows:

Social security tax	@	7.65%
State unemployment tax	@	4.50% (with earning limits)
Federal unemployment tax	@	0.80% (with earning limits)
Workers' compensation	@	15.00%
Contractor public liability	@	3.00%
Contractor property liability	@	2.00%
Total		32.95%

The base wage determination, as specified by regional wage information, is as follows:

CS = crew supervisor	@	$15.00
OP = equipment operators	@	$12.00
CF = cement finishers	@	$11.00
LB = laborers	@	$ 9.00

The average wage per hour is determined next by calculating the standard and premium time wages, by the appropriate hours, to arrive at an average wage per hour.

	Average wage/hour
CS @ [($15.00 * 8)+($15.00 * 1.5 * 0.5)]/8.5 =	$15.44
OP @ ($12.00 * 8)/8 =	$12.00
CF @ ($11.00 * 8)/8 =	$11.00
LB @ ($9.00 * 8)/8 =	$ 9.00

From the given information, the labor crew can be cost allocated for the wage concern appropriate for the project.

Trade type	Number of personnel	Shift duration	Total hours	Average wage/hour	Crew cost wages
CS-Supervisor	1.0	8.5	8.5	$15.44	$131.24
OP	1.0	8.0	8.0	$12.00	$ 96.00
CF	2.0	8.0	16.0	$11.00	$176.00
LB	2.0	8.0	16.0	$ 9.00	$144.00
Total	6.0		48.5		$547.24

From this calculation, the payroll tax and insurance factor are allocated. This is applied by multiplying the burden factor by the total base wages, then summarizing the two amounts to arrive at the budget wage concern.

```
$547.24 * 0.3295 (32.95%)              = $180.32
$547.24 + $180.32                      = $727.56 wage concern
Average hours per shift: 48.5/6        =    8.083
Average wage per crew hour: $727.56/48.5 = $ 15.00
```

The average cost per crew hour for the determined labor crew will become the multiplier for the craft-hour production factor that determines the unit labor component cost.

The labor crew format must be given consideration and realistic production factors. Many labor crews may be structured in a similar manner but utilize different trade allocations and appropriations.

Equipment Crew Determination

As with labor crew structuring, the formulation and sizing of equipment crews will be based on the component operation for which they are required. The required component operation will determine the number of units and the equipment type and classification needed. Based on the production requirements, condition demands, and personal preference, the variation and structuring of these equipment crew determinations will differ.

There are many types of equipment with similar functions and performance, and there are many types with special and unique features and performances. As with labor crews, these equipment crew determinations and performance requirements must be fully researched and understood prior to the bid, cost analysis, and structuring.

The equipment crew structure should be formed in a manner equal to the operation for which the labor crew is structured. Each labor crew should have a comparable equipment crew. This method forms a systematic estimating structure that allows the estimator to quickly form the crew performance, both labor and equipment, for the project.

The equipment crews will be compiled of the various types and capacities required to adequately perform the defined activity. The equipment will be categorized first by group, then by component size and capacity. The various group category definitions will include:

TK = trucks

MC = maintenance equipment

CR = cranes and hoisting

EM = earthmoving equipment

AC = air compressors

PD = piling equipment

CF = concrete finishing equipment

MS = miscellaneous support equipment

Along with the primary classifications of the general equipment type, there will be secondary-level descriptions for each category. The secondary classifications will include unit type and capacity by class and group levels for each unit.

Each equipment crew must be properly sized and formed to provide the most productive and cost-efficient grouping possible. The equipment selected for the specific crew should be classified by the controlling component function of the crew: cranes for hoisting requirements, concrete finishing equipment for concrete placement crew, etc. The production base for the operation will be prorated and factored utilizing certain key component units within the crew along with the required support units. It is recommended to assign the structured equipment crew with an identifier code.

The unit cost for each piece of equipment with the crew, for the shift duration required, will be the total "cost concern" associated with the unit as described in the section on Equipment Ownership. This cost concern must include the ownership cost and operating costs required for each piece. The ownership cost must include the purchase cost, financing cost, depreciation allowances, and trade revenue. The operating costs are considered as fuels, parts, oils, tires and tracks, and other component units required for the normal daily operation of the equipment unit.

The craft-hour factor utilization and productivity factor is based solely on the quantities performed by the chosen crew, both labor and equipment. The crew must be determined for the actual equipment units required to perform the operation, an example being that of a concrete placement crew, performing at a placement rate of 20 cubic yards of concrete per hour, for a cycle of an 8-hour shift; it will be formatted as follows:

Equipment crew function, concrete placement crew, footer

Equipment type	Number of units	Shift duration	Total hours
$\frac{1}{2}$-ton pickup	1.0	8.5	8.5
5-ton flat-bed	1.0	8.0	8.0
30-ton crane (hydraulic)	1.0	8.0	8.0
Loader/backhoe	1.0	2.0	2.0
$\frac{3}{4}$ cy concrete bucket	1.0	8.0	8.0
Tools	1.0	8.0	8.0
Total	6.0		42.5

The crew consists of the necessary and required equipment to perform the operation. This equipment crew was formed by visualizing the required operation or structuring it based on past experience and requirements.

The production is based on the equipment crew and assumes it will be productive for the full shift. The pickup truck for the crew supervisor is allotted an additional $\frac{1}{2}$ hour per shift, as with the supervisor. The usage and performance are factored on a full shift with an equal proportionate rate applied to each hour worked.

Once the crew is structured, the associated cost is applied and calculated to arrive at a total cost for each designated shift, relative to the specific crew.

This is calculated using the methods to determine the total cost concern for equipment ownership and operating, discussed in the section on Equipment Ownership.

The required cost allocations are the owning and operating allocation for each unit. The equipment cost concern is based on the defined operating hours, or hours worked, with no additional compensation for overtime. From the equipment ownership factoring, each unit is allocated an ownership rate based on an 8-hour shift. Any variation above or below this should be prorated accordingly.

An example follows, using assumed owning and operations costs for each unit described, based on an 8-hour shift.

Equipment type	Units	Owning and operating cost	Cost/hour
½-ton pickup	1.0	$ 24.00	$ 3.00
5-ton flat-bed	1.0	$ 42.00	$ 5.25
30-ton crane (hydraulic)	1.0	$240.00	$30.00
Loader/backhoe	1.0	$ 60.00	$ 7.50
¾ cy concrete bucket	1.0	$ 10.00	$ 1.25
Tools	1.0	$ 62.00	$ 7.75

From the given information, the specific equipment crew can be cost allocated for the cost concern appropriate for the project.

Equipment type	Number of units	Shift duration	Total hours	Average cost/hour	Crew cost concern
Pickup	1.0	8.5	8.5	$ 3.00	$ 25.50
Flat-bed	1.0	8.0	8.0	$ 5.25	$ 42.00
Crane	1.0	8.0	8.0	$30.00	$240.00
Backhoe	1.0	2.0	2.0	$ 7.50	$ 15.00
Concrete bucket	1.0	1.0	8.0	$ 1.25	$ 10.00
Tools	1.0	8.0	8.0	$ 7.75	$ 62.00
Total	6.0				$394.50 shift

To determine the component activity budget, the total operation cost concern for the equipment requirement can be calculated by multiplying the shift cost by the required shifts for each activity.

For a placement operation requiring 1.30 shifts, the cost concern of $394.50 would be multiplied by 1.30, producing an operation budget of $512.85.

The equipment crew format must be given consideration for realistic ownership and operating rates. Many equipment crews may be structured in a similar manner but utilize different capacity unit allocations and appropriations.

The equipment cost associated with the entire project should be evaluated by recouping the direct cost associated with each piece to determine that a full budget is allocated and included for the project. There also may be an over budget based on time required; therefore, a reduction adjustment can be made in the bid closeout process.

Production Unit Factoring

The theory of this method, production unit factoring, is calculating the production performance for a defined time period for a specific crew within an operation. The key identifier within the labor crew which controls the production performance is the skilled personnel. The support and unskilled personnel are designated to aid the skilled personnel and are not calculated as actual production promoters.

The second method of production factoring is computing historical data of craft-hours consumed for a defined unit of measured quantity for each specific activity. Both methods develop a factor that defines the required craft-hours for each unit of component measurement; i.e.,

Required craft-hours per square foot

Required craft-hours per square yard

Required craft-hours per cubic foot

Required craft-hours per cubic yard

Required craft-hours per each unit

The craft-hour determination must be translated for a specific unit of measure. The calculation is performed by dividing the designated craft-hours by the operation quantity producing the production factor. This factor, multiplied by the required quantity, allocates the required craft-hours needed to perform the operation. This method can be performed separately for similar operations throughout the project that require differing performance time durations.

Developing primary production factors

To perform the primary method, first develop the required labor crew in number of personnel (*crew manpower*). The next step is to define the *shift duration* and unit of measurement (*production unit*). From this the total *quantity requirement* for this specific operation, component, or project must be defined. This quantity is determined from the operation quantity takeoff procedure. When applicable, the *formwork type* and *forming condition* must be identified. This will determine the production performance required for certain form systems.

The following procedures will develop the actual production defining and performance calculations. The first procedure will be to define the duration, in minutes, required to perform one unit of the component for the operation, identified as *unit cycle time*. Following this, the *productivity factor* will be established, or the amount of productive time the crew will perform under normal working conditions. This procedure is determined as minutes per hour. Once these two performance components have been determined, the *cycles per hour* can be established. This is accomplished by dividing the productivity factor by the cycle time, producing the related *production performance per hour* for the component unit.

The subsequent group of procedures will develop the shift production, the required shifts, and the total required craft-hours, producing the production operation factor. The *production per shift* will be established by multiplying the unit production per hour by the defined hours per shift, allocated by specific unit of measure of performed quantity per shift. To determine the *required number of shifts* needed to perform the specified activity, divide the defined activity quantity by the allocated production per shift. From this the *total craft-hours* required will be determined by multiplying the designated number of shifts by the designated hours per shift by the required personnel per crew.

The final result, the *activity production factor,* is determined by dividing the total craft-hours by the operation quantity. The cost allocation for the labor is allocated by multiplying the average wage cost per craft-hour by the total craft-hours required to perform the operation. The unit labor cost is derived by multiplying the operation production factor by the average wage cost per craft-hour. To verify the accuracy of the calculation method, multiply the operation production factor by the operation quantity defining the required operation craft-hours.

The example shows the primary factoring method for activity production factor calculation.

Operation: interior deck, formwork placement, conventional

Crew members	= 10.0 each
Shift duration	= 8.0 hours
Production unit	= square feet
Project quantity requirement	= 2400.0
Formwork type	= wood
Forming condition	= good, 4
Unit cycle time	= 0.70 min/sf
Productivity factor	= 50 min
Cycles per hour: 50/0.70	= 71.43 cycles
Production per hour	= 71.43 sf/hour
Production per shift: 71.43 * 8	= 571.44 sf/shift
Required shifts: 2400.0 sf/571.44 sf	= 4.20 shifts
Total craft-hours: 4.20 * 8 * 10	= 336.0 craft-hours
Operation production factor: 336.0/2400.0	= 0.140
Operation factor =	= 0.140 craft-hour/sf
Verification: 0.140 * 2400.0	= 336.0 craft-hours

Developing historical production factors

Developing operation production factors from historical data entails an accurate and precise record-keeping system. This method is developed from the account tracking of specific operations for the expended craft-hours required to perform the quantity specified. Each operation should be identified, normally from the original takeoff procedure, and accurately tracked and monitored throughout the course of the work.

Once the operation has been completed, the total craft-hours expended can be divided by the component quantity, field verified, to define the actual operation production factor (45 craft-hours divided by 250 sf will define a factor of 0.180 craft-hours per square foot required to perform the work).

The craft-hour usage must be the total expended for the performance of the identified operation for the limits and parameters established. These operation production factors can be compiled in a database spreadsheet and categorized by similar component operations, geographic regions, demographic labor areas, or any definable method required by the company.

From these records of establishing the production performance factors from historical projects, the factor method can be applied to a current similar project to determine the required craft-hours for specific operations. This can be performed in lieu of or as a verification and check to the method of developing a primary production factor.

Part

2

Construction Systems, Methods, and Materials

George Westinghouse Memorial Bridge, Pittsburgh, Pa. Structure type, concrete spandrel arch. Historical landmark (built 1931). (*Courtesy of Federal Highway Administration.*)

Chapters 3, 4, and 5 contain the three primary types of equipment categories in bridge and concrete structure construction. Forming systems, pouring and placing systems, and lifting systems play an important role in a company's advantages to their estimators' applications to a project.

Chapter 3, "Formwork and Falsework," shows how to determine, analyze, and cost the various types of systems used in today's market along with proper application procedures, the most advantageous and the most cost-effective to use. Ownership costing is also discussed.

Chapters 4 and 5 follow the same scenario only with concrete placing and lifting equipment. There is much discussion along with detailed illustrations showing types, capacities, and application procedures. Other minor and specialty equipment are thoroughly discussed in other chapters dealing with their individual use.

Chapter 3

Formwork and Falsework

Applications and Systems

Technical

The formwork and falsework systems are the elements that form, support, shape, and contain the concrete components of a bridge and concrete structure.

Form systems

The form system must be able to withstand, not only the load of the fresh concrete, but also the working load of any personnel and equipment. This activity consists of designing, constructing, and removing the form systems and falsework that temporarily support structural concrete, cast-in-place girders, and other structure elements and components. The form system will remain in place until the concrete has sufficiently cured and is able to support itself according to the project plans and specifications.

For the form system and falsework selected, the drawings and design must detail the safe construction of the system and provide the adequate and necessary rigidity, the support required for the imposed loads, the finished structure within the required lines and grades specified, and the desired finished surface required. The following list details a selective procedure for the appropriate form system and falsework units.

1. Define and identify the maximum applied structural load to the form system and/or falsework.

2. Describe and precisely detail all proposed and intended materials.

3. Define all recommended working loads imposed on the form system and falsework.

4. Evaluate the form system and falsework and ascertain whether the physical properties and material conditions are able to support the designed loads.

5. Furnish calculations and designs to assure the proposed system will support the imposed initial concrete pressures and long-term loads.
6. Provide an outline of the intended concrete placement procedures and identify the equipment, labor, and procedures anticipated for duration of the operation, along with the placing sequence and joint locations.
7. Detail the concrete volume placement rate and design pressures for each placement.
8. Identify and provide design calculations of the structure falsework with all proposed stresses and deflections within the system and load-supporting members.
9. Define and include the anticipated settlement and "crush" of the falsework and form system. The falsework design must assure there will be no differential settlement of the supporting deck slabs and overhangs between the girders and deck forms during the concrete placement procedure. The deflection and required camber must be designed to compensate all imposed loads.
10. The embedded support systems for the form system must be designed and detailed for the required support of the concrete deck slabs and overhang members.
11. The required falsework protection and additional strengthening members must be detailed for areas needing the identified system.
12. The intended form system and falsework calculations must include the steel erection procedures along with sufficient detail to substantiate the loads and geometry.
13. The proposed anchorages and ties must be detailed.
14. Removal and dismantling procedures must be accurately detailed and outlined, and sequenced in a manner to accommodate the concrete curing and loading details.

A form system is an integral part of bridge construction. A properly chosen form system will affect not only the cost but also the production duration. Form systems and falsework can be of different materials and components. The most common types of form systems are wood-framed conventional, metal-framed handset, and metal gang panels. These form systems are versatile and durable and lend themselves to many configurations and uses.

The two basic uses of a functional form system are to support and form vertical structural members or to support and form horizontal structural members. The formwork system for the vertical members is mostly used in the substructure portion, consisting of footers, abutment units, pier units, and walls. The horizontal bridge member formwork system is most commonly used in the superstructure portion, consisting of interior and cantilevered deck units, cast-in-place units, and beam structure diaphragm units. The design and load transfer of these systems differ greatly depending on the use and construction intent.

Vertical. For vertical form systems and falsework, horizontal loads of differing nature will exist. To initiate the design for horizontal loads, an assumed value will be used on falsework towers, bents, frames, and other vertical structures to verify lateral stability. The assumed horizontal load force is the sum of the actual horizontal loads implied by equipment, construction sequence, other industry causes, and allowances for wind. A minimum load, however, should be not less than 2% of the total supported dead load at the specific location under consideration.

The minimum wind allowance for vertical heavy-duty falsework members is calculated by the sum of the products or components having the wind impact area. From this, the applicable wind pressure value for each height zone and shape factor is applied. The wind impact area is defined as the total projected area of all elements, facing the normal applied wind.

The minimum wind allowance for all other types of form systems and falsework, including that supported by shoring, is the sum of the members of the wind impact area and the applicable wind pressure value for each height zone. The wind impact area will be defined as the gross projected area of the form system or falsework, and the unrestrained portion of the permanent structure, excluding the areas previously calculated for the heavy-duty units (see Tables 3.1 and 3.2)

The design and use of the vertical form system and falsework will have sufficient rigidity to resist the assumed horizontal load without a vertical dead load (without frictional resistance).

TABLE 3.1   Minimum Design Wind Pressure—Heavy-Duty Shoring and Falsework

Height zone, ft	Rated at pounds per square foot	
	Normal location	Adjacent to traffic
0	15	20
30–50	20	25
50–100	25	30
Over 100	30	35

TABLE 3.2   Minimum Design Wind Pressure—Normal Formwork and Falsework

Height zone, ft	Rated at pounds per square foot	
	Normal location	Adjacent to traffic
0	1.5 Q	2.0 Q
30–50	2.0 Q	2.5 Q
50–100	2.5 Q	3.0 Q
Over 100	3.0 Q	3.5 Q

Note: To determine Q, tabulate as $Q = 1 + 0.2W$, but not more than 10. In the preceding formula, W is the width of the formwork and falsework system in feet measured in the direction of the wind force being considered.

Horizontal. Within the horizontal form system exist vertical loads, defined as dead loads and live loads. The dead loading of a form system includes the weight of the concrete, reinforcing steel, the form system, and the falsework. In addition to this, any increased or readjusted loads and forces caused by procedures such as prestressing will be included within the design of the form system and falsework.

For the proper design and evaluation of the horizontal form system, the following criteria must be met. The first area to define is the *rated load,* or the maximum load that can be applied to a hanger assembly or overhang bracket. From this, a *safe working load* (or SWL) must be established.

The *ultimate load* is the average load, or force, at which an item fails or will no longer support or carry a load.

The next criteria to define are the *dead load* and *live load* weights. The dead load will be defined as the weight of the concrete and reinforcing steel combined with the formwork itself. The concrete mass being supported by the horizontal form system and falsework will be considered a fluid dead load with no ability to support itself. The assumption of a unit weight of a minimum of 150 pounds per cubic foot will be used for the density of normal concrete and reinforcing steel. The concrete and steel load calculation is performed using the assumed weight of 150 pounds per cubic foot. The unit weight of the formwork must then be calculated.

The live load will be defined as the additional loads imposed during construction such as material storage, crew, and equipment. Consider the live loads to be the maximum actual weight of the equipment, material stored, and crew, to be supported by the form system and falsework. This calculation will be applied as concentrated loads at the point of contact; plus a uniform load of (recommended) not less than 20 pounds per square foot applied over the supported area, and 75 pounds per lineal foot applied at the outside edge of the deck cantilevered section, plus the finish machine screed load ratio.

The proportioned finish machine and apparatus weight is calculated by identifying the total operating weight of the deck surface finish machine and distributing it equally to the exterior loading point or area of the deck form system.

The *design load* is defined as the dead load plus the live load per square foot of net form contact area.

The *impact load* is defined as the resulted impact loading of discharged concrete, or the starting and stopping of construction equipment on the formwork. This load can be several times the dead load.

A *safety factor* is a term denoting the theoretical capability which has been determined by dividing the ultimate load of the product by its rated load. This is expressed as a ratio, for example, 2 to 1 factor.

The *bay span* width is the actual dimension the formwork is required to carry a load, between beams. The *overhang dimension* is the dimension the proposed deck extends past the exterior beam.

Horizontal formwork charts

The following charts were supplied by Dayton Superior, *The Bridge Deck Forming Handbook.*

Formwork and Falsework 33

Safety Notes

Factors Affecting A Products Rated Load

Failure of coil bolts and coil rods is generally caused by excessive wear on threads, field modification or bending and restraightening. It shall be the responsibility of the user to continually inspect coil bolts, coil rods and other related items for wear and to discard the parts when wear is noted. Do not straighten bent coil bolts or coil rods, rather discard and replace them. Also discard any coil bolt or coil rod known to have been used at loads of 70% of ultimate strength or more. Such coil bolts or coil rods may have stretched sufficiently to become brittle hard. Every user must establish a control program that replaces coil bolts or coil rods after a predetermined number of uses, regardless of their exterior appearance.

All reuseable items shown in this handbook are subject to wear, misuse, overloading, corrosion, deformation, intentional alteration and other factors which may affect the products safe working load. Therefore, all reuseable items must be regularly inspected by the user to determine their overall condition and whether the item may be used at the rated safe working load or if it should be removed from service. The frequency of inspection is dependent upon factors such as, frequency of use, period of use, environment and is best determined by the user consistent with good construction practice.

When in doubt about the proper use or installation of Dayton Superior's bridge deck forming accessories and hardware, contact Dayton Superior for clarification.

Failure to do so may result in exposure of workers to safety hazards, resulting in the possibility of injury or death to workers in the vicinity of the bridge site.

C-50 Bridge Overhang Bracket and C-60 Pres-steel Hangers

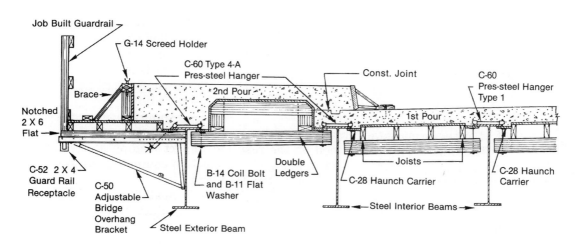

Typical Section Through Bridge Deck

4-83B

34 Construction Systems, Methods, and Materials

General and Technical Information

Technical Data—Plywood

Data based on information supplied by the American Plywood Association. For plywood used with face grain parallel to spacing, the tabulated spacings may be used with Plyform Class 1, Structural 1, Exterior A-B, Exterior B-B, and Exterior B-C or equivalent grades of plywood. For face grain perpendicular to spacing, tabulated spacings are suitable only for use with Plyform Class 1 or Structural 1 grades or their equivalent.

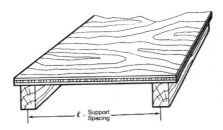

Plywood used the strong way
(Face Grain parallel to spacing)

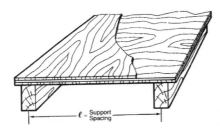

Plywood used the weak way
(Face Grain perpendicular to spacing)

	PLYWOOD DATA				
Thickness	Approximate Weight lbs.		Plies	Minimum Bending Radii	
	4x8 Sheet	Square Foot		Across Grain	Parallel to Grain
½"	48	1.5	3	6 Ft.	12 Ft.
⅝"	58	1.8	5	8 Ft.	16 Ft.
¾"	70	2.2	5	12 Ft.	20 Ft.
1"	96	3.0	5	—	—

Safe Spacing in Inches of Support for Plywood Sheathing, Continuous Over Three or More Spaces.								
Design Load of Concrete Pounds per Square Foot	$f = 1{,}900$ psi; Rolling Shear = 72 psi $E = 1{,}500{,}000$ psi							
	Sanded, Face Grain Parallel to Spacing				Sanded, Face Grain Perpendicular to Spacing			
	½" (3 Ply)	⅝" (5 Ply)	¾" (5 Ply)	1" (5 Ply)	½" (3 Ply)	⅝" (5 Ply)	¾" (5 Ply)	1" (5 Ply)
75	20"	23"	26"	31"	10"	14"	18"	25"
100	18"	21"	24"	29"	9"	13"	17"	23"
125	16"	20"	23"	27"	8"	12"	15"	22"
150	15"	18"	21"	26"	8"	11"	14"	21"
175	15"	17"	20"	25"	7"	10"	14"	20"
200	14"	17"	19"	24"	7"	10"	13"	19"
225	13"	16"	18"	23"	7"	10"	13"	18"
250	13"	15"	17"	23"	7"	10"	12"	17"
275	12"	15"	17"	22"	6"	9"	11"	16"
300	12"	15"	17"	22"	6"	9"	11"	16"

Maximum deflection is $\ell/360$, but not more than 1/16".

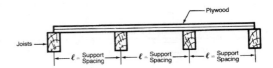

Safe Spacing in Inches of Supports for Plywood Sheathing, Single Spacing.								
Design Load of Concrete Pounds per Square Foot	$f = 1{,}900$ psi; Rolling Shear = 72 psi $E = 1{,}500{,}000$ psi							
	Sanded, Face Grain Parallel to Spacing				Sanded, Face Grain Perpendicular to Spacing			
	½" (3 Ply)	⅝" (5 Ply)	¾" (5 Ply)	1" (5 Ply)	½" (3 Ply)	⅝" (5 Ply)	¾" (5 Ply)	1" (5 Ply)
75	16"	19"	22"	26"	8"	11"	15"	21"
100	14"	17"	20"	25"	7"	10"	13"	19"
125	13"	16"	18"	23"	7"	9"	12"	18"
150	12"	15"	17"	22"	6"	9"	12"	17"
175	12"	14"	16"	21"	6"	8"	11"	16"
200	11"	14"	16"	20"	6"	8"	11"	15"
225	10"	12"	15"	19"	6"	8"	10"	14"
250	10"	12"	14"	18"	6"	8"	10"	14"
275	10"	12"	14"	18"	5"	7"	9"	14"
300	10"	12"	14"	18"	5"	7"	9"	13"

Maximum deflection is $\ell/360$, but not more than 1/16".

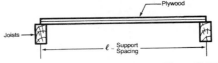

REV. 4-85B 12-84B

General and Technical Information

Formwork and Falsework 35

Technical Data — Joists

Safe Spacing, Inches, of Supports for Joists, Continuous Over Three or More Spans

$f = 1,500$ psi $E = 1,700,000$ psi $H = 140$ psi

Uniform Load, Pounds per Lineal Foot (Equals Design Load, Pounds per Square Foot, Times Spacing of Joists in Feet.)	Nominal Size Lumber, bxh S4S 19% Maximum Moisture Content					
	2x4	2x6	2x8	4x2	4x4	6x2
75	84"	119"	147"	52"	105"	60"
100	74"	111"	137"	47"	98"	55"
125	66"	104"	129"	43"	92"	51"
150	60"	95"	123"	39"	88"	48"
175	56"	88"	116"	36"	85"	45"
200	52"	82"	108"	34"	80"	43"
225	49"	77"	102"	32"	75"	40"
250	46"	72"	95"	30"	71"	38"
275	42"	67"	88"	29"	68"	36"
300	39"	62"	82"	28"	65"	35"
325	37"	58"	76"	26"	62"	33"
350	35"	55"	72"	25"	60"	32"
375	33"	52"	68"	25"	58"	31"
400	31"	49"	65"	24"	56"	30"
450	28"	45"	59"	22"	53"	28"
500	26"	41"	55"	21"	50"	27"
550	24"	39"	51"	20"	48"	25"
600	23"	36"	48"	19"	45"	24"

Spacings are governed by bending, shear or deflection. Maximum deflection is $\ell/270$, but not more than 1/4".

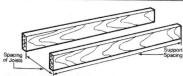

Safe Spacing, Inches, of Supports for Joists, Single Span

$f = 1,500$ psi $E = 1,700,000$ psi $H = 140$ psi

Uniform Load, Pounds per Lineal Foot (Equals Design Load, Pounds per Square Foot, Times Spacing of Joists in Feet.)	Nominal Size Lumber, bxh S4S 19% Maximum Moisture Content					
	2x4	2x6	2x8	4x2	4x4	6x2
75	72"	102"	125"	42"	89"	49"
100	66"	95"	116"	38"	83"	44"
125	59"	89"	110"	35"	79"	41"
150	54"	85"	105"	33"	75"	39"
175	50"	73"	101"	31"	72"	37"
200	46"	69"	97"	30"	70"	35"
225	44"	66"	91"	28"	67"	34"
250	41"	62"	87"	27"	64"	33"
275	40"	60"	82"	26"	61"	32"
300	38"	57"	79"	25"	58"	31"
325	36"	55"	76"	24"	56"	30"
350	35"	53"	73"	23"	54"	29"
375	34"	52"	71"	22"	52"	28"
400	33"	50"	68"	21"	50"	27"
450	31"	49"	64"	20"	47"	25"
500	29"	46"	61"	19"	45"	24"
550	28"	44"	58"	18"	43"	23"
600	26"	41"	55"	17"	41"	22"

Spacings are governed by bending, shear or deflection. Maximum deflection is $\ell/270$, but not more than 1/4".

REV. 4-85B   12-84B

General and Technical Information

Technical Data — Ledgers

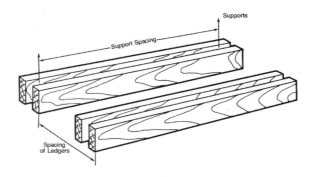

Uniform Load, Pounds per Lineal Foot (Equals Design Load, Pounds per Square Foot, Times Spacing of Ledgers in Feet.)	Safe Spacing, Inches, of Supports for Double Ledgers, Single Span				
	f = 1,500 psi E = 1,700,000 psi H = 140 psi				
	Nominal Size Lumber, bxh S4S 19% Maximum Moisture Content				
	Double 2x4	Double 2x6	Double 2x8	Double 2x10	Double 2x12
600	38"	60"	79"	101"	123"
800	33"	52"	68"	87"	106"
1,000	30"	46"	61"	78"	95"
1,200	26"	41"	55"	70"	85"
1,400	23"	37"	49"	62"	76"
1,600	21"	34"	44"	57"	69"
1,800	20"	31"	41"	53"	64"
2,000	18"	29"	38"	49"	60"
2,200	18"	27"	36"	46"	56"
2,400	16"	26"	34"	44"	54"
2,600	16"	25"	33"	42"	51"
2,800	15"	24"	31"	40"	49"
3,000	14"	23"	30"	39"	47"
3,200	14"	22"	29"	37"	46"
3,400	14"	21"	28"	36"	44"
3,600	13"	21"	28"	35"	43"
3,800	13"	20"	27"	34"	42"
4,000	12"	20"	26"	34"	41"

Spacings are governed by bending, shear or deflection.
Maximum deflection is $\ell/270$, but not more than 1/4".

REV. 6-88A   12-84B

General and Technical Information

Formwork and Falsework

Technical Data — Lumber

X—X = Neutral Axis

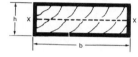 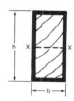

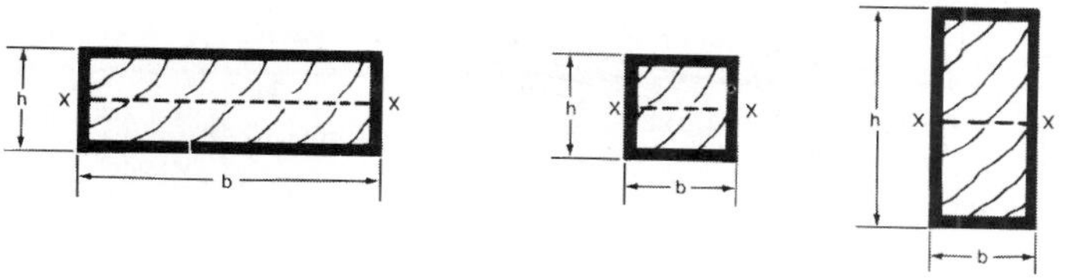

Lumber Properties

Nominal Size in Inches, $b \times h$	American Standard Sizes in Inches, $b \times h$ S4S* 19% Maximum Mositure	Area of section $A = bh$, sq. in.		Moment of Inertia, in.^{4} $I = \dfrac{bh^3}{12}$		Section Modulus, in.^{3} $S = \dfrac{bh^2}{6}$		Board Feet per Linear Foot of Piece
		Rough	S4S	Rough	S4S	Rough	S4S	
4x1	3½x¾	3.17	2.62	0.20	0.12	0.46	0.33	⅓
6x1	5½x¾	4.92	4.12	0.31	0.19	0.72	0.52	½
8x1	7¼x¾	6.45	5.44	0.41	0.25	0.94	0.68	⅔
10x1	9¼x¾	8.20	6.94	0.52	0.32	1.20	0.87	⅚
12x1	11¼x¾	9.95	8.44	0.63	0.39	1.45	1.05	1
4x2	3½x1½	5.89	5.25	1.30	0.98	1.60	1.31	⅔
6x2	5½x1½	9.14	8.25	2.01	1.55	2.48	2.06	1
8x2	7¼x1½	11.98	10.87	2.64	2.04	3.25	2.72	1½
10x2	9¼x1½	15.23	13.87	3.35	2.60	4.13	3.47	1⅔
12x2	11¼x1½	18.48	16.87	4.07	3.16	5.01	4.21	2
2x4	1½x3½	5.89	5.25	6.45	5.36	3.56	3.06	⅔
2x6	1½x5½	9.14	8.25	24.10	20.80	8.57	7.56	1
2x8	1½x7¼	11.98	10.87	54.32	47.63	14.73	13.14	1⅓
2x10	1½x9¼	15.23	13.87	111.58	98.93	23.80	21.39	1⅔
2x12	1½x11¼	18.48	16.87	199.31	177.97	35.04	31.64	2
3x4	2½x3½	9.25	8.75	10.42	8.93	5.75	5.10	1
3x6	2½x5½	14.77	13.75	38.93	34.66	13.84	12.60	1½
3x8	2½x7¼	19.36	18.12	87.74	79.39	23.80	21.90	2
3x10	2½x9¼	24.61	23.12	180.24	164.89	38.45	35.65	2½
3x12	2½x11¼	29.86	28.12	321.96	296.63	56.61	52.73	3
4x4	3½x3½	13.14	12.25	14.39	12.50	7.94	7.15	1⅓
4x6	3½x5½	20.39	19.25	53.76	48.53	19.12	17.65	2
4x8	3½x7¼	26.73	25.38	121.17	111.15	32.86	30.66	2⅔
4x10	3½x9¼	33.98	32.38	248.91	230.84	53.10	49.91	3⅓
6x3	5½x2½	14.77	13.75	8.48	7.16	6.46	5.73	1½
6x4	5½x3½	20.39	19.25	22.33	19.65	12.32	11.23	2
6x6	5½x5½	31.64	30.25	83.43	76.26	29.66	27.73	3
6x8	5½x7½	42.89	41.25	207.81	193.36	54.51	51.56	4
8x8	7½x7½	58.14	56.25	281.69	263.67	73.89	70.31	5⅓

*Roughdry sizes are ⅛ in. larger, both dimensions.

Properties of American Standard Board, Plank Dimension and Timber Sizes Commonly used for Formwork Construction.
Based on data supplied by the National Lumber Manufacturers Association.

12-84B

Construction Systems, Methods, and Materials

General and Technical Information

Representative Working Stress Values (PSI) for Lumber at 19 Percent Moisture Content, Continuing or Prolonged Reuse

LUMBER SPECIES AND GRADE / PROPERTIES	Extreme Fiber Bending	Compression ⊥ to Grain	Compression ∥ to Grain	Horizontal Shear	Modulus of Elasticity
CALIFORNIA REDWOOD					
Range, all grades	225-2300	270-425	500-2150	120	900,000-1,400,000
No. 2, 4x4 and smaller	1400	425	1100		1,250,000
Constr., 4x4 and smaller	825	270	925		900,000
DOUGLAS FIR-LARCH					
Range, all grades	275-2450	385-455	600-1850	140	1,500,000-1,900,000
No. 2, 4x4 and smaller	1450	385	1000		1,700,000
Constr., 4x4 and smaller	1050	385	1150		1,500,000
EASTERN SPRUCE					
Range, all grades	200-1785	255	425-1200	105	1,100,000-1,400,000
No. 2, 4x4 and smaller	1050	255	700		1,200,000
Constr., 4x4 and smaller	775	255	800		1,100,000
HEM-FIR					
Range, all grades	225-1650	245	500-1300	110	1,200,000-1,500,000
No. 2, 4x4 and smaller	1150	245	825		1,400,000
Constr., 4x4 and smaller	825	245	925		1,200,000
SOUTHERN PINE					
Range, all grades	275-2600	405-475	575-2000	135	1,400,000-1,800,000
No. 2, 4x4 and smaller	1250	405	975		1,600,000
Constr., 4x4 and smaller	1050	405	1100		1,400,000
ADJUSTMENT FOR MOISTURE CONTENT GREATER THAN 19 PERCENT: Use percentage shown (also applies to wood used wet)	86**	67	70	97*	97**
INCREASE FOR LOAD DURATION OF 7 DAYS OR LESS	25%	0%	25%	25%	0%

Note: Derived from National Design Specification for Wood Construction.
*For redwood use 94 percent; for southern pine use 90 percent.
**For redwood and southern pine, use 80 percent of bending stress, 93 percent of E.

Formulas Used to Calculate Safe Support Spacings of Joists and Ledgers

To Check	for Single Span Beam	for Two-Span Beam	for Three or More Span Beam
$\triangle_{max} = \ell/360$	$\ell = 1.37\sqrt[3]{\frac{EI}{w}}$	$\ell = 1.83\sqrt[3]{\frac{EI}{w}}$	$\ell = 1.69\sqrt[3]{\frac{EI}{w}}$
$\triangle_{max} = \ell/270$	$\ell = 1.51\sqrt[3]{\frac{EI}{w}}$	$\ell = 2.02\sqrt[3]{\frac{EI}{w}}$	$\ell = 1.86\sqrt[3]{\frac{EI}{w}}$
$\triangle_{max} = 1/16$ in.	$\ell = 2.75\sqrt[4]{\frac{EI}{w}}$	$\ell = 3.43\sqrt[4]{\frac{EI}{w}}$	$\ell = 3.23\sqrt[4]{\frac{EI}{w}}$
$\triangle_{max} = 1/8$ in.	$\ell = 3.27\sqrt[4]{\frac{EI}{w}}$	$\ell = 4.08\sqrt[4]{\frac{EI}{w}}$	$\ell = 3.84\sqrt[4]{\frac{EI}{w}}$
$\triangle_{max} = 1/4$ in.	$\ell = 3.90\sqrt[4]{\frac{EI}{w}}$	$\ell = 4.85\sqrt[4]{\frac{EI}{w}}$	$\ell = 4.57\sqrt[4]{\frac{EI}{w}}$
BENDING	$\ell = 9.80\sqrt{\frac{fS}{w}}$	$\ell = 9.80\sqrt{\frac{fS}{w}}$	$\ell = 10.95\sqrt{\frac{fS}{w}}$
HORIZONTAL SHEAR	$\ell = \frac{16Hbh}{w} + 2h$	$\ell = \frac{192Hbh}{15w} + 2h$	$\ell = \frac{40Hbh}{3w} + 2h$

ℓ = safe spacing of supports, in.
h = depth of section, in.
I = moment of inertia, in.^{4}
$\triangle$ = deflection, in.
w = load, lbs. per lineal ft.
E = modulus of elasticity, psi
b = width of section, in.
S = section modulus, in.^{3}
f = extreme fiber stress, psi
H = horizontal shear stress, psi

12-84B   REV. 6-88A

Spacing Tables

Formwork and Falsework

How to Use Spacing Tables

The spacing tables shown on the following pages indicate the maximum hanger and overhang bracket spacings for the various slab thicknesses and screed loads. The type of hanger and overhang bracket required, as well as the proper bracket "A" and "D" dimensions which must be used to safely obtain the spacings shown, are listed.

When selecting a trial hanger and overhang bracket spacing and the selected spacing is:

    Equal to or less than D_1, multiply Wheel Load W_1 by a Screed Load Factor of 1.0;
    Over D_1 and up to $2D_1$, multiply Wheel Load W_1 by a Screed Load Factor of 1.5;
    Over $2D_1$ and up to $3D_1$, multiply Wheel Load W_1 by a Screed Load Factor of 1.7;
    Over $3D_1$ and up to $4D_1$, multiply Wheel Load W_1 by a Screed Load Factor of 1.9;
    Greater than $4D_1$, multiply Wheel Load W_1 by a Screed Load Factor of 2.3;

to determine a close approximation of the total Screed Load (S_1) that will be applied to an individual overhang bracket. Use this value or next highest incremental value for the total Screed Load (S_1) per bracket when using the spacing tables.

The two basic types of bridge deck finishing/screed machines in use today are illustrated below.

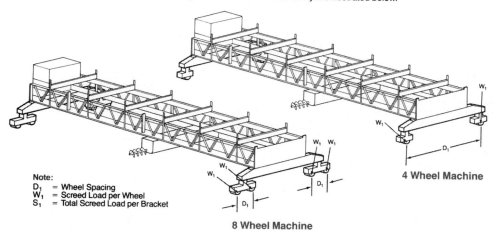

Note:
D_1 = Wheel Spacing
W_1 = Screed Load per Wheel
S_1 = Total Screed Load per Bracket

Warning: Check with Dayton Superior if concrete conveyors are to be supported on overhang formwork.

Example
40" Deep Plate Girder with 1" Thick Flanges
3'-0" Overhang
8" Thick Overhang Slab (150 PSF)
C-49 Bridge Overhang Bracket
C-62 Type 6-A Pres-Steel Hanger

8 Wheel Screed Machine
D_1 = 1'-6"
W_1 = 525 lbs. Wheel Load

As we are using the C-49 Overhang Bracket in the above example to support a 3'-0" overhang from a plate girder, the spacing table on page 53 should be used. The correct "D" dimension (30") is determined by subtracting away from the girders 40" depth, both flange thicknesses, the overall thickness of the form lumber plus a clearance allowance of 2" to 6".

For the above example it has been decided to use a trial hanger and bracket spacing of 6'-0". This results in a total screed load (S_1) per bracket of 998 lbs.

$$\frac{6.0' \text{ Trial Spacing}}{1.5' \, D_1} = 4$$ which means the Screed Load Factor (SLF) as shown above is 1.9.

$$S_1 = (W_1)(SLF) = 525 \text{ lbs.} \times 1.9 = 998 \text{ lbs.}$$

Enter the spacing table at 150 PSF design load. (8" slab thickness), "D" = 30" and upper row for a Type 6-A Pres-steel Hanger. Follow this row until it intersects with the vertical column having a total screed load (S_1) per bracket of 1,000 lbs. The allowable hanger and bracket spacing is 5'-3".

12-84B

Spacing Tables for C-49 Bracket

C-49 Bridge Overhang Bracket and Exterior Hanger Spacings
For Use on Up to 2'-0" Overhangs on Steel Beams or Girders

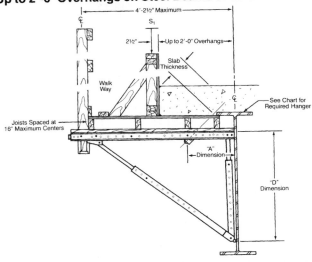

Design Load PSF	Slab Thickness	Bracket "D" Dimension	Screed Load Per Bracket = S_1							Hanger Type
			1,500 lbs.	1,250 lbs.	1,000 lbs.	750 lbs.	500 lbs.	250 lbs.	0 lbs.	
			Bracket "A" Dimension = 12⅛" or 14⅝"							
125	6"	30"	7'-6"	8'-0"	8'-0"	8'-0"	8'-0"	8'-0"	8'-0"	6A
			8'-0"	8'-0"	8'-0"	8'-0"	8'-0"	8'-0"	8'-0"	4A
125	6"	40"	7'-6"	8'-0"	8'-0"	8'-0"	8'-0"	8'-0"	8'-0"	6A
			8'-0"	8'-0"	8'-0"	8'-0"	8'-0"	8'-0"	8'-0"	4A
125	6"	50"	8'-0"	8'-0"	8'-0"	8'-0"	8'-0"	8'-0"	8'-0"	6A
			8'-0"	8'-0"	8'-0"	8'-0"	8'-0"	8'-0"	8'-0"	4A
150	8"	30"	6'-9"	7'-6"	8'-0"	8'-0"	8'-0"	8'-0"	8'-0"	6A
			8'-0"	8'-0"	8'-0"	8'-0"	8'-0"	8'-0"	8'-0"	4A
150	8"	40"	6'-9"	7'-6"	8'-0"	8'-0"	8'-0"	8'-0"	8'-0"	6A
			8'-0"	8'-0"	8'-0"	8'-0"	8'-0"	8'-0"	8'-0"	4A
150	8"	50"	7'-6"	8'-0"	8'-0"	8'-0"	8'-0"	8'-0"	8'-0"	6A
			8'-0"	8'-0"	8'-0"	8'-0"	8'-0"	8'-0"	8'-0"	4A
175	10"	30"	6'-0"	6'-9"	7'-3"	8'-0"	8'-0"	8'-0"	8'-0"	6A
			8'-0"	8'-0"	8'-0"	8'-0"	8'-0"	8'-0"	8'-0"	4A
175	10"	40"	6'-0"	6'-9"	7'-6"	8'-0"	8'-0"	8'-0"	8'-0"	6A
			8'-0"	8'-0"	8'-0"	8'-0"	8'-0"	8'-0"	8'-0"	4A
175	10"	50"	6'-9"	7'-6"	8'-0"	8'-0"	8'-0"	8'-0"	8'-0"	6A
			8'-0"	8'-0"	8'-0"	8'-0"	8'-0"	8'-0"	8'-0"	4A
200	12"	30"	5'-6"	6'-0"	6'-9"	7'-3"	7'-9"	8'-0"	8'-0"	6A
			8'-0"	8'-0"	8'-0"	8'-0"	8'-0"	8'-0"	8'-0"	4A
200	12"	40"	5'-6"	6'-0"	6'-9"	7'-3"	8'-0"	8'-0"	8'-0"	6A
			8'-0"	8'-0"	8'-0"	8'-0"	8'-0"	8'-0"	8'-0"	4A
200	12"	50"	6'-3"	6'-9"	7'-6"	8'-0"	8'-0"	8'-0"	8'-0"	6A
			8'-0"	8'-0"	8'-0"	8'-0"	8'-0"	8'-0"	8'-0"	4A

Note: Design includes 50 PSF live load on walk way area.
Overhang form lumber must be checked to make sure it will span the selected spacing.
Warning: Contact our Technical Service Department for recommended spacings, when conditions on your specific project vary from those shown.

12-84B

Spacing Tables for C-49 Bracket

Formwork and Falsework   41

C-49 Bridge Overhang Bracket and Exterior Hanger Spacings
For Use on Over 2'-0" to 3'-0" Overhangs on Steel Beams or Girders

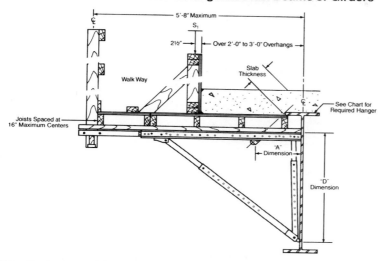

Design Load PSF	Slab Thickness	Bracket "D" Dimension	Screed Load Per Bracket = S₁							Hanger Type
			1,500 lbs.	1,250 lbs.	1,000 lbs.	750 lbs.	500 lbs.	250 lbs.	0 lbs.	
			Bracket "A" Dimension = 12⅛" or 14⅝"							
125	6"	30"	4'-9"	5'-3"	6'-0"	6'-9"	7'-6"	8'-0"	8'-0"	6A
			5'-6"	6'-3"	7'-3"	8'-0"	8'-0"	8'-0"	8'-0"	4A
125	6"	40"	5'-6"	6'-0"	6'-9"	7'-6"	8'-0"	8'-0"	8'-0"	6A
			7'-3"	8'-0"	8'-0"	8'-0"	8'-0"	8'-0"	8'-0"	4A
125	6"	50"	5'-9"	6'-6"	7'-3"	7'-9"	8'-0"	8'-0"	8'-0"	6A
			8'-0"	8'-0"	8'-0"	8'-0"	8'-0"	8'-0"	8'-0"	4A
150	8"	30"	4'-0"	4'-9"	5'-3"	6'-0"	6'-6"	7'-0"	7'-9"	6A
			4'-9"	5'-6"	6'-3"	7'-3"	8'-0"	8'-0"	8'-0"	4A
150	8"	40"	4'-9"	5'-3"	6'-0"	6'-6"	7'-0"	7'-9"	8'-0"	6A
			6'-6"	7'-6"	8'-0"	8'-0"	8'-0"	8'-0"	8'-0"	4A
150	8"	50"	5'-0"	5'-9"	6'-3"	6'-9"	7'-6"	8'-0"	8'-0"	6A
			8'-0"	8'-0"	8'-0"	8'-0"	8'-0"	8'-0"	8'-0"	4A
175	10"	30"	3'-9"	4'-3"	4'-9"	5'-3"	5'-9"	6'-3"	6'-9"	6A
			4'-3"	5'-0"	5'-9"	6'-6"	7'-3"	7'-9"	8'-0"	4A
175	10"	40"	4'-3"	4'-9"	5'-3"	5'-9"	6'-3"	6'-9"	7'-3"	6A
			6'-0"	6'-9"	7'-6"	8'-0"	8'-0"	8'-0"	8'-0"	4A
175	10"	50"	4'-6"	5'-0"	5'-6"	6'-0"	6'-6"	7'-3"	7'-9"	6A
			7'-6"	8'-0"	8'-0"	8'-0"	8'-0"	8'-0"	8'-0"	4A
200	12"	30"	3'-3"	3'-9"	4'-3"	4'-9"	5'-3"	5'-9"	6'-3"	6A
			4'-0"	4'-6"	5'-3"	5'-9"	6'-6"	7'-3"	7'-9"	4A
200	12"	40"	3'-9"	4'-3"	4'-9"	5'-3"	5'-6"	6'-0"	6'-6"	6A
			5'-6"	6'-0"	6'-9"	7'-6"	8'-0"	8'-0"	8'-0"	4A
200	12"	50"	4'-0"	4'-6"	5'-0"	5'-6"	6'-0"	6'-6"	7'-0"	6A
			6'-9"	8'-0"	8'-0"	8'-0"	8'-0"	8'-0"	8'-0"	4A

Note: Design includes 50 PSF live load on walk way area.
Overhang form lumber must be checked to make sure it will span the selected spacing.
Warning: Contact our Technical Service Department for recommended spacings, when conditions on your specific project vary from those shown.

Spacing Tables for C-49 Bracket

C-49 Bridge Overhang Bracket and Exterior Hanger Spacings
For Use on Over 3'-0" to 4'-0" Overhangs on Steel Beams or Girders

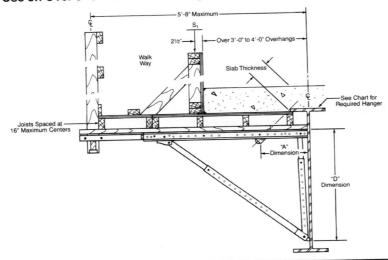

Design Load PSF	Slab Thickness	Bracket "D" Dimension	Screed Load Per Bracket = S₁							Hanger Type
			1,500 lbs.	1,250 lbs.	1,000 lbs.	750 lbs.	500 lbs.	250 lbs.	0 lbs.	
			Bracket "A" Dimension = 12⅛" or 14⅝"							
125	6"	30"	2'-9"	3'-6"	4'-3"	4'-9"	5'-6"	6'-0"	6'-9"	6A
			2'-9"	3'-6"	4'-3"	5'-3"	6'-0"	6'-9"	7'-6"	4A
125	6"	40"	3'-6"	4'-3"	4'-9"	5'-3"	6'-0"	6'-6"	7'-0"	6A
			3'-6"	4'-3"	5'-3"	6'-0"	6'-9"	7'-9"	8'-0"	4A
125	6"	50"	3'-9"	4'-6"	5'-0"	5'-9"	6'-3"	6'-9"	7'-6"	6A
			3'-9"	4'-9"	5'-6"	6'-3"	7'-3"	8'-0"	8'-0"	4A
150	8"	30"	2'-3"	3'-0"	3'-6"	4'-0"	4'-6"	5'-3"	5'-9"	6A
			2'-3"	3'-0"	3'-9"	4'-6"	5'-3"	6'-0"	6'-6"	4A
150	8"	40"	3'-0"	3'-6"	4'-0"	4'-6"	5'-0"	5'-6"	6'-0"	6A
			3'-0"	3'-9"	4'-6"	5'-3"	5'-9"	6'-6"	7'-3"	4A
150	8"	50"	3'-3"	3'-9"	4'-3"	4'-9"	5'-3"	5'-9"	6'-3"	6A
			3'-3"	4'-0"	4'-9"	5'-6"	6'-3"	6'-9"	7'-6"	4A
175	10"	30"	2'-0"	2'-6"	3'-0"	3'-6"	4'-0"	4'-6"	5'-0"	6A
			2'-0"	2'-9"	3'-3"	4'-0"	4'-6"	5'-3"	5'-9"	4A
175	10"	40"	2'-9"	3'-0"	3'-6"	4'-0"	4'-6"	4'-9"	5'-3"	6A
			2'-9"	3'-3"	4'-0"	4'-6"	5'-3"	5'-9"	6'-6"	4A
175	10"	50"	2'-9"	3'-3"	3'-9"	4'-3"	4'-9"	5'-0"	5'-6"	6A
			3'-0"	3'-6"	4'-3"	4'-9"	5'-6"	6'-0"	6'-9"	4A
200	12"	30"	1'-9"	2'-3"	2'-9"	3'-0"	3'-6"	4'-0"	4'-6"	6A
			1'-9"	2'-3"	3'-0"	3'-6"	4'-0"	4'-6"	5'-0"	4A
200	12"	40"	2'-3"	2'-9"	3'-0"	3'-6"	4'-0"	4'-3"	4'-9"	6A
			2'-3"	3'-0"	3'-6"	4'-0"	4'-6"	5'-0"	5'-9"	4A
200	12"	50"	2'-6"	3'-0"	3'-3"	3'-9"	4'-0"	4'-6"	5'-0"	6A
			2'-6"	3'-3"	3'-9"	4'-3"	4'-9"	5'-3"	6'-0"	4A

Note: Design includes 50 PSF live load on walk way area.
Overhang form lumber must be checked to make sure it will span the selected spacing.
Warning: Contact our Technical Service Department for recommended spacings, when conditions on your specific project vary from those shown.

12-84B

Spacing Tables for C-49-D Bracket

C-49-D Bridge Overhang Bracket and Exterior Hanger Spacings
For Use on Up to 3'-0" Overhangs on Steel Girders

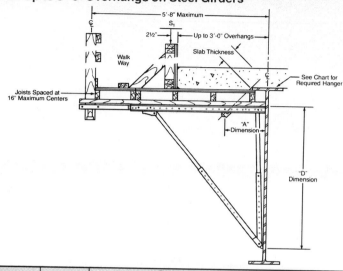

Design Load PSF	Slab Thickness	Bracket "D" Dimension	Screed Load Per Bracket = S_1							Hanger Type
			1,500 lbs.	1,250 lbs.	1,000 lbs.	750 lbs.	500 lbs.	250 lbs.	0 lbs.	
			Bracket "A" Dimension = 14⅝" or 17⅛"							
125	6"	50"	6'-0"	6'-9"	7'-6"	8'-0"	8'-0"	8'-0"	8'-0"	6A
			8'-0"	8'-0"	8'-0"	8'-0"	8'-0"	8'-0"	8'-0"	4A
125	6"	60"	6'-6"	7'-0"	7'-9"	8'-0"	8'-0"	8'-0"	8'-0"	6A
			8'-0"	8'-0"	8'-0"	8'-0"	8'-0"	8'-0"	8'-0"	4A
125	6"	70"	6'-9"	7'-6"	8'-0"	8'-0"	8'-0"	8'-0"	8'-0"	6A
			8'-0"	8'-0"	8'-0"	8'-0"	8'-0"	8'-0"	8'-0"	4A
150	8"	50"	5'-3"	5'-9"	6'-6"	7'-0"	7'-6"	8'-0"	8'-0"	6A
			8'-0"	8'-0"	8'-0"	8'-0"	8'-0"	8'-0"	8'-0"	4A
150	8"	60"	5'-6"	6'-0"	6'-6"	7'-3"	7'-9"	8'-0"	8'-0"	6A
			8'-0"	8'-0"	8'-0"	8'-0"	8'-0"	8'-0"	8'-0"	4A
150	8"	70"	5'-9"	6'-6"	7'-0"	7'-6"	8'-0"	8'-0"	8'-0"	6A
			8'-0"	8'-0"	8'-0"	8'-0"	8'-0"	8'-0"	8'-0"	4A
175	10"	50"	4'-6"	5'-0"	5'-6"	6'-0"	6'-6"	7'-3"	7'-9"	6A
			8'-0"	8'-0"	8'-0"	8'-0"	8'-0"	8'-0"	8'-0"	4A
175	10"	60"	4'-9"	5'-3"	5'-9"	6'-3"	6'-9"	7'-3"	7'-9"	6A
			8'-0"	8'-0"	8'-0"	8'-0"	8'-0"	8'-0"	8'-0"	4A
175	10"	70"	5'-0"	5'-6"	6'-0"	6'-6"	7'-0"	7'-9"	8'-0"	6A
			8'-0"	8'-0"	8'-0"	8'-0"	8'-0"	8'-0"	8'-0"	4A
200	12"	50"	4'-0"	4'-6"	5'-0"	5'-6"	6'-0"	6'-3"	6'-9"	6A
			7'-3"	8'-0"	8'-0"	8'-0"	8'-0"	8'-0"	8'-0"	4A
200	12"	60"	4'-3"	4'-9"	5'-3"	5'-6"	6'-0"	6'-6"	6'-9"	6A
			7'-9"	8'-0"	8'-0"	8'-0"	8'-0"	8'-0"	8'-0"	4A
200	12"	70"	4'-6"	5'-0"	5'-6"	6'-0"	6'-3"	6'-9"	7'-3"	6A
			8'-0"	8'-0"	8'-0"	8'-0"	8'-0"	8'-0"	8'-0"	4A

Note: Design includes 50 PSF live load on walk way area.
Overhang form lumber must be checked to make sure it will span the selected spacing.
Warning: Contact our Technical Service Department for recommended spacings, when conditions on your specific project vary from those shown.

12-84B

C-50 Bridge Overhang Bracket and Exterior Hanger Spacings
For Use on Up to 2'-0" Overhangs on Steel Beams or Girders

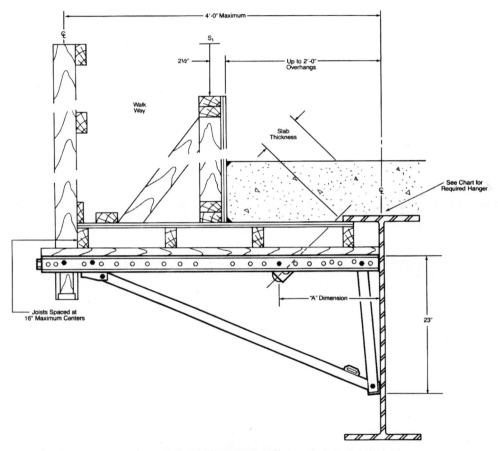

Design Load PSF	Slab Thickness	Screed Load Per Bracket = S_1							Hanger Type
		1,500 lbs.	1,250 lbs.	1,000 lbs.	750 lbs.	500 lbs.	250 lbs.	0 lbs.	
		Bracket "A" Dimension = 12⅛" or 14⅝"							
125	6"	1'-3"	2'-0"	3'-0"	4'-9"	3'-9"	5'-6"	6'-6"	6A
		4'-6"	5'-0"	5'-6"	8'-0"	7'-0"	8'-0"	8'-0"	4A
150	8"	1'-0"	1'-9"	2'-6"	4'-3"	3'-3"	5'-0"	5'-9"	6A
		4'-0"	4'-9"	5'-6"	7'-0"	6'-3"	7'-9"	8'-0"	4A
175	10"	1'-0"	1'-6"	2'-3"	3'-9"	3'-0"	4'-6"	5'-3"	6A
		3'-6"	4'-3"	5'-0"	6'-3"	5'-9"	7'-0"	7'-9"	4A
200	12"	—	1'-6"	2'-0"	3'-6"	2'-9"	4'-0"	4'-9"	6A
		3'-3"	3'-9"	4'-6"	5'-9"	5'-3"	6'-6"	7'-0"	4A

Note: Design includes 50 PSF live load on walk way area.
Overhang form lumber must be checked to make sure it will span the selected spacing.
Warning: Contact our Technical Service Department for recommended spacings, when conditions on your specific project vary from those shown.

12-84B

Spacing Tables for C-50 Bracket

C-50 Bridge Overhang Bracket and Exterior Hanger Spacings
For Use on Over 2'-0" to 3'-0" Overhangs on Steel Beams or Girders

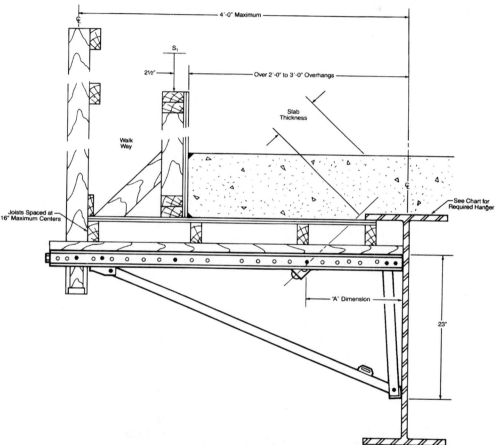

Design Load	Slab Thickness	Screed Load Per Bracket = S_1							Hanger Type
		1,500 lbs.	1,250 lbs.	1,000 lbs.	750 lbs.	500 lbs.	250 lbs.	0 lbs.	
		Bracket "A" Dimension = 12⅛" or 14⅝"							
125	6"	—	—	1'-6"	2'-6"	3'-3"	4'-3"	5'-3"	6A
		2'-3"	3'-3"	4'-3"	5'-0"	6'-0"	6'-6"	7'-9"	4A
150	8"	—	—	1'-3"	2'-0"	2'-9"	3'-6"	4'-6"	6A
		2'-0"	2'-9"	3'-6"	4'-3"	5'-0"	5'-9"	6'-9"	4A
175	10"	—	—	1'-0"	1'-9"	2'-6"	3'-3"	3'-9"	6A
		1'-9"	2'-6"	3'-0"	3'-9"	4'-6"	5'-0"	5'-9"	4A
200	12"	—	—	1'-0"	1'-6"	2'-3"	2'-9"	3'-6"	6A
		1'-6"	2'-0"	2'-9"	3'-3"	4'-0"	4'-6"	5'-3"	4A

Note: Design includes 50 PSF live load on walk way area.
Overhang form lumber must be checked to make sure it will span the selected spacing.
Warning: Contact our Technical Service Department for recommended spacings, when conditions on your specific project vary from those shown.

Spacing Tables for C-49 Bracket

C-49 Bridge Overhang Bracket and Exterior Hanger Spacings
For Use on Up to 2'-0" Overhangs on Precast/Prestressed Concrete Beams

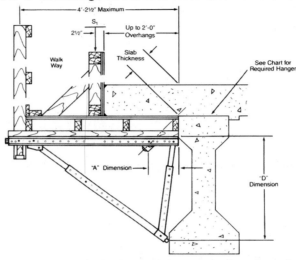

Design Load PSF	Slab Thickness	Bracket "D" Dimension	Screed Load Per Bracket = S_1							Hanger Type
			1,500 lbs.	1,250 lbs.	1,000 lbs.	750 lbs.	500 lbs.	250 lbs.	0 lbs.	
			Bracket "A" Dimension = 7⅛" or 9⅝"							
125	6"	30"	5'-9"	6'-3"	7'-0"	7'-9"	8'-0"	8'-0"	8'-0"	6A
			5'-9"	6'-6"	7'-3"	8'-0"	8'-0"	8'-0"	8'-0"	4A
125	6"	40"	6'-3"	7'-3"	8'-0"	8'-0"	8'-0"	8'-0"	8'-0"	6A
			6'-3"	7'-3"	8'-0"	8'-0"	8'-0"	8'-0"	8'-0"	4A
125	6"	50"	7'-0"	8'-0"	8'-0"	8'-0"	8'-0"	8'-0"	8'-0"	6A
			7'-0"	8'-0"	8'-0"	8'-0"	8'-0"	8'-0"	8'-0"	4A
150	8"	30"	5'-0"	5'-9"	6'-3"	7'-0"	7'-6"	8'-0"	8'-0"	6A
			5'-0"	5'-9"	6'-6"	7'-3"	7'-9"	8'-0"	8'-0"	4A
150	8"	40"	5'-6"	6'-3"	7'-0"	7'-9"	8'-0"	8'-0"	8'-0"	6A
			5'-6"	6'-3"	7'-0"	7'-9"	8'-0"	8'-0"	8'-0"	4A
150	8"	50"	6'-0"	6'-9"	7'-9"	8'-0"	8'-0"	8'-0"	8'-0"	6A
			6'-0"	6'-9"	7'-9"	8'-0"	8'-0"	8'-0"	8'-0"	4A
175	10"	30"	4'-6"	5'-0"	5'-9"	6'-3"	6'-9"	7'-6"	8'-0"	6A
			4'-6"	5'-3"	5'-9"	6'-6"	7'-0"	7'-9"	8'-0"	4A
175	10"	40"	4'-9"	5'-6"	6'-3"	6'-9"	7'-6"	8'-0"	8'-0"	6A
			4'-9"	5'-6"	6'-3"	6'-9"	7'-6"	8'-0"	8'-0"	4A
175	10"	50"	5'-6"	6'-0"	6'-9"	7'-6"	8'-0"	8'-0"	8'-0"	6A
			5'-6"	6'-0"	6'-9"	7'-6"	8'-0"	8'-0"	8'-0"	4A
200	12"	30"	4'-0"	4'-6"	5'-3"	5'-9"	6'-3"	6'-9"	7'-3"	6A
			4'-3"	4'-9"	5'-3"	6'-0"	6'-6"	7'-0"	7'-6"	4A
200	12"	40"	4'-3"	5'-0"	5'-6"	6'-0"	6'-9"	7'-3"	8'-0"	6A
			4'-3"	5'-0"	5'-6"	6'-0"	6'-9"	7'-3"	8'-0"	4A
200	12"	50"	4'-9"	5'-6"	6'-0"	6'-9"	7'-3"	7'-9"	8'-0"	6A
			4'-9"	5'-6"	6'-0"	6'-9"	7'-3"	7'-9"	8'-0"	4A

Note: Design includes 50 PSF live load on walk way area.
Overhang form lumber must be checked to make sure it will span the selected spacing.

Warning: Contact our Technical Service Department for recommended spacings, when conditions on your specific project vary from those shown. Beam must be checked against overturning before concrete is placed.

Spacing Tables for C-49 Bracket

Formwork and Falsework

C-49 Bridge Overhang Bracket and Exterior Hanger Spacings
For Use on Over 2'-0" to 3'-0" Overhangs on Precast/Prestressed Concrete Beams

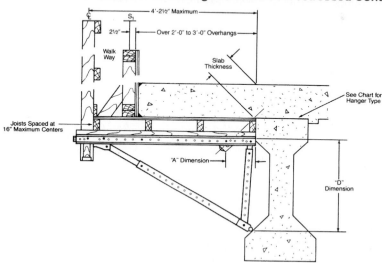

Design Load PSF	Slab Thickness	Bracket "D" Dimension	Screed Load Per Bracket = S_1							Hanger Type
			1,500 lbs.	1,250 lbs.	1,000 lbs.	750 lbs.	500 lbs.	250 lbs.	0 lbs.	
			Bracket "A" Dimension = 7⅛" or 9⅝"							
125	6"	30"	2'-3"	3'-6"	4'-9"	6'-0"	6'-9"	7'-9"	8'-0"	6A
			2'-3"	3'-6"	4'-9"	6'-0"	7'-0"	7'-9"	8'-0"	4A
125	6"	40"	4'-6"	5'-6"	6'-9"	7'-6"	8'-0"	8'-0"	8'-0"	6A
			4'-6"	5'-6"	6'-9"	8'-0"	8'-0"	8'-0"	8'-0"	4A
125	6"	50"	5'-3"	6'-3"	7'-3"	8'-0"	8'-0"	8'-0"	8'-0"	6A
			5'-3"	6'-3"	7'-6"	8'-0"	8'-0"	8'-0"	8'-0"	4A
150	8"	30"	1'-9"	3'-0"	4'-0"	5'-0"	5'-9"	6'-6"	7'-3"	6A
			1'-9"	3'-0"	4'-0"	5'-0"	6'-0"	6'-6"	7'-3"	4A
150	8"	40"	3'-9"	4'-9"	5'-9"	6'-3"	7'-0"	7'-6"	8'-0"	6A
			3'-9"	4'-9"	5'-9"	6'-9"	7'-9"	8'-0"	8'-0"	4A
150	8"	50"	4'-6"	5'-6"	6'-3"	6'-9"	7'-6"	8'-0"	8'-0"	6A
			4'-6"	5'-6"	6'-6"	7'-6"	8'-0"	8'-0"	8'-0"	4A
175	10"	30"	1'-9"	2'-6"	3'-6"	4'-6"	5'-0"	5'-9"	6'-3"	6A
			1'-9"	2'-6"	3'-6"	4'-6"	5'-0"	5'-9"	6'-6"	4A
175	10"	40"	3'-3"	4'-3"	5'-0"	5'-6"	6'-0"	6'-6"	7'-0"	6A
			3'-3"	4'-3"	5'-0"	6'-0"	6'-9"	7'-9"	8'-0"	4A
175	10"	50"	3'-9"	4'-9"	5'-6"	6'-0"	6'-6"	7'-0"	7'-6"	6A
			3'-9"	4'-9"	5'-9"	6'-6"	7'-6"	8'-0"	8'-0"	4A
200	12"	30"	1'-6"	2'-3"	3'-3"	4'-0"	4'-6"	5'-0"	5'-6"	6A
			1'-6"	2'-3"	3'-3"	4'-0"	4'-6"	5'-0"	5'-9"	4A
200	12"	40"	3'-0"	3'-9"	4'-3"	4'-9"	5'-3"	5'-9"	6'-3"	6A
			3'-0"	3'-9"	4'-6"	5'-3"	6'-0"	7'-0"	7'-9"	4A
200	12"	50"	3'-6"	4'-3"	4'-9"	5'-3"	5'-9"	6'-3"	6'-9"	6A
			3'-6"	4'-3"	5'-0"	5'-9"	6'-6"	7'-6"	8'-0"	4A

Note: Design includes 50 PSF live load on walk way area.
Overhang form lumber must be checked to make sure it will span the selected spacing.

Warning: Contact our Technical Service Department for recommended spacings, when conditions on your specific project vary from those shown. Beam must be checked against overturning before concrete is placed.

48 Construction Systems, Methods, and Materials

Spacing Tables for C-50 Bracket

C-50 Bridge Overhang Bracket and Exterior Hanger Spacings
For Use on Up to 2'-0" Overhangs on Precast/Prestressed Concrete Beams

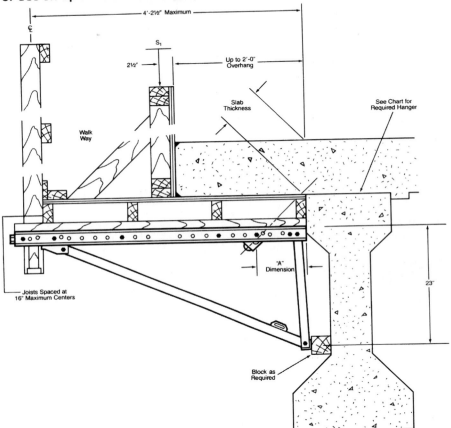

Design Load PSF	Slab Thickness	Screed Load Per Bracket = S_1							Hanger Type
		1,500 lbs.	1,250 lbs.	1,000 lbs.	750 lbs.	500 lbs.	250 lbs.	0 lbs.	
		Bracket "A" Dimension = 7⅛" or 9⅝"							
125	6"	1'-3"	2'-0"	2'-6"	3'-3"	4'-0"	4'-9"	5'-6"	6A
		1'-3"	2'-0"	2'-6"	3'-3"	4'-0"	4'-9"	5'-6"	4A
150	8"	1'-0"	1'-9"	2'-3"	3'-0"	3'-6"	4'-3"	4'-9"	6A
		1'-0"	1'-9"	2'-3"	3'-0"	3'-6"	4'-3"	4'-9"	4A
175	10"	—	1'-6"	2'-0"	2'-6"	3'-3"	3'-9"	4'-3"	6A
		—	1'-6"	2'-0"	2'-6"	3'-3"	3'-9"	4'-3"	4A
200	12"	—	1'-3"	1'-9"	2'-3"	2'-9"	3'-3"	3'-9"	6A
		—	1'-3"	1'-9"	2'-3"	2'-9"	3'-3"	3'-9"	4A

Note: Design includes 50 PSF live load on walk way area.
Overhang form lumber must be checked to make sure it will span the selected spacing.
Warning: Contact our Technical Service Department for recommended spacings, when conditions on your specific project vary from those shown. Beam must be checked against overturning before concrete is placed.

12-84B

Spacing Tables For C-50 Bracket

Formwork and Falsework 49

C-50 Bridge Overhang Bracket and Exterior Hanger Spacings
For Use on Over 2'-0" to 3'-0" Overhangs on Precast/Prestressed Concrete Beams

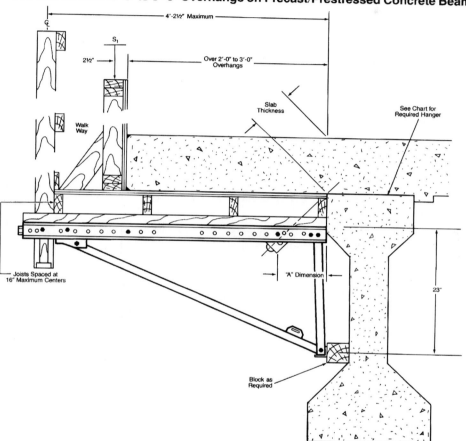

Design Load PSF	Slab Thickness	Screed Load per Bracket = S_1							Hanger Type
		1,500 lbs.	1,250 lbs.	1,000 lbs.	750 lbs.	500 lbs.	250 lbs.	0 lbs.	
		Bracket "A" Dimension = 7⅛" or 9⅝"							
125	6"	—	—	1'-3"	2'-0"	2'-9"	3'-6"	4'-3"	6A
		—	—	1'-3"	2'-0"	2'-9"	3'-6"	4'-3"	4A
150	8"	—	—	1'-0"	1'-9"	2'-6"	3'-0"	3'-9"	6A
		—	—	1'-0"	1'-9"	2'-6"	3'-0"	3'-9"	4A
175	10"	—	—	—	1'-6"	2'-0"	2'-9"	3'-3"	6A
		—	—	—	1'-6"	2'-0"	2'-9"	3'-3"	4A
200	12"	—	—	—	1'-3"	1'-9"	2'-3"	2'-9"	6A
		—	—	—	1'-3"	1'-9"	2'-3"	2'-9"	4A

Note: Design includes 50 PSF live load on walk way area.
Overhang form lumber must be checked to make sure it will span the selected spacing.

Warning: Contact our Technical Service Department for recommended spacings, when conditions on your specific project vary from those shown. Beam must be checked against overturning before concrete is placed.

12-84B

Interior Hangers

Interior Hanger Spacing Charts

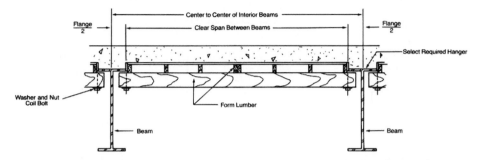

The following charts list the maximum safe hanger spacing for the various types of interior hangers produced by Dayton Superior. When the clear span on a project is not an even foot, the next larger clear span, from the chart, should be used in determining the maximum hanger spacing.

In many cases, the form lumber may control the maximum spacing between hangers, therefore the lumber must always be checked before a hanger spacing is determined for actual use.

These charts are based on the following formula:

$$\text{Maximum Hanger Spacing (Limited at 8'-0" Maximum Centers)} = \frac{\text{S.W.L. per Side of Hanger}}{\text{Design Load, PSF} \times \left(\frac{\text{Clear Span, Feet}}{2}\right)}$$

2,000 lbs. per Side Hanger Safe Working Load											
Design Load PSF	Slab Thickness	Clear Span Between Beams									
		3'-0"	4'-0"	5'-0"	6'-0"	7'-0"	8'-0"	9'-0"	10'-0"	11'-0"	12'-0"
		Maximum Interior Hanger Spacing									
125	6"	8'-0"	8'-0"	6'-4"	5'-4"	4'-6"	4'-0"	3'-6"	3'-2"	2'-10"	2'-8"
150	8"	8'-0"	6'-8"	5'-4"	4'-5"	3'-9"	3'-4"	2'-11"	2'-8"	2'-5"	2'-2"
175	10"	7'-7"	5'-8"	4'-6"	3'-9"	3'-3"	2'-10"	2'-6"	2'-3"	2'-0"	1'-10"
200	12"	6'-8"	5'-0"	4'-0"	3'-4"	2'-10"	2'-6"	2'-1"	2'-0"	1'-9"	1'-8"

2,375 lbs. per Side Hanger Safe Working Load											
Design Load PSF	Slab Thickness	Clear Span Between Beams									
		3'-0"	4'-0"	5'-0"	6'-0"	7'-0"	8'-0"	9'-0"	10'-0"	11'-0"	12'-0"
		Maximum Interior Hanger Spacing									
125	6"	8'-0"	8'-0"	7'-7"	6'-4"	5'-5"	4'-9"	4'-2"	3'-9"	3'-5"	3'-2"
150	8"	8'-0"	7'-11"	6'-4"	5'-3"	4'-6"	3'-11"	3'-6"	3'-2"	2'-10"	2'-7"
175	10"	8'-0"	6'-9"	5'-5"	4'-6"	3'-10"	3'-4"	3'-0"	2'-8"	2'-5"	2'-3"
200	12"	7'-11"	5'-11"	4'-9"	3'-11"	3'-4"	2'-11"	2'-7"	2'-4"	2'-1"	1'-11"

2,500 lbs. per Side Hanger Safe Working Load											
Design Load PSF	Slab Thickness	Clear Span Between Beams									
		3'-0"	4'-0"	5'-0"	6'-0"	7'-0"	8'-0"	9'-0"	10'-0"	11'-0"	12'-0"
		Maximum Interior Hanger Spacing									
125	6"	8'-0"	8'-0"	8'-0"	6'-8"	5'-8"	5'-0"	4'-5"	4'-0"	3'-7"	3'-4"
150	8"	8'-0"	8'-0"	6'-7"	5'-6"	4'-9"	4'-2"	3'-8"	3'-4"	3'-0"	2'-9"
175	10"	8'-0"	7'-1"	5'-8"	4'-9"	4'-1"	3'-6"	3'-2"	2'-10"	2'-7"	2'-4"
200	12"	8'-0"	6'-3"	5'-0"	4'-2"	3'-6"	3'-1"	2'-9"	2'-6"	2'-3"	2'-1"

Interior Hangers

Formwork and Falsework

3,000 lbs. per Side Hanger Safe Working Load

Design Load PSF	Slab Thickness	Clear Span Between Beams									
		3'-0"	4'-0"	5'-0"	6'-0"	7'-0"	8'-0"	9'-0"	10'-0"	11'-0"	12'-0"
		Maximum Interior Hanger Spacing									
125	6"	8'-0"	8'-0"	8'-0"	8'-0"	6'-10"	6'-0"	5'-4"	4'-9"	4'-4"	4'-0"
150	8"	8'-0"	8'-0"	8'-0"	6'-8"	5'-8"	5'-0"	4'-5"	4'-0"	3'-7"	3'-4"
175	10"	8'-0"	8'-0"	6'-10"	5'-8"	4'-10"	4'-3"	3'-9"	3'-5"	3'-1"	2'-10"
200	12"	8'-0"	7'-6"	6'-0"	5'-0"	4'-3"	3'-9"	3'-4"	3'-0"	2'-8"	2'-6"

3,500 lbs. per Side Hanger Safe Working Load

Design Load PSF	Slab Thickness	Clear Span Between Beams									
		3'-0"	4'-0"	5'-0"	6'-0"	7'-0"	8'-0"	9'-0"	10'-0"	11'-0"	12'-0"
		Maximum Interior Hanger Spacing									
125	6"	8'-0"	8'-0"	8'-0"	8'-0"	8'-0"	7'-0"	6'-2"	5'-7"	5'-0"	4'-8"
150	8"	8'-0"	8'-0"	8'-0"	7'-9"	6'-8"	5'-10"	5'-2"	4'-8"	4'-2"	3'-10"
175	10"	8'-0"	8'-0"	8'-0"	6'-8"	5'-8"	5'-0"	4'-4"	4'-0"	3'-7"	3'-4"
200	12"	8'-0"	8'-0"	7'-0"	5'-10"	5'-0"	4'-4"	3'-10"	3'-6"	3'-2"	2'-11"

4,500 lbs. per Side Hanger Safe Working Load

Design Load PSF	Slab Thickness	Clear Span Between Beams									
		3'-0"	4'-0"	5'-0"	6'-0"	7'-0"	8'-0"	9'-0"	10'-0"	11'-0"	12'-0"
		Maximum Interior Hanger Spacing									
125	6"	8'-0"	8'-0"	8'-0"	8'-0"	8'-0"	8'-0"	8'-0"	7'-2"	6'-6"	6'-0"
150	8"	8'-0"	8'-0"	8'-0"	8'-0"	8'-0"	7'-6"	6'-8"	6'-0"	5'-5"	5'-0"
175	10"	8'-0"	8'-0"	8'-0"	8'-0"	7'-4"	6'-5"	5'-8"	5'-1"	4'-8"	4'-3"
200	12"	8'-0"	8'-0"	8'-0"	7'-6"	6'-5"	5'-7"	5'-0"	4'-6"	4'-1"	3'-9"

6,000 lbs. per Side Hanger Safe Working Load

Design Load PSF	Slab Thickness	Clear Span Between Beams									
		3'-0"	4'-0"	5'-0"	6'-0"	7'-0"	8'-0"	9'-0"	10'-0"	11'-0"	12'-0"
		Maximum Interior Hanger Spacing									
125	6"	8'-0"	8'-0"	8'-0"	8'-0"	8'-0"	8'-0"	8'-0"	8'-0"	8'-0"	8'-0"
150	8"	8'-0"	8'-0"	8'-0"	8'-0"	8'-0"	8'-0"	8'-0"	8'-0"	7'-3"	6'-8"
175	10"	8'-0"	8'-0"	8'-0"	8'-0"	8'-0"	8'-0"	7'-7"	6'-10"	6'-2"	5'-8"
200	12"	8'-0"	8'-0"	8'-0"	8'-0"	8'-0"	7'-6"	6'-8"	6'-0"	5'-5"	5'-0"

12-84B

Interior Hangers

C-46 and C-47 Con-Beam Hangers
Designed for use on Concrete Interior Beams

Field welding of the Type D Hanger is shown for information only. Since field welding may alter the hanger's strength, a safe working load has not been shown.

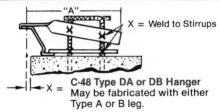

X = Weld to Stirrups

C-48 Type DA or DB Hanger
May be fabricated with either Type A or B leg.

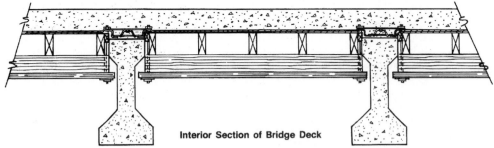

Interior Section of Bridge Deck

The Type DA or DB Hanger is ideal for use as an extra support on one side of a concrete beam.

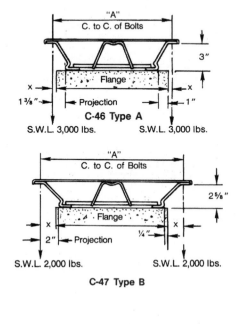

C-46 Type A
S.W.L. 3,000 lbs.   S.W.L. 3,000 lbs.

C-47 Type B
S.W.L. 2,000 lbs.   S.W.L. 2,000 lbs.

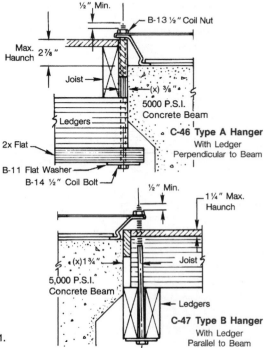

C-46 Type A Hanger
With Ledger Perpendicular to Beam

C-47 Type B Hanger
With Ledger Parallel to Beam

NOTE: Dimension "A" = Flange Width Plus 2 (x).

S.W.L. shown have a safety factor of approximately 2 to 1.

4-83B

Haunch Forming

Formwork and Falsework 53

C-28 Haunch Carrier

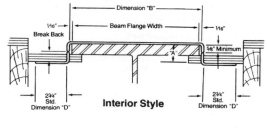

Interior Style

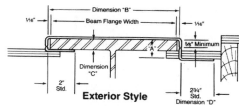

Exterior Style

"A" = Flange Thickness + Plywood Thickness Less 1/8" for Tightness (1 3/8" = Minimum "A")

**Safe Working Load
100 lbs. per Side**

S.W.L. provides a factor of safety of approximately 2 to 1.
Warning: Do not weld haunch carrier to beam!

To Order:
Specify: (1) quantity, (2) type, (3) "A", "B", "C" and "D" dimensions, (4) break back
Example:
500 pcs. C-28 Exterior Haunch Carrier "A" = 1 3/8" "B" = 12 1/8" "C" = 3/4", Break Back 3/4".

C-29 Fillet Clip

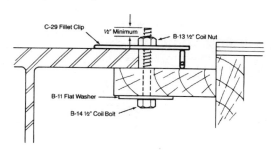

For use in supporting 3" to 4" wide fillets.

**Safe Working Load
350 lbs. per Clip**

S.W.L. provides a factor of safety of approximately 2 to 1.

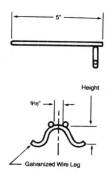

Fillet Clip Detail

12-84B

Vertical formwork charts

The following charts were supplied by Dayton Superior, *The Formwork Accessory Book*.

General and Technical Information

The safety factor to be applied to a particular product is a variable, depending on the degree of hazard of risk involved in the application of that product. In concrete construction various job site conditions can often increase the degree of risk. Concentrated loads of reinforcing bars, storage of construction materials on the formwork, unsymmetrical placement of concrete, uplift, impact of machine delivered concrete, use of motorized carts and formwork height are some of the conditions that have high risk factors. Safety factors must be increased accordingly by the user to reduce these risks. Dayton Superior recommends that the provisions of the American National Standards Institute (ANSI A10.9), OSHA (Occupational Safety and Health Administration Act, Part 1910) and the American Concrete Institute's "Guide to Formwork for Concrete," (ACI 347) be strictly followed when considering safety factors. It is for this reason that we state the safe working loads of our products and only the approximate minimum safety factor. We especially advise that the minimum safety factors as listed below be adhered to. When there are unusual job conditions, such as shock, impact or vibration, these minimum safety factors must be increased by the user.

Minimum Safety Factors of Formwork Accessories Per ACI 347-89		
Accessory	Factor of Safety	Type of Construction
Form Tie	2.0 to 1	All applications
Form Anchor	2.0 to 1	Formwork supporting form weight and concrete pressures only.
Form Anchor	3.0 to 1	Formwork supporting weight of forms, concrete, construction live loads and impact
Form Hangers	2.0 to 1	All applications
Anchoring Inserts (Used as Form Ties)	2.0 to 1	Precast concrete panels when used as formwork

If a larger safety factor than the one shown in this handbook is required for any reason, a product's safe working load must be changed accordingly by the user. The following equation is used to reduce a safe working load when a larger factor of safety is required:

$$\frac{\text{Published SWL} \times \text{Published Factor of Safety}}{\text{Required Factor of Safety}} = \text{New SWL}$$

THE USER OF DAYTON SUPERIOR PRODUCTS MUST EVALUATE THE PRODUCT APPLICATION, DETERMINE THE APPROPRIATE SAFETY FACTOR, CALCULATE THE SAFE WORKING LOAD AND CONTROL ALL FIELD CONDITIONS TO PREVENT APPLICATION OF LOADS IN EXCESS OF THE SAFE WORKING LOADS.

Factors Affecting a Product's Safe Working Load

Failure of coil bolts, coil rods, she-bolts and similar forming accessories is generally caused by excessive wear on the threads, field modification or bending and restraightening. It shall be the responsibility of the user to continually inspect coil bolts, coil rods, she-bolts and other related items for wear and to discard parts when wear is noted. Do not restraighten bent forming accessories, rather discard and replace them. Also discard any reuseable item that has been used at 70% of ultimate load or more. Such items may have been stretched sufficiently to become brittle hard. Every user must establish a control program that replaces reuseable forming accessories after a predetermined number of uses, regardless of their exterior appearance. All reuseable forming accessories shown in this handbook are subject to wear, misuse, overloading, corrosion, deformation, intentional alteration and other factors which may affect the products safe working load. Therefore, all reuseable items must be inspected by the user to determine their overall condition and whether the item may be used at its rated load or if it should be removed from service. The frequency of inspection is dependent upon factors such as, frequency of use, period of use, environment and is best determined by the user consistent with good construction practice.

When in doubt about the proper use or installation of Dayton Superior's forming accessories and hardware, contact Dayton Superior for clarification. Failure to do so may result in exposure of workers to safety hazards, resulting in the possibility of injury or death to workers in the vicinity of the job site.

Formwork and Falsework

DAYTON SUPERIOR®

Wall Formwork Hardware Selection

There is a large variety of wall formwork hardware available from Dayton Superior for concrete formwork construction and that variety may, perhaps, make form hardware selection a difficult decision. However, by classifying this hardware in two ways, it will help simplify the selection process.

The first method is to classify form ties by their use as listed below:

Through Ties

These ties are so named because the entire tie rod extends through the wall as well as through both form sides. There are four types of through ties offered by Dayton Superior—Rod clamps and pencil rod, snap ties and wedges, loop ties and wedge bolts or taper ties which use nut washers as the clamping member.

Advantages—Inexpensive, easy to install, readily available and carpenters and laborers are generally familiar with the required installation procedures. Pencil rod is a field emergency tie for unusual or unexpected forming situations.

Limitations—Pencil rod and taper ties do not have a form spreader feature. Pencil rod must be chiseled back for the break back, taper ties leave a hole through the wall and patching costs can be expensive. Pencil rod and snap ties may cause rust stains on exposed walls after a period of time. Loop ties require the use of special forms.

She-Bolts

The she-bolt has external threads on the outer end to provide adjustment for variable form thicknesses. The inside rod is threaded at each end and fits into the threaded tapered end of the she-bolt. Various sizes of washers for bearing on the wales are available, as well as wing nuts or hex nuts to fit on the running thread of the she-bolt.

Advantages—She-bolts, inside rods and hardware can be bench assembled; both sides of the form can be erected and the she-bolt assembly fed through the forms. Inside rods are available with machine threads (NC) or coil threads and the contractor can make up emergency inside rods by cutting all thread or coil rod to the required length. The she-bolt system can be used for various form grips or form thicknesses and can be used to re-anchor to an inside rod in a previous placement. Coil threads have a faster pitch than machine threads, are self-cleaning and will not cross thread.

Limitations—No positive internal spreader action unless adaptor cones are used, then it requires seven pieces to make up an assembly. Also, if adaptor cones are used the she-bolt assembly can not be fed through both sides of the forms. Reinforcing steel may be in the way when feeding the tie through the forms. She-bolt has a high initial cost but can be reused hundreds of times depending on care; danger of cross threading with machine thread. Also, it is difficult to inspect the nose of the she-bolt for wear.

Coil Ties

The coil tie system consists of he-bolts, washers, screw-on cones and a coil tie. The he-bolts are fed through the round or flat washers then through one of the form sides. The plastic screw-on cone is an accessory that acts as an internal form spreader and assures a proper setback of the coil tie from the face of the form. Continous coil threaded rod can be cut in the field to handle emergency conditions, and a wing nut would then be the final clamping device.

Advantages—Coil threads are self-cleaning, no chance of cross threading, easy to examine hardware, positive attachment to one side of the forms so iron workers cannot place reinforcing steel in front of the tie hole, working parts can usually be rented, excellent re-anchor tie.

Limitations—High initial cost for the external hardware, cannot be used as a feed through tie assembly, one side of the forms must be erected, form ties attached, reinforcing steel positioned and the closure form erected, and the remaining hardware installed.

The second method of classifying form ties is by their load carrying capacity:

Light Duty

Light duty ties have safe working loads of 3,750 lbs. or less. Types of light duty ties are pencil rod, snap ties, loop ties and flat ties.

Forms for light duty ties are generally built up at the job site with ¾" plyform, 2" x 4" studs and double 2" x 4" wales. The **Jahn** system, however, is a single liner system and is built with ¾" plyform and single 2" x 4" horizontal wales or vertical studs. Forms for flat ties and loop ties are generally factory built.

Heavy Duty

Heavy duty ties have safe working loads of 4,500 lbs. or more and consist of she-bolts, coil ties and taper ties. When ties having safe working loads of 9,000 lbs. or less are used, the forms for this type of hardware will consist of ¾" plyform, 2" x 4" or 2" x 6" studs and double 2" x 6" or 3" x 6" wales. Tie and wale spacing will usually be spaced between 24"-48" on center. Typical form designs for heavy duty ties with safe working loads of 12,000 lbs. or more usually have ¾" plyform, 2x studs, 2x or 3x wales or steel/aluminum channels. Sometimes these forms are built completely as an all steel form.

Lumber and Form Tie Cost Analysis

Assume your project contains 100,000 sq. ft of form contact area, 12" thick walls x 14'-0" high and that 10,000 sq. ft. of form will be constructed. Concrete schedule will be 6 months with form reuse based on 3 uses per month.

Assume that working parts are purchased. Experience has shown that working parts for Example A have a life of 10 uses and for Example B a life of 50 uses. Form lumber in Example A has a salvage value of 25% while Example B has a salvage value of 60%.

For this analysis we have omitted the cost of nails, band iron, connecting bolts, lifting devices, etc. Example B was calculated in the same manner as Example A, with the exception of the number of uses of working parts and higher lumber salvage value as noted.

Example A	Example B
¾" Plywood	¾" Plywood
2"x4" Studs @ 12" o.c.	2"x4" Studs @ 6 o.c.
2-2"x4" Wales @ 24" o.c.	2-3"x6" Wales @ 24" o.c.
A-3, A-4 or A-44 Standard	B-1 Heavy Coil Ties @ 32" o.c.
Snap Ties @ 24" o.c.	D-1 or D-18 Inside Rod with She-Bolts @ 32" o.c.
Rate of Placement: 50° F. = 2¼ ft./hr. 70° F. = 3¼ ft./hr.	**Rate of Placement:** 50° F. = 10 ft./hr. 70° F. = 10 ft./hr.

Refer to "Typical Formwork Designs for Wall Forms" footnotes for data regarding allowable stresses for plywood and lumber, concrete temperature and short term loading conditions.

The two formwork design examples shown present average costs for lumber and form ties. They are intended as examples only, serving the reader an opportunity to prepare similar cost analyses of specific formwork designs. Labor cost is a factor that must be added to the material costs, remembering that for Example A 12,500 ties must be installed and removed and that 25,000 tie holes must be patched, while in Example B only 9,375 ties must be installed and removed, with 18,750 holes requiring patching.

These comparative figures illustrate the advantage of "balanced" formwork designs; proper capacity form ties matched with appropriate lumber size and strength result in an efficient and economical formwork design. Also evident is the small material cost difference of building a "strong" form compared to a "light-duty" form. Users must account for the significant labor cost difference of installing, removing and patching of the additional form ties.

Note also that the placement rate for Example B is 4½ times greater than Example A. The placing crew savings in cost must also be considered to arrive at the total in place cost per unit of measure.

Calculations for Formwork Costs

Description	Example A	Example B
a) Form Contact Area per Tie = $\frac{\text{Wale Center (in.)}}{12} \times \frac{\text{Tie Centers (in.)}}{12} \times 2$	8 sq. ft.	10.67 sq. ft.
b) Unit Cost of Tie	$0.49	$0.65
c) Tie Cost per sq. ft of Form Contact Area = (b ÷ a)	$0.061	$0.061
d) Working Part Cost/Tie	$1.60	$21.90
e) Working Part Cost/Tie per sq. ft. of Form Contact Area per Use = (d ÷ a)	$0.02	$0.041
f) Total Tie Cost per sq. ft. of Form Contact Area per Use = (c+e)	$0.081	$0.102
g) Board Feet of Lumber per sq. ft. of Form, Excluding Plywood. See note below.	1.43	2.76
h) Material Cost of Lumber per sq. ft of Form. See note below.	$1.13	$1.62
i) Lumber cost per sq. ft. of Form Contact Area = $\frac{(10{,}000 \text{ sq. ft.})(\ h\ )}{100{,}000 \text{ sq. ft. of Form Contact Area}}$	$0.113	$0.162
j) Salvage Value per sq. ft. of Form Contact Area = (i x .25) or (i x .60)	= $0.028	$0.065
k) Net Lumber Cost per sq. ft. of Form Contact Area = (i—j)	= $0.08	$0.097
l) Total Formtie and Lumber Cost per sq. ft. of Form Contact Area = (f+k)	= $0.166	$0.199
m) Total Number of Ties Required	12,500 pcs.	9,375 pcs.

Check (Example A):
Total Tie Cost = (0.081)(100,000) = $ 8,100
Total Lumber Cost = (1.13)(10,000)(.75) = 8,500
TOTAL COST = $16,600

Cost per sq. ft. of Form Contact Area = $\frac{\$16{,}600}{100{,}000}$ = $0.166 This is okay as it checks out with line l.

Note: Depending upon local prices, the plywood and structural lumber costs may be separated as follows:

¾" Plyform Class 1, Grade B-B = $0.59/sq. ft.
1.43 bd. ft. @ $340/M = 0.49
Bracing Lumber @ 10% = 0.05
Total Lumber Cost/sq. ft. = $1.13/sq. ft.

Chart for Determining Required Quantities of Form Ties

Form Tie Calculator Based on 10,000 sq. ft. of Wall Area or 20,000 sq. ft. of Form Contact Area	
Form Tie Spacing	Form Ties Required
16" x 16" = 1.77 sq. ft.	5,650
24" x 24" = 4.0 sq. ft.	2,500
24" x 32" = 5.33 sq. ft.	1,877
32" x 32" = 7.11 sq. ft.	1,407
32" x 48" = 10.67 sq. ft.	938
48" x 48" = 16 sq. ft.	625
60" x 60" = 25 sq. ft.	400

Typical Formwork Designs for Wall Forms

The table below lists several of the most common form lumber sizes and spacings that are being used in the industry today. For each formwork design the appropriate form tie is shown.

Typical Formwork Designs

Recommended Form Ties	Form Tie Safe Working Load (lbs.)	Maximum Rate of Placement Vertical Feet per Hour		Maximum Form Tie Spacings		Form Design			
						Single Vertical Studs		Double Horizontal Wales	
		50°F.	70°F.	Vertical	Horizontal	Size	Centers	Size	Centers
A-3, A-4 or A-44 Snap Ties, Standard	2,250	2¼	3¼	24"	24"	2"x4"	12"	2"x4"	24"
		5¾	10	16"	16"	2"x4"	8"	2"x4"	16"
A-3 Snap Tie, Heavy	3,350	2⅔	3⅔	24"	32"	2"x4"	12"	2"x4"	24"
		3¾	5⅓	24"	24"	2"x4"	8"	2"x4"	24"
B-1 Coil Tie, Standard D-19 Taper Tie	4,500	2⅔	3¾	32"	32"	2"x6"	12"	2"x6"	32"
		5⅓	8¾	24"	24"	2"x4"	8"	2"x6"	24"
D-19 Taper Tie	6,000	3¾	5⅓	32"	32"	2"x6"	8"	2"x6"	32"
		5⅓	8¾	24"	32"	2"x4"	8"	2"x8"	24"
		5⅔	10	24"	24"	2"x4"	8"	2"x4"	24"
B-1 Coil Tie, Heavy	6,750	2⅔	3¾	32"	48"	2"x6"	12"	2"x8"	32"
		4⅓	6¼	32"	32"	2"x4"	8"	2"x6"	32"
		6	10	24"	32"	2"x4"	8"	2"x6"	24"
B-1 Coil Tie, Heavy D-1 or D-18 Inside Rod with She-Bolts	9,000	3⅓	4⅔	32"	48"	2"x4"	8"	2"x8"	32"
		3⅔	5	24"	32"	2"x4"	8"	2"x6"	24"
		10	10	24"	32"	2"x4"	6"	3"x6"	24"
D-9 Taper Tie	18,000	5⅓	8¾	48"	48"				
B-2 Coil Tie, Standard	18,000	5⅓	8¾	48"	48"				
D-1 or D-18 Inside Rod with She-Bolts	18,000	5⅓	8¾	48"	48"	Aluminum or steel studs and wales are normally used for these conditions.			
B-2 Coil Tie, Standard	27,000	5	7	60"	60"				
D-9 Taper Tie	34,000	4⅔	6¾	60"	60"				
D-1 or D-18 Inside Rod with She-Bolts	37,500	5	7	72"	72"				
D-9 Taper Tie	40,500	5⅓	8¾	72"	72"				

Note: The above table is based on the following conditions:

- **Concrete** — Made with type 1 cement weighing 150 pcf, contains no admixtures, slump of 4" or less and normal internal vibration to a depth of 4 ft. or less. If your conditions vary contact Dayton Superior for additional recommendations.
- **Concrete Temperature** — For practical purposes, 50°F. is used by many form designers as the temperature of fresh concrete during winter, with 70°F. being used as the summer temperature. This "rule of thumb" appears to work satisfactory unless the concrete has been heated or cooled to a controlled temperature.
- **Plywood Sheathing** — ¾" plyform class 1 or structural 1 used the strong direction. Experience has shown that ¾" plywood is more economical in form usage than other thicknesses even though initial cost may be slightly more. Deflection has been limited to $\ell/360$ or $1/16$" whichever is less and plyform is supported by four or more studs.
- **Studs** — Fiber stress in bending = 1,500 psi, modulus of elasticity = 1,700,000 psi, horizontal shear = 238 psi, deflection limited to $\ell/270$ or ⅛" whichever is less with studs continuous over four or more wales.
- **Double Wales** — Fiber stress in bending = 1,500 psi, modulus of elasticity = 1,700,000 psi, horizontal shear = 238 psi, deflection limited to $\ell/270$ or ¼" whichever is less with wales continuous over four or more ties.
- **Short Term Loading Conditions** — Allowable stresses, except for modulus of elasticity include a 25% increase for short term loading.
- **Form Ties** — Safe working loads are based on a factor of safety of approximately 2 to 1 (ultimate to SWL).

Vertical Formwork Design Loads

The selection of the proper sheathing, studs and/or wales for concrete formwork requires a knowledge of the maximum lateral pressure which will be exerted by the concrete. Dayton Superior is in agreement with the **Lateral Pressure Design Formulas** contained in the American Concrete Institute's, "Guide to Formwork for Concrete", (ACI 347 latest revision). Designers of formwork for concrete walls or columns will find the following information useful:

- For general purpose conditions and unless the special conditions listed below are met, all formwork should be designed for the lateral pressure of the newly placed concrete using the formula of;

    $P = W \times H$

    Where P = lateral pressure, pounds per square foot;

    W = unit weight of fresh concrete, pounds per cubic foot or 150 pcf for normal weight concrete;

    H = depth of fluid or plastic concrete in feet. (Normally height of wall or column form.)

    Please note that the maximum and minimum values given for the formulas under the special conditions do not apply to the above lateral pressure formula.

- **Special Condition No. 1** — For concrete made with type 1 cement, weighing 150 pounds per cubic foot, containing no pozzolans or admixtures, having a slump of 4″ or less and normal internal vibration to a depth of 4 ft. or less. Then the formwork may be designed for a lateral pressure as follows:

    For columns;

    $P = 150 + \dfrac{9{,}000 \times R}{T}$

    with a maximum of 3,000 pounds per square foot, a minimum of 600 pounds per square foot, but in no case greater than $W \times H$.

    For walls with a rate of placement less than 7 ft. per hour;

    $P = 150 + \dfrac{9{,}000 \times R}{T}$

    with a maximum of 2,000 pounds per square foot, a minimum of 600 pounds per square foot, but in no case less than $W \times H$.

    For walls with a rate of placement of over 7 ft. per hour but less than 10 ft. per hour;

    $P = 150 + \dfrac{43{,}400}{T} + \dfrac{2800 \times R}{T}$

    with a maximum of 2,000 pounds per square foot, a minimum of 600 pounds per square foot, but in no case less than $W \times H$.

    Where P = lateral pressure, pounds per square foot;

    R = rate of placement, feet per hour, and

    T = temperature of concrete in the form, degree fahrenheit. For practical purposes, 50°F. is used by many form designers as the temperature of fresh concrete during the winter, with 70°F. being used as the summer temperature. This "rule of thumb" appears to work satisfactorily unless the concrete has been heated or cooled to a controlled temperature.

- **Special Condition No. 2** — If concrete is to be pumped from the base of the form, the form should be designed for a full hydrostatic head of concrete ($W \times H$) plus a minimum allowance of 25% for pump surge pressure. In certain instances pressures may be as high as the face pressure of the pump piston.

- **Special Condition No. 3** — Caution must be taken when using external vibration or concrete made with shrinkage compensating or expansive cements. Pressure in excess of equivalent hydrostatic may occur.

Wall forms should be designed to meet wind load requirements of American National Standards Institute A-58.1 or of the local building code, whichever is more stringent. The minimum wind design load should be 15 pounds per square foot. Bracing for wall forms should also be designed for a horizontal load of at least 100 pounds per lineal foot of wall, applied at the top of the form.

Lateral Pressure of Concrete for General Purpose Conditions

Depth of Fluid or Plastic Concrete In Feet	Pounds Per Square Foot
4	600
5	750
6	900
7	1,050
8	1,200
9	1,350
10	1,500
12	1,800
14	2,100
16	2,400
18	2,700
20	3,000

Lateral Pressure of Concrete for Special Condition No. 1 — Walls

Rate of Placement Feet Per Hour	Pounds per Square Foot for Indicated Temperature	
	50°F.	70°F.
2	600	600
3	690	600
4	870	664
5	1,050	793
6	1,230	921
7	1,410	1,050
8	1,466	1,090
9	1,522	1,130
10	1,578	1,170

Note: Do not use lateral pressures in excess of 150 x height of fluid or plastic concrete in forms.

12-88A

Points to Remember

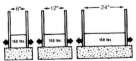

Fluid or plastic concrete exerts the same side pressure on forms regardless of their width.

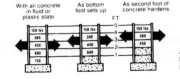

As concrete hardens, lateral pressure on forms decreases.

As you add more fluid or plastic concrete to forms, the pressure will build up toward the bottom at about the rate of 150 pounds per foot of depth. This will be true as long as all concrete remains in a plastic state.

Example: Eight feet of fluid or plastic concrete bears on the bottom foot of forms with a pressure of 8 x 150 pounds or 1200 pounds per square foot.

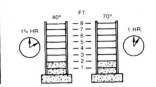

Concrete sets up or hardens faster with an increase in temperature.

Example: At 70°F. concrete sets in approximately 1 hour. At 40°F. concrete will set up in about 1¾ hours.

Slab Formwork Design Loads

The loadings used in the design of slab formwork consists of a dead load and a live load. The weight of the formwork plus the concrete is considered dead load while the live load is made up of the weight of workers, equipment, material storage and other like items which is supported by the formwork. The tables below tabulate design loads based on the concrete weight for the thicknesses indicated, and includes 10 pounds per square foot for the weight of forms and a live load of 50 or 75 pounds per squre foot as indicated. A live load of 75 pounds per square foot is generally used when motorized carts are used to transport concrete during the placing operation.

Slab Formwork Design Load for Uniform Slab Thickness (Includes 50 psf Live Load)									
Pounds per Square Foot for Indicated Thickness									
2"	4"	6"	8"	10"	12"	14"	16"	18"	20"
100	110	135	160	185	210	235	260	285	310

Note: Chart is based on a concrete weight of 150 pounds per cubic foot.

Slab Formwork Design Load for Uniform Slab Thickness (Includes 75 psf Live Load)									
Pounds per Square Foot for Indicated Thickness									
2"	4"	6"	8"	10"	12"	14"	16"	18"	20"
110	135	160	185	210	235	260	285	310	335

Note: Chart is based on a concrete weight of 150 pounds per cubic foot.

For a complete explanation of general objectives in formwork design, planning, materials and accessories, loads and pressures, design tables and much more, it's recommended that a copy of ACI publication SP-4 "Formwork for Concrete" be obtained. The current edition is available from American Concrete Institute, P.O. Box 19150, Redford Station, Detroit, Michigan 48219.

12-88A

Formwork and Falsework   61

DAYTON SUPERIOR®

Technical Data—Plywood

Data is based on information supplied by the American Plywood Association (APA). The recommended spacings listed in the following table are for Plyform Class 1 or STRUCTURAL 1 Plyform. Plyform is a special exterior type of plywood designed by APA for use in formwork for concrete construction.

Though not manufactured specifically for concrete forming, grades other than Plyform have been used in formwork. The spacings shown in the table give a good estimate of performance for sanded grades such as APA A-C Exterior, APA B-C Exterior and unsanded grades such as APA RATED SHEATHING Exterior and Exposure 1 (CDX) (marked PSI), provided the plywood is used in the strong direction only.

For additional information on APA Plyform, please contact the American Plywood Association, PO Box 11700, Tacoma, WA 98411.

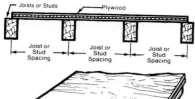

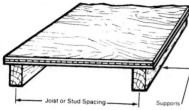

Plywood Used Strong Way
Face Grain Across Supports

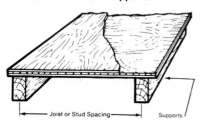

Plywood Used Weak Way
Face Grain Along Supports

Safe Spacing in inches of Support for Plyform Sheathing Continuous Over Four or More Supports

$F_b = 1{,}930$ psi; Rolling Shear $= 72$ psi
$E = 1{,}500{,}000$ psi

Design Load of Concrete Pounds Per Sq. Ft.	Plyform Used Weak Way				Plyform Used Strong Way			
	19/32″	5/8″	23/32″	3/4″	19/32″	5/8″	23/32″	3/4″
100	12″	14″	17″	18″	20″	21″	23″	24″
125	12″	13″	16″	17″	19″	19″	21″	22″
150	11″	12″	15″	16″	17″	18″	20″	21″
175	10″	11″	14″	15″	16″	17″	19″	20″
200	10″	11″	13″	14″	16″	16″	18″	19″
225	9″	10″	13″	14″	15″	16″	18″	18″
250	9″	10″	12″	13″	15″	15″	17″	17″
275	9″	10″	12″	13″	14″	15″	16″	17″
300	9″	9″	12″	13″	14″	14″	16″	16
350	8″	9″	11″	12″	13″	13″	15″	16″
400	8″	8″	11″	11″	12″	13″	14″	15″
500	7″	8″	10″	11″	11″	12″	14″	14″
600	6″	7″	9″	9″	11″	11″	12″	13″
700	5″	6″	7″	8″	10″	10″	12″	12″
800	5″	5″	6″	7″	9″	10″	11″	11″
900	4″	4″	6″	6″	9″	9″	10″	10″
1,000	4″	4″	5″	5″	8″	9″	10″	10″
1,200	3″	3″	4″	4″	7″	7″	8″	8″
1,400	—	3″	3″	4″	6″	6″	7″	7″
1,600	—	—	3″	3″	5″	5″	6″	6″
1,800	—	—	3″	3″	4″	5″	5″	5″
2,000	—	—	—	3″	4″	4″	5″	5″

Support spacings are governed by bending, shear or deflection. Maximum deflection 1/360 of spacing, but not more than 1/16″. Contact Dayton Superior for safe spacing of supports when plyform is used over two or three supports.

Curved Forms: Plyform can be used for building curved forms. However, the following radii have been found to be appropriate minimums for mill run panels of the thicknesses shown, when bent dry. An occasional panel may develop localized failure at these radii.

Plywood Data

Plywood Thickness	Approximate Weight, lbs.		Minimum Bending Radii, Ft.	
	4 x 8 Sheet	Sq. Ft.	Across Grain	Parallel to Grain
1/4″	26	.8	2	5
5/16″	32	1.0	2	6
11/32″ or 3/8″	35	1.1	3	8
15/32″ or 1/2″	48	1.5	6	12
19/32″ or 5/8″	58	1.8	8	16
23/32″ or 3/4″	70	2.2	12	20

12-88A

Technical Data—Lumber

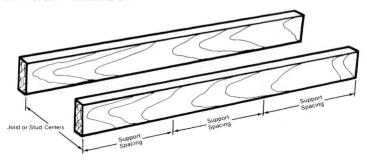

Safe Spacing of Supports for Joists or Studs
Continuous Over Four or More Supports

F_b = 1,500 psi   E = 1,700,000 psi F_v = 238 psi

Uniform Load, Pounds per Lineal Foot (Equals Design Load, Pounds per Sq. Ft., Times Joist or Stud Centers in Feet.)	Nominal Size Lumber, bxh (S4S) at 19% Maximum Moisture					
	2 x 4	2 x 6	2 x 8	3 x 6	4 x 2	4 x 4
100	66"	93"	115"	106"	43"	82"
200	56"	78"	96"	89"	34"	69"
300	42"	67"	87"	80"	28"	62"
400	37"	58"	76"	75"	24"	56"
500	33"	52"	68"	67"	21"	50"
600	30"	47"	62"	61"	19"	46"
700	28"	44"	58"	56"	18"	42"
800	26"	41"	54"	53"	17"	40"
900	24"	38"	51"	50"	16"	37"
1,000	23"	36"	48"	47"	15"	35"
1,100	22"	34"	45"	45"	14"	34"
1,200	20"	32"	43"	43"	14"	32"
1,300	19"	31"	40"	41"	13"	31"
1,400	18"	29"	39"	40"	12"	30"
1,500	18"	28"	37"	38"	12"	29"
1,600	17"	27"	36"	37"	12"	28"
1,700	16"	26"	34"	36"	11"	27"
1,800	16"	25"	33"	35"	11"	26"
1,900	15"	24"	32"	33"	11"	26"
2,000	15"	24"	31"	32"	10"	25"
2,200	14"	22"	30"	30"	10"	24"
2,400	13"	21"	28"	29"	9"	23"
2,600	13"	21"	27"	27"	9"	21"
2,800	12"	20"	26"	26"	8"	20"
3,000	12"	19"	25"	25"	8"	19"

Note: F_b and F_v shown above includes a 25% increase because of short term loading conditions.

Support spacings are governed by bending, shear or deflection. Maximum deflection 1/270 of spacing, but not more than ⅛". Contact Dayton Superior for safe spacings of supports for joists or studs used over two or three supports.

12-88A

Technical Data—Lumber

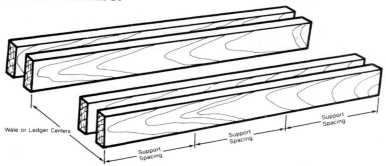

Safe Spacing of Supports for Double Ledgers or Wales Continuous Over Four or More Supports

F_b = 1,500 psi   E = 1,700,000 psi F_v = 238 psi

Nominal Size Lumber, bxh (S4S) at 19% Maximum Moisture

Uniform Load, Pounds per Lineal Foot (Equals Design Load, Pounds per Sq. Ft., Times Ledger or Wale Centers in Feet.)	Double 2 x 4	Double 2 x 6	Double 2 x 8	Double 3 x 6	Double 3 x 8
1,000	33"	52"	68"	67"	88"
1,100	31"	49"	65"	64"	84"
1,200	30"	47"	62"	61"	81"
1,300	29"	45"	60"	59"	77"
1,400	28"	44"	58"	56"	75"
1,500	27"	42"	56"	55"	72"
1,600	26"	41"	54"	53"	70"
1,700	25"	40"	52"	51"	68"
1,800	24"	38"	51"	50"	66"
1,900	24"	37"	49"	48"	64"
2,000	23"	36"	48"	47"	62"
2,200	22"	34"	46"	45"	59"
2,400	20"	32"	44"	43"	57"
2,600	19"	31"	42"	41"	55"
2,800	18"	29"	41"	40"	53"
3,000	18"	28"	39"	38"	51"
3,200	17"	27"	38"	37"	49"
3,400	16"	26"	37"	36"	48"
3,600	16"	25"	36"	35"	46"
3,800	15"	24"	35"	33"	45"
4,000	15"	24"	34"	32"	43"

Note: F_b and F_v shown above includes a 25% increase because of short term loading conditions.

Support spacings are governed by bending, shear or deflection. Maximum deflection 1/270 of spacing, but not more than ¼". Contact Dayton Superior for safe spacings of supports for ledgers or wales used over two or three supports

12-88A

Technical Data—Lumber

Formulas for Calculating Safe Support Spacings of Lumber Formwork Members

To Check	for Single Span Beam	for Two-Span Beam	for Three or More Span Beam
$\Delta_{max} = \ell/360$	$\ell = 1.37 \sqrt[3]{\frac{EI}{w}}$	$\ell = 1.83 \sqrt[3]{\frac{EI}{w}}$	$\ell = 1.69 \sqrt[3]{\frac{EI}{w}}$
$\Delta_{max} = \ell/270$	$\ell = 1.51 \sqrt[3]{\frac{EI}{w}}$	$\ell = 2.02 \sqrt[3]{\frac{EI}{w}}$	$\ell = 1.86 \sqrt[3]{\frac{EI}{w}}$
$\Delta_{max} = 1/16$ in.	$\ell = 2.75 \sqrt[4]{\frac{EI}{w}}$	$\ell = 3.43 \sqrt[4]{\frac{EI}{w}}$	$\ell = 3.23 \sqrt[4]{\frac{EI}{w}}$
$\Delta_{max} = 1/8$ in.	$\ell = 3.27 \sqrt[4]{\frac{EI}{w}}$	$\ell = 4.08 \sqrt[4]{\frac{EI}{w}}$	$\ell = 3.84 \sqrt[4]{\frac{EI}{w}}$
$\Delta_{max} = 1/4$ in.	$\ell = 3.90 \sqrt[4]{\frac{EI}{w}}$	$\ell = 4.85 \sqrt[4]{\frac{EI}{w}}$	$\ell = 4.57 \sqrt[4]{\frac{EI}{w}}$
Bending	$\ell = 9.80 \sqrt[2]{\frac{F_b S}{w}}$	$\ell = 9.80 \sqrt[2]{\frac{F_b S}{w}}$	$\ell = 10.95 \sqrt[2]{\frac{F_b S}{w}}$
Horizontal Shear	$\ell = \frac{16 F_v bh}{w} + 2h$	$\ell = \frac{192 F_v bh}{15w} + 2h$	$\ell = \frac{40 F_v bh}{3w} + 2h$

Notation:

A = area of cross section, sq. in.
b = width of section, in.
E = modulus of elasticity, psi
F_b = design value for extreme fiber in bending, psi
F_v = design value in horizontal shear, psi
F_c = design value in compression parallel to grain, psi
$F_{c\perp}$ = design value in compression perpendicular to grain, psi
h = depth of section, in.
I = moment of interia, in.[4]
ℓ = safe spacing of supports, in.
S = section modulus, in.[3]
w = load, lbs. per lineal ft.
Δ = deflection, in.

Design Stress Values (psi) for Lumber at 19 Percent Moisture Content, Continuing or Prolonged Reuse

Grade	Size	F_b	F_v	$F_{c\perp}$	F_c	E
Dense No. 1	2" x 4" or 4" x 4"	2,050	190	730	1,450	1,900,000
No. 1		1,750	190	625	1,250	1,800,000
Dense No. 2		1,700	190	730	1,150	1,700,000
No. 2		1,450	190	625	1,000	1,700,000
No. 3		800	190	625	600	1,500,000
Appearance		1,750	190	625	1,500	1,800,000
Stud		800	190	625	600	1,500,000
Dense No. 1	2" x 6" and wider or 4" x 6" and wider	1,800	190	730	1,450	1,900,000
No. 1		1,500	190	625	1,250	1,800,000
Dense No. 2		1,450	190	730	1,250	1,700,000
No. 2		1,250	190	625	1,050	1,700,000
No. 3		725	190	625	675	1,500,000
Appearance		1,500	190	625	1,500	1,800,000
Stud		725	190	625	675	1,500,000
Increase for load duration of 7 days or less (short term loading)		25%	25%	0%	25%	0%

Species: Douglas Fir — Larch

Formwork and Falsework

DAYTON SUPERIOR®

Technical Data—Lumber

Species: Southern Pine

Grade	Size	F_b	F_v	$F_{c\perp}$	F_c	E
No. 1 Dense	2" x 4" or 4" x 4"	2,000	200	660	1,450	1,800,000
No. 2		1,400	180	565	975	1,600,000
No. 2 Dense		1,650	180	660	1,150	1,600,000
No. 3		775	180	565	575	1,400,000
No. 3 Dense		925	180	660	675	1,500,000
Stud		775	180	565	575	1,400,000
Construction		1,000	200	565	1,100	1,400,000
No. 1 Dense	2" x 6" and wider or 4" x 6" and wider	1,700	180	660	1,450	1,800,000
No. 2		1,200	180	565	1,000	1,600,000
No. 2 Dense		1,400	180	660	1,200	1,600,000
No. 3		700	180	565	625	1,400,000
No. 3 Dense		825	180	660	725	1,500,000
Stud		725	180	565	625	1,400,000
Increase for load duration of 7 days or less (short term loading)		25%	25%	0%	25%	0%

Species: Hem — Fir

Grade	Size	F_b	F_v	$F_{c\perp}$	F_c	E
No. 1	2" x 4" or 4" x 4"	1,400	150	405	1,050	1,500,000
No. 2		1,150	150	405	825	1,400,000
No. 3		650	150	405	500	1,200,000
Appearance		1,400	150	405	1,250	1,500,000
Stud		650	150	405	500	1,200,000
Construction		825	150	405	925	1,200,000
No. 1	2" x 6" and wider or 4" x 6" and wider	1,200	150	405	1,050	1,500,000
No. 2		1,000	150	405	875	1,400,000
No. 3		575	150	405	550	1,200,000
Appearance		1,200	150	405	1,250	1,500,000
Stud		575	150	405	550	1,200,000
Increase for load duration of 7 days or less (short term loading)		25%	25%	0%	25%	0%

Species: Eastern Spruce

Grade	Size	F_b	F_v	$F_{c\perp}$	F_c	E
No. 1	2" x 4" or 4" x 4"	1,200	140	390	825	1,500,000
No. 2		975	140	390	650	1,400,000
No. 3		550	140	390	400	1,200,000
Appearance		1,200	140	390	1,000	1,500,000
Stud		550	140	390	400	1,200,000
Construction		700	140	390	750	1,200,000
No. 1	2" x 6" and wider or 4" x 6" and wider	1,000	140	390	825	1,500,000
No. 2		825	140	390	700	1,400,000
No. 3		475	140	390	450	1,200,000
Appearance		1,000	140	390	1,000	1,500,000
Construction		475	140	390	450	1,200,000
Increase for load duration of 7 days or less (short term loading)		25%	25%	0%	25%	0%

The above charts have been derived from the "National Design Specification For Wood Construction" published by National Forest Products Association, 1619 Massachusetts Ave., N.W., Washington, D.C. 20036

12-88A

DAYTON SUPERIOR®

Technical Data—Lumber

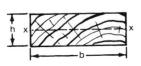

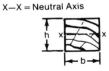

X—X = Neutral Axis

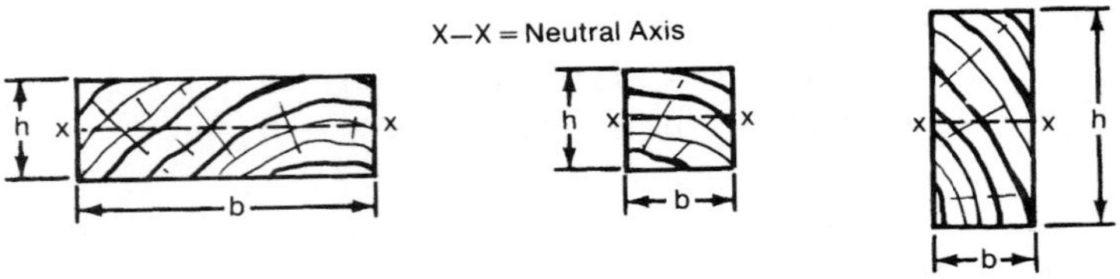

Properties of Structural Lumber										
Nominal Size in Inches, *bxh*	American Standard Sizes in Inches, *bxh* S4S* 19% Maximum Moisture	Area of section $A = bh$, sq. in.		Moment of Inertia, in.^{4} $I = \frac{bh^3}{12}$		Section Modulus, in.^{3} $S = \frac{bh^2}{6}$		Board Feet per Lineal Foot of Piece	Approx. Weight per Lineal Foot (lbs.) of S4S Lumber
		Rough	S4S	Rough	S4S	Rough	S4S		
4x1	3½x¾	3.17	2.62	0.20	0.12	0.46	0.33	⅓	.7
6x1	5½x¾	4.92	4.12	0.31	0.19	0.72	0.52	½	1.0
8x1	7¼x¾	6.45	5.44	0.41	0.25	0.94	0.68	⅔	1.4
10x1	9¼x¾	8.20	6.94	0.52	0.32	1.20	0.87	⅚	1.7
12x1	11¼x¾	9.95	8.44	0.63	0.39	1.45	1.05	1	2.1
4x2	3½x1½	5.89	5.25	1.30	0.98	1.60	1.31	⅔	1.3
6x2	5½x1½	9.14	8.25	2.01	1.55	2.48	2.06	1	2.0
8x2	7¼x1½	11.98	10.87	2.64	2.04	3.25	2.72	1⅓	2.7
10x2	9¼x1½	15.23	13.87	3.35	2.60	4.13	3.47	1⅔	3.4
12x2	11¼x1½	18.48	16.87	4.07	3.16	5.01	4.21	2	4.1
2x4	1½x3½	5.89	5.25	6.45	5.36	3.56	3.06	⅔	1.3
2x6	1½x5½	9.14	8.25	24.10	20.80	8.57	7.56	1	2.0
2x8	1½x7¼	11.98	10.87	54.32	47.63	14.73	13.14	1⅓	2.7
2x10	1½x9¼	15.23	13.87	111.58	98.93	23.80	21.39	1⅔	3.4
2x12	1½x11¼	18.48	16.87	199.31	177.97	35.04	31.64	2	4.1
3x4	2½x3½	9.52	8.75	10.42	8.93	5.75	5.10	1	2.2
3x6	2½x5½	14.77	13.75	38.93	34.66	13.84	12.60	1½	3.4
3x8	2½x7¼	19.36	18.12	87.74	79.39	23.80	21.90	2	4.4
3x10	2½x9¼	24.61	23.12	180.24	164.89	38.45	35.65	2½	5.7
3x12	2½x11¼	29.86	28.12	321.96	296.63	56.61	52.73	3	6.9
4x4	3½x3½	13.14	12.25	14.39	12.50	7.94	7.15	1⅓	3.0
4x6	3½x5½	20.39	19.25	53.76	48.53	19.12	17.65	2	4.7
4x8	3½x7¼	26.73	25.38	121.17	111.15	32.86	30.66	2⅔	6.2
4x10	3½x9¼	33.98	32.38	248.91	230.84	53.10	49.91	3⅓	7.9
6x3	5½x2½	14.77	13.75	8.48	7.16	6.46	5.73	1½	3.4
6x4	5½x3½	20.39	19.25	22.33	19.65	12.32	11.23	2	4.7
6x6	5½x5½	31.64	30.25	83.43	76.26	29.66	27.73	3	7.4
6x8	5½x7½	42.89	41.25	207.81	193.36	54.51	51.56	4	10.0
8x8	7½x7½	58.14	56.25	281.69	263.67	73.89	70.31	5⅓	13.7

*Roughdry sizes are ⅛ in. larger, both dimensions.

Properties and weights of American Standard Board, Dimension and Timber sizes commonly used for formwork construction are based on data supplied by the National Forest Products Association.

Approximate weights listed are based on lumber weighing 35 lbs. per cubic foot.

Total vertical load for horizontal form systems

The total vertical load for horizontal form systems and falsework is designed for the sum of the dead load plus all live loads and impact loads. When concrete is placed, the lateral fluid pressure that exists must be accounted for. For concrete containing retarding admixture and fly ash or pozzolan cement replacement, the formwork, form ties, and bracing must be calculated for the fluid pressure of the mix to have a density of 150 pcf. For concrete containing no pozzolans or admixtures which affect the initial set time, the lateral fluid pressure will be determined on the concrete temperature and rate of placement, at time of placement (ACI Standard 347, Formwork for Falsework).

The following criteria of allowable maximum stresses and loads for formwork assume that high-quality material, undamaged, and normally maintained formwork is used and that it has not exceeded its life expectancy or normal reuse factor. Any deviation from this, below or above, will reduce or overstress the allowable stresses and loads.

The placement of concrete will follow a detailed placement procedure to assure the proper displacement of weight within the form system and falsework.

Timber criteria

$$\text{Compression perpendicular to the grain} = 450 \text{ psi}$$

$$\text{Compression parallel to the grain*} = \frac{480,000 \text{ psi}}{(L/d)^2}$$

where L = unsupported length
d = least dimension of a square or rectangular column, or the width of a square of equivalent cross-sectional area for round columns

$$\text{Flexural stress†} = 1800 \text{ psi}$$

$$\text{Horizontal shear} = 190 \text{ psi}$$

$$\text{Axial tension} = 1200 \text{ psi}$$

Deflection due to the weight of concrete may not exceed $\frac{1}{500}$ of the span, even if the deflection is compensated for by camber strips.

$$\text{Modulus of elasticity } E \text{ for timber} = 1.6 * 10^n \text{ psi}$$

where $n = 6$

$$\text{Maximum axial loading on timber piles} = 45 \text{ tons}$$

---

*NOTE: Not to exceed 1600 psi.

†NOTE: Reduced to 1450 psi for members with a nominal depth of 8 in or less.

The timber connections will be designed according to the stresses and loads allowed in the National Design Specification for Wood Construction (published by the National Forest Products Association) with the exception of:

1. Reductions in allowable loads required therein for high moisture condition of the lumber and service conditions do not apply.
2. Use 75% of the tabulated design value as the design value of bolts in two member connections (single shear).

Steel criteria. For identified grades of steel, do not exceed the design stresses (other than stresses due to flexural compression) specified in *The Manual of Steel Construction,* published by the AISC.

When the grade of steel cannot be positively identified, do not exceed the design stresses, other than stresses due to flexural compression, either specified in the *AISC Manual for ASTM A36 Steel,* or the following:

$$\text{Tension, axial and flexural} = 22{,}000 \text{ psi}$$

$$\text{Compression, axial} = 16{,}000 - 0.38(L/r)^2 \text{ psi}[110{,}320 - 2.62(L/r)^2 \text{ kPa}]$$

NOTE: L/r shall not exceed 120.

$$\text{Shear on the web gross section of rolled shapes} = 14{,}500 \text{ psi}$$

$$\text{Web crippling for rolled shapes} = 27{,}000 \text{ psi}$$

For all grades of steel, do not exceed the following design stresses and deflection:

$$\text{Compression, flexural*} = \frac{12{,}000{,}000 \text{ psi}}{Ld/bt}$$

where L = unsupported length
d = least dimension of a square or rectangular column, or width of a square of equivalent cross-sectional area for round columns, or the depth of beams
b = width of compression flange
t = thickness of compression flange
r = radius of gyration of member
F_y = specified minimum yield stress for the grade of steel used

Deflection due to the weight of concrete may not exceed $1/500$ of the span, even if the deflection is compensated for by camber strips.

---

*NOTE: Not to exceed 22,000 psi for unidentified steel or steel conforming to ASTM A36. Not to exceed $0.6 F_y$ for other identified steel.

$$\text{Modulus of elasticity } E \text{ for steel} = 30*10^n \text{ psi}$$

where $n = 6$.

Falsework criteria. Falsework spans supporting T-beam girder bridges are limited to 14 ft plus 8.5 times the overall depth of the T-beam girder.

Assembly criteria. Jacks, brackets, columns, joists, and other manufactured devices used shall not exceed the manufacturer's recommendations, or 40% of the ultimate load-carrying capacity of the assembly. This criterion will be based on the manufacturer's tests. The maximum allowable dead load deflection of joists will be $1/500$ of their spans. The components submitted or tested will be the components used, with no substitutions.

A steel-framed falsework tower component exceeding two or more tiers in height shall not exceed a 4 to 1 differential leg loading within the steel tower unit.

Falsework and foundation criteria. For the initial design of a ground-supported falsework system, the existing ground elevations at the proposed foundation locations will be verified.

Where spread-footing-type foundations are specified, determine the bearing capacity of the soil. The recommended maximum allowable bearing capacity for foundation material, other than rock, will be 2 tons/ft^2. Based on soil conditions and local practices, it is recommended not to locate the footing edges closer than 12" from the intersection of the bench and top of slope. Unless the excavation for the footer is adequately supported by properly designed shoring, it is recommended not to locate the edge of the footer closer than 4 ft from the edge of the excavation, or the depth of the excavation, whichever is greater.

With pile-supported foundation for falsework, the criteria design for piling, and the recommended soil analysis will be followed. Spread footers used to support the falsework will be designed at the assumed load-bearing capacity of the soil, and with use of reinforcing steel within the concrete if required.

For individual falsework tower components having a maximum leg load exceeding 30 kips, it is recommended to provide for uniform settlement under each leg for all loading conditions.

The foundation for falsework systems and components will be protected from adverse conditions, erosion, and traffic concerns.

The falsework system will be constructed and designed to be protected from vehicle impact and restrictions. This would include falsework posts and support members. The distance from the roadway and the height clearance above the roadway will be the controlling criteria, then secondly the load impacts and spanning criteria resulting from the required span distances.

To assure the falsework system remains stable, the following minimum design loads for falsework posts, columns, and towers will be followed:

1. 150% of the ultimate design load is calculated.

2. The increased or readjusted loads caused by the prestressing forces are calculated.

Falsework column components that are steel should have a minimum section modulus about each axis of 9.5 inches^n; where $n = 3$. Column components that are wood should have a section modulus about each axis of 250 inches^n; where $n = 3$. The base of each column or tower should be mechanically connected to the footing or piling substantial enough to withstand any lateral movement or force in excess of 2000 pounds in any direction. The falsework cap of each column or tower should be mechanically connected to the column or tower substantial enough to withstand any lateral movement or force in excess of 1000 pounds in any direction.

The exterior stringer girders of the falsework system for which overhang deck brackets are supported from, should be braced, tied, or mechanically connected to the adjacent interior girders to prevent rotation or overstressing to the exterior web. The mechanical connections should be capable of resisting a load from any direction, including uplift, of not less than 500 pounds.

False decking will be provided for areas over, or impeding traffic areas. The minimum required vertical clearances will be maintained for all roadways, railways, pedestrians, and marine waterways.

Falsework systems for beam structures. The criteria for falsework components required for the support of a beam structure will be calculated and identified as follows:

1. The calculation of falsework design loading will consist of the weight of the beam and girder members, the load imposed by supported erection equipment, the permanent deck loads including impact and design loads, concrete placing equipment, and any incidental load supported by the falsework.
2. The falsework and formwork for concrete is supported on a steel structure so that loads are applied to the girder webs within 6 inches of a flange or stiffener. The load will be distributed in a manner that does not produce local distortion of the web.
3. Strut, tie, and/or brace exterior beam members, which support cantilevered deck formwork and falsework components, to adjacent interior beam members to prevent the distortion and overstressing of the exterior member.
4. Refrain from applying loads to existing, new, or partially completed structures that exceed the load-carrying capacity of any portion of the structure. Reference to this is The Load Factors Design Methods of the AASHTO Bridge Design Specifications, Load Group IB.
5. Design and construct supporting falsework and formwork that will accommodate the proposed method of erection without overstressing the beam member, as required, and that will produce the required final structural geometry, intended continuity, and structural action.

The falsework and formwork will be constructed to the following criteria and standards:

1. The required camber will be built into the falsework and formwork system to compensate for the falsework system deflection and the anticipated structure deflection.
2. Provide telltale markers in systematically placed locations to determine the actual settlement and crush of the system and components during concrete placement.
3. Do not apply any dead loads, formwork, reinforcing steel, or live loads unevenly that will cause unbalance to the falsework system, differently than the design placement procedure details and identifies.
4. If an unanticipated event or if an uneven deviation or settlement occurs, cease the placement or erection procedure and mandate corrective action.

Suitable jacks and wedges will be incorporated into the falsework system to perform any adjustment of settlement of the formwork either before or during the placement of concrete. The falsework and formwork supporting deck slabs and cantilevered overhangs will be designed so there will be no appreciable differential settlement between the beam structure and the deck structure during the placement of concrete.

The formwork and falsework system will show the stresses and deflections in load-supporting members in a clear and concise manner. The allowable stresses are calculated upon the use of high-quality undamaged materials. The design must assure that the proper evaluation of the formwork and falsework system has been completed to safely carry the actual imposed loads.

Formwork, wood system. Wood formwork and panels, for exposed concrete surfaces, will be U.S. Product Standard PS 1 for Exterior Grade B-B (Concrete Form) Class I Plyform that will produce a smooth and uniform surface. The form panels will be in good structural condition, free from defects on the exposed surfaces. The form material will have the required flexural strength, modulus of elasticity, and other physical properties mandated for the intended use.

The form panels should be placed in uniform widths and lengths that will produce a structurally sound wall unit that interlocks together and that aligns with any back panel system for embed and tie matching locations.

The form panels will be arranged in a symmetrical pattern that conforms to the general lines and grades of the structure. Panels for vertical surfaces will be placed with the long dimension horizontal, and with horizontal joints level and continuous. In walls that are sloped or not symmetrical, the panels will be placed parallel to the footing in long dimension.

The form panel units will be precisely aligned on each side of the panel joint by supports or fasteners which are common to both panels. Triangular fillets or chamfer material will be used at all intersections and corners of the structure to deform the sharp edge.

Mechanical devices or embedded inserts may be cast into the concrete structure for later use in supporting formwork systems. Ties such as form bolts, clamps, taper ties, and yokes will be utilized to prevent the spreading or buckling of the form system during concrete placement. Tie devices such as wire loops that may expand should not be used. The tie devices that are required to be removed should be done without causing damage to the concrete surface, and will be removed to at least 1″ below the surface face. The cavities and crevices will be patched with cement mortar and finished to a smooth and sound, uniform-colored finish (rub and patch all the required surfaces).

The formwork system will be of sufficient strength so the concrete surface does not undulate more than 0.1 inch, or $\frac{1}{360}$ of the center-to-center distance between studs, joists, form stiffeners, form fasteners, or wales. Construct all formwork, for all exposed surfaces, for each element and component of a concrete structure, with the same forming material or with a material that will produce a similar surface texture, color, and appearance.

Concrete form panels and formwork will be mortartight, true to dimensions, lines, and grades for the structure, and of sufficient strength to prevent appreciable deflection during the placement of concrete.

All material required for embedment will be placed prior to the concrete placement.

Formwork supporting the slab of a box girder structure will be supported on wales or similar supports fastened, as nearly as possible, to the top web of the wall unit.

The inside surface of the formwork must be clean from all dirt, mortar, and foreign material. All loose material will be removed from the formwork within the placement area, prior to concrete placement.

The formwork system will be coated thoroughly with form oil and a form release agent prior to concrete placement. This material must not cause any discoloration to the concrete surface.

Formwork, stay-in-place metal. Fabricated permanent steel bridge deck forms and supports are designed and manufactured of steel conforming to ASTM A446 (Grades A–E), having a coating class of G165, which is specified by ASTM A525.

Corrugated metal permanent steel formwork. The following charts were supplied by Topikal, *Wheeling Bridge Forms*.

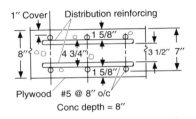
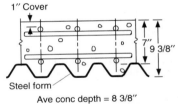

(a)

PROJECT DESCRIPTION
Rebars—#5 @ 6" o.c.
Stringers—W36 × 160 at 8'2" o.c.
Slab design thickness—8"
Bottom of slab to be located near centroid of form.

SLAB THICKNESS FOR DEAD LOAD CALCULATION
Average slab thickness shall be the conventional slab thickness minus the bottom transverse bar diameter plus one inch.

Conventional slab	=	8"
− #5 Rebar Dia.	=	− 5/8"
		7 3/8"
+ Distance to centroid	=	1"
		8 3/8" Slab design thickness

DESIGN LOADS
Deflection load shall be the wet concrete weight (150 POF or 12½ lb/inch of depth per AASHTO) plus 4 PSF for weight of form. If lightweight concrete is used, the actual weight shall be calculated plus 4 PSF for form dead load. Stress load shall be the total dead load weight plus 50 PSF for construction live load and impact.

Wt. of Slab 8.375 × 12.5	= 105 PSF
Wt. of Form	= 4
Deflection Load	= 109 PSF*
Construction Live Load	= 50
Stress Load	= 159 PSF

FORM SPAN
Design span shall be clear span of form sheet plus two inches.

Stringer Spacing	=	8' 2"	
Flange Width	=	−1' 0"	W36 × 160
		7' 2"	
Haunch Angle Flange	=	−0' 4"	2 @ 2"
Required End Bearing	=	0' 2"	
Design Span	=	7' 0'	

REQUIRED PHYSICAL PROPERTIES
Maximum Applied Moment for single span uniformly loaded sheet.

$$M = \frac{12 WL^2}{8} \text{ in.-lb}$$

$$M = \frac{12 \times 159 \times 7^2}{8} = 11,664 \text{ in.-lbs.}$$

Section Modulus Required equals Maximum Applied Moment divided by the Allowable Working Stress,

*If application is on USFHWA Project, see deflection load limitation under "Required Physical Properties."

which is 29 KSI for gages 14, 15 & 16, and 36 KSI for gages 17 thru 22.

$S_{Req'd}$ (29 KSI) = 11,664 ÷ 29,000 = 0.4022 in.³
$S_{Req'd}$ (36 KSI) = 11,664 ÷ 36,000 = 0.3240 in.³

Allowable Deflection Limitation equals L/180 (where L = design span of form sheet) or ½" whichever is less under Slab Dead Loads. If USFHWA specification applies. Mimimum Slab Weight shall not be less than 120 PSF.

NOTE: L/180 limitation will govern on spans less than 7'6" and ½" limitation will govern spans of 7'6" or greater.

For this application

$$\text{Allowable } \Delta = \frac{L}{180} = \frac{7.0 \times 12}{180} = 0.466 \text{ in.}$$

$$I_{Req'd} = \frac{5WL^4}{384 \, E\Delta}$$

Where W = Actual Slab Dead Loads for non-FHWA Projects or Actual Slab Dead Load or 120 PSF. whichever is greater for FHWA Projects

$$I_{Req'd} = \frac{5 \times 109 \times 1728 \times 7^4}{384 \times 29.5 \times 10^6 \times 4.66 \times 10^{-1}} = 0.427 \text{ in.}^4$$

if USFHWA Specification does not apply.
or

$$I_{Req'd} = \frac{5 \times 120 \times 1728 \times 7^4}{384 \times 29.5 \times 10^6 \times 4.66 \times 10^{-1}} = 0.473 \text{ in.}^4$$

if USFHWA Specification applies.

SELECT PANS
Selection of the correct gage is made from the Physical Property Tables using the required physical properties previously determined. Pitch Selection is governed by bottom transverse bar pitch if the bottom of the slab is near the centroid of the form as it is in this case. Otherwise 6" pitch (60N) pans are used.

Refer to Table I for form properties. Select 60N 19 pans

f = 36 KSI
I = 0.4734 in.⁴ 0.473 req'd.
S = 0.4057 in.³ 0.324 req'd.

ACTUAL STRESS AND DEFLECTION
Actual Stresses (f) and Deflections (Δ) can be determined by ratios of physical properties required to those furnished

Actual Stress
$$f = \frac{0.324}{0.406}(36,000) = 28,750 \text{ PSI}$$

Actual Deflection
$$\Delta = \frac{0.473}{0.4734}(0.466) = 0.466 \text{ in}$$

(b)

Metal stay-in-place design criteria.

Formwork and Falsework 75

SLAB CROSS-SECTION

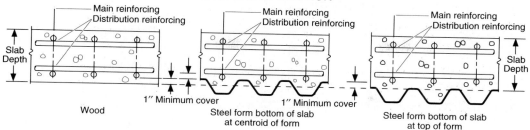

Wheeling Bridge Form physical properties have been derived in accordance with all provisions of the latest edition of the AISI Specifications for the Design of Light Gage, Cold Formed Structural Members.

Wheeling Bridge Forms have specially designed closures and pans that seal off all areas not covered by the forms. These accessory items are zinc coated in the same class as the forms when exposed in finished construction.

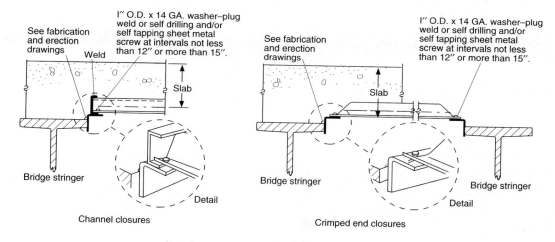

BOTTOM OF SLAB ABOVE STRINGER

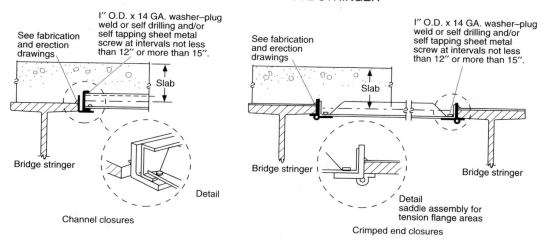

BOTTOM OF SLAB AT TOP OF STRINGER

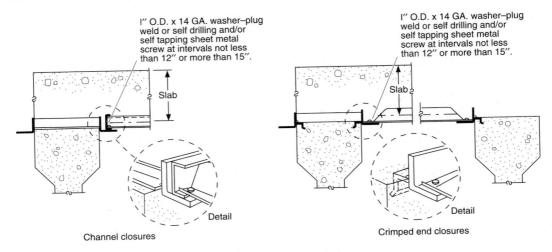

PRECAST CONCRETE BRIDGE BEAMS

HAUNCH ANGLES

2 3/16" x 3", 12 gage angles are supplied as standard. These have been found to meet the broadest range of loading requirements. Haunch angles may be supplied in 10 gage for extremely heavy loadings or in 14 gage for light loading at the discretion of Wheeling Corrugating Company.

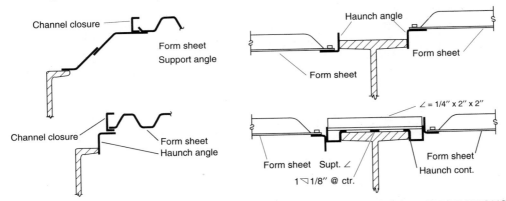

TYPICAL DIAPHRAGM HAUNCH SECTIONS   TYPICAL STRINGER HAUNCH SECTIONS

The steel stay-in-place form system will be installed according to approved and detailed fabrication and erection drawings, for each specific structure. The form system will be supported from, and fastened to, the designated form supports (metal angle). The metal form supports will be in direct contact with, and supported from, the individual structure beam member. The support angle will be attached to the beam member according to project specifications. These support angles will designate and profile the haunch of the structure as designed.

Void formwork. Void form components are designed to provide a hollow section within a structural member. Void form systems will be stored in a dry and flat location to prevent distortion. The void form system will be secured using embedded anchors, supports, and ties which leave a minimum of metal or other supports exposed at the bottom of the finished slab.

The exterior of the form system will be waterproof, and the ends will be covered with waterproof, mortartight caps. PVC vents are normally required at each end of the void form.

Metal formwork systems. Metal form panels are designed for volume forming situations and conditions. The metal skin of the panel will produce a smooth and uniform surface. The form panels will be in good structural condition, free from defects on the exposed surfaces. The form material will have the required flexural strength, modulus of elasticity, and other physical properties mandated for the intended use.

The following charts were supplied by Economy Forms Corp., *Forming Systems Catalog*.

20'-0" RADIUS & LARGER

Height Length	2'-0R Wt.	2'-0R Item No.	3'-0R Wt.	3'-0R Item No.	4'-0R Wt.	4'-0R Item No.	5'-0R Wt.	5'-0R Item No.	6'-0R Wt.	6'-0R Item No.
Inside Form 12'-0"	552	73357	702	73359	846	77357	994	72357	1151	71357
Outside Form 12'-0"	552	83357	702	83359	846	87357	994	82357	1151	81357
Inside Form 4'-0"	300	76357	250	74359	300	78357	353	75357	404	74357
Outside Form 4'-0"	200	86357	250	84359	300	88357	353	85357	404	84357
2" Filler Form	15	18357	22	66359	29	60357	35	14357	42	10357
4" Filler Form	19	19357	28	67359	36	61357	44	15357	52	11357
6" Filler Form	23	20357	33	68359	43	62357	52	16357	61	12357
8" Filler Form	27	21357	38	69359	50	63357	60	17357	71	13357
Pour Window Panel							1180	71359	1366	70359

(a)

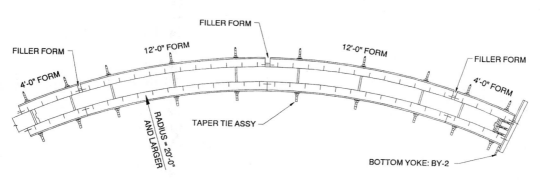

(b)

STANDARD *PLATE GIRDER FORM SYSTEM* PANEL SIZES

Length	1'-0"		2'-0"		4'-0"		8'-0"		12'-0"		20'-0"	
Width "R"	Wt.	Item No.	Wt.	Item No.	Wt.	Item No.	Wt.	Item No.	Wt.	Item No.	Wt.	Item No.
2'-0R	67	61800	104	62800	178	64800	299	68800	504	72800	651	80800
3'-0R	82	31801	133	32801	215	34801	369	38801	540	42801	862	50801
4'-0R	102	01802	168	02802	279	04802	494	08802	728	12802	1161	20802
5'-0R	133	61802	225	62802	377	64802	687	68802	1020	72802	1600	80802
6'-0R	155	31803	255	32803	437	34803	820	38803	1178	42803	1860	50803
7'-0R	181	01804	313	02804	539	04804	976	08804	1372	12804	2230	20804
8'-0R	234	61804	382	62804	662	64804	1190	68804	1740	72804	2817	80804
9'-0R	250	31805	431	32805	733	34805	1550	38805				
10'-0R	283	01806	473	02806	803	04806	1619	08806				
12'-0R	434	31807	712	32807	1211	34807	2158	38807				

(a)

(b)

Formwork and Falsework

STANDARD EFCO LITE PANEL SIZES

Item No.	Size	Area	Wt.
01300	8'-0"R x 6'-0"	48	512
03300	8'-0"R x 4'-0"	32	347
05300	8'-0"R x 2'-0"	16	202
06300	8'-0"R x 1'-0"	8	130
71300	6'-0"R x 6'-0"	36	462
73300	6'-0"R x 4'-0"	24	203
75300	6'-0"R x 2'-0"	12	177
76300	6'-0"R x 1'-0"	6	116

TAKE-UP PANELS

Item No.	Size	Area	Wt.
11301	8'-0"R x 18"	12	159
09301	8'-0"R x 16"	10.7	149
05301	8'-0"R x 10"	6.7	99
03301	8'-0"R x 8"	5.3	89
01301	8'-0"R x 6"	4	81
31301	6'-0"R x 18"	9	117
29301	6'-0"R x 16"	8	108
27301	6'-0"R x 12"	6	91
25301	6'-0"R x 10"	5	82
23301	6'-0"R x 8"	4	67
21301	6'-0"R x 6"	3	60

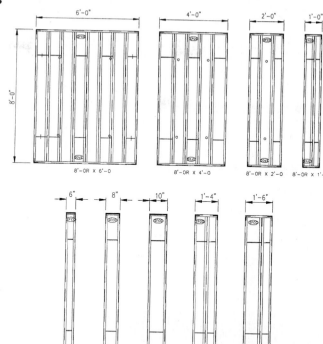

STANDARD COLUMN FORM PANEL SIZES

Width \ Height	8'-0" Wt.	8'-0" Item No.	4'-0R Wt.	4'-0R Item No.	2'-0R Wt.	2'-0R Item No.	1'-0R Wt.	1'-0R Item No.
30/24	196	02372	105	03372	56	04372	38	13372
24/18	159	05372	86	06372	45	07372	31	12372
16/10	127	08372	68	09372	36	10372	24	11372
3/4" Alum. Chamfer	4	51176	2	52176	1	53372	.5	62176

(a)

(b)

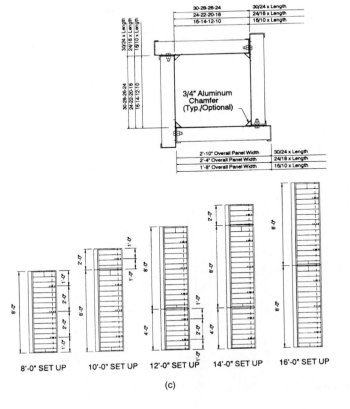

(c)

ROUND COLUMN FORM SIZES

Diameter \ Length	1'-0"		2'-0"		4'-0"		8'-0"		12'-0"	
	Wt.	Item No.	Wt.	Item No.	Wt.	Item No.	Wt.	Item No.	Wt.	Item No.
D16" x 360°	60	P0230	92	P0231	162	P0232	304	P0233	424	P0234
D18" x 360°	64	P0235	102	P0236	180	P0237	340	P0238	470	P0239
D20" x 360°	74	P0240	124	P0241	212	P0242	396	P0243	552	P0244
D24" x 360°	84	P0245	140	P0246	252	P0247	428	P0248	674	P0249
D30" x 360°	112	P0250	184	P0251	308	P0252	574	P0253	810	P0254
D36" x 360°	121	P0255	200	P0256	338	P0257	630	P0258	892	P0259
D42" x 360°			230	P0261	400	P0262			1020	P0264
D48" x 360°			268	P0266	460	P0267			1222	P0269
D54" x 360°			320	P0271	560	P0272			1468	P0274
D60" x 360°			344	P0276	604	P0277			1580	P0279
D72" x 360°			400	P0281	672	P0282			1800	P0284
D84" x 360°			436	P0286	752	P0287			2020	P0289
D96" x 360°			512	P0291	892	P0292			2408	P0294

Item Numbers and Weights Indicate Complete 360° of Forms and Required Bolts

(a)

(b)

HANDSET EFCO Forming System

Width-Description / Length	48" Item No.	Wt.	24" Item No.	Wt.	12" Item No.	Wt.
24" Wide Form Panel	01900	48.5	01902	25.6	-	-
16" Wide Form Panel	05900	35.0	05902	18.5	05903	9.6
12" Wide Form Panel	09900	27.0	09902	13.2	09903	7.8
8" Wide Form Panel	13900	19.0	13902	9.8	13903	6.0
7" Wide Form Panel	14900	16.3	14902	9.2	14903	5.4
6" Wide Form Panel	15900	14.5	15902	8.7	15903	4.4
5" Wide Form Panel	16900	13.3	16902	7.0	16903	4.0
4" Wide Flexible Form Panel	01980	8.0	15980	4.0	22980	2.0
3 1/2" Wide Flexible Form Panel	02980	7.4	16980	3.7	23980	1.9
3 Wide Flexible Form Panel	03980	7.0	17980	3.5	24980	1.8
2 3/4" Wide Flexible Form Panel	04980	6.8	18980	3.4	25980	1.7
2 1/2" Wide Flexible Form Panel	05980	6.6	19980	3.3	26980	1.7
2 1/4" Wide Flexible Form Panel	06980	6.4	20980	3.2	27980	1.6
2" Wide Flexible Form Panel	07980	6.2	21980	3.1	28980	1.6
2 1/4" x 2 1/4" 90° Outside Corner	06985	6.5	08985	3.5	09985	1.7
1 1/2" x 1" Insert Angle	06984	3.5	03984	1.7	04984	.8
4" x 4" 90° Inside Corner	01940	15.9	03940	8.1	04940	4.6
3" x 5" 90° Inside Corner	01944	16.5	03944	8.3	04944	4.5
2" x 6" 90° Inside Corner	01942	16.5	03942	8.3	04942	4.5
6" x 6" 90° Inside Corner	80629	22.7	82629	11.6	83629	6.0
8" x 8" 90° Inside Corner	80628	33.9	82628	16.7	82628	8.5
12" x 12" 90° Inside Corner	80627	42.5	82627	22.1	83627	11.5
2" x 1" x 1/2" Rad. 90° Outside Corner	01954	14.3	03954	7.0	04954	3.5
4" Rad. 90° Inside Corner	01955	13.5	03955	6.8	04955	3.4
2 1/4" x 2 1/4" x 90° Outside Tie Angle Corner	01986	7.0	03986	4.0	-	-
3" x 3" Outside Flex. Angle Corner	01981	9.4	03981	4.7	04981	2.4
3" x 3" Outside Rigid 135° Angle Corner	05981	10.0	07981	5.1	-	-
3" x 3" Inside Flex. Angle Corner	01982	9.4	03982	4.7	04982	2.4
3" x 3" Inside Rigid 225° Angle Corner	05982	10.0	07982	5.1	-	-
12" x 12" Inside Fillet Corner	01946	36.7	03946	18.0	04946	9.5
10" x 10" Inside Fillet Corner	01947	34.0	03947	17.1	04947	8.6
8" x 8" Inside Fillet Corner	01948	28.9	03948	14.2	04948	7.7
6" x 6" Inside Fillet Corner	01949	23.0	03949	11.0	04949	6.2
4" x 4" Inside Fillet Corner	01950	18.5	03950	8.5	04950	4.7
3" x 3" Inside Fillet Corner	01951	15.8	03951	8.0	04951	4.2
4" x 4" x 2" x 2" Inside Fillet Corner	10968	15.2	12968	7.8	13968	5.0

(a)

(b)

Placement of formwork. The form panels should be placed in uniform widths and lengths that will produce a structurally sound wall unit that interlocks together and will align with any back panel system for embed and tie matching locations.

The form panels will be arranged in a symmetrical pattern that conforms to the general lines and grades of the structure. Panels for vertical surfaces will be placed with the long dimension horizontal and with horizontal joints level and continuous. In walls that are sloped or not symmetrical, the panels will be placed parallel to the footing in long dimension.

The metal form panel units will be precisely aligned on each side of the panel joint by supports or fasteners which are common to both panels. Triangular fillets or chamfer material will be used at all intersections and corners of the structure to deform the sharp edge.

Mechanical devices or embedded inserts may be cast into the concrete structure for later use in supporting formwork systems. Ties such as form bolts, clamps, taper ties, and yokes will be utilized to prevent the spreading or buckling of the form system during concrete placement. Tie devices such as wire loops that may expand will not be used. The tie devices that are required to be removed will be done without causing damage to the concrete surface and will be removed to at least 1″ below the surface face. The cavities and crevices will be patched with cement mortar and finished to a smooth and sound, uniform-colored finish (rub and patch required surfaces).

The formwork system will be of sufficient strength so the concrete surface does not undulate the panel skin or webbing. Construct all formwork, for all exposed surfaces, for each element and component of a concrete structure, with the same forming material or with a material that will produce a similar surface texture, color, and appearance.

Metal form panels and formwork will be mortartight, true to dimensions, lines, and grades for the structure, and of sufficient strength to prevent appreciable deflection during the placement of concrete. All material required for embedment will be placed prior to the concrete placement.

The inside surface of the formwork will be clean from all dirt, mortar, and foreign material. All loose material must be removed from the formwork within the placement area prior to concrete placement.

The formwork system will be coated thoroughly with form oil and a form release agent prior to concrete placement. This material must not cause any discoloration to the concrete surface.

Removal of forms and falsework. The form system and falsework will not be removed until such time the concrete has reached sufficient strength, or the minimum time specified for the formwork to remain in place. The formwork removal will not cause damage to the concrete structure. All formwork and falsework will be completely removed except for the following:

1. Interior soffit panels for the deck slabs of cast-in-place box girder structures

2. Formwork for the interior voids of precast members
3. Formwork for members or components where no permanent access is available into the cells or voids

The removal of forms that do not support the dead load of concrete members, other than railings and barriers, will not begin prior to at least 24 hours after the concrete for the component has been placed and the concrete has sufficient strength to resist damage to the surface. The exposed concrete surface will remain protected from damage.

Table 3.3 will be a guideline for formwork removal.

TABLE 3.3 **Formwork Removal Guide**

Structure component	Percent of 28-day strength	Minimum time since placement, days
Columns and walls nonsupportive	50	3
Mass placements nonsupportive	50	3
Box girder structure	80	14
Bridge decks, pier caps, supportive	80	14
Trestle slabs	70	10
Slabs and overhangs (supported)	70	10
Pier caps continuously supported	60	7
Arches and continuous spans	90	21

Supporting falsework, formwork, and shoring systems will be removed uniformly, systematically, and gradually, according to the removal sequence, beginning at the crown and working progressively toward the springing. Falsework and formwork systems for adjacent spans will be removed simultaneously. Falsework for cast-in-place box girder structures will not be removed until the posttensioning procedures have been completed.

Falsework supporting rigid framed structures will not be removed until the backfill material has been placed and compacted.

The contractor will be responsible for designing and constructing a safe and adequate falsework system that provides the necessary rigidity, supports the imposed loads, and produces the lines and grades indicated on the plans.

NOTE: The system for all supporting falsework, formwork, and shoring systems, foundation pads, and piling should be designed by a licensed professional engineer proficient in structural design. The engineer will also inspect the completed falsework system prior to any placement of concrete or load-supporting equipment, and before the removal operation begins for the supporting falsework or formwork system.

Applications for Conventional Formwork Systems

Conventional wood formwork systems are hand-built panels, either prebuilt in common-sized dimensions or field-built for specific custom and unique uses, or wall units built on site with individual components.

Prebuilt and job-built wood systems. The prebuilt panels normally have a skeletal framework of dimensional lumber and are skinned with a structural-grade plyform sheet. The conventional wood formwork panels can be fabricated for virtually any reasonable size and shape and can be somewhat easily altered in size and shape.

A job-built system is individually and specifically built for a one-time use, then dismantled. This system has components similar to the prebuilt panel system but is normally more labor-intensive. However, this system has its advantages as well as the prebuilt system and in more cases than not is used in conjunction with the prebuilt system.

The key to any form system is time savings with the best and most efficient and effective result. The prebuilt wood panels can be utilized where a gang-type or bulk system requirement exists, and the job-built or custom panels would be utilized in the interim or in conjunction. If prepared, maintained, and properly stored, this type of form system can be long-lasting and cost-effective. (See illustrations on pages 86 and 87.)

The main components of a prebuilt wood form system are the wood panels, double wales, wood spreaders or yokes, braces, and metal ties. The components of a job-built wood form system are wood studs, plyform or wood board sheathing, sill plates (these components constitute a prebuilt panel), double wales, wood spreaders or yokes, braces, and metal ties.

The wood panel design allows a combination of both structural webbing and a formed surface face, but with multiple components to create a modular unit. The stud components of a wood panel act as stiffeners to transfer and distribute the horizontal pressures of liquid concrete from the form face, throughout the panel framework to the top and bottom flanges. These panels can then be attached to form a continuous, modular sectional form unit. The wood panel design requires vertical and horizontal supports, walers and spreaders to increase their structural capability. Without these, the wood panel has limited capabilities. The wood form panel system normally is the most time-consuming and costly type of form system.

The benefits of a wood panel form system are: (1) Built for custom or standard-sized uses, (2) flexibility of continuous and irregular forming, (3) little or no lifting equipment, (4) more versatile with smaller, nonproductive uses, and (5) ease of stripping and removal.

Vertical. Vertical form systems are constructed with the components listed in the illustration. The vertical forming systems must contain the concrete placement horizontal fluid pressures but normally require little vertical support, unless the use is a multitier, second, third, etc., lift placement. A vertical form system will at a minimum be required to support its own weight, plus the scaffolding, crew, and equipment intended for it (live load).

A vertical wood form system is most commonly used in wall, pier, abutment, and miscellaneous structure components.

Horizontal. Horizontal form systems are constructed with the components listed in the illustration. The horizontal form systems must contain the con-

Vertical Forming System

Column and Pilaster Forming Ideas

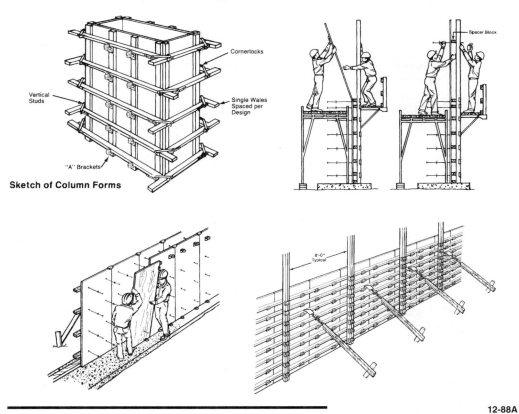

Job-built form systems. (*Dayton Superior*).

12-88A

Typical Section Horizontal Forming System

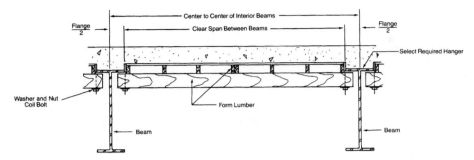

Interior Section

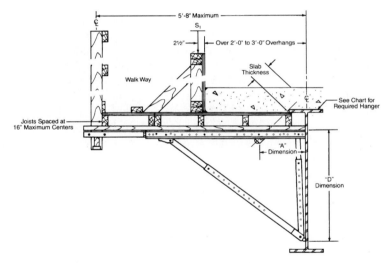

Cantilevered Section

Job-built form system (*Dayton Superior*).

crete placement and vertical pressures but normally require little horizontal support. A horizontal form system will at a minimum be required to support its own weight, plus the scaffolding, crew, and equipment intended for it, and the imposed loads of reinforcing steel and concrete (dead load plus live load).

The horizontal wood form system is most commonly used in deck, cantilevered overhangs, and soffit components.

Hand-set panels. Metal hand-set form panels consist of a lightweight metal framework with a thin, structural skin of steel or wood sheeting. This system is considered a hand-set forming system owing to its lightweight design. These panels, even though a lightweight system, are versatile and durable. The hand-set form system is manufactured in standard-sized panels and custom-sized panels.

The main components of a hand-set form system are the panels, wales, spreaders or yokes, braces, and metal ties. The hand-set panel design allows a combination of both structural webbing and a formed surface face. The steel ribs of a steel panel act as stiffeners to transfer and distribute the horizontal pressures of liquid concrete from the form face throughout the panel framework to the top and bottom flanges. These panels can then be bolted together to form a continuous, modular sectional form unit. The hand-set panel design requires vertical and horizontal support to increase the structural capability.

The benefits of a hand-set panel form system are: (1) Precision-built for dimensional accuracy, (2) flexibility of continuous forming, (3) little or no lifting equipment, (4) reduced use of stiffbacks and walers, (5) more consistent use of ties and embedded components, (6) smoother and more cosmetic concrete surface, (7) more versatile with smaller, nonproductive uses, and (8) longer-lasting form system.

Hand-set panels are normally vertical form systems and are constructed with the components listed in the illustration. The vertical forming systems must contain the concrete placement horizontal fluid pressures but normally require little vertical support, unless the use is a multitier, second, third, etc., lift placement. A vertical form system (p. 82) will at a minimum be required to support its own weight, plus the scaffolding, crew, and equipment intended for it (live load).

A vertical hand-set form system is most commonly used in wall, pier, abutment, and miscellaneous structure components.

Gang form system. A gang form system is one of steel constructed panels, intended for large common size and repetitious applications. This type of form system is long-lasting, versatile, and durable. This system is considered a heavy-duty forming system owing to its structural design and unit weight. These panels are versatile and durable. The metal gang or heavy-duty form system is manufactured in standard- and custom-sized panels (pp. 77–81).

The main components of a steel, heavy-duty form system are the panels, spreaders or yokes, braces, and metal ties. This type of form system is structurally designed to give both horizontal and vertical support, thus eliminating or decreasing the requirement of wales and intermediate support components required by other form systems.

The steel panel design allows a combination of both structural webbing and a formed surface face. The steel ribs of a steel panel act not only as stiffeners but as structural beams to transfer and distribute the horizontal pressures of liquid concrete from the form face, throughout the panel framework to the top and bottom flanges. These panels can then be bolted together to form a continuous, modular sectional form unit. The steel panel design also allows a self-spanning support capability, which decreases interspan shoring.

The benefits of a steel panel form system are:

1. Precision built for dimensional accuracy
2. Flexibility of continuous forming
3. Self-spanning capability
4. Elimination of stiffbacks and walers
5. Fewer ties and embedded components
6. Smoother and more cosmetic concrete surface
7. Reduced cycle time of form relaying
8. Longer-lasting form system

Steel form panels are normally vertical form systems and are constructed with the components listed in the illustration. The vertical forming systems must contain the concrete placement horizontal fluid pressures but normally require little vertical support, unless the use is a multitier, second, third, etc., lift placement. A vertical form system will, at a minimum, be required to support its own weight, plus the scaffolding, crew, and equipment intended for it (live load).

A vertical steel form system is most commonly used in wall, pier, and abutment components.

Horizontal steel form systems are constructed with the components listed in the illustration. The horizontal steel forming systems must contain the concrete placement and vertical pressures but normally require little horizontal support. A horizontal steel form system will, at a minimum, be required to support its own weight, plus the scaffolding, crew, and equipment intended for it, and the imposed loads of reinforcing steel and concrete (dead load plus live load).

The horizontal steel form system is most commonly used in deck, cantilevered overhangs, and soffit components.

Form reuse

The formwork cost of a cast-in-place structure accounts for a substantial part of the total direct cost. The form system cost can be reduced and prolonged by form care, type of use, and good planning with regard to sizing, standardizing, and obtaining the most efficient reuses of the formwork.

The ability to reuse the formwork depends on the nature of the intended use. The repetitive, standard, or continuous structural features will favor multiple reuses, whereas complicated and custom components will lend to a one-time or less reuse factorability. Along with these variables, site conditions, phasing, fabricated formwork components, availability of lifting equipment, and long-term cash investments of form systems are important and critical factors.

Of the three most common types of form systems discussed, the plyform wood system remains the most vulnerable for reuse consideration. The metal steel form panels offer the highest reuse capability. The wood system normally is the most labor-intensive, and the heavy-duty metal panel is normally the highest initial investment system. The wood panel system is also the most demanding, with respect to unit cost and embed item and form accessory cost. These components will also affect the reuse factorability.

Plyform wood system. An exterior-type plyform material is recommended for wood panel formwork, but the plyform that is manufactured for concrete use will perform with best results and provide the most efficient reuse factor. This type of plyform has special treatment to the wood and has more plies for a stronger, dimensional stability.

With proper care and normal intended functions, five to eight uses out of wood form panels are common. The wood form panel life can be increased by proper handling, sealed edges, quality materials, proper cleaning and oiling after each use, and the nonoverstressing of the form panel. Unnecessary cutting of large panels, cutouts, and differing phasing conditions will take from the reuse life factor of a form panel also. It is wise to utilize specialty panels, older panels, or previously altered panels for these conditions, provided the form panel still retains its structural capability.

When not in use, it is recommended that the form panels, regardless of type, be properly stored from exposure to weather and physical damage and stacked to allow for air circulation. The form panels should be stored in a flat position to eliminate warping, bending, and twisting; and for large panels, interim supports may be required.

The life of the formwork, regardless of wood or metal, depends greatly on the care and maintenance given to the system. The form panels and accessories must be properly moved, cleaned, oiled, prepared, and stacked flat immediately after the removal from the structure component. All concrete laitance will be removed and any damage areas will be repaired prior to reuse.

A well-maintained and oiled form panel will be more efficient to prepare, cause less rubbing and patching to the concrete surface, and be easier to clean after the next use. The form panel utilized under these guidelines will prove a more cost-efficient investment and system, and provide more reused square feet of formed surface area.

Form panels will have a tendency to adhere to the freshly placed concrete from the curing process; therefore, for ease of stripping and nondamage to the form panel, a form release agent will be applied to the panel prior to placement.

Eventually, form panels suffer from fatigue and stress. The more common uses, evenly distributed loading, and care in construction of bracing and supports will add to the reuse life factor of the form system, whether wood or metal.

A low reuse form system normally consists of wood panels, of common material components, such as plyform and dimensional lumber, having a low initial cost. This type of form system is versatile, easy to modify, and readily adaptable.

Hand-set and steel gang forms. A high reuse form system, including hand-set and steel gang forms, will have more than eight uses as categorized with the wood panel formwork. These larger systems, however, can require heavy lifting capabilities and advance fabrication, and are not generally well suited for small and modification-demanding uses.

Formwork requirements will require the use of multiple systems within a singular component. This application would mandate the use of high reuse formwork in the large, repetitive, continuous section, and low reuse formwork in the complicated or one-time-use section.

Form systems are available to purchase, rent, or job-built considerations, or a combination dependent on custom components and long-term needs.

The hardware, accessory, and form embed items are a continual consideration in the choice of formwork. Some accessories have an initial low cost but require a higher stripping and patching cost. Some hardware, such as snap ties for a wood form panel, are a nonrecoverable embedded item. The larger steel form panels offer a reusable form tie, she bolt, or coil rod system.

Form tie spacing will also be a consideration and will impact the final cost of the chosen form system. The increased tie spacing to minimize hardware cost, installation, and patching will result in a heavier, more structurally designed, and costly form panel. For the most efficient form system cost, all variables, conditions, reuses, removal, embed and hardware demands, and equipment needs must be analyzed.

In order to maximize form reuse efficiency, increase the form tie spacing, maintain a lightweight form system, minimize the number of individual panels, and increase the reuse factor.

Reuse factor in initial cost distribution

The reuse factor of formwork will allow the user to distribute the initial cost of the form panel, or system, over more uses, resulting in a lower per job square foot cost of the formwork.

A wood form system will be categorized by two methods; the normal built-up wood form panel system, including component deck formwork and dimension lumber, and consumable formwork. The built-up panel system will be sized by dimension thickness, being nominal 4", 6", or 8" panels of various width and length.

The reuse consideration for a wood panel or wood decking form system will be factored on a scale of 1 to 5, with a recommendation of not considering more than five reuses for ownership purposes.

Wood form system reuse factor

1. Severe = 1 use
2. Poor = 2 uses
3. Average = 3 uses
4. Good = 4 uses
5. Excellent = 5 uses

Consumable lumber will be an incidental job-related consumed component and not considered for any reuse value.

Dimensional lumber is measured and identified by a unit of required board feet, and structural plyform material is measured and identified by a unit of required square feet. The purchase of these materials will be based on the unit cost for each, board foot by dimensional size, and square foot by thickness.

Required board feet. The calculation required to determine board feet will be performed by the following procedure. For estimating purposes, dimensional lumber will be identified by its nominal size, not the actual milled size ($2'' \times 6''$ nominal size; $1\frac{1}{2}'' \times 5\frac{1}{2}''$ actual).

Definition: 1 board foot = $1''$ thick $\times 12''$ wide $\times 1'$ long.

To determine the required board feet of a specific component of dimensional lumber, perform the following calculation:

$$\text{Board feet} = \frac{t \times w \times l}{12}$$

where t = dimensional thickness, inches
w = dimensional width, inches
l = length, feet
12 = constant (l = feet, therefore, $12''$ per foot)

The calculation will be reduced to its lowest number to arrive at the required board feet for the component.

Example The board feet of a $2'' \times 6'' \times 12'$ dimensional component is

$$\frac{2 \times 6 \times 12}{12} = \frac{144}{12} = 12.0 \text{ board feet}$$

Required square feet. To determine the required square feet of a specific component of plyform lumber, perform the following calculation:

$$\text{Square feet} = w \times l$$

where w = dimensional width, feet
l = length, feet

Example The square feet of a $\frac{3}{4}'' \times 4' \times 8'$ surface component is

Formwork and Falsework

$$4 \times 8 = 32.0 \text{ square feet of } \tfrac{3}{4}''$$

The support components for the form panels will be calculated and listed by quantity required, and designated as

$$r = \text{reuse, factor } x$$
$$c = \text{consumable}$$

Application: 6″ Wood Form System

The cost and ownership of a job-built wood form system, and accessories, will be distributed by the reuse factor.

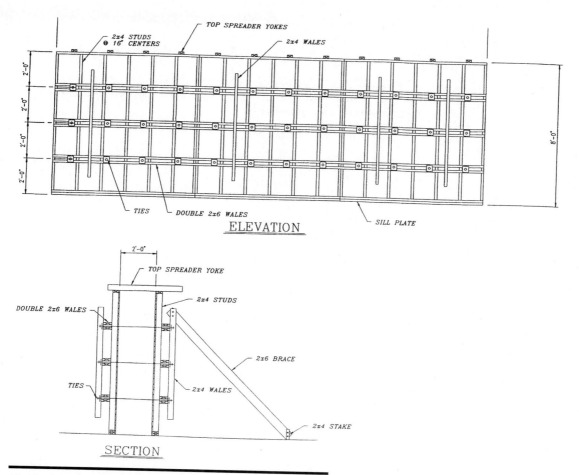

6″ wall panel 8′h×24′l @ 2′ thick.

94   Construction Systems, Methods, and Materials

Example: 6″ wood panel form system

The example conventional form system consists of wood, prebuilt panels; with support components of walers, sill plates, spreaders, and braces (both sides); and accessories of metal snap tie hardware.

The scope and structural needs of this wall panel will require a dimensional panel thickness of 6″. The condition will allow standard-sized form panels 4′ wide by 8′ long, with an excellent reuse factor of 5.

The individual panel construction specification will require dimensional 2″ × 6″ construction with ¾″ plyform sheathing, and the studs to be placed at 16″ centers within the 8′ dimension.

Material list. The material list will be as listed for the reusable component form panels:

Panels. Required: 12 each 6″ dimensional 4′ × 8′ panels.

Each panel
Side plates	2 ea 2″×6″ @ 8′	= 16 lf
Studs	7 ea 2″×6″ @ 45″ (say 4′)	= 28 lf
¾″ plyform	1 ea ¾″×4′×8′	= 32 sf
Form oil	2 oz per sf (32*2)/128	= 0.5 gal
Nails	0.25 lb per sf (32*0.25)	= 8 lb

Lumber. Total required: 12 each panels.

Side plates	12 ea*16 lf	= 192.0 lf
Studs	12 ea*28 lf	= <u>336.0 lf</u>
	2″×6″ dimensional	= 528.0 lf
Board feet	$\dfrac{2 \times 6 \times 528}{12}$	= 528.0 bd ft

¾″ plyform	12 ea*32 sf	= 384.0 sf
Form oil	2 oz per sf (384*2)/128	= 6.0 gal
Nails	12 ea*8 lb	= 96.0 lb

The support components for the form panels will be calculated and listed by quantity required, and designated as:

$$r = \text{reuse factor } x$$

$$c = \text{consumable}$$

The double 2″ × 6″ walers will span the wall panel horizontally (perpendicular to the studs) at 2′ centers vertically, front and back face, and will extend past the wall panel form system a distance of 1′ on each end.

Walers. Total number required: 2″ × 6″ double walers, 6 each.

Wale each	2 ea 2″×6″ @ 26′	= 52 lf ea
	6 ea*52 lf	= 312 lf r<u>5</u>

Board feet $\quad \dfrac{2 \times 6 \times 312}{12} \quad = 312$ bd ft

The sill plate will consist of a dimensional 2″×6″ component under each front and back panel unit.

Sill plates. Total number required: 2″×6″ sill plate, 2 each.

Sill plate $\quad$ 2 ea 2″×6″ @ 24′ $= 48$ lf r5

Board feet $\quad \dfrac{2 \times 6 \times 48}{12} \quad = 48$ bd ft

Because of the horizontal seam between the individual form panels, it is recommended to install vertical support members (perpendicular to the horizontal double walers) consisting of a double 2″ × 4″ waler component, spaced at 4′ centers horizontally at both front and back faces. These can be standard stock length lumber (8′).

Double walers. Total number required: 2″×4″ double walers, 14 each.

Wale each $\quad$ 2 ea 2″×4″ @ 8′ $= 16$ lf ea
$\qquad\qquad\quad$ 14 ea∗16 lf $\quad= 224$ lf r5

Board feet $\quad \dfrac{2 \times 4 \times 224}{12} \quad = 149.3$ bd ft

The top spreader yokes will consist of a dimensional 2″ × 4″ component spaced at 4′ center across the top of the form unit.

Spreader yokes. Total number required: 2″ × 4″ spreader yokes, 7 each.

Spreader yokes $\quad$ 7 ea 2″ × 4″ @ 3′ $= 21$ lf c

Board feet $\quad \dfrac{2 \times 4 \times 21}{12} \quad = 14$ bd ft

Because the form system is anchored to the footer component by the sill plates, and the back panel unit is attached to the front by the top spreader yokes and the metal snap ties, the requirement of bracing will be on only one side, consisting of dimensional 2″×6″ lumber. They will be anchored to the vertical 2″×4″ waler members.

Braces. Total number required: 2″×6″ braces, 7 each.

Support braces $\quad$ 7 ea 2″×6″ @ 12′ $= 84$ lf r5

Board feet $\quad \dfrac{2 \times 6 \times 84}{12} \quad = 84$ bd ft

Ground stakes. Total number required: 2″ × 4″ stakes, 7 each.

Ground stakes	7 ea 2″×4″ @ 2′	= 14 lf c
Board feet	$\dfrac{2 \times 4 \times 14}{12}$	= 9.3 bd ft

Material summary: form panels

Form panels 2″×6″. Total lumber required: 12 each panels.

2″×6″ dimensional lumber	= 528.0 lf
Board feet	= 528.0 bd ft r5
¾″ plyform	= 384.0 sf r5
Form oil	= 6.0 gal c
Nails	= 96.0 lb rc

Support components

Horizontal waler	2″×6″	6 ea = 312 lf r5	= 312 bd ft
Sill plate	2″×6″	2 ea = 48 lf r5	= 48 bd ft
Vertical waler	2″×4″	14 ea = 224 lf r5	= 149.3 bd ft
Spreader yokes	2″×4″	7 ea = 21 lf c	= 14 bd ft
Braces	2″×6″	7 ea = 84 lf r5	= 84 bd ft
Ground stakes	2″×4″	7 ea = 14 lf c	= 9.3 bd ft

The material listed is composed of the required lumber to construct the form system specified. This scenario will assume the reuse factor to be equally applied for all the material, and the form system will provide the same term result, except for the components which are listed as consumable. These consumable components, which are classified as part of the form system, not consumable project components, will be disbursed within the cost of the ownership for the form system, 6″ wood panel, and allocated as an asset.

Ownership cost. The ownership cost procedure for this system will provide for the fabrication of the form panels and the purchase of the listed lumber on the material list. The material listed will be considered the complete requirement for the form panels, including the walers, spreaders, and bracing material.

The labor and equipment required will be only that associated with the form panels, waler, and component fabrication and furnishing, as with the lumber requirement. All labor, equipment, and material required for the erection and placement of the form system is budgeted with the project's structural components, as outlined in the quantity takeoff section.

The lumber and material cost will be allocated by the specific unit of measure relative to the component, board feet with lumber, square feet with plyform, pounds with nails, etc.

The labor and equipment cost will be allocated against the time duration required to fabricate the formwork and make ready of the walers and support components.

The lumber cost per measured unit will be calculated from the average local unit, or index price market, from which the material will be purchased, including applicable sales and use tax.

Formwork and Falsework

Price index

2″×4″ dimensional lumber	S4S common grade	@ $0.540 bd ft
2″×6″ dimensional lumber	S4S common grade	@ $0.533 bd ft
¾″ plyform	B-B concrete grade	@ $0.822 bd ft

Material cost

Summary of required material by size

6″ wood panel form:	384.0 sf	Reuse factor: 5	
2″×6″ dimensional	972.0 bd ft	@ $0.533 =	$518.08
¾″ plyform	384.0 sf	@ $0.822 =	$315.65
2″×4″ dimensional	149.3 bd ft	@ $0.540 =	$ 80.62
Nails	96.0 lb	@ $0.750 =	$ 72.00 c
			$986.35

Consumable:

2″×4″ dimensional	23.3 bd ft	@ $0.540 =	$12.58
Form oil	6.0 gal	@ $4.500 =	$27.00
			$39.58

Labor calculation. The labor and equipment will be calculated for the time required to construct the required formwork and components and have them available for use.

Operation. Construct 6″ wood form panels and associated components.

Crew members	= 2.0 ea
Shift duration	= 8.0 hours
Production unit	= sf
Project quantity requirement	= 384.0
Formwork type	= wood
Form panel size	= 4′ × 8′
Unit cycle time	= 1.25 min/sf
Productivity factor	= 50 min
Cycles per hour: 50/1.25	= 40.0 cycles
Production per hour	= 40.00 sf/hour
Production per shift: 40.00∗8	= 320.00 sf/shift
Required shifts: 384.0 sf/320.0 sf	= 1.20 shifts
Total craft-hours: 1.20∗8∗2	= 19.20 craft-hours
Operation production factor: 19.20/384.0	= 0.050
Operation factor	= 0.050 craft-hours/sf

Example: Crew costing and formulating

The formula detailed under this heading will show the distribution and application of the production factoring method. The cost breakdown will summarize the unit labor and equipment, along with the total required labor and equipment cost, independently and total for the operation.

The application of cost will be derived from the production factors calculated by an average wage per craft-hour, and the equipment crew cost per operation. The labor crew and equipment crew cost, per shift, will be the total cost expended, derived from the labor and equipment formulations.

The cost per shift, for both labor and equipment, is calculated by the production per shift, times the number of required shifts. This format produces the total labor and equipment cost required to perform the given activity.

To arrive at the total labor cost, multiply the unit production factor by the average wage per hour by the required operation quantity. To arrive at the unit labor cost, multiply the unit production factor by the average wage per hour. For the total equipment cost, multiply the cost per shift by the required shifts. For the unit equipment cost, divide the total equipment cost by the required quantity for the operation. To determine the average wage per craft-hour, multiply the labor cost per shift by the required shifts divided by the required craft-hours.

Operation. Construct 6" wood form panels and associated components.

```
Required quantity: 12.0 ea           384.00 sf
Labor cost/shift                     $240.00
Equipment cost/shift                 $ 80.00
Total cost/shift                     $320.00
Required shifts                         1.20
Required craft-hours                   19.20
Average wage/craft-hour              $ 15.00
Labor production factor                0.050
Production/shift                     320.00 sf
Unit labor: 0.050*$15.00           = $   0.750 sf
Unit equipment: $80.00/320.0       = $   0.250 sf
  Unit production cost             = $   1.000 sf
  Labor: 1.20*$240.00              = $288.00
  Equipment: 1.20*$80.00           = $  96.00
  Total cost, labor/equipment      = $384.00
```

Cost distribution. 6" wood form panels and associated components.

```
Quantity: 384.0 sf, reuse factor 5.
Reuse material                  $  986.35
Labor                           $  288.00
Equipment                       $   96.00
Reusable cost                   $ 1370.35
Consumable material cost        $   39.58
```

Form asset cost. To arrive at a form asset cost, per square foot, apply the reuse applicator factor to the reusable cost and divide by the applicable unit of form panel surface square feet (384.0 sf). Then divide the consumable material item cost by the form surface square feet (384.0) and add the two numbers together, giving the asset value for the formwork, based on the reuse factor of 5.

Calculation

```
Reuse cost ($1370.35/5)/384.0 sf     = $0.714 sf
Consumable cost $39.58/384.0 sf      = $0.103 sf
                                     = $0.817 sf
```

Required. 384.0 sf form system with appurtenances @ $0.817 = $313.73.

The listed example provides for the procedure to allocate the cost of fabricating and supplying the required form system, with supports and bracing, for a defined reuse cost factor.

Application: Metal Panel Form System

A steel form system, when purchased, will be considered a long-term asset with respect to ownership cost distribution.

Cost distribution. The cost will include the form panels along with the necessary hardware and accessory items that are standard and required for the system. The cost will not include any specialty, short-term, or specific project-related expenses.

Since the metal system is purchased as a complete and fabricated system, the cost distribution will include only the form panel and direct material accessory cost. The associated labor and equipment for erection will be distributed within the job budget analysis.

Depreciation. The metal form system cost will be allocated over a life expectancy and use period. There are several methods of performing the depreciation factor. One method of allocation is the time allocation. This is performed by disbursing the normal expected life of the form panel over a specified number of years and uses within each year. The interim yearly use can be realized by number of days per month and number of months per year. The daily or monthly cost will then be allocated by the number of days required for the project, providing an ownership cost per day, or for the number of square feet of production unit requiring the form system, providing an ownership cost per square feet of formed surface required by the component.

A second method of allocating the form cost is by realizing a specified unit use, or number of square feet of surface formed within a year or normal life cycle. This method is a defined life cycle of predetermined unit use.

The ownership of form systems will not be distributed over a reuse factor allocation as the wood system. Normally, a quality metal form system used under normal conditions and well maintained can be allocated over an average ownership period, and the asset will be realized well before the life of the system has expired.

For a *time depreciation method,* an example would be the unit cost disbursed over a 5-year, 8-month per year, 15-day per month usage factor:

$d = \underline{5,8,15}$

d = form cost/(5 years*8 months) = 40 months (duration cost of 40 months)

d = cost per month/15 days = unit cost per day (duration cost of 600 days)

Example

Form system cost of $6000 over a duration of 600 days equals an allocation cost of $10 per day of use.

For a *unit of measurement use allocation,* an example would be the system cost disbursed over a specified quantity of total use:

$u = \underline{50,000.0}$ sf

u = form cost/a specified number of allocation square feet

Example

Form system cost of $6000 over square foot use quantity of 50,000 sf equals an allocation cost of $0.12 per square foot of use.

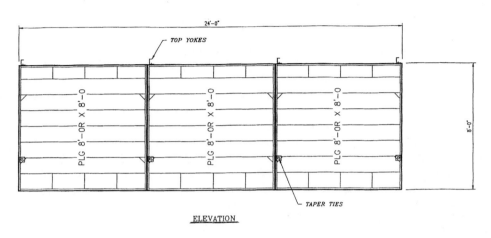

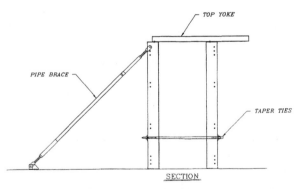

8" wall panel 8'h×24'l @ 2' thick.

Example: Metal panel form system

The example form system consists of fabricated structural metal panels, with support braces and accessories of metal tie hardware.

The scope and structural needs of this wall panel will require a dimensional panel thickness of 8". The condition will allow standard-sized form panels 8' wide by 8' long.

The individual panel construction specification will require a structurally constructed, steel-skinned, form panel capable of supporting and stabilizing itself, with braces required for the vertical support of horizontal movement.

Material list. The material list will be as required for the steel component form panels:

Panels. Required: 6 each 8″ dimensional steel 8′ × 8′ panels.

The *support components* for the steel form panels will be calculated and listed by quantity required, and designated as:

$$d = \text{depreciation factor } x$$

$$a = \text{accessory}$$

$$c = \text{consumable}$$

The *form bolt requirement* will consist of the required number of form bolts required for the form unit assembly. Required: ¾″ × 4″.

Total number required: 17 per vertical joint
                                     4 per brace
                                     4 per top yoke

Vertical joints	4 joints*17 ea	= 68.0 ea
Pipe braces	7 braces*4 ea	= 28.0 ea
Top yokes	7 yokes*4 ea	= 28.0 ea
		124.0 4″ bolts a

The *top spreader yokes* will consist of a dimensional 4″ metal angle component spaced at 4′ center across the top of the form unit.

Total number required: 4″ spreader yokes, 7 each.
Spreader yokes 7 ea   4″ angle @ 3′   = 21 lf a

The *metal taper ties* will consist of a tapered metal bolt assembly component spaced at 4′ center across the center of the form unit that will tie the front panel to the back panel.

Total number required: 24″ taper ties = 7 ea a

The form panel system will be anchored to the footer component by *anchor assemblies,* the back panel unit will be attached to the front by the top spreader yokes and the metal taper ties, and the requirement of bracing will be on only one side, consisting of 2″ structural pipe bracing. They will be anchored to the vertical steel form panels (front face).

Braces

Total number required: 2″ pipe braces, 12′ long; 7 each
Support braces 7 ea 2″ pipe braces @ 12′   = 84 lf a
Ground stakes 7 ea 2′ stakes   = 7 ea a

NOTE: The end closure bulkheads will be calculated as consumable components within the project, or as a reusable wood form panel, and not allocated within the metal system. The metal form system allocation will be the components and support accessories directly related for the square feet of main panel designated.

Material summary: form panels

8" steel form panels:
Total required: 6 each panels
    8'×8' panels   = 384.0 sf 5,8,15 d
    Form bolts   = 124.0 ea 5,8,15 d
Support components
    Pipe braces, 2"×12', 7 each   = 7.0 ea 5,8,15 d
    Ground stakes   = 7.0 ea 5,8,15 d
    Taper ties 24"   = 7.0 ea 5,8,15 d
    Top yoke 3" angle   = 21.0 lf 5,8,15 d

The material listed is composed of the required panels and accessories required to construct the form system specified. This scenario will assume the time allocation depreciation factor to be equally applied for all the components within the form system and allocated as an asset.

The ownership cost procedure for this system will provide for the purchase of the form panels and components. The material listed will be considered the complete requirement for the form panels, including the support and bracing accessories.

All labor, equipment, and material required for the erection and placement of the form system are budgeted with the project's structural components, as outlined in the quantity takeoff section.

Material cost

Summary of required material components by type:
8" steel panel form: 384.0 sf, depreciation factor: 5,8,15

8' × 8' panels	384.0 sf	@ $ 22.00 sf =	$ 8,448.00
Form bolts/nuts	124.0 ea	@ $ 4.50 sf =	$ 558.00
Pipe braces	7.0 ea	@ $175.00 ea =	$ 1,225.00
Ground stakes	7.0 ea	@ $ 25.00 ea =	$ 175.00
Taper ties 24"	7.0 ea	@ $ 85.00 ea =	$ 595.00
Top yoke 3" angle	21.0 lf	@ $ 3.00 lf = $	63.00
			$11,064.00

Cost distribution: 8" steel form panels and associated components
Quantity: 384.0 sf, depreciation factor 5,8,15
Form panel material: $11,064.00

To arrive at a form asset cost, per day, apply the daily depreciation applicator factor to the initial cost. To arrive at the unit square foot ownership cost for the specific use, determine the total time the form system will be required, multiply the daily ownership cost by the time duration, and divide by the applicable unit of form panel surface square feet (384.0 sf). The time duration required will include the setup and make ready time, the erection time, the concrete placement and cure time, the dismantling and removal time, and the form cleaning, preparation, and storage time.

Calculation

$$d = \underline{5, 8, 15}$$

$$d = 5 \text{ years} * 8 \text{ months} = 40\text{-month payoff}$$

$$d = \$11{,}064.00/40 \text{ months} = \$276.60 \text{ per month}$$

$$d = \$276.60/15 \text{ days} = \$18.44 \text{ per day ownership}$$

Time duration required	
Make ready, set up	2 days
Form system erection	2 days
Placement time	1 day
Cure time	3 days
Form removal time	1 day
Clean and prep time	<u>1 day</u>
	10-day duration

$$d = \$ \ 18.44 * 10 \text{ days} = \$184.40$$

$$\$184.40/384.0 \text{ sf} = \$0.48 \text{ sf}$$

The listed example provides the procedure to allocate the cost of furnishing and supplying the required form system, with supports and bracing, for a defined allocation cost factor.

For a short-term condition requirement of a form system, a rental or lease may be appropriate, rather than fabricating or purchasing a form system. When this condition exists, it is necessary to identify and define the components required for the specific structural use, arrive at a firm cost including shipping and handling, and distribute this cost within the required use, by either the square feet or a lump-sum unit per item.

For specialty formwork that is specific to one use, one job, or one location, it will be required to absorb the entire end result cost within the component requiring the special component.

Normally, for ownership form systems, the surface calculation of concrete contact area required for the formwork will be assumed at a 1 to 1 ratio. In other words, if the contact surface area of a structure requires 384 sf of formwork and the actual formwork panel area is 400 sf, the concrete surface contact area of 400 will be used for bidding purposes. This will hold true for most scenarios other than special or large irregular areas.

For the form rental of specialty conditions, the actual form surface area of the required panel that is rented will be used in the budget. In other words, if the contact surface area of a structure requires 384 sf of formwork and the actual formwork panel area is 400 sf, the form panel surface contact area of 400 will be used for bidding purposes.

Chapter 4

Concrete Mixes, Technical Aspects

Introduction

Concrete is the material component which is most dominantly used in structures. It contains the strength needed to support and withstand many and great implied forces. Depending on their intended use, concrete mixes can be designed by varying their basic components and by the addition of other nonbasic components, to obtain greater *compressive strengths* needed in today's technology. The compiling of a *concrete mix design* is a scientific determination of blending proper materials together to achieve a desired and useful product. (See Table 4.2, page 111.)

Owing to the *flowability* state of concrete at the time of mixing, it can be molded, shaped, or formed into a variety of shapes.

Concrete is a widely used product. It is used in bridges, roadways, buildings, and many other structures. In conventional bridge construction its use extends from foundation materials such as pilings and caissons, to substructure supports and containments, to the superstructure for precast beams and the roadway surface decks.

Concrete can be custom-designed as to shape and strength to serve in a wide variety of bridge construction techniques. This work will consist of constructing major and minor structures with the use of concrete.

The placement and finishing of concrete must be thoroughly planned and followed as outlined by the project specifications. Concrete must be placed in such a manner as not to cause segregation to the mix. Care must also be taken during placing procedures not to cause damage to the form system in which the concrete is placed and by which it is supported.

The finishing of concrete to its specified form and texture must be completed during the time of which the mix is still in its *plastic* state. The concrete must be properly consolidated prior to this, usually by a form of mechanical vibration or other acceptable means. The concrete must then be finished and/or textured prior to its initial set, as per specifications. Some variables that may dictate the final finishing time may be temperature, weather, and cement content or special accelerating or retarding admixtures.

Improper mixing, placing, finishing, and curing of concrete can be detrimental to the product with respect to both design strength and consolidated mass.

Work with structural concrete will consist of *furnishing, placing,* and *finishing* concrete mixes in bridges, culverts, and other structures in accordance with specific project specifications and details. Also with the end result of concrete, it must be constructed in a reasonably close conformity to the design plans and details of the project as established by the owner and engineer.

Mix Design

The basic components of concrete will consist of a mixture of *portland cement, fine aggregate, coarse aggregate, chemical admixtures* when required, and water. *Pozzolan* or fly ash can, in most cases, be added to a mix design as a partial cement replacement. These materials must be mixed in the exact proportions as detailed on the mix design as specified to assure the stability and strength of the desired mix.

Concrete is specified by classes which govern the strength and type of materials used to produce a given mix. Materials used within a concrete mix design must be of an approved source and conform to project specifications. Concrete materials will be proportioned in accordance with the given project specifications and guidelines. Mix designs will be furnished for each class or strength of concrete specified by the project.

The proposed mix designed will be suitable for all uses and conditions for which the individual classes will be used. Concrete mixes will be designed using an absolute volume analysis. Each mix proposed will be approved and verified by a certified laboratory. This is accomplished by material proportioning by weight and volume, *slump tests, air tests, uniformity, workability,* and *compressive strength tests.* The results of these tests along with the materials certification of each detailed mix design will be submitted to the owner for final approval based on conformity with the project specifications. This mix design, once approved, will always remain the responsibility of the producer to produce the mix to the conforming specifications of the mix design and the minimum acceptance criteria to the project guidelines.

Testing requirements

Throughout the project the raw materials and the mixed concrete will be continually tested for conformity to specifications. A change of source of raw materials will not be permitted owing to the possibility of change in *specific gravity* of these materials which would alter the design of concrete mix either by weight and absolute volume or by chemical properties. If an instance existed to necessitate a change, a new mix design would need to be developed. A contractor must also keep in mind the time needed to produce a mix design with respect to the curing requirement of a 28-day compressive strength test.

Outside of additional admixtures to regulate strength and durability of concrete, there are controlling items within the basic components and materials

CONCRETE MIX DESIGN CRITERIA
MIX DESIGN #1:   100% Cement

Source of Raw Materials:

			Specific Gravity Property
	Cement	Cement Products Inc.	3.150 (Bulk Sp.Gr.)
#23	Fine Aggregate	Rockstone Aggregate Co.	2.591 S.S.D.
#8	Coarse Aggregate	Rockstone Aggregate Co.	2.569 S.S.D.
	Air Entrainment	Concrete Admixture Co.	
	Admixture 2	n/a	
	Admixture 3	n/a	

Mix Criteria:

Minimum Cement Content	564	pounds
Water/Cement Ratio	0.488 = 275.2	lbs
Fine/ Coarse aggregate Ratio	0.41 / 0.59	
Air Entrainment Target	6.5% (+/- 1.5%)	
Slump Target	2.5"	
Unit Weight Target	138	lbs/c.f.
Flexural Strength 28 day	650	PSI

Specific Gravity Calculation:

1. Volume of air per cubic yard   6.5% * 27 = 1.755 c.f.
2. Volume of cement   564.0 / 3.15 * 62.4 = 2.869 c.f.
3. Volume of water   275.2 / 62.4 = 4.411 c.f.
4. Volume- Air, Cement, Water   9.035 c.f.
5. Volume available for aggregate   27.0 - 9.035 = 17.965 c.f.
6. Volume of fine aggregate   0.41 * 17.965 = 7.366 c.f.
7. Fine aggregate weight SSD   7.366 * 62.4 * 2.591 = 1191 lbs
8. Volume of coarse aggregate   0.59 * 17.965 = 10.599 c.f.
9. Coarse aggregate weight SSD 10.599 * 62.4 * 2.569 = 1699 lbs

Batch Weights:

	Pounds	Volume-C.F.
Cement	564	2.87
Water	275	4.41
Fine Aggregate	1191	7.37
Coarse Aggregate	1699	10.60
Air Entrainment	6.5%	1.76
Totals	3729	27.00
Unit Weight / cubic foot	3729 / 27.0 = 138.1 lbs/c.f.	

Sample mix design.

used. The composition of concrete is controlled by achieving *cement content,* which is a ratio by pounds of cement per cubic yard of mixed concrete; *water-cement ratio,* which is a ratio by pound of water to pound of cement within a cubic yard of mixed concrete; *consistency* and uniformity of the mix, which is measured by a range in *slump* inches and unit weight; entrained *air content* for workability and durability, which is a ratio by a range of percent; and by varying gradation sizes of coarse aggregate based on AASHTO specifications M43, which is gradation size by pounds per cubic yard of concrete mix.

TABLE 4.1 AASHTO Gradation for Coarse Aggregates

Size	Number designation
2 inch to number 4	357
$1\frac{1}{2}$ inch to number 4	467
1 inch to number 4	57
$\frac{3}{4}$ inch to number 4 or 8	67 or 68
$\frac{1}{2}$ inch to number 4 or 8	7 or 78

Within these controlling items of measurement there are maximum and minimum limitations of the guideline specifications and target ranges to follow. This allows somewhat of a window for control owing to field conditions that prevent exact control but will maintain the designed consistency of the desired mix owing to the altered field conditions.

Table 4.1 represents the standard AASHTO gradation sizes and number designation for coarse aggregate.

Aggregates both fine and coarse must be clean, durable, and uniformly graded. The fine aggregates or sand must conform to the outlined specifications and gradations. They can be of a natural source or of a manufactured source. Coarse aggregates can be of natural gravel, crushed slag, or crushed stone passing specified *sieves*. They must be of sound, durable particles and be of specified freeze-thaw requirements. Aggregates of a carbonate substance, or any known to polish, will not be permitted.

Concrete composition. The requirements set forth in concrete composition for designed acceptance are as follows.

The concrete mix shall contain the specified percent of entrained air within the limitations set forth by specifications. Concrete shall consist of the required slump in inches within the range specified of the mix design.

The concrete mix shall contain the minimum amount of cement in pounds and shall not exceed the maximum amount of water specified in pounds, referred to as the water-cement ratio. This controls the slump which ultimately controls the compressive strength. If fly ash is substituted as a partial cement replacement, it may only be replaced up to a maximum percentage specified and by the ratio of replacement pounds specified. This combination of cement and fly ash would then be used and considered total *cementatious* material in conjunction with the water-cement ratio. Also, after this substitution, the aggregate volumes shall be adjusted by an amount equal to the net adjustment in volume of the combined cementatious material. This is required to assure an absolute volume yield of 27 cf.

For purposes of controlling the maximum water-cement ratio as specified, the water-cement ratio of fly ash modified concrete will be the ratio of the weight of water to the combined weights of portland cement and/or percentage of the weight of fly ash.

Portland cement is classified by types depending on the mixture of raw materials and types of materials used within the manufacturing process. Types of cement are determined by end result strengths obtained, and use characteristics. They are classified as type 1, type 1A, type 2, type 2A, type 3, or type 3A.

Fly ash or pozzolan is obtained from a by-product of coal-fired plants. It is classified, depending on the chemical composition, as types F and C.

The concrete mix shall develop a minimum compressive strength as specified by class, in 28 days.

The concrete must be subject to acceptance by visual inspections, field controlled air and slump tests, and intermediate specified compressive strengths.

Within the process of establishing an approved concrete mix design the contractor must submit, as discussed, whether self-produced or produced by a commercial redimix supplier:

1. The type and source of aggregate
2. The type and sources of cement, blended cement, and fly ash as applicable
3. The scale weights of each aggregate proposed as pounds per cubic yard of total concrete mix
4. The quantity of water proposed as pounds per cubic yard of total concrete mix
5. The quantity of cementatious materials proposed as pounds per cubic yard of total concrete mix
6. The percent of entrained air content
7. The slump of the designed mix
8. The unit weight per cubic foot

Each mix design will designate the target slump at which the mix will be produced. This target will be selected based on the expected conditions of placements.

The importance of quality control and adhering to the proportioned quantities of material controls the exact volume to produce 27 cf within specific gravities. This will also produce the required compressive designed strength of the specific class of concrete.

Each individual component within the concrete mix contains a *specific gravity* or exact uniform weight and a known volume. With respect to the aggregates this is obtained as a *dry weight* or moisture-free. This information is vital to obtaining a mix design to assure an exact *yield*. With regard to absorption of moisture within the aggregate, a calculation must be made to calculate the SSD or saturation of moisture for the water calculation of the mix, for both slump and yield control.

Admixtures. Concrete admixtures contain a wide variety of uses and functions which control a desired use of a concrete mix design and enable functions and strength that are not obtainable otherwise. They are also used for flowability and retarding and accelerating concrete.

Water-reducing admixtures allow for the reduction of water and cement content to a mix but produce the same desired strength result. These are normally used for an economic result but are sometimes required for other additions such as fly ash. The *dosage* rate of water-reducing agents is a ratio of ounces of material to pounds of cementatious material.

Retarding admixtures are chemicals which retard the set of concrete for a specified amount of time for performance of certain placement procedures or other construction requirements. The dosage rate of retarding agents is a ratio of ounces of material to pounds of cementatious material.

Air-entraining admixtures or chemicals give the concrete mix some workability and allow for an expansive design to the concrete with respect to freeze-thaw requirements. The dosage rate of air-entraining agent is a ratio of ounces of material to pounds of cementatious material.

Calcium chloride admixtures are used as accelerators and heat-generating agents. The dosage rate of calcium chloride is a ratio of percent of pounds to total weight of the mix with maximum limitations.

Nonchloride admixtures are also designed for accelerators and early-strength gainers; they develop workability without the harshness of calcium chloride. The dosage rate of these agents is a ratio of ounces of material to pounds of cementatious material.

Corrosion-inhibiting admixtures are designed for concrete to retard or inhibit corrosion to the embedded steel within the structure of the concrete. The corrosion-inhibiting agents are a ratio of ounces of material to pounds of cementatious material.

Fly ash or pozzolan admixtures allow for the partial replacement of raw cement for economic purposes without surrendering final strength requirements. Fly ash admixtures are a ratio of percentage of pounds of admixed material to a replacement pound of cement.

Superplasticizing admixtures allow for increased workability and higher early and final compressive strengths, ease placements and produce higher slumps with some water reduction. Superplasticizing admix agents are a ratio of ounces of material to pounds of cementatious material.

Latex admixtures are used for increased durability under freeze-thaw conditions, salt and deicing exposure protection, increased strength, both tensile and flexural, aids in bonding, and durability. Latex admix agents are a ratio, usually of liquid gallons to either a percentage of water content or by weight of cementatious material.

Microsilica admixtures are used for increased compressive strength, increased abrasion resistance, and increased chemical resistance and durability. Microsilica admix agents are a ratio of percent of weight of material to pounds of cementatious material.

Fiber reinforcing materials are enhancement admixtures to concrete for the reducing of cracking due to shrinkage and expansion, and an aid to durability. Fibers are a ratio of pounds of material added per cubic yard of mixed concrete.

Construction requirements. All materials to be used within a concrete mix must be stored and handled in a manner that prevents segregation, contamination, or any other harmful or detrimental effects.

TABLE 4.2 AASHTO Specifications for Concrete

Component	AASHTO designation
Cement	M85
Fly ash	C618 (ASTM)
Fine aggregate (natural)	M6
Fine aggregate (manufactured)	T176
Coarse aggregate	M80
Lightweight aggregate	M195
Entrained air	T152 (ASTM C260)
Slump	T119
Aggregate gradation	M43
Admixture water reducer	M194 (ASTM C494) (type for G)
Membrane curing compound	M148
Burlap curing materials	M182
Sheet curing materials	M171
Water	T26

All cementitious materials such as cement and fly ash must be stored in weathertight containers and be kept moisture-free at all times.

Aggregates, during storage and handling, must be maintained in such a manner to assure a constant uniform moisture content at the time of batching. Aggregates must be stockpiled and handled to prevent segregation, contamination, and crushing.

Concrete must be mixed in accordance with the project specifications and the approved mix design. All components will be measured separately either by weight or by a calibrated volumetric system. The tolerances of measurement must be accurately maintained with respect to the categories of cement, water, aggregates, and additives.

Production and delivery rate. The rate of production and the delivery of the concrete must be sufficient to provide a continuous placement operation without the concrete obtaining any initial set until completion, as per specific project specifications. In some conditions a minimum pour rate per hour will be outlined within the specifications, and proper control to achieve this must be maintained. These specifications must be facilitated with a minimum of rehandling of the mix and without damage to the structure. Certain specifications may require a placement schedule of both equipment and time to assure the known procedure.

Concrete will be mixed only in quantities required for immediate use. The concrete mix shall not be used if it has developed initial set or exceeds the established time limit set forth from the point of mixing to the point of placement. The *retempering* of concrete by adding additional amounts of water will not be permitted, as also any concrete that does not conform to the mix design.

Concrete may be mixed in a central mix plant or by truck-mounted mixers. The equipment used must be operated within its specified capacity and produce a concrete mix of uniform consistency within the mix design specifications. Precise mixing sequences must be followed with respect to individual components entering the mix drum within certain intervals. The inside of the

TABLE 4.3 Normal Time Limitations for Concrete Delivery

Vehicle type	Time, hours
Truck Mixer	1
Agitator	3/4
Open-top nonagitating	1/2

mixing drum will be kept free of debris and other accumulations which may restrict the free flow of materials into the drum and the mixing sequence. Batching procedures must be maintained to contain the raw materials from the point of weigh-up to the entrance of the mixing drum to assure a proper mix, strength, and yield. The entire operations and mixing procedures must follow the manufacturer's recommendations as to drum speed and volume capacity and the project specifications.

When concrete is transported from the mixing site, it must be delivered and discharged, after the point of water entry, within the time limits set forth by the project specifications. Table 4.3 shows normal limits.

Certain circumstances with regard to special cements and admixtures may alter these time limitations, which the project specifications will outline and govern.

The bodies of concrete hauling equipment will be kept clean and free of debris and buildup of accumulated concrete and cleaning water. They will be kept smooth and mortartight and be capable of discharging the concrete mix at a controlled rate and without segregation.

The admixtures must be dispersed into the mix at the proper time of mixing and at a controlled, measurable dosage rate per batch.

Truck mixers will be operated within the limits of their rated capacity and rotation speed for mixing as designated by the manufacturer. Mixing within a truck mixer is controlled by cycle revolutions rather than mix time. After initial mixing, as with agitator trucks, any revolutions will be at agitating speed. The mixing count for revolutions will begin after all materials, including water, have entered the drum.

Concrete that is mixed in a central mix plant will be measured for mixing by time as outlined in the project specifications. The blending of materials, including water and admixtures, into the mixing drum will be outlined and defined within the specifications. All mixing must be complete prior to the concrete's being delivered into a hauling unit. All contents of the drum must be discharged or removed before a succeeding batch is charged into the mixing drum.

When a truck mixer or truck agitator is used to transport concrete which has been mixed in a stationary central mixing unit, the speed of rotation of the drum or agitator must be maintained at agitating speed as recommended by the specific equipment manufacturer.

Volume proportioning devices. When volume proportioning is employed, devices such as counters, calibrated gates, or flowmeters will be incorporated within the process for precise measurement and control of determining quantities of the ingredients. In operation, the entire measurement and dispersing mechanisms will produce the specified proportions of the mix design.

All measurement devices, indicators, proportioning equipment, scales, and other mechanisms will be accurately calibrated, maintained, and in full view of the qualified operator. The water flowmeter shall be of sufficient quality to assure an accurate measurement without any effect from varying pressures within the supply line.

All necessary ingredients needed for the concrete mix must be contained in separate compartments within the batcher-mixer. The unit must be equipped with calibrated proportioning devices and be able to vary the mix proportions by specifications and approved mix designs.

Producer's responsibilities. The concrete producer shall maintain a competent workforce and experienced operators and technicians, not only for the mixing operation but for overall quality assurance of the product. These responsibilities will include:

1. Maintenance and proper storage and handling procedures of all components of the mix.
2. Assurance, maintenance, and cleanliness of plant, tanks, and other related equipment.
3. Gradation testing of both fine and coarse aggregates at frequencies necessary to assure specification compliance.
4. Moisture testing of aggregates to assure proper adjustments of free water within the mix design, daily or as often as necessary due to conditions. This is to maintain the water-cement ratio.
5. Batch weight computations for each production day or plant calibration checks necessary for accurate plant operation.
6. Accurate batching procedures of concrete in accordance with approved mix designs and specifications.
7. Accurate production of batch tickets with the following information:
   - Name of supplier
   - Number of ticket
   - Date and truck number
   - Name of agency controlling work
   - Designation or location of placement
   - Identification and class of mix design
   - Quantities of all components within the mix and total volume
   - Moisture corrections for aggregate content
   - Total water in mix at plant and added prior to discharge
   - Time at batching and at discharge

After mixing and prior to placement, the concrete will be randomly sampled and tested by a qualified technician for the following:

- Assurance that the final adjustments to the mix were within specifications
- Completion of the batch ticket along with a computation of the proper water-cement ratio
- Performance of tests such as temperature, air content, slump, and other screening tests to verify compliance with specifications in regard to mix design and plant operation
- Obtaining random samples of concrete for test cylinders for future strength tests and proper curing procedures

The producer may be required to furnish and complete these or other procedures, a testing laboratory facility, and curing facilities along with heated conditions as specified by project controls.

Temperature. Minimum and maximum temperatures of both the concrete and ambient air must be maintained both at the time of placement and throughout the curing time specified by project requirements. Cement content and admixture are controlling items that may also factor in on curing time and temperature control regulations.

A cold-weather and hot-weather concrete plan must be maintained and controlled by the contractor to assure a quality product.

Specifications will govern the need for continuous recording thermometers, heating and maintaining, and the cooling of the concrete mixing and placement operations, along with any housing requirements set forth to protect the structure until the curing limitations have expired.

The heating of a concrete mix can be aided by the heated enclosure of the plant, heated water, along with covered or enclosed and protected aggregate storage. A placement or structure can be protected by a heated housing or enclosure and the maintaining of heat within a specified limit of temperature throughout the curing process.

The cooling of concrete in adverse hot weather conditions can be aided by refrigerated water, shading, or enclosing aggregate and plant, underground storage, and cooling aggregates by sprinkling or by the addition of ice to the mix under controlled conditions. The aid of windbreaks, enclosures, fog sprayers, or other means of protection may be employed.

Depending on specific conditions and placements, other specifications with regard to air temperature, relative humidity, concrete temperature, and wind velocity for evaporation at placement may govern.

After proper placement has been achieved and prior to initial set, the pour must be protected against any rain that may be forecasted.

Volume calculations. Concrete mixes are produced and measured by volume relating to a cubic yard or cubic meter.

Unit weight is the traditional terminology used to describe the properties of materials within the mix design. The *mass* is the quantity of matter in a body, the *unit weight* is the weight per unit volume, the *weight* is the force exerted on a body of gravity, and *voids* within the volume of a specific material are the space between particles not occupied by the solid matter.

The calculation to determine unit weight, as detailed by ASTM C29, will be

$$M = \frac{G-T}{V}$$

where M = unit weight of the aggregate, pcf
G = mass of the aggregate plus the measure, lb
T = mass of the measure, lb
V = volume of measure, cf

The measure shall be defined as a cylindrical metal, watertight container.

The *absolute volume method* for determining concrete mix designs will establish the limits for the ingredients within the concrete mix. The design mix will be calculated for 1 cubic yard with the specified cement content and workability characteristics desired as specified for the particular use. The following guideline method will illustrate a typical concrete mixture.

Required:

Cement content: 6 sacks (94 lb ea) per cy
Air content: 6.5% (target)
Water-cement ratio: 5.5 gal per sack maximum
   (8.33 lb/gal, 62.4 lb/cf)
   [(5.5*6)*8.33]/(6*94) = 0.487 water-cement ratio
Target slump: 3″
Design for 38% fine aggregate

Properties

Component	Bulk specific gravity	Bulk specific gravity, SSD	Absorption percent	Moisture percent
Cement	3.15	n/a	n/a	n/a
Aggregate:				
Fine	2.59	2.63	1.1	5.0
Coarse	2.63	2.67	1.6	3.0

For the purpose of computing absolute volumes, the specific gravity may be considered as the ratio of the weight in air of a given volume of material to the weight in air of an equal volume of water.

The bulk specific gravity is based on the dry unit weight of the material, which includes both the permeable and impermeable voids. The bulk specific gravity saturated surface dry basis (SSD) is based on the saturated surface dry unit weight. The unit volume, including both the permeable and impermeable voids, is the same as for the bulk specific gravity.

The bulk specific gravity (SSD) is used in computing the absolute volume of aggregates. The SSD can be obtained by multiplying the bulk specific gravity by a factor equal to 1.0 plus the absorption percent expressed as a decimal.

116 Construction Systems, Methods, and Materials

Given the quantity of cement, air, and water, the absolute volume of cement paste binder can be calculated. The remaining volume will be aggregate.

For 1 cy
A. Volume of air: 0.065*27   = 1.755 cf
B. Volume of cement: 6 sacks*94   = 564 lb

$$\frac{564}{3.15*62.4} = 2.869 \text{ cf}$$

C. Volume of water: 6*5.5   = 33 gal
33*8.33   = 274.89 lb

$$\frac{274.89}{62.4} = 4.405 \text{ cf}$$

Absolute volume of cement paste binder   = A + B + C in cf
1.755 + 2.869 + 4.405   = 9.029 cf

If a fly ash product is used, the calculation will be performed in the same manner with the fly ash being added as a fourth component.

To calculate the remaining volume, which will be filled by aggregate, subtract the cement binder paste volume from 27, the absolute cubic feet within a cubic yard; 27 − 9.029 = 17.971 cf.

The specific gravity and absorption for the fine and coarse aggregates will be different; therefore, 38% of the aggregate volume will not provide exactly 38% of fine aggregates based on dry weights, but suitable for determining batch weights.

Volume of fine aggregate 38%: 0.38*17.971   = 6.829 cf
Fine aggregate, weight SSD: 6.829*2.63*62.4   = 1120.72 lb
Volume of coarse aggregate: 0.62*17.971   = 11.142
Coarse aggregate weight SSD: 11.142*2.67*62.4   = 1856.37

NOTE: A rule of thumb is for coarse aggregate to be 0.60, or 60% of the total aggregate weight, for initial evaluation.

Computation for dry weights:

Absorption defined as:

$$\frac{\text{Saturated surface dry weight} - \text{dry weight} * 100}{\text{Dry weight}}$$

Fine aggregate dry weight $\dfrac{1120.72 - D(100)}{D} = 1.1$

$$D = \frac{1120.72}{1.011} = 1108.53$$

20.44*1.011 = 20.66 lb SSD material

21.87 − 20.66 = 1.21 lb free water

Volume of material replaced:

$$\frac{20.66}{62.4*2.63} + \frac{1.21}{62.4} = 0.126 + 0.019 = 0.145 \text{ cf}$$

Volume of material added:

$$\frac{21.87}{62.4} = 0.350$$

Increase in volume:

$$0.350 - 0.145 = 0.205 \text{ cf, or } 1\%$$

For a mix design to function properly under normal field operating conditions, the amount of mixing water will require periodic adjustment. When the amount of mixing water is adjusted, the aggregate weights must maintain a yield of 1 cy in volume. An example would be a decrease of 2 gal for mixing water:

$$\frac{2*8.33}{62.40} = 0.27 \text{ cf}$$

To restore the volume of 0.27 cf with aggregate, the criteria of the aggregate must be defined:

Bulk specific gravity SSD = 2.63

Moisture content = 5%

Absorption = 1.1%

Both absorption and moisture contents are based on dry weight; for this it is mathematically correct to use the difference between the percentage of moisture and absorption as the percent of free water. Therefore,

W = required aggregate weight replacement (from stockpile)

0.05 moisture content

0.11 absorption

0.039 free water

$$\frac{W}{1.05} = \text{aggregate dry weight}$$

$$\frac{W}{1.05}*1.011 = \text{aggregate SSD weight}$$

$$\frac{W}{1.05}*0.039 = \text{amount of free water}$$

$$\frac{W*1.011}{1.05*2.62*62.4} + \frac{W*0.039}{1.05*62.4} = 0.27$$

$$W = 41.8 \text{ lb}$$

Ratio of aggregate weight to water weight:

$$\frac{41.8}{16.66} = 2.51$$

Normally, a guideline to follow for adjustments is 2½ lb of aggregate for each 1 lb change in water. If water is added, aggregate should be removed.

Placing and Finishing

Concrete will not be placed until the necessary forms and reinforcing steel have been properly installed, checked, and approved. The formwork and reinforcing steel must be cleaned of all debris prior to the placement of concrete. All forms will be treated with an approved form release agent prior to placing reinforcing steel. In addition, wood forms will be moistened with water immediately before placing concrete. No material or treatment that will be detrimental to, adhere to, or discolor the concrete will be used.

During placement procedures, the concrete must be thoroughly consolidated to avoid air pockets and honeycombs. The exposed surface will be worked during placement to bring the mortar to and against the formwork and to the surface to produce a smooth and desired finish.

Equipment for placing

Concrete will be placed to avoid segregation of the material components and displacement of the reinforcement steel and also the formwork. The use of *chutes, tremies,* and *troughs* may be required to aid in proper placement from the discharge point.

Concrete will not be dropped into the formwork or final area a distance of more than 5 ft unless confined by closed chutes or pipes. Care will be taken to fill each part of the form area by depositing the concrete as near to the final position as possible. The coarse aggregate shall be worked back from the forms and around the reinforcing steel without any displacement of the steel. After the consolidation and initial set of the concrete mix, the forms shall not be jarred or moved and no strain will be placed on any projecting reinforcing steel.

The use of *pneumatic* placing equipment will be arranged such that no vibrations would result that may damage the freshly placed concrete. When concrete is conveyed and placed by pneumatic methods, the equipment used will be of suitable kind and adequate in capacity to handle the work as outlined. The machine will be located as close as practicable to the point of deposit and placement of the concrete. The position of the discharge end will be not more than 10 ft or as specified from the point of deposit. The discharge lines will be horizontal or inclined upward from the machine.

When *pumping* concrete, the equipment used shall be so arranged that no vibrations would result that may damage the freshly placed concrete. The operation of the pump will be such that a continuous flow of concrete without air pockets is maintained. Upon completion of the pumping operation, any remaining concrete within the pipeline, if it is to be used, will be ejected in such a manner as not to contaminate the mix or that no segregation will occur. Any remaining concrete within the limits of the pump line that is not used must be disposed of so that fresh concrete mix will be maintained without contamination. Sampling of the concrete mix, when pumped, will be at the discharge flow end of the pipeline at the point of placement.

Concrete will be vibrated and consolidated by approved mechanical vibrations that assure proper consolidation without segregation of the volume of concrete. The vibrators will transmit vibration of the frequencies of impulses per minute as specified and within the radius of the vibrator as specified. Care must be taken not to overvibrate the mixture to the point that segregation occurs. Vibration should occur when the concrete mix is in its theoretical final position and should not be used to move or transport the flow of concrete in the forms.

Underwater placement. Depositing concrete underwater will be done in a manner as specified by project requirements. The concrete mix will meet the composition and requirements of *seal or tremie concrete* and will be placed in one continuous operation within the pour limits.

Tremies. To prevent segregation and the entrapment of water, the concrete will be placed in a consolidated mass, in its final position, by means of a *tremie* or by other approved methods. The concrete will not be disturbed after being deposited. Concrete will not be placed in moving water. The method of placement of the concrete will be regulated to the confinement of the mass within the limits with the top surface kept as level as possible.

When using a tremie, the means of supporting the tremie will permit free movement of the discharge and over the entire top of the concrete area and permit a rapid lowering when necessitated to choke off or retard the flow of mix. The tremie will be filled by methods that prevent the washing of the concrete mix. During placement procedures, it is imperative that the discharge end remain completely submerged within the concrete and the tremie tube be kept full at all times. This will prevent segregation and the trapping of water within the placement.

When the concrete is placed with a bottom-dump bucket, the bucket will be of such size and capacity specified and shall be equipped with loose-fitting top covers. The bucket shall be lowered to either the prepared foundation or upon previously placed concrete gradually. It will then be slowly raised during the discharge procedure with the intent to maintain, as nearly as possible, a still water environment to avoid agitating the concrete mixture.

Pumping. When underwater concrete is placed by pumping, the pump pipe shall be equipped with a bottom valve at the point of discharge to avoid and prevent the mixing of water with the concrete within the pipe. The pump pipe, as with the tremie, will be withdrawn slowly during the placement operation as the concrete rises but always maintaining to keep the discharge end submerged below the concrete.

Concrete placed by conventional methods of either direct discharge from the truck or by means of a bucket should always be maintained in a controlled and planned procedure. Place the concrete mix as near to its final position as possible with the least amount of handling. This will be both economical and nondetrimental to the concrete mix.

Columns. When concrete is placed in columns, it will be in one continuous operation within the limits of the placement and construction joints as specified by project conditions. The period of time between adjacent or additional placements will be specified to permit proper curing procedures and time.

Concrete which is placed in supported forms, cantilever areas, girder spans, caps, arches, etc., will be done in a manner to continually equalize the load over the form system and not to unbalance the placement. Any falsework used along with the form system should be constantly checked for shrinkage, creep, or settlement which may occur causing undue stress and danger to the placement.

No loads will be placed upon nor should any formwork or supporting falsework be removed until the concrete has achieved the proper strength required under the project specifications relating to these conditions.

The delivery of concrete to the placement site must be detailed to permit a somewhat continuous operation as not to disrupt the plastic state of the mix prior to its initial set. Failure to maintain a constant adhering flow of concrete within the placement may result in a *cold joint.* This would be detrimental to the design of the structure with regard to imposed loads and reinforcement shear value.

Weather conditions also could greatly control the placement of concrete and could cause and adversely affect the economical and structural conditions of the concrete. Careful planning must be an important decision on a placement schedule.

Finishing exposed surfaces

Exposed concrete surfaces must be finished in a manner not only to seal and close the surface for durability but for aesthetic value and texturing.

Finishing of surfaces of concrete is categorized by method, appearance, and location of area on a structure. The normal types of surface finishes are as listed:

- Ordinary surface finish
- Rubbed finish
- Tooled finish
- Sandblasted finish
- Wire brush or scrubbed finish
- Roadway surface finish
- Sidewalk finish

- Color finish
- Spray finish

Ordinary surface finish. Normally, an ordinary surface finish is specified for undersurfaces of slab spans, box girders, arch spans, and floor slabs between the girders of the superstructure. Areas and surfaces that are to be buried underground, covered with embankment, or surfaces above the finished ground that are not normally visible from a traveled roadway are not required to be finished. The ordinary surface type of finish usually requires that any holes, honeycombs, or other minor defects that are not harmful to the structure be patched for sealing purposes.

Immediately following the removal of forms, fins and irregular projections will be removed from surfaces that are to be exposed. With any surface, exposed or not, the cavities produced by form ties, bolt holes, broken corners or edges, honeycombs, or other defects will be thoroughly cleaned and prepared, then pointed and patched with mortar or cement and fine aggregate. This material will be mixed in proportions equal to the grade and mix of the concrete surface being finished. This rub patching and curing should be done according to project specifications. Care should be taken when patching around construction or expansion joints to produce a carefully tolled edge which is free from mortar and concrete for the full length of the joint. The resulting surfaces from the rubbing and patching procedure will be true and uniform to line, grade, and evenness of surface. A straightedge will be used on areas not protected by forms.

The tops of caps in the area of bridge bearing beam seats will be finished by rubbing or grinding to a smooth finish and true to slope at the specified elevation. Bearing areas will be finished to a true plane that does not vary more than $\frac{1}{16}$ inch per foot tolerance in any direction. The use of additional mortar to provide a grout finish will not be permitted. In bearing areas where a buildup is required, it will be necessary to apply a bonding agent or form of epoxy to assure adherence of the mixes and stability to the bearing area.

Rubbed finish. A rubbed finish is that of which the entire specified surface receives a hand-rubbed effect with mortar of the same cement content of the mix for color, to uniformly finish a surface. This normally requires the chipping and patching of form irregularities prior to the application of the mortar paste.

The normal areas of structures to receive this type of finish are all surfaces of bridge superstructures outside the areas as specified for an ordinary finish: all exposed above-ground surfaces of the substructure such as piers, piles, columns, abutments, and retaining walls. This area of finish should extend below the ground level at least 1 foot. Areas of culverts, headwalls, and endwalls which are exposed and visible by the traveled roadway are required to receive the application. All concrete railing, pedestrian crossings, arches, or other parts of structures which are either above-ground or within visible distances will be finished as outlined by the specifications.

After the removal of forms, the rubbing of concrete shall be started as soon as conditions permit. In preparing for this and immediately before rubbing, the concrete will be saturated with water. Sufficient time shall have elapsed prior to this wetting down to allow any patching mortar used in pointing of bolt holes, honeycombs, and defects to have properly set. The surfaces to be finished will be rubbed with a medium coarse carborundum stone, using a small amount of mortar on its face. The mortar will be composed of cement and fine sand mixed in the same proportions used in the concrete to be finished. Rubbing should be continued until the form marks, projections, and irregularities have been removed, voids have been filled, and a uniform surface has been obtained. The paste produced by this procedure should be left in place.

After all the concrete above the surface being treated has been cast, the final finish will be obtained by rubbing with a fine carborundum stone and water. This procedure should continue until the entire surface is of a smooth texture and uniform color.

The surface can then be final rubbed with burlap to remove any laitance and loose powder. This will be a final check and observance for any unsound patches and objectionable marks.

Tooled finish. A tooled finish is an architectural finish of certain character which is secured by the use of a bushhammer, pick, crandall, or other approved tool. No tooling of this nature will be done until the concrete has achieved a minimum strength as detailed in the project specifications, as not to structurally damage the concrete. The finished surface will show a grouping of broken aggregate particles in a matrix of mortar with each aggregate particle in slight relief.

Sandblasted finish. A sandblasted finish should commence after the concrete has thoroughly cured. The surface will be blasted with a hard, sharp sand to produce an even, fine-grained surface in which the surface mortar has been cut away, leaving the aggregate in whole exposed.

Wire brush or scrubbed finish. A wire brush or scrubbed finish is achieved as soon as the forms are removed, and while the concrete is yet comparatively green, the surface will be thoroughly and evenly scrubbed with a stiff wire or fiber brush. This procedure is done using a solution of muriatic acid proportioned of 1 part acid to 4 parts water until the cement film or surface is completely removed and the aggregate particles are exposed. This will leave the effect of an even, pebbled texture presenting an appearance grading from that of fine aggregate to coarse, depending on the size and gradation of aggregate used. Once the desired effect has been achieved by scrubbing, the area should be washed with water to rinse the acid traces away.

Roadway surface finish. The roadway surface finish shall be applied to all roadway surfaces of bridge decks, approach slabs, and other traveled concrete surfaces of structures according to specific project specifications.

All structure surfaces will be finished by mechanical, power-driven machinery traveling on rails which are adjusted to conform to the profile or cross section of the roadway. The machine will be equipped with oscillating transverse or longitudinal screeds and shall be adjusted to conform to the profile or the required cross section of the roadway surface. The screeds will have sufficient strength and stability to retain their shape and profile.

Rails and headers. The rails for the support and operation of the finishing machines and headers for hand-operated strike-off devices will be completely in place and firmly secured for the scheduled length of the concrete placement. These devices will be checked and approved prior to the placing of the concrete. The rails for the finishing machines will extend beyond both ends of the scheduled length of the concrete placement area for a sufficient distance that will permit the floating devices of the finish machine to fully clear the concrete surface to be placed.

The rails or headers shall be adjustable for elevation and will be set to allow for anticipated settlement *camber*, and *deflection* of falsework. This will be deemed as necessary to obtain a finished surface true to the required grade and cross section. These rails and headers, along with their supports, will be of such type and will be so installed that no springing or deflection will occur under the weight of the finishing equipment. These guides will also be located that the finishing equipment may operate without interruption over the entire area of surface being finished. The rails and headers will be able to be adjusted as necessary to correct for unanticipated settlement or deflection that may occur during the finishing operations.

Prior to the placement of concrete, the finishing machine shall make a *dry run* over the entire area of surface to be finished to determine and substantiate that the elevations of grade, profile, cross section, camber, and depth are correct as per specifications and plans.

During placing operations, a slight excess of concrete will be kept in advance of the cutting edge of the screed at all times. This excess of concrete will be carried the entire distance of the pour and will then be wasted and not worked into the slab placement.

Excess water, laitance, or foreign materials brought to the surface during the course of the finishing operations will not be reworked into the concrete surface and will be immediately removed upon appearance. This can be accomplished by means of a squeegee or straightedge, drawing it from the center of the slab toward either edge.

The addition of foreign water to the surface of the concrete to aid in workability and assist in the finishing operations will not be permitted. There are specific specifications that allow a misting or fogging procedure to aid in the finishing of the slab placement.

Check after finishing. After finishing, as described above by the use of a mechanical device, the entire surface will be checked with a metal straightedge of such length as outlined by specifications. This will be operated parallel to the centerline of the placement and shall show no deviation in excess of that specified, normally $1/8$ inch from the testing edge of the straightedge. Deviations in excess of this requirement will be corrected by a strike-off or a transverse float prior to the initial set of the concrete.

This checking operation will progress by overlapping the straightedge tool by at least one-half the length of the preceding pass. Once corrections have been made in the surface, a burlap drag shall be dragged transversely across the placement in a uniform manner traveling longitudinally with the finishing machine. This procedure is done to seal the placement surface and remove any slight deviation markings left by the straightedge.

Tined texture. All bridge and roadway surfaces will be final finished with a tined texture. Depending on governing specifications, this procedure can either be done while the concrete is in its plastic state or, after the curing process of the slab, be cold cut by means of a mechanical power-driven saw. This procedure produces an even texture to the slab to prevent skidding and hydroplaning that a smooth surface may emit.

The tined grooves will be transverse, of such size, dimension, and spacing as specified, usually between $1/16$ and $1/8$ inch wide by $1/8$ and $3/16$ inch deep and spaced $1/2$ to $3/4$ inch on centers.

Sidewalk finish. A sidewalk finish to a concrete surface will be such that the concrete will be deposited in place, consolidated, and the surface be struck by means of a float or strike board with the surface tolerance deviations being as specified by the contract. All edges and expansion joints will be tooled.

The surface will be broomed once the concrete has hardened sufficiently to prevent any tearing or segregation of the surface by the broom. The texturing strokes will be transversely square across the slab, from edge to edge, with adjacent strokes being slightly overlapped. This will proceed producing uniform and regular corrugations as outlined by specifications. Prior to the surface being finished, it will be free from porous spots, irregularities, depressions, small pockets or honeycombs, rough spots, or small particles of aggregate that may inhibit or disturb the brooming operation.

Color finish. A color finish is that of which a uniform permanent color is applied to the concrete surface as outlined by specifications.

Usually, the surface that is to receive a color finish will meet the requirements of an ordinary surface finish or rubbed finish specification. In addition to this, the surface will be free from all dust, foreign matter, curing compound, form oil, and grease. The surface may need a washing to properly prepare the area for coloring application. This color coating must be applied as per specifications regarding temperature limitations, moisture content, and other weather conditions that may inhibit or damage the final result. Also, the color finish must be applied following the manufacturer's recommendations or as otherwise specified. This finish will not be applied until all concrete operations pertaining to the unit have been completed.

Spray finish. The application of a spray finish shall be such as to provide a uniform, fine-grained, textured surface with a permanent color to all concrete surfaces designated by specifications.

Normally, the concrete surfaces to receive a spray finish shall meet the requirements of an ordinary surface finish or a rubbed surface finish. In addition to these requirements, the surface will be free from all dust, foreign matter, curing compound, and form oil or grease. The surface may need a washing to properly prepare the area for this application procedure.

At the time of application, the area must be in such condition consistent with both the manufacturer's recommendations and the project specifications. The weather conditions and temperature limitations must be such as to permit application. The spray finish will be applied in a uniform rate and texture and will be tightly bonded to the surface. The surface finish must adhere tightly to the structure to attain the desired and specified appearance.

Curing Concrete

All newly placed concrete will be properly cured. Curing will begin immediately after placing and finishing to protect the mix as per specifications.

Importance and time limits

The curing process will continue until such time, or as procedures outlined by specifications. The curing of concrete is a vitally important operation to assure the quality and design strength are maintained within the concrete mix. Without proper curing procedures, the concrete may become unsatisfactory to specifications and be of poor quality for its intended use. Curing retains the moisture within the concrete mix which aids in the designed strength gain and also protects the placement with regard to some weather conditions and temperatures. The moisture retention is an integral part of concrete operations. Improperly cured concrete will be considered defective.

The curing process must be maintained until the concrete has achieved the compressive strength outlined by specifications. The normal time requirement may be waived provided that the concrete has achieved a minimum strength required by ratio. Also, the curing process may be extended but only to the maximum time as allowed by strength design. In other words, the concrete must achieve the design strength within 28 days as specified by maximum time allowed.

If a formed surface is to require a rubbed finish, the concrete will be kept moist during the rubbing procedure. The curing method should be initiated immediately following this, while the concrete surface is still moist.

Additional moisture and coverings. When additional moisture is needed to be applied or water curing is specified, the method used will be done by *ponding, sprinkling,* or *fogging.* Coverings such as burlap should be used to retain the water so supplied. Covering materials that may cause discoloration to the concrete surface may not be used. The curing method used will maintain an even and constant supply of water to the surface being cured. The type of covering used to retain the moisture will be placed as soon as possible after the finishing operations have been completed to assure no damage to the finished surface. These coverings should be kept continuously moist throughout the specified time for curing.

A covering method will be used to prevent moisture loss to the finished concrete surface. This may be accomplished by using approved materials such as waterproof paper, plastic sheets, or a liquid membrane curing compound as permitted by specifications. The water curing method will be applied no later than 4 hours after the completion of the surface finishing or as specified. The surface that is covered by burlap, cotton mats, or other suitable moisture-retaining material will be saturated with water and the entire surface area will remain covered throughout the curing period, maintaining a constant supply and even application of water.

Waterproof paper and plastic will be of sufficient size and unit weight to completely form a waterproof cover over the entire concrete surface. It will be secured so wind will not displace it throughout the recommended curing time period.

Membrane curing compound. Membrane curing compound, as specified for use under the contract terms and meeting AASHTO M148 specifications, may be used as the initial and final curing agents on structural concrete surfaces. If the membrane coating becomes damaged during the curing period, it must be immediately repaired.

The curing compound will be applied in even, uniform applications as soon as the water sheen has practically dissipated from the concrete surface, as soon as the formwork has been removed, or immediately after the rubbing has been completed as specified. Should there be any delay in the application of the membrane curing compound, the surface area will receive moist curing until the compound can be applied. The surface area of the concrete cannot be permitted to dry out until the curing period has been completed.

The curing compound will be applied with approved equipment that will produce a fine spray and applied in an even and uniform coating. Any mixing or agitation needed must comply with recommendations. The rate of applications as to surface area square feet per gallon of compound must be as specified. Care must be taken in areas where future concrete bonding is required or where reinforcement is exposed, as the membrane compound will act as a bond breaker. The use of liquid membrane curing compounds on surfaces of construction joints and on areas to receive a spray finish will be prohibited.

Sampling and testing

Initially, throughout the placement, periodic samples will be taken and field tests will be performed for concrete and air temperature, air content, and slump. These tests will be done in accordance with the specifications and AASHTO T141 with regard to temperature, air content, and slump; and also AASHTO T152 or AASHTO T196 and AASHTO T119. Tests for controlling the maximum water-cement ratio, as specified by the mix design, shall be the ratio of the weight of the water to the combined weights of portland cement and a percentage of the weight of fly ash, if used.

Compressive strength of concrete will be derived by samples of the mix taken and molded into cylinders according to testing specifications. The results of these cylinder tests for compressive strength will be relative to the compliance with the 28-day compressive strength as outlined by specifications to each mix design. This is done in accordance with AASHTO T22 and AASHTO T23. For purposes of acceptance for compressive strengths, this test may be modified by a set ratio to extend the time for fly ash modified concrete mixes.

A unit weight test will be performed according to AASHTO T121 to comply with uniformity and the approved mix design.

Any concrete that does not comply to these test procedures may be rejected and deemed intolerably detrimental with respect to the effect on the structure.

Chapter 5

Equipment Requirements

The equipment needs and requirements for structure-related projects place demanding financial commitments on the contractor. The long-term investment for owned equipment, and the short-term cost for leased or rented equipment require careful planning, research, and cost analysis. The decision must be made to purchase the required equipment, commit to a long-term lease, or merely rent the unit for a short-term duration; or a combination of these. Regardless of the choice or needs, equipment is the most valuable and costly asset.

There are four main categories of equipment: the primary major equipment units, secondary support equipment, minor equipment, and specialty equipment. Within these categories, there exist many different types having multiple classifications and capacities.

Primary and Major Equipment Units

Within the category of the primary major equipment, there exist the types of *lifting* and *hoisting, concrete placing* and *finishing,* and *pile driving.* For a structure contractor, these will be considered the most required, demanding, and largest investment equipment units.

Cranes

The lifting and hoisting units (cranes) are required for most component portions of a concrete structure, from hoisting and setting formwork, to placing concrete, to the heavy-duty requirement of beam erection. Cranes are designed by lifting capacity, style, and mobility features, which lends itself to the specific and unique feature requirements.

The three major classes of cranes are crawler, truck-mounted carrier, and rough terrain. The crawler and truck-mounted are available with conventional lattice boom, in varying lifting and reach capacities. The truck-mounted and rough terrain types also are available in hydraulic type, telescoping boom, with varying lifting and reach capacities. These crane units range in capacity from 15 to 250 ton, for the most commonly used units, and with capacities to 750 ton and up for larger, specialty units. As they increase in lifting capacity, they also increase in reach capacity.

The maximum capacity of any crane is based on the crane's lifting capability with the base boom (shortest) section, full part line, and greatest (fully raised) boom angle. As the boom length increases and the boom angle lowers or becomes more horizontal, the lifting capacity decreases to the crane's minimum lift capacity.

Lifting charts are furnished with every crane so that instant analysis can be made for each specified lift. These charts reference three required criteria: the distance the hoisted object is from the center pin of the crane, the length of boom required for the lift, and the weight of the component to be lifted.

The distance requirement will be the maximum reach needed within the cycle. The boom length requirement will be the maximum length required within the cycle. By defining the maximum reach and maximum boom length, the minimum boom angle will be determined, which will identify the maximum allowable weight to be safely lifted. By comparison, the charted weight is referenced to the desired component to be hoisted.

This will control the lifting capacity for the crane unit. If the component is heavier than the lifting chart allows, one of three decisions can be made. Move the crane closer to the component, thus closing the radius and increasing the boom angle, and possibly shorting the boom if permissible for the height required. Utilize two crane units to equally lift the component. Use a larger-capacity crane that meets the lifting criteria. Regardless of the choice, the required lifting criteria will have to be reevaluated for each component lift.

The total weight of the component to be hoisted will include the component weight, any required lifting apparatus, slings and cables, spreader beams, or any other additional item that adds weight to the hook load of the crane.

In order for a crane unit to operate safely and to its designed optimum capacity, the ground on which it sets must be stable, suitable, and level. For soil conditions that are somewhat unstable, wood crane mats can be utilized. However, this procedure should be done with the utmost care and by the crane manufacturer's recommendations, following the specific soil criteria for stability and safety.

Crawler-mounted crane units

Crawler-mounted crane units normally are more stable and are more maneuverable for off-road, heavy-type construction projects. The ground conditions must be level and stable for the crawler crane to function properly.

Crawler-mounted cranes have normal ranges of lifting capacities from 25 to 750 ton, and have boom lengths from 40 to 500′. A crawler-mounted crane, once moved into a site and assembled, normally is planned to be present for a longer-duration project. The load being hoisted, provided the soil is level and

LS-138H Lift Crane Capacities

Boom – angle:
48" (1.22 m) wide and 48" (1.22 m) deep with open throat top section; with or without 24' (7.32 m) live mast, 1-1/4" (32 mm) diameter boom pendants.

Mounting – crawler:
extended gauge: 13' 0" (3.96 m)
retracted: 8' 11" (2.72 m)
overall length: 19' 5" (5.92 m)

Counterweights:
Ctwt. "A": 19,600 lbs.
Ctwt. "AB": 39,750 lbs

Length	Boom					Side Frames Extended			
	Radius		Angle	Boom Pt. Height①		Ctwt. "A"		Ctwt. "AB"	
	feet	meters	degree	feet	meters	pounds	kilograms	pounds	kilograms
40' (12.19 m)	12	3.66	77.4	44' 8"	20.3	125,500	56 927	150,000*	68 040*
	13	3.96	75.9	44' 6"	20.2	105,700	47 946	138,100*	62 642*
	14	4.27	74.5	44' 2"	20.1	91,300	41 414	128,200*	58 152*
	15	4.57	73.0	43' 11"	19.9	80,200	36 379	114,200	51 801
	20	6.10	65.3	42' 0"	19.1	49,400	22 408	70,800	32 115
	25	7.62	57.1	39' 3"	17.8	35,200	15 967	50,800	23 043
	30	9.14	48.1	35' 5"	16.1	27,000	12 247	39,300	17 826
	35	10.67	37.5	30' 0"	13.6	21,600	9 798	31,800	14 424
	40	12.19	23.4	21' 6"	9.8	17,800	8 074	26,400	11 975
50' (15.24 m)	13	9.96	78.8	54' 8"	16.7	106,000	48 082	138,100*	62 642*
	14	4.27	77.6	54' 6"	16.6	91,400	41 459	128,200*	58 152*
	15	4.57	76.4	54' 3"	16.6	80,300	36 424	114,300	51 846
	20	6.10	70.5	52' 9"	16.1	49,400	22 408	70,800	32 115
	25	7.62	64.3	50' 8"	15.5	35,100	15 921	50,800	23 043
	30	9.14	57.7	47' 11"	14.6	26,900	12 202	39,200	17 781
	35	10.67	50.6	44' 4"	13.5	21,600	9 798	31,800	14 424
	40	12.19	42.7	39' 7"	12.1	17,900	8 119	26,500	12 020
	50	15.24	20.9	23' 6"	7.2	12,800	5 806	19,400	8 800
60' (18.29 m)	13	9.96	80.7	64' 10"	19.8	106,200	48 172	129,900*	58 923*
	15	4.57	78.7	64' 6"	19.7	80,400	36 469	114,400	51 892
	20	6.10	73.8	63' 3"	19.3	49,400	22 408	70,800	32 115
	25	7.62	68.8	61' 7"	18.8	35,100	15 921	50,700	22 998
	30	9.14	63.6	59' 5"	18.1	26,800	12 156	39,200	17 781
	35	10.67	58.1	56' 7"	17.3	21,500	9 752	31,600	14 334
	40	12.19	52.3	53' 1"	16.2	17,700	8 029	26,400	11 975
	50	15.24	38.9	43' 4"	13.2	12,800	5 806	19,400	8 800
	60	18.29	19.0	25' 3"	7.7	9,600	4 355	15,000	6 804
70' (21.34 m)	15	4.57	80.4	74' 8"	22.8	80,400	36 469	112,000*	50 803*
	20	6.10	76.2	73' 8"	22.4	49,200	22 317	70,700	32 070
	25	7.62	71.9	72' 2"	22.0	34,900	15 831	50,500	22 907
	30	9.14	67.6	70' 4"	21.5	26,700	12 111	39,000	17 690
	35	10.67	63.1	68' 1"	20.8	21,300	9 662	31,500	14 288
	40	12.19	58.4	65' 3"	19.9	17,500	7 938	26,200	11 884
	50	15.24	48.1	57' 9"	17.6	12,600	5 715	19,200	8 709
	60	18.29	35.9	46' 8"	14.2	9,500	4 309	14,900	6 759
	70	21.34	17.6	26' 10"	8.2	7,300	3 311	11,800	5 352
80' (24.38 m)	16	4.88	80.9	84' 8"	25.8	71,500	32 432	101,100*	45 859*
	20	6.10	77.9	83' 11"	25.6	49,100	22 272	70,500	31 979
	25	7.62	74.2	82' 8"	25.2	34,700	15 740	50,400	22 861
	30	9.14	70.5	81' 1"	24.7	26,500	12 020	38,200	17 328
	35	10.67	66.6	79' 1"	24.1	21,500	9 752	31,200	14 152
	40	12.19	62.7	76' 9"	23.4	17,300	7 847	26,100	11 839
	50	15.24	54.3	70' 7"	21.5	12,400	5 625	19,000	8 618
	60	18.29	44.8	62' 2"	18.9	9,300	4 218	14,700	6 668
	70	21.34	33.5	49' 10"	15.2	7,100	3 221	11,700	5 307
	80	24.38	16.5	28' 4"	8.7	5,500	2 495	9,400	4 264
90' (27.43 m)	18	5.49	80.6	94' 5"	28.8	58,200	26 400	83,300	37 785
	20	6.10	79.3	94' 2"	28.7	48,900	22 181	70,400	31 933
	25	7.62	76.0	93' 0"	28.4	34,500	15 649	50,200	22 771
	30	9.14	72.7	91' 7"	27.9	26,300	11 930	38,600	17 509
	35	10.67	69.4	89' 11"	27.4	20,900	9 480	31,000	14 062
	40	12.19	65.9	87' 10"	26.8	17,100	7 757	25,700	11 658
	50	15.24	58.7	82' 7"	25.2	12,100	5 489	18,800	8 528
	60	18.29	50.9	75' 6"	23.0	9,000	4 082	14,400	6 532
	70	21.34	42.2	66' 1"	20.1	6,900	3 130	11,400	5 171
	80	24.38	31.5	52' 9"	18.3	5,300	2 404	9,200	4 173
	90	27.43	15.5	29' 9"	9.1	4,000	1 814	7,500	3 402
100' (30.48 m)	19	5.79	81.0	104' 5"	31.8	53,000	24 041	76,200	34 564
	25	7.62	77.5	103' 3"	31.5	34,300	15 558	50,000	22 680
	30	9.14	74.5	102' 0"	31.1	26,000	11 794	38,300	17 373
	35	10.67	71.5	100' 6"	30.6	20,600	9 344	30,800	13 971
	40	12.19	68.5	98' 8"	30.1	16,900	7 666	25,500	11 567
	50	15.24	62.1	94' 1"	28.7	11,900	5 398	18,500	8 392
	60	18.29	55.4	88' 0"	26.8	8,800	3 992	14,200	6 441
	70	21.34	48.1	80' 2"	24.4	6,600	2 994	11,200	5 080
	80	24.38	39.9	69' 10"	21.3	5,000	2 268	9,000	4 082
	90	27.43	29.9	55' 5"	16.9	3,800	1 724	7,300	3 311
	100	30.48	14.7	31' 1"	9.5	2,800	1 270	5,900	2 676

① Measured vertically from center of boom head sheave to ground

Sample lifting chart. (*Courtesy of Linkbelt Construction Equipment.*)

Crawler crane. (*Courtesy of Linkbelt Construction Equipment.*)

stable, is distributed over the track contact area of ground surface. A crawler-mounted crane has equal lifting capacity within the radius of the unit.

Truck-mounted crane units

Truck-carrier-mounted crane units are more mobile and roadway maneuverable. The stability of this type of crane unit is established by the use of outrigger stabilizers. If the terrain is somewhat uneven or not level, the outrigger stabilizers can adjust the crane into a level position for lifting.

Truck-mounted crane units are classified as conventional lattice boom and hydraulic boom. The lattice boom type, which requires assembly of the boom section and other components, is best suited for mobile conditions within a project of heavy-type construction. The hydraulic boom truck cranes are best suited for a quick setup and mobility-type projects owing to their self-contained features.

Conventional lattice boom, truck-mounted cranes have normal ranges of lifting capacities from 35 to 250 ton, and have boom lengths from 40 to 390′.

Truck-mounted crane, conventional. (*Courtesy of Linkbelt Construction Equipment.*)

Hydraulic boom, truck-mounted cranes have normal ranges of lifting capacities from 14 to 100 ton and have boom lengths from 32 to 115′. A lattice boom truck-mounted crane, once moved into a site and assembled, normally is planned to be present for a longer-duration project than that of the hydraulic boom truck-mounted crane.

The load being hoisted is distributed over each of the outrigger pad contact areas of ground surface. A truck-mounted crane has different lifting capacity within the radius of the unit, hoisting over the rear of the unit, hoisting over the outriggers, or hoisting over the front of the unit. These limitations and capacities, if applicable, will be outlined by the lifting charts.

Rough terrain crane units

Rough terrain crane units are designed for mobility and maneuverable within the normal and rugged conditions of a construction project. The stability of this type of crane unit is established by the use of outrigger stabilizers. If the terrain is somewhat uneven or not level, the outrigger stabilizers can adjust the crane into a level position for lifting.

Truck-mounted crane, hydraulic. (*Courtesy of Linkbelt Construction Equipment.*)

Rough terrain crane units are classified as hydraulic boom type. The hydraulic boom rough terrain cranes are best suited for a quick setup and mobility within a construction project owing to their self-contained features.

Rough terrain hydraulic boom cranes have normal ranges of lifting capacities from 22 to 90 ton, and have boom lengths from 113 to 208′. A rough terrain crane, once moved into a site, normally is ready for work and with its maneuverability has quick and easy access around the project. These types of crane units are normally utilized for lighter-duty support functions but are capable of heavy-duty performance.

The load being hoisted is distributed over each of the outrigger pad contact areas of ground surface. A rough terrain crane has different lifting capacity within the radius of the unit, hoisting over the rear of the unit, hoisting over the outriggers, or hoisting over the front of the unit. These limitations and capacities, if applicable, will be outlined by the lifting charts.

Concrete placing and finishing

Concrete placing equipment consists of concrete pumps, belt placing conveyor units, and buckets. Concrete finishing equipment consists of bridge deck finish screeds.

Concrete pump units can be truck-mounted, trailer-mounted, or pedestal placing booms. They are classified and rated by their reach capability and

Rough terrain crane, hydraulic. (*Courtesy of Grove Worldwide.*)

concrete volume pumping capacity. Concrete pumps range in size and capacity from 42 cubic yards per hour for trailer-mounted pumps to 195 cubic yards per hour for truck-mounted pumps. Trailer-mounted pumps can pump a horizontal distance of 4000′ and a vertical distance of 1200′. Truck mounted pumps have a reach capacity from 23 to 52 meters.

Concrete pumping is a procedure which allows more flexibility and distance from the placement area. A pump truck unit is a quick-setup mobile unit which can maneuver within the work area and has the reach capability required with some conditions that prevent adjacent access to the placement area. Pumping also gives a more consistent flow of concrete to the placement area, provided that the delivery source of concrete is consistent to the pumping unit.

A feeder conveyor placement system is an interim carrier transfer device that relays the concrete product from one fixed point to another.

These are most commonly used for conditions that prevent the concrete delivery unit access to the placement area. Concrete conveying and placing systems are classified by the volume capacity with respect to cubic yards

Concrete pump truck.

Concrete pump pedestal. (*Courtesy of Schwing America.*)

Concrete conveyor. (*Courtesy of Morgen Manufacturing Company.*)

relayed or conveyed per hour. The conveyor units are categorized by length of unit and width of belt.

Once the concrete product has been delivered to the placement point by the conveyor, a placing spreader unit has the ability to spread it uniformly. This procedure is most commonly used for a deck placement in conjunction with a deck finish machine. The spreader unit saves on the labor power required to spread the concrete evenly in front of the finish machine and provides a greater volume in concrete pour rate capability.

The concrete bucket is the conventional method of concrete placement. The bucket procedure requires crane capability to hoist and position the bucket from the delivery point to the placement point.

Concrete buckets are rated and categorized by the size in volume of concrete. Normal concrete buckets range in size from $3/8$ to 4 cubic yards. The larger the bucket, the larger the crane capacity requirement.

The deck finish machinery is a self-propelled unit that finishes the concrete mix for a bridge deck, or horizontal surface, to the established profile. The finish machine has ranges in width from 36 to 130′.

The finish machine travels on steel screed rails that are preset to the desired elevation which the finish machine will perform to. Normally, the finish machine travels longitudinally with the deck surface, and the roller fin-

Concrete side discharge conveyor placer. (*Courtesy of Morgen Manufacturing Company.*)

Concrete bucket. (*Courtesy of Camlever.*)

Bridge deck finish machine. (*Courtesy of Bid-well: a Division of CMI Corp.*)

isher travels and finishes transversely across the deck surface. The finish machine will strike off, and with proper attachments consolidate, the concrete to the desired profile and uniform depth. The finish machine is capable

Bridge deck screed machine. (*Courtesy of Shugart Manufacturing Inc.*)

of profiles of flat, parabolic, rooftop, and inverted crowns and will finish on a skew, transitions, and superelevated slab surfaces.

A hydraulic screed is another type of slab finish machine that works excellently for flat slab decks or for crowned decks finishing longitudinally.

Pile driving

Pile-driving equipment has many classifications, types, and capacities. The pile hammer, which is the actual driving unit, is classified by types of air/steam, diesel, and vibratory. Accessory and support equipment to the hammer includes leads, bonnets, cushion material, driving heads, air compressors, spotters, and templates. For crane requirements, the weight of the pile, hammer, leads, and accessories or components will be totaled and accounted for the lift.

Pile hammers are rated by the driving energy produced in foot-pounds, blows per minute, and the stroke of the driving mechanism and are classified

Air/steam pile hammer. (*Courtesy of Vulcan Iron Works Inc.*)

by the propulsion type. Dependent on soil conditions, piling type, and specified bearing, the choice of hammer may vary somewhat for productivity and performance.

Air/steam powered pile hammers are classified as single-acting and differential-acting units. With single-acting hammers, the propulsion (air/steam) raises the ram, allowing it to free fall by gravity to impact. With differential acting hammers, the air/steam power assists the ram in a downward motion to gain striking energy.

Air/steam powered hammers have energy ratings from 15,000 to 150,000 foot-pounds, with weight of striking parts ranging from 5000 to 30,000 pounds. The blows per minute vary from 41 to 60, and the stroke varies from 36 to 60″.

Diesel pile hammer. (*Courtesy of International Construction Equipment, Inc.*)

Diesel-powered pile hammers are classified as double-acting variable units. With double-acting hammers, the propulsion force (diesel firing) raises and lowers the ram, to create impact. With double-acting variable hammers, the diesel power assists the ram in a downward motion to gain striking energy. The variable cycle allows for a controlled adjustment to the energy produced for added performance to the hammer, and flexibility to the pile-driving operation.

Diesel-powered hammers have variable energy ratings from 8100 to 4000 foot-pounds, up to a range of 100,000 to 40,000 foot-pounds. The weight of striking parts ranges from 1725 to 20,000 pounds. The blows per minute vary from 90 to 95 for the smaller hammers to a range of 53 to 70 for the larger hammers. The stroke varies from 4′8″ to 10′5″. The developed bearing obtained from these hammers ranges from 40.5 to 500 tons, based on the following formula:

Vibratory driver/extractors. (*Courtesy of International Construction Equipment, Inc.*)

$$\text{Pile bearing (tons)} = 2E/(S + 0.1)/2000$$

where E = hammer energy, ft-lb
S = pile set, inches per blow

Vibratory hammers are hydraulically powered units that are designed for driving and extracting. Eccentric weights work in vibrating motion to control the operation. These hammers are well suited for sheet-piling operations, extracting piling, timber and steel piling, and soils that offer no or little resistance. These hammers are suspended from the hook of the crane and require no leads, monkey stick, or guides.

Hydraulic vibratory hammers have centrifugal force energy ratings from 36 tons up to a range of 164 tons, and a line pull extraction force from 30 to 100 tons. The suspended weight of vibratory hammers ranges from 5350 to 26,900 pounds. The pile clamping forces range from 25 tons for the smaller hammers to 244 tons for the larger hammers.

Secondary and Support Equipment Units

The secondary and support equipment classification consists of the units of trucks, minor earthwork equipment, air compressors, screeds, and incidental components associated with them.

This equipment will be categorized within the type and functional capability of the unit. The units shall be grouped with the major equipment for which they are supportive or with which they are associated.

The truck category will include pickup-type vehicles, light-duty trucks, dump body, mechanic and maintenance vehicles, flat-deck units, tool vans and trailers, and concrete trucks.

The minor earthwork units will include small track dozers, front-end loaders, backhoe and excavating units, roller and compaction equipment, and hauling units.

The minor support equipment units will include air compressors, miscellaneous concrete equipment, welding machines, mortar mixers and small concrete screeds, personnel baskets and lifts, scaffolding, and incidental support equipment.

All required equipment shall be categorized by component type and use.

Minor equipment units

The minor equipment category will consist of hand tools, minor motorized equipment units such as generator units, pumps, compactors, and small concrete support equipment. The minor equipment grouping will include a complete supply of equipment required for support and operation of a crew. Normally, for a structure crew, this would include the following group:

144   Construction Systems, Methods, and Materials

Assorted hand tools	Slings and cables
Johnson bar	Sledgehammers
Hard hats	Concrete tremie and chutes
Rain gear	Water hose
Concrete and mud boots	Mortar bucket
Portable generator	Compactor and vibro plate
Water pump and hose	Carpenter tools
Circular saw	Bolt cutters
Chop saw	Transit and level instruments
Extension cords	Shovels and picks
Banding tool and material	Concrete vibrators
Tie wire	Cutting torches
Concrete floats and tools	Wrenches
Safety equipment	Drill motors and bits
Air tools	Hilti tools
Water cans	Chains, binders, shackles

This is general list for comparison purposes and evaluation. The actual requirement for the specific use intended shall be prepared in a similar manner.

Specialty equipment units

The specialty equipment category classification consists of the equipment units that are not commonly utilized. These specialty units include large cranes, special or uncommon pumping equipment, certain piling components, drilling and augering units, or any equipment units not owned by a specific contractor. In some conditions, contractors may desire to rent certain equipment rather than maintain the piece within their fleet. This may be simply due to the nonoccurrence of commonly performing the work, or performance of smaller component-type units. With this scenario, acquiring a larger equipment unit on an as-needed condition would constitute a specialty requirement.

Part

3

Fundamentals of Estimating

Ohio River Bridge, Wheeling, W.Va. Structure type, suspension bridge (built 1849). Owner, West Virginia Department of Highways. (*Courtesy of Federal Highway Administration.*)

Part 3 covers areas of corporate structure, establishing procedures, compiling data, formulating guidelines, and obtaining a system of costing and factoring. The five chapters of this part show how to utilize this knowledge.

Chapter 6 deals with scoping and locating potential projects, establishing contacts, acquiring specific project advertisements, and obtaining project plans and specifications.

On to Chapters 7 and 8 where the detailing of an actual project is defined. A contract proposal, contract specifications, and the actual project drawings and plans are discussed in detail. The general data along with special provisions, areas of potential concern, contract changes, and addendum are thoroughly detailed.

Chapter 9 concentrates on project structure. It outlines the owner's profile and payment schedule, the qualifications needed to bid and build the project, any risks or advantages involved, and the hands-on performance of a company's being able to construct the project. The labor force and equipment needed and a complete but precise overview of the project conditions must be developed into a routine function.

In Chapter 10, the importance of scheduling is discussed. An experienced estimator must understand not only the schedule and completion of a certain project but the full impact of the entire company's workload. This is a key factor in regard to financing, cash flow, implementing equipment and labor, and future bidding and bonding power.

Chapter 6

Scoping the Market

Locating potential work and future projects requires a vast network of information and contacts. In today's market, there exist basically two sources for construction projects, the private sector and the public, government sector.

The private sector consists of negotiated and cost-plus projects. This manner of work performance entails an owner and a contractor, mostly chosen, selected, or invited, based on their past performance and reputation, working together and negotiating to an agreed amount for a specified quantity of work. The negotiated work can even be between a group of contractors, all dealing with the same owner and general scope but differing slightly on the specific details. Each offers the best solutions and possibilities. With this type of work, the lowest-priced contractor may not always be successful in obtaining the awarded project.

Private sector work is usually more open-ended with regard to qualifications of contractors, specifications, and material requirements.

The public sector work is mainly comprised of government, federal, state, and local; municipality; and military. These types of projects are publicly advertised by various methods. They usually consist of itemized and unitized predetermined work items for a specific project. The project can be of different categories of turnkey or component segments. A turnkey project entails the complete construction of various sectors and multiple segments, from the start to the finish of a particular project. A component segment project consists of the specific categories within a project, being built separately under individual contracts.

The public sector projects normally require the contractor, in advance, to be qualified with the owner or agency in order to participate in the bidding process for a specific project. These types of projects are more stringent with regard to specifications, performance, and the selection criteria for obtaining and prequalifying of contractors. The prequalifying process is defined by the owner and establishes requirements, standards, procedures, and limitations, to which the contractor must adhere.

Projects, both private and public, are advertised by prescribed and predetermined methods. Contractors for private sector contracts can be procured by select invitation by the owner or owner's representative, or by public advertisement. Public sector projects are procured by public advertisements and broadcasts, mailing lists by the owner to prequalified, preapproved contractors, and advertisements from private companies that specialize in contract notification, such as the *Dodge Reports*, a subsidiary of The McGraw-Hill Companies.

A project is normally advertised by category, building-related for general contractors; mechanical-related for specialty trade contractors; street and roads for highway-type contractors; and heavy and civil for structure contractors. Bridge and structure projects will normally fall within the heavy and civil or highway category. This allows contractors to follow the type of work, by the type of advertisement, in which they specialize or have an interest.

Potential projects are advertised by specialty magazine publications, owner mailing lists, state and local highway departments, private engineering and consulting firms, federal agencies, military district offices such as the Army Corps of Engineers and the Bureau of Reclamation, private advertising agencies, contractor associations, private development companies, and local newspaper listings.

Specialty magazines and private advertising agencies such as the *Dodge Reports* are regionalized by specified demographic areas. By this feature, more related projects and detailed specifics are available for defined geographic regions and contractor locations.

One important fact for this chapter: A contractor can always perform the first job for an owner; the key to success and repeated projects is quality, on time, on budget, and contractual performance.

Listed here is a sampling of sources of project advertisements.

City and municipal highway departments, city offices

State Highway Department of Transportation, state capital

Federal Highway Administration, Washington, D.C.

The *Dodge Reports,* The McGraw-Hill Companies, New York, New York

The *Engineering News-Record* magazine, The McGraw-Hill Companies, New York, New York

Chapter 7

Contracts and Specifications

Every project is governed and controlled by specifications, both standard specifications and special provisions. These specify the control, scope, and specific requirements relative to each project.

The standard specifications are normally considered as general requirements, more detailed and defined as generic conditions specific to all projects which are under the jurisdiction of the owner or specified by a governing body such as the Federal Highway Administration or the State Highway Department.

The standard specifications contain the general scope which the contractor must follow. They contain the general material requirements, the specifications by which the contractor will construct the work, the definitions and terms of the contract requirements, a detail of the unit quantity identifications and determinations, and standard payment schedules for the defined items of work. The provisions define the controlling permanent items of work and detail the incidental and secondary components within each item.

The special provisions of the contract define specific components and requirements of a particular project, those which are not considered standard by the owner definitions. These are items such as the time duration and completion requirements for the project, specific material components, special construction requirements, labor and specific wage determinations, and unique unit quantity construction items. They outline coordination clauses, special schedule demands, and the phasing and sequencing of the project.

The special provisions and requirements are detailed for each specific project that they are relevant for. The contract requirements, standard and special provisions, and the detailed plans are included within the request for proposals and bidding documents.

Special requirements such as the qualifications which the contractor must follow, prequalification of contractors, financial requirements, and general bid information are outlined within the request for proposals and general guidelines of the contractor's responsibilities.

Each contract proposal identifies the items of work to be performed and how the project will be bid, either unit-price component as defined by the owner or lump-sum as defined by the owner (see page 151).

The contract proposal outlines not only how the work will be performed but when the progress payments will be generated. The proposal outlines the specific requirements for the submission of the bid, the date and time at which the proposal is due to the owner, and if a sealed bid is specified, the parameters and requirements of that procedure.

Prior to these requirements, the contractor must first be capable and qualified to perform the work, as detailed by the owner's requirements. The qualifications include the expertise and management staff to perform the work, equipment assets required by the scope of work, the financial capabilities to fund the project, and the bonding limits set forth.

Public sector projects normally require the contractor to be qualified with the owner or agency in advance of the notice to bid on a specific project. The prequalifying process is defined by the owner establishing requirements, standards, procedures, and limitations, which the contractor must adhere to.

Contractors must first prove their financial capability to the owner. This procedure consists of cash on hand, assets, financial commitments and backing, lines of credit, and bonding capacity. Most projects today require the contractor to be bonded by a reputable and qualified insurance company. This ensures that if the contractor defaults or fails to complete the project, the bonding company of record must finish the remaining work at no additional cost to the owner.

The bonding of a project requires multiple types of bonds. First, a bid bond is assurance to the owner that the bid the contractor submits to perform the work will be honored. In other words, if a contractor fails to execute a contract with the owner for which the contractor was tabulated as the lowest successful and responsible bidder, the bid bond will be forfeited and the project will be awarded to the second bidder. This bond is normally a percentage of the total bid.

Once the project has been awarded and contracts have been issued, the owner may, and most likely will, require a performance bond, labor bond, material bond, and maintenance bond.

A performance bond will assure to the owner that the project will be completed for the contract amount, even in default of the contractor. The labor and material bonds will assure the owner, and any workers and material suppliers, that all commitments with regard to wages and purchases will be honored if the contractor defaults.

The maintenance bond is a warranty to the owner, for a specified period of time, that all workmanship, materials, and components will be repaired or replaced upon failure within the scope of the specifications.

BID ITEMS

PROPOSAL
93 04018

THIS INFORMATION IS REQUIRED BY MICHIGAN STANDARD SPECIFICATIONS FOR HIGHWAY CONSTRUCTION IN ORDER TO COMPLETE A BID FOR A CONSTRUCTION PROJECT.

MDOT 1337 (9,79) ALL ENTRIES MADE ON THIS PAGE SHALL BE HANDWRITTEN IN INK.

ITEMS OF WORK	QUANTITY	UNIT PRICE	AMOUNT
65. 5020017 SPLICE, STEEL SHELL	336.00 EACH		
266. 5030010 CONCRETE GRADE 35S-SUBFOOTING	26.00 CYD		
267. 5030023 SUBSTRUCTURE CONCRETE	969.00 CYD		
268. 5030024 SUPERSTRUCTURE CONCRETE	945.00 CYD		
269. 5030030 STEEL REINFORCEMENT	71811.00 LBS		
270. 5030031 STEEL REINFORCEMENT, EPOXY COATED	179468.00 LBS		
271. 5030052 PENETRATING WATER REPELLENT TREATMENT	42.00 SYD		
272. 5030099 EXPANSION JOINT DEVICE	115.00 LFT		
273. 5030171 CONDUIT, 3"	952.00 LFT		
274. 5040064 ELASTOMERIC BEARING, 1 1/2"	88.00 SFT		
275. 5040065 ELASTOMERIC BEARING, 1 3/4"	30.00 SFT		
276. 5050037 PRESTRESSED CONCRETE I-BEAM, 70" FURNISHED	3278.00 LFT		
277. 5050038 PRESTRESSED CONCRETE I-BEAM, 70" ERECTED	3278.00 LFT		
278. 5060001 JOINT WATERPROOFING	702.00 SFT		

CHECK UNIT PRICE COLUMN FOR OMISSIONS-BEFORE ENTERING BID TOTAL.

Sample of unit price bid item schedule. (*Courtesy of Michigan Department of Transportation.*)

Once the contractor and owner accept all the criteria set forth within the proposal, the contract remains a binding agreement for the duration of the project.

During the course of a project, there will exist the possibility of required changes and alterations to the original contract proposal and drawings. These are negotiated between the two parties, or in conjunction with a subcontractor and prime contractor, and agreed to. Once this has been accomplished, a formal change to the contract is executed prior to the commencement of the work. Revised or new specifications and/or construction drawings are issued for the work.

For extra work or changed conditions that cannot be agreed to prior to the performance of work, the work is performed by the force account method, a cost of time, equipment, and material record of the actual and specific work requested, as it is performed. This method itemizes all labor, equipment, and materials expended for the change ordered work, allows a profit and overhead allocation, and computes the actual cost of the work. This is authorized by a formal change to the contract, after completion of said work.

Prior to the quantity takeoff and formal bidding of a project, the contract specifications, both standard and special, must be fully understood. There may be components within these that determine whether the project will be bid by the contractor, such as schedules conflicting with current work, specific items of the contract that are of concern, anticipated value of the project, or other controlling factors.

Chapter 8

Contract Construction Drawings and Plans

As with the contract specifications, every project is governed by construction drawings, both standard and specific. These specify and identify each component of the project, location, size and dimension, sequencing, and special details of construction.

The standard drawings are normally considered as general details that are common for certain components found within multiple projects. Standard drawings can be applied as specific complete components, or by formula application, workable for any situation. These drawings are designed and utilized for commonly occurring components that are usable for the owner's requirements. Standard drawings, along with standard specifications, are utilized as cost- and timesaving features and components.

An example of a standard drawing application would be a structure approach slab, where the bridge width is a standard defined size and the approach slab will always be specified as 25′ in length. This design can be drawn, implemented, and utilized for many application projects.

The standard drawings contain the general scope, dimensions, calculations, and requirements which the contractor must follow. They contain the general material requirements and the specifications by which the contractor will construct the work. The drawings show a typical cross-section view, plan view, and elevation view of the component that is specified and required.

The specific drawings of the contract define actual structure, and components and details of a particular project. The drawings detail specific cross-section views, plan and elevation views, and special dimensional details for which the component will be constructed.

The project drawings and requirements are detailed for each specific project that they are relevant for. The contract requirements, standard and special provisions, and the detailed plans are included within the request for proposals and bidding documents.

154　Fundamentals of Estimating

Quantity descriptions, itemized component lists, fabrication and construction details, and general construction notes are formulated for each project and included within the plans.

Each drawing identifies the items of work to be performed and how the component will be built.

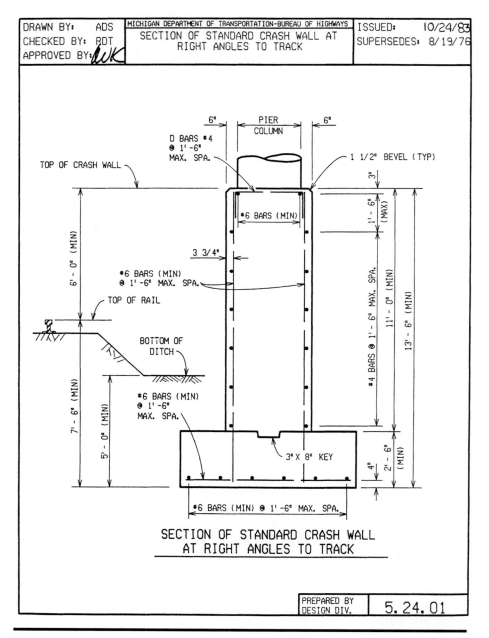

Sample of drawing. (*Courtesy of Michigan Department of Transportation.*)

In order for a detailed and accurate quantity takeoff to be performed, the contract construction drawings, both standard and special, must be fully understood. The drawings will contain all the required items and components, along with specific dimensions, elevations, and details, necessary for the construction and execution of the project.

Chapter 9

Project Structuring and Format

The estimating and takeoff of a specific project must be well formatted, detailed, and structured to assure that the proper analysis has been performed. The project plans and specifications must be thoroughly researched and understood. A complete understanding of every function and component must be reached.

The three most critical questions to be applied: does the project fit the company, is the project constructable, and will the anticipated profit margin be acceptable and in reason?

The project must fit the needs of the company in both scope and schedule. A project may be suited perfectly as to the type of work, conditions and nature of the work, and possible area or regional concerns but may not be suited at the specific time for current workload, personnel and equipment requirements, and schedule.

On the other side, the intended project may not be well suited with regard to type of work, equipment and personnel skill demands, region, financial burdens, or overlapping schedule impacts. The project may apply a restrictive capacity to the company with regard to unrealized or unwanted bonding restrictions, prequalification limitations and impacts, bank line of credit concerns, or demands to personnel and equipment. The analysis of a project must take these factors into account, not only the field construction reasoning decisions.

The project must lend itself to the company's expertise, equipment capabilities and resources; financial concerns, both banking and bonding; and within the realm of the specific company, be constructable, both within the allotted time schedule based on current and future commitments, and with the owner's anticipated schedule being realistic.

A well-detailed thorough bid analysis and quantity takeoff will outline and identify each component of the project, list the requirements of materials, equipment demands, and time schedules; research the specification requirements; identify the labor requirements and agreements; and identify the specialty requirements for the listed work. The scope of performance, subcontract requirements, and other special details must be highlighted within the bid analysis. A complete material list, permanent, indirect, and incidental, along with a component listing for each item of work will be performed. The labor and equipment crew requirements must also be identified for each component unit.

The project, in the minds of the estimators and managers, must be constructable and fit their needs. If the principals of a company do not feel comfortable and confident of a particular project, don't expend the time and money pursuing it any further. The initial scope, plan and specification investigation, or actual site visit normally will provide the alert signals.

The anticipated profit margin and markup must be the prime concern to bid any project, not for the fact of applying your name to it merely to say it was built. We all are in business to make money. If a project is bid well and within reason, anticipated for quality performance, maintaining the anticipated schedule time and budget, and providing the owner with the end result specified by contract, the project normally will go well.

Economic conditions, workload of competition, and the amount of work anticipated within a certain region may have an effect on the anticipated or deserved margin. The decision to bid or not to bid should be a prime factor with this in mind. Regardless of this, the responsibility of the estimator will be the performance of the most realistic, accurate, dependable quantity takeoff and material summary, and labor and equipment crew determination for the project analysis.

Chapter 10

Project Schedule

The schedule or anticipated schedule will have a significant impact on a project. It will be a primary factor from the beginning bid analysis and formatting through to the actual planning and construction phases and operations. A well-planned and realistic schedule will aid in the performance of the bid analysis and will prove to be a useful and profitable tool for the management and construction of the project.

The estimator must thoroughly understand and acknowledge the schedule of events, operations, and components in order to compile an accurate and realistic bid. The procedure of the quantity takeoff, material summary, and crew determination is only a portion of a detailed and accurate bid preparation.

The project must be addressed, at the time of bid, as to schedule concerns for construction: during the component takeoff, the time required for the specific and individual component construction time, along with mobilization and setup, form erection and dismantling, concrete placement, curing time, and all incidental work associated with each component. The individual schedules will then be overlaid within one another, along with multiple crew impacts, to determine the overall schedule and construction time required.

Along with the specific component construction duration, the initial project setup and mobilization, interim phasing time requirements, special material acquisition impacts, weather delays, holiday shutdown periods, and project cleanup and finishing requirements will be realistically factored in with the project schedule.

By thoroughly analyzing the schedule component duration of a project at the onset, during the takeoff procedure, the bidding procedure will flow more realistically and evenly. A well-thought-out and planned schedule will determine the number of crews required, equipment demands, the daily shift durations needed, the weekly workday requirements, and the overhead and management staff required. An accurate schedule will also design the needs of future project requirements and commitments.

Part

4

Takeoff and Cost Analysis Techniques

Smith Avenue High Bridge, St. Paul, Minn. Structure type, welded steel tied deck arch. Owner, Minnesota Department of Transportation. (*Courtesy of Federal Highway Administration.*)

This part involves the actual quantity takeoff methods of this system along with developed work sheets. This is a very detailed, systematic section. It starts at the foundation level and moves upward through the finished structure. It gives concise methods of measurement and quantity summary of specific parts and components of a structure along with detailed illustrations. This shows how to apply certain job conditions to a given area and how to work with a formulated system to end up with accurate and detailed results. Within this section, an enhancing feature for each bridge component operation is discussed, titled the Technical Section.

This section gives a technical explanation and view of each topic to help readers better understand the operation and component they are to learn about. It gives a concise definition along with a detailed specification of the type of work involved for the estimator, with both the direct permanent work involved and the indirect incidental work involvement. It also explains the various designs and functions of the intended use.

Each technical section discusses the following:

1. *The item involved*
2. *A constructive definition of the corresponding item*
3. *The relative specifications in general form*
4. *The type of material from which the item is derived*
5. *The intended use of the item*
6. *The design features of each item*
7. *The basic equipment needed to perform the work*
8. *The standard construction requirements that are involved*

Following each technical section, the chapters detail the precise takeoff and quantity measurement procedures involved with each item.

Chapter 11

Foundation Preparation

TECHNICAL SECTION

Introduction

Foundation preparation is a key ingredient to bridge construction. A bridge structure must bear on a suitable foundation. In cases where an excavated foundation cannot obtain suitable bearing, foundation piling will be required to support the bridge structure within a prepared foundation area.

Depending on specific soil conditions and requirements, the foundation work can differ in regard to excavation and backfilling requirements.

The preparation of suitable foundation limits within a project could include a work area of water, rock, unsuitable earth, suitable soils, or combinations of the above. A wide variety of options exist to perform this work.

It is essential to obtain satisfactory foundations within these specifications for all components of a bridge structure.

Definition

This work will consist of foundation excavation, preparation, and backfill for bridge structures. It will be performed within the limits specified by the project plans and requirements. Within these guidelines reasonably close conformity of the line, grade, and section will be followed and established.

Excavation

This work includes all necessary excavation and the proper use or disposal of bracing, sheeting, shoring, cribbing, cofferdams, pumping, and dewatering; suitable base preparation, backfilling with suitable material, and the subsequent removal of any temporary support systems.

All required line and grade elevations must be properly established and approved. Overexcavation of width and length may be required to permit and facilitate the construction process.

Excavation of structure foundations is generally classified by the type of material encountered, which is generally noted on project soil boring reports.

Unclassified excavation within the limits specified is a common type of excavation, normally encountering a mixture of sound and some unsuitable soils, with the possibility of some rock. This is normally considered as a variety of materials and will not specify one type of earth in general. This would be material either above or below the established datum line as defined within the project.

Dry excavation is defined as non- (or subsurface) water-containing soils. This is normally represented as dry soil that can be excavated by conventional methods. This would be excavation performed above the *datum line* as outlined on the plans but not classified as rock excavation.

Wet excavation, excavation performed below the specified datum line, is classified as outlined. This also would not be determined or defined as rock excavation within the limits.

Rock excavation is classified as material, either above or below the established and defined datum line, that cannot be economically excavated without the use of explosives, also any boulder, slab, or rock fragment normally greater than 0.5 cubic yard in size. This may not include cemented gravel, shut, shale, or slate which is not necessarily of a solid formation but may require explosives or ripping. This will be determined and outlined by individual project specifications.

Rock drilling within a specified or prepared foundation may be used to determine the actual underlining strata and formation soundness, or the pilot holes for permanent foundation piling.

Unsuitable excavation within the foundation limits is defined as earthen soils not suited or stable for a load-bearing stratum or backfill. This material requires proper disposal outside the structure limits.

After completion of all excavations, the foundation must be inspected and approved to assure its stability and load-bearing capability. There will be no areas of undisturbed or noncompacted soils within the limits of the foundation.

In areas of rock encounters or areas of unsuitable, unyielding materials, it may be necessary for overexcavation or undercutting to be performed. This would be material below the specified bottom of footing elevation. Normally this extra depth would be outlined and directed. The area would then be backfilled and compacted with a suitable specified material, normally of a granular source, to restore the foundation to its proper elevation and stability.

All final foundation surfaces will provide a firm foundation of uniform density throughout the entire planned area as defined by project design.

The excavated materials must be properly handled and disposed. Excavated material used for suitable backfill must be utilized and handled according to specifications. The contractor shall handle and deposit such materials in a manner as to furnish proper protection to any material to be used within the project limits.

Protection of the excavated area is required in areas of noncontrol, unsuitable or unstable soils, and water conditions. Precaution must be taken to maintain the stability and defined limits of the area, keeping it intact.

Sheet-piling wall. (*Courtesy of Bethlehem Steel Corporation.*)

In these conditions, the use of cofferdams, cribs, sheeting, or other means of shoring will be required to properly construct the foundation.

Cofferdams, or cribs, used in the preparation and protection of the foundation normally will be carried well below the bottom of the footing elevation to establish a safe *toe* and should be adequately braced and supported in all directions. They will be as watertight as practicable and will be constructed in such a manner to permit proper dewatering and pumping and be sufficient to maintain a watertight environment. The interior of the cell will provide sufficient clearance, when applicable, for the intended forming system and construction of the foundation. Cofferdams should be constructed to protect the foundation from future and/or unexpected or anticipated conditions, as best as can be determined. A *seal* of specified concrete can be used with a cofferdam in areas where dewatering the cell is impracticable. Special excavation methods, such as *airlifting* the unsuitable material from the foundation bottom, must be implemented.

Airlift excavation is performed in a water-filled environment. Normal *clamshell* excavation is performed within the cofferdam cell within its capabilities. Once the clamming proves to be insufficient, the airlift operation is performed to clean the remaining unwanted material from the foundation. This consists of a vacuum system which extracts the material from the underwater environment, thus preparing the foundation for the seal.

Once the foundation has been completed, the cofferdam will be removed as to not disturb the completed portion of the structure, unless otherwise noted, for it to remain permanently.

Takeoff and Cost Analysis Techniques

In the case of landward excavation instability, sheeting and bracing is required to contain the earth and the stability of the foundation limits. Sheeting will be constructed with proper toe and suitable bracing for substantial protection to the foundation area, or cantilevered within the excavation.

Preparation

Once the foundation area is properly excavated and stabilized, it must be prepared in accordance with the following specifications.

When the foundation has been completed to its initial planned elevation, an inspection is performed to its stability and bearing status. Unless permanent foundation piles are indicated, each foundation must be tested to verify the apparent conditions.

Should unsuitable conditions exist, the excavation will be carried lower until a satisfactory foundation elevation is acquired or other methods are designed to assure a suitable foundation. When rock is encountered in the foundation area, it will be cleared off to have a determination made of its line of demarcation, classification, and stability of the rock. The excavation will continue until an approved foundation is reached.

Rock formations used as a foundation will be stripped, cleaned, and detailed of all overlying materials. All loose, disintegrated, or slabby portions of the rock will be removed. If the rock is shattered below the foundation elevation, the shattered material will be removed and the exposed area rebuilt with an approved material.

Unsatisfactory material within a foundation area will be removed and replaced with satisfactory bearing material and compacted according to specifications.

The foundation area must be kept free from water and will be in a dry condition prior to the concrete construction. If water cannot be controlled, measures will be taken to install a concrete foundation seal as specified previously.

When foundation piling is indicated on the plans to be driven, the necessary foundation excavation and preparation must be completed first. Any upheaval or bulge of spoil excavation caused by the pile-driving operation must be removed and the foundation area restored to its condition and elevation prior to the driving operation.

Prior to the placement of any permanent components, including seal concrete, the foundation area must be free and cleaned of all objectionable material and laitance and be prepared for further construction activity.

Backfill

Backfilling will take place once the component or structure has been constructed and has achieved proper designed strength. Backfill material will consist of an approved source meeting project specifications.

All backfill that becomes a permanent part of the roadway structure will be placed in designated layers and compacted according to density specifications. The backfill material will be constructed to the designated line and grade as shown. This will include all areas that have been excavated of which the volume is not occupied by the permanent structure. The mechanical compaction equipment used must be capable of achieving the designed density. When backfilling is to take place within a structure or on both sides, the material must be placed simultaneously and in equal horizontal lifts.

Any designed drainage filters, piping, or blankets must be installed prior to the permanent backfill.

Measurement and Payment

Measurement and payment for structure excavation is normally by the cubic yard by volume of material excavated for the type of material or method excavated. This will be calculated using the *neat line method* to the line and grade shown on the plans unless an approved change is authorized.

Any material excavated outside the neat line area or any rehandling of such will normally be at the contractor's expense. Incidental items such as dewatering, blasting, bracing, and shoring should be thoroughly researched by the estimator to verify the means of payment, if any. Sheeting for retention walls or cofferdams should be calculated along with necessary bracing and verified for method of separate payment.

Foundation preparation will include all work necessary to prepare the foundation for the permanent structure component. This work should be calculated and determined for the method of payment—separate or incidental to the excavation.

Foundation backfill material is normally calculated and paid for by the volume of in-place cubic yards placed to the neat line designed dimensions as shown and the type specified on the plans, or any approved changes thereto.

Rock drilling, if measured and paid separately, will be quantified by the lineal foot method and the diameter of the hole required.

Cofferdams will be measured and paid for as the specific work details. The work outlined should include all sheeting, bracing, excavation, bottom preparation, and the removal of the cell.

Sheeting will normally be measured by the square foot at the wall surface installed as per neat line dimensions. This may or may not include the required toe. All bracing and shoring required for the sheeting will be included in this item and the removal of same.

Shoring, cribbing, and related work will be measured and paid for at the appropriate unit specified by the project. This should include all work to furnish, construct, maintain, and remove the system.

Any specific work not outlined should be accounted for as incidental to an appropriate construction item for relative work demands.

Individual project plans and specifications will clearly outline the required process to perform the excavation, preparation, and backfill. The estimator should thoroughly research this prior to bidding a project.

Specific projects may contain *differing site conditions* or materials and areas which are different from that proposed and detailed by the owner. The conditions must be thoroughly researched and studied by the estimator as to their fullest detail and scope.

TAKEOFF QUANTIFICATION

Introduction

This section defines the parameters and techniques needed to perform an accurate and detailed quantity takeoff and summary. This helps in building a formulated, procedural system of consistent methods for estimating. The section gives the proper guidelines required for an estimator to learn the required work activities associated with foundation preparation and to locate and call out from the plans and specifications incidental items found within this work.

Definition

The foundation preparation area for structures is defined as the area required for the bearing of or stability of bearing for the structure to be seated. This contains the area of excavation, backfill, grading, embanking, dewatering, and bottom preparation.

Foundation preparation is the detailing and preparing of the bottom surface of the excavated or fill area on which the structure will be founded. It is the manicuring to prepare a suitable and sound surface.

Foundation preparation may consist of various components required to construct and complete the defined area to project specifications. It is the responsibility of the estimator to define the work activities and compile them into a detailed and accurate takeoff.

Takeoff

The conditions of the foundation preparation and type of takeoff required are outlined by the project specifications or normally, given aspects of the industry, are required in some form merely owing to construction procedures. In other words, regardless of, and over and above the project conditions, some form of preparation will be required for the construction of the structure.

The variables of the preparation required are dependent on regional soil conditions, the presence of foundation piling, rock formations or fragments thereof, special or required stabilization, any drainability requirements, and earthwork manipulation.

The foundation preparation takeoff procedure for the required quantity and type of work activity is derived by determining the controlling function of work to be performed. The main activity will be driver, and all detailing and formulating of the takeoff should be controlled by this. The incidental or subsequent work items that are necessary to complete the entire work activity should then be itemized as secondary items to this, but still maintained as relative work.

An example is a scenario of rough excavation being performed in a foundation area becoming the control to the work activity. The excavation item is itemized by a volume measurement of cubic yards of material excavated. The secondary work activity associated with this item is the fine grading and detailing of the bottom surface area after the main excavation is completed. The operation is quantified by a measurement of square yards of surface area prepared.

This comparison shows differing units of measure being summarized. Each work activity requires a ratio of units to be performed against the time duration and crew size required to complete the activity. This is derived by applying a crew-hour production rate to the appropriate unit of work. In this case, the excavation requirement is formulated by determining the required craft-hours needed to excavate the specified cubic yards, and the fine grading is the required craft-hours to prepare the specified square yards of surface area.

The formulating of the costing procedure for the controlling activity, that being the excavation, is done by accumulating the total craft-hours required to complete both operations, the excavation and fine grading. By doing this you are working with like units of measure of which are craft-hours.

When working with different activities of work that is required to complete a defined item as determined by the project specifications, the controlling work item is always determined first; the subsequent items, along with their appropriate units of measure, are detailed next. A thorough understanding of the scope of work, the project requirements, and general procedures of construction must be a requirement for formulating a responsible and detailed estimate.

Earthwork and Excavation Takeoff

Procedure

Earthwork is calculated in a volume (three-dimensional) of cubic yard measurement. The area that is considered has a defined dimension of width, length, and depth, in feet and inches.

Cubic yard calculation (in feet)

$$W = \text{width}$$

$$L = \text{length}$$

$$D = \text{depth}$$

$$(W \times L \times D) = \text{volume of cubic feet}$$

$$\text{Cubic feet}/27 = \text{cubic yards}$$

The method of the excavation is controlled by site conditions, logistics of the approach to perform the work, and the time duration allowed by the schedule. The site excavation must be performed in an organized manner to control the ease of the operation.

WORK SHEET

Project: BRIDGE Estimator: JDN Item No. 4
Type of Work: FOUNDATION EXCAVATION Date: 10-13-94 Sheet No. 2a

FOOTER EXCAVATION: PIER 1

25.0' W × 30.0' L × 10.0' D = NEAT LINE DIMENSIONS SHOWN

MATERIAL — CLAY

SCALE 1/8" = 1.0'

[Diagram of rectangular excavation: 25.0' × 30.0' × 10.0' deep]

25.0' × 30.0' × 10.0 = 7500.0 CUBIC FEET

7500.0 CF ÷ 27 = 277.78 CUBIC YARDS (NEAT LINE)

NOTE: 1. DOES NOT INCLUDE ANY REQUIRED "OVER EXCAVATION."
2. DOES NOT SHOW EXCAVATED OR "LAYED BACK" SLOPES FOR TRENCH SAFETY.

Volume calculation.

Questions of estimate

When estimating and detailing a takeoff of required excavation, the following questions must be raised and, if positive, formulated into a unit of productivity of time.

1. What type of material is required to be excavated?
2. Will the material be suited for reuse within the project?
3. Does the material need to be wasted or removed from the project limits?
4. What processing needs to be done to the excess material?
5. If blasting is required, is it allowable, and what are the procedures?
6. Is any shoring, sheeting, or bracing required within the limits of excavation to stabilize the work area?

7. Are there utilities or obstructions, either underground or overhead, that may restrict or cause additional and required work?
8. Will any dewatering or stabilizing of the work area be required?
9. Is there a shrink or swell factor of the material that may be of issue?
10. How is the item of work specified for measurement and payment, *direct* or *incidental?*

This list aids the formation of the labor and equipment crew required to perform the item of work detailed and specified.

Once the area has been defined and calculated, it will be decided which type and condition of material is present. The normal types of excavation that exist are unclassified, dry, wet, rock, and unsuitable. These differing conditions require different methods and equipment to perform the work required. Each of these must be approached and reasoned with a view to perform the work required as efficiently and economically as possible.

The above-mentioned types of excavation require a different labor analysis as well as equipment needs to perform the work as outlined. Excavation is normally computed and measured by the cubic yard. The takeoff and price computation must be detailed and presented in this manner.

The estimator must determine the primary or controlling item component of the prevailing work item initially, then the subsequent or secondary activities that correspond with this, thus preparing an outline of takeoff and pricing structure required for this work item.

A detailed summation of required excavation includes these minimum categories and components. First and foremost is the type of excavation present at the site (i.e., earth, rock, wet, etc.). This determines the primary controlling component. Next determined is whether the soils are suitable or unsuitable and whether the project can utilize the excess excavation or if it is to be disposed of off site. Access to the specific work site must next be addressed to assess the type and method of approach to the takeoff.

Once these items have been identified, a crew can be formulated for the performance of the work required. This includes but is not limited to the actual excavation and disposing of the material. If the material can be utilized within the specific work area, it must be stockpiled out of the area of construction for future use. If the material is not suited for use at the site, it must somehow be removed. This is normally accomplished by trucking. When this method is encountered, a secondary crew may need to be realized at the dump end to process the material in a manner outlined by the project specifications.

Another characteristic of soils to determine for excavation is the shrink and swell factor of the material being removed. This is a determining factor that will control the material when it is disposed of or placed elsewhere. Normally rock swells and earth shrinks. In other words, when a solid formation of rock is fragmented or fractured, the excavated material will not embank back to its original volumetric state. Earth, on the other hand, when compacted under normal specifications, will compress to a volume smaller than its original state.

For the excavation and removal of rock, blasting is usually the preferred method, but one must research the specifications to determine if this is permissible owing to local ordinances, the proximity of dwellings, etc. The method determined by this analysis will greatly impact the takeoff, pricing structure, and modeling of the bid.

When trucking of the excavated material is required, three factors must be determined: the type of hauling unit needed, the amount or weight that is permissible to haul, and the distance the material is to be hauled. The type of hauling unit may be determined on whether the material can be hauled within the limits of the project or if it has to be hauled on local and public roads and thoroughfares.

The weight or amount of material hauled in the unit is then determined by the capacity of the individual haul unit and the restriction or gross weight allowed by the local roadway that the unit transports on.

The distance of the haul then determines the number of haul units required to maintain a continuous excavating operation at the site, keeping in mind that the excavating unit is the primary activity. The round-trip haul time includes the loading time, the normal transport and return elapsed time, the time required to dump, and some bank time for contingency. The time usage factor for productivity is a 50- to 55-minute hour.

Application Example: Hand Unit Cycle Time

Operation time duration analysis

Haul unit	Cycle time, minutes
Load time	5
Travel time	12 (average)
Dump time	7
Return time	13 (average)
Total cycle time	37
Productivity factor = 55-minute hour (55/60)	
Trips per hour per haul unit = 55/37	= 1.49 trips per hour
Haul unit capacity	
Haul unit *gross* vehicle weight	80,000 lb
Haul unit *tare* vehicle weight	26,000 lb
Haul unit *net* vehicle weight	54,000 lb
Unit weight of excavated material	125 lb per cf
Volume calculation	
lb per cf	125.0
cf per cy	27.0
Total average material weight per cy	
125 * 27 = 3375 lb/cy	
Haul unit *payload* = 4,000/3375	= 16 cy
Total net weight = 3375/2000 lb (ton)	= 1.69 ton per cy
1.69 tons per cubic yard * 16	= 27.04 tons
27.04 tons * 2000 lb	= 54,080 lb (check)

Given the above example, the haul unit is capable of making 1.49 trips per hour and hauling a payload of 16 cubic yards of material; thus for each hour worked the haul unit can move 23.84 cubic yards (16∗1.49).

Under normal excavation methods, the cycle time and capacity of the primary excavating unit is the control for measurement of production. The excavating unit has two functions that must be realized by the estimator: the cycle time or the elapsed time the unit takes to excavate, travel or swing to the discharge point, dispose of the material within the bucket, and return to the area for the next pass (dig; travel, swing; dump; return). The capacity is the amount of material the excavating unit is designed to handle or dig under normal operating conditions, i.e., bucket size in cubic yards. This can be valued as "level" or "heaped." Again this must be realized under normal operating conditions and consistent use for the entire shift.

Based on the excavating unit's digging capacity, in cubic yards, this determines the production per hour of material excavated. Also, if trucking is needed, this will be the controlling factor for the number of haul units required to maintain the excavator's production.

In determining this production, one must always consider an average production per shift.

Two main items to always consider with cycle time production and the type of unit required are the type of material being excavated and the size of the excavating unit relative to the nature of the work. In other words, the excavating unit should be sized in both digging capacity, with regard to reach and bucket size, and in operating weight relative to the size of the piece of equipment. Do not oversize or undersize the unit. This will only either compound operating expense with a larger than needed unit or restrict productivity with a smaller unit.

Also, the type of material being excavated will greatly impact the productivity, i.e., granular soils, rock, clay, wet material or muck, etc.

Application Example: Excavator Cycle Time

Operation time duration analysis

Activity	Time, seconds
Dig or load	10
Swing or travel	8
Discharge	7
Return	6
Excavator cycle time	31 seconds
Productivity factor	= 55-second minute (55/60)
	= 55-minute hour (55/60)
Cycles per hour	= (55/31)
	= cycles/minute = 1.77
	= (55/1.77)
	= cycles/hour = 31.07
Bucket capacity (normal)	= 1.5 cubic yards
Shift duration	= 8 hours
Production per hour	
31.07∗1.5	= 46.6 cy per hour

Given that the excavation cycle time has been determined along with the haul cycle time per haul unit, a determination of the number of haul units can now be established.

Application Example: Production Factoring

Production unit	Cubic yards
Haul units required?	
Excavated material per hour	= 46.6
cy hauled per hour per unit	= 23.84
Haul units required (46.6/23.84)	= 1.95
Haul units rounded	= 2.0 units

The number of haul units required must be rounded up to the nearest whole number, and with some conditions, additional units may be added for a contingency factor for traffic, dump restrictions, site conditions, etc. The main objective is not to have the excavating unit waiting for a haul unit, nor is it to have too many haul units waiting for the excavator.

Once the equipment requirements are determined, a labor force can be formulated for the work activity required.

This example provides for a consistent production of excavation which may normally not occur during small construction activities. The estimator must apply the determination of high, normal, and low production of the required job activity, and formulate a factor to cover the time incurred for that specific type of work activity encountered.

With differing conditions that may restrict or alter the production sequence, the unit of productivity can be greatly affected from the formulated method listed above. This would impact the units produced per hour, normally to a reduced scale, but may still maintain the same equipment usage and labor force detailed. A condition of this nature would in turn affect the end result unit cost to perform the work activity.

Given the above example, one can realize the effect of differing productivity factors affecting the end result cost.

Earthwork, Backfill

Procedure

Backfill material is normally calculated by a volume method of cubic yards, in place. The term *in place* refers to the neat line area of backfill limits detailed by the project plans and includes the factor of compaction, or the compressed loss of material required to fill the void, referred to as *shrink*. There is also the possibility that the backfill material may not return to its virgin state of volume. It then would have an expansive factor, referred to as *swell*.

Backfill material is specified by the type and characteristic of properties which it contains or, in some cases, of what it needs to contain to give it specific properties. Backfill material is intended to perform certain design functions along with merely filling a void. It is required, by design, to be of such *density* as to support and maintain support of a structure or other load-bearing surfaces.

A *proctor* must first be analyzed of the intended specific material to obtain the physical characteristics and *unit weight* of the material. This determines how the material will react to compaction, loads implied, moisture absorption and loss, and the percent of compaction required to obtain a stable area. The proctor establishes the measurement of optimum unit weight, with regard to the minimum and maximum moisture content, which controls the compaction density results, unit weight plus moisture.

Soils and aggregate materials react in many different ways in regard to manipulative manners of placement. One must fully realize the intent, design, and material to be used in order to formulate a competitive but constructable estimate for the work to be performed. Some materials need to be dried in order to be placed in a compacted state, and some need a certain amount of moisture added. The thickness of the *lift* or layer also varies with specific materials owing to their density requirements and compaction limitations.

Example: If an area is calculated to require, by volume, 100 cubic yards of a specified material, and the shrink is determined to be 8%, the total required material by volume needed to fill the void is 108 cy.

Questions of estimate

When estimating and detailing the furnishing and placement of backfill material, these questions must be raised and, if positive, formulated into a unit of productivity of time.

1. What type of backfill material is required?
2. What is the local source, availability, and haul distance to the site?
3. What are the physical characteristics of the specified material and proctor?
4. In the procedure of placement, what manipulation, processing, or special handling must be done to the material in order for it to be placed and compacted to project specifications?
5. In addition, what is the thickness of the allowable lift?
6. Does the material need to be processed after each lift, or can it be processed once at the source?
7. With regard to the processing, how much time will elapse during the processing and when can the next lift be placed?
8. What is the shrink or swell factor?
9. Do any foreign materials, such as lime, cement, blended soils, or aggregates, need to be added in the processing by specifications to make the backfill material constructable?

10. How is the item of work specified for measurement and payment, direct or incidental?

This list aids with the formation of the labor and equipment crew required to perform the work detailed and outlined by the project specifications.

To determine the quantity of backfill material required, the area of fill must be calculated to a volume of cubic yard measurement. This must be calculated to the neat line dimensions shown by the plans, and an allowance must be made for areas of overexcavation and overfill that may exist, in other words, a yield factor to compensate for some overrun, or possibly an underrun.

The next procedure is to identify the type of backfill material specified by the project documents. This determines and identifies the specific conditions that will occur for procuring the sources and placement techniques required along with developing and obtaining the proctor. The project specifications outline the specific guidelines and compaction requirements needed.

For the placement of backfill material, a cycle time of production must be established. This will be the control for the delivery of the material as well as the placing and manipulation of it. With some conditions it may be prudent to take delivery of the material and stockpile within the site area for future use or ease of operations. This is referred to as *double handling*. Of course, this may add to the overall cost but aid in the production due to availability.

To establish a cycle of production, these key factors must be addressed. The first is the material characteristics, which determine the handling required. The next is the thickness of the lift. This establishes and controls the volume of material relative to the surface area required. The third factor is the compaction requirements. Does the site embankment area allow for a continuous operation of placement, spreading, and compaction? The fourth factor is how the material will be placed. The normal choices may include a bucket-type placement, either a backhoe or a loader; a crane and bucket; or a more conventional method of dump and spread, being a haul unit and a spreading device such as a dozer. The last key factor is does the backfill material need processing during or upon the completion of each lift.

In determining these factors, one can realize the production availability for the work items required, and choose whether the work is classified as high, normal, or low production. This formulation must be relative to the unit of measurement detailed and, as always, must be given a unit of time to formulate the task.

The controlling productivity item should be the placement of the backfill material, making all subsequent and incidental items subordinate to it. In conjunction with this, the secondary item of control is the demand of material or delivery of it to the placing unit. This is a circular function of which a balance must be made for the amount of delivered material per placing unit. That should equal the capacity or optimum requirements of that unit. All other operations are, by direct result, controlled by these components. The next item of control is the *lift* requirements, or layers of fill. This is normally a specification based on the design of the project with respect to the characteristics of the fill material. This determines the surface area to be filled at one time and the volume of material required, which may affect the equipment capacities.

Foundation Preparation

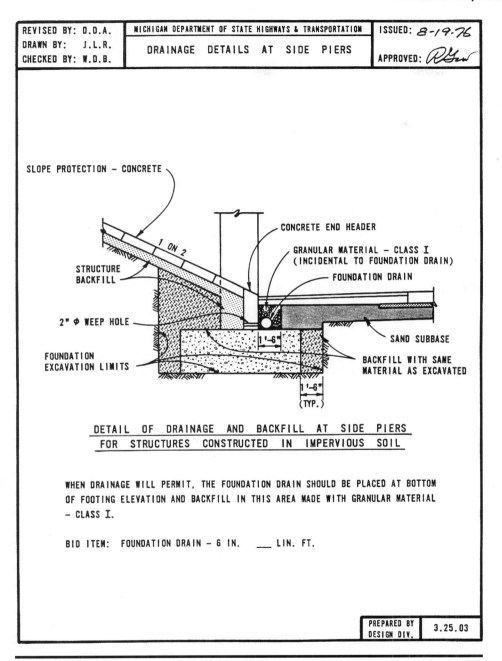

Typical excavation/backfill detail.

178 Takeoff and Cost Analysis Techniques

Application Example: Placement, Spreading, and Compaction

Operation time duration analysis

Quantity of material delivered	50 cy/hour
Placement capacity (area)	40'w * 60'l * 10'h
Lift thickness	8"
Number of lifts required (10'×12")/8"	= 15
Surface area/lift (sf) 40 * 60	= 240 sf
Cubic yards/lift ((8/12) * (40 * 60))/27	= 59.26
Total volume required (40 * 60 * 10)/27	= 888.89 cy
Placement capacity (equipment unit)	45 cy/hour
Type of material	Granular
Proctor @ 9% optimum	111.18 lb/cf
Unit weight (dry)	102 lb/cf
Compaction requirement	96%
Optimum moisture content	4%
Actual material moisture	7%
Compaction capacity (equipment unit)	60 cy/hour
Manipulation: Required *disking*	= None

The above example shows the differing conditions that must be first determined with the takeoff in order to define the parameters required for the specific operation. Once this has been outlined, the production specific to this actual condition can be formulated.

Example: Production factoring, time duration

Operation: Placement, spreading, and compaction

Crew members	= 3.0
Production unit	= cy
* Material delivery	= 50 cy/hour
Lifts required	= 15
Volume per lift	= 59.26 cy neat line
* Placement capacity	= 45 cy/hour
* Compaction capacity	= 60 cy/hour
**Control production	= 45 cy/hour
Time duration per lift: 59.26/45	= 1.32 hours/lift
Production duration: 15 * 1.32	= 19.8 hours
Shift duration	= 8 hours
Production per shift: 45 cy/hour * 8 hours	= 360 cy
Total volume required: 59.26 * 15	= 888.9 cy
Required shifts: 888.89/360.0	= 2.47 shifts
Total craft-hours: 2.47 * 8 * 3	= 59.28 labor-hours
Operation production factor: 59.28/888.89	= 0.067 craft-hour/cy

---

*Denotes control items for operation.
**Denotes determining control item.

This example is shown by neat line volume without any yield loss, shrink, or swell adjustment for compaction, overrun, or waste accounted for. The formulated listing is adjusted for these conditions.

One must note that the controlling factor of this operation, given the above criteria, is the equipment placement capacity. The chosen or logistic equipment can handle and manipulate only 45 cubic yards per hour; therefore, the delivery required must be equal. This must be realized for two important reasons: (1) If the delivery overrides the placement, the fill area will be mismanaged and the delivery vehicles will be delayed. (2) If the delivery falls behind the placing equipment, the fill operation will be nonproductive to its capacity to place fill.

When condition 1 exists, the delivery of material must be cut back or the type of placing equipment may be changed to accommodate the increase in delivered fill, thus increasing productivity. This can be accomplished only if the area of placement is able to accommodate the increased production and larger equipment. When condition 2 exists, the delivery source must be increased to accommodate the need for the volume capacity of the placing equipment. If this cannot be accomplished, the effect will alter the production factor of the operation, thus increasing the cost to perform the work activity.

The above example shows the required information and format normally needed to quantify a backfill embankment operation. It produces the checks and balances to assure a productive operation.

Foundation Detailing

Procedure

Foundation detailing exists when some corrective or preparatory work is required within a specified foundation area. Most commonly, this item exists when the bearing foundation structure is founded on rock. This *dentil* work is done in order to prepare the foundation to a more uniform and sound surface.

Another type of detailing exists if the bottom of the foundation was specified to be undisturbed and in fact it became disturbed. With this condition, a specified granular select backfill material or concrete may be required.

Also, an undercut condition may exist in a portion of the foundation area requiring some additional attention. The foundation surface area of a load-bearing structure is crucial to the design load points intended. The estimator must fully realize the project specifications concerning this area of work.

Questions of estimate

When estimating and detailing the foundation detail component, these questions must be answered to formulate a justifiable response and determine a unit of productivity:

1. What type of material is being encountered?
2. Is the required work productive or nonproductive?
3. Can equipment be utilized or will it be performed by raw labor?

4. Is the material to be exported from the foundation area, or is there a need to import some select material?
5. Will any shoring or protection be required to give stability to the foundation area?
6. How will this work function be compensated, direct or incidental?

This list aids in the determination of the labor and equipment crews needed to perform the work that is required and specified. This work type usually exists within an excavated area.

To determine the work required and the area in which the work is to be performed, a referral of the project plans and soil boring reports is necessary. When the soil analysis and plans indicate a solid formation of rock evident at the foundation line, some type of bottom preparation may be required. To determine the surface area to be prepared, a calculation of the bottom bearing area of the foundation must be made, plus any additional perimeter work area that may be required.

The preparation of a rock foundation requires a variety of differing conditions and specifications. First, one must determine the requirement that is specified. These conditions normally include (1) undercut and removal of the rock to a specified depth with a selected backfill material, (2) hand chipping the irregular surface area to a more uniform and flat surface, and (3) removal and chipping of primarily unsound areas of the surface with sound or solid protrusions remaining.

The listed conditions require a variety of methods which the estimator must realize and quantify to compile a detailed and accurate takeoff and cost estimate.

Application: Method 1. Undercut and Removal

This condition requires a measurement of the surface area to be undercut along with a volume calculation of removal in cubic yards. The next determination is that of the soil condition that is present and is the undercut required for the entire area and the depth.

The quantifying of this condition is in cubic yards of removed material within the work area defined.

Example: Operation time duration analysis

Operation. Foundation surface area, undercut.

Bearing surface area (sf): $30'w * 50'l$	= 1500 sf
Added perimeter required	= say 4'
Calculation: $(30 + 4 + 4) * (50 + 4 + 4)$	= 2204 sf
Total surface area	= 2204 sf
Area to be treated for preparation	= entire
Depth of undercut	= 1.5'
Volume calculation: $(1.5' * 38' * 58')/27$	= 122.44 cy
Type of material: limestone and shale (loose)	

Foundation Preparation

Example: Production factoring

Operation. Undercut.

Crew members	= 2.0
Production unit	= cubic yards
Unit cycle time (½ cy bucket)	= 0.75 minute/cy
Productivity factor	= 55-minute hour
Cycles per hour: (55/0.75)	= 73.33 cycles/hour
Production per hour: 73.33 * 0.5 cy	= 36.67 cy/hour
Shift duration	= 8 hours
Production per shift: 8 * 36.67 cy	= 293.36 cy/shift
Required shifts: 122.44/293.36	= 0.417 shift
Total craft-hours: 0.417 * 8 * 2	= 6.67 craft-hours
Operation production factor: 6.67/122.44	= 0.054 cy/craft-hour

This example shows an undercut operation with a two-member crew. The anticipated production base is ½ cubic yard of material to be excavated every ¾ minute with a productivity factor of a 55-minute hour. This produces 73⅓ cycles per crew hour worked. With the excavating unit having a ½ cubic yard capacity, the hourly production is 36.67 cubic yards per crew hour.

The crew consists of two personnel for which the yardage is factored for costing, thus equating to 18.34 cubic yards per craft-hour, or an operation production factor of 0.054 craft-hour expended for each cubic yard of material excavated.

Application: Method 2. Hand Chip and Dentil Work

This condition requires a calculation of the entire surface bearing area, any perimeter work area required, and a determination of the approximate or average depth of chipping or dentil work specified. Because this condition is a heavy labor user, the need of the additional perimeter area should be analyzed for the added cost. Next would be a report of the material type encountered. Since this is a somewhat unknown condition with regard to the intensity of labor to be expended, some assumptions must be made. The entire bearing surface area should be included in the area of performance, and an average depth of cut should be assumed. The factoring for craft-hours used should then be derived by the actual condition that exists, normally by a classification of light, 0 to 2″; medium, 2 to 4″; and heavy, 4 to 6″, preparation. Any condition that exists greater than 6″ in average depth should be considered as an undercut area. This quantifying is done in square feet of affected surface area with a factoring condition.

Example: Operation time duration analysis

Operation. Foundation surface area, hand chip.

Bearing surface area (sf): 30′w * 50′l	= 1500 sf
Added perimeter required	= none
Total surface area	= 1500 sf
Average chip depth	= 3″
Factoring condition	= medium
Area to be prepared	= entire
Volume calculation: (30 * 50 * 3/12)	= 375.0 cf
Type of material	= limestone

Example: Production factoring

Operation. Hand chip.

Crew members	= 1.0
Production unit	= square foot surface
Unit cycle time	= 15 minutes/sf
Productivity factor	= 50-minute hour
Cycles per hour: (50/15)	= 3.33 cycles/hour
Production per hour: 3.33 * 1	= 3.33 sf/hour
Shift duration	= 8 hours
Production per shift: 8 * 3.33 sf	= 26.64 sf/shift
Required shifts: 1500 sf/26.64	= 56.31 shifts
Total craft-hours: 56.31 * 8 * 1.0	= 450.45 craft-hours
Operation production factor: 450.45/1500	= 0.30 craft-hours/sf

Operation. Removal (hand shovel) 3″ deep.

Crew members	= 1.0
Production unit	= cubic feet
Unit cycle time	= 10 minutes/cf
Productivity factor	= 50-minute hour
Cycles per hour (50/10)	= 5.0 cycles/hour
Production per hour	= 5.0 cf/hour
Shift duration	= 8 hours
Production per shift: 8 * 5.0 cf	= 40.0 cf/shift
Required shifts: 375 cf/40.0	= 9.38 shifts
Total craft-hours: 9.38 * 8 * 1.0	= 75.04 craft-hours
Operation production factor: 75.04/375.0	= 0.20 craft-hour/cf
Convert to square feet: (40/(3/12))	= 160.0 sf/shift
Combined production factor	
Required hours chipping: 1500 sf/3.33	= 450.45 hours
Required hours removal: 375 cf/5.0	= 75.04 hours
Total operation hours	= 525.49 hours
Production factor: 525.49 hours/1500 sf	= 0.3503 sf/crew hour
Shift production: 1500/65.69	= 22.835 sf/shift

This example gives a condition where two separate work operations are comprised into one costing function. It combines the activity of a hand chipping operation which will include the removal of the material being hand chipped within the foundation area. The total work operation is formulated using two separate units of measure for control. The hand chipping is measured by the square foot of surface area prepared with a given average depth, and the removal is accounted for by the cubic feet of the given material removed.

Each of the work operations must be analyzed using its own form of criteria, with the end result producing the total craft-hour consumption for that item. Once this has been answered, the craft-hours for each function of work are summed together and factored into the controlling unit of measure for overall costing. With this scenario, the unit control is the square feet of foundation surface area prepared. The craft-hour production factor is then based on this operation.

This example shows that even though multiple items may be providing for the costing of one operation, it can be easily accomplished by following this procedure, assuring that all the incidental items of work are accounted for.

Application: Method 3. Selected Removal

With the condition of selected preparation and removal, the entire surface bearing area of the foundation also must be calculated. The perimeter work area of the foundation must be acknowledged if required for additional work area, and the amount of selected preparatory work and chipping must be realized. As with method 2, this condition requires a heavy labor usage factor. The need for an additional perimeter preparation must be determined and accounted for.

This condition of work also requires some assumptions based on the plan specifications, soil boring logs, and actual site conditions. With this type of preparation classified as selected, areas of the foundation bearing surface are not affected. This condition is classified by a percentage calculation area with a labor factoring determination of light, medium, and heavy, as utilized with method 2. The control calculation summary is in square feet of surface area along with an average depth of chipping. The quantifying of work to be performed with this condition is the entire foundation surface area along with a factoring condition of difficulty, and some realization of the total area that will be affected.

Example: Operation time duration analysis

Operation. Foundation surface area: Select preparation.

Bearing surface area (sf): 30'w * 50'l	= 1500 sf
Added perimeter required	= none
Total surface area	= 1500 sf
Percent of prepared area	= 40%
Average chip depth	= 3″
Factoring condition	= light
Total affected area	= 1500 sf * 40%
Area calculation (1500 × 0.40)	= 600 sf
Type of material	= loose and fractured shale

Example: Production factoring

Operation. Select preparation.

Crew members	= 1.0
Production unit	= square foot surface
Unit cycle time	= 20 minutes/sf
Productivity factor	= 50-minute hour
Cycles per hour: (50/20)	= 2.50 cycles/hour
Production per hour: 2.50 * 1	= 2.50 sf/hour
Shift duration	= 8 hours
Production per shift: 8 * 2.50	= 20.0 sf/shift
Required shifts: 600/20	= 30 shifts
Total craft-hours: 30 * 8 * 1	= 240 craft-hours
Operation production factor: 240/1500	= 0.16 craft-hour/gross sf

The example for method 3 gives a condition where there may be a requirement for only a partial area of the foundation to be prepared. This area must be estimated to a percentage of the total surface area of the foundation. The example allows for 40 percent of the area to be prepared. The production factoring is calculated for the entire surface area of 1500 square feet, but the unit cycle time of actual production is derived by prorating the 40 percent of area over the entire area. In other words, the unit productive measurement of 2.5 square feet per hour equates to 600 craft hours if 100 percent of the foundation area was required to be prepared.

By formulating the production factors in this manner, the controlling unit of measure, the total square feet of foundation surface area, remains accounted for.

Final Grading

Procedure

The final grading of the foundation surface area is the finishing touches of the preparation to the surface. It normally exists within the limits of earth excavated areas and areas of embankment or suitable backfill. This procedure should be the last item of work done within the foundation area, usually hand type or small equipment usage, and provides the bottom line dimensional configuration of the structure outline and template as detailed by the drawings. This procedure also ensures proper dimensions for no overyields or underyields with respect to the structure footings.

A final confirmation of compaction and density is also completed during this procedure.

Questions of estimate

When compiling a takeoff of final grading for the foundation area, these questions should be answered to formulate and detail the estimate and to arrive at a justifiable productivity factor.

1. What type of material is to be encountered?
2. Will this operation be considered productive or nonproductive?
3. Will this operation be performed by raw labor or with the aid of equipment?
4. Was the area constructed on fill or virgin excavation?
5. How will this work operation be compensated, direct or indirect?

By compiling this list, one can define the work operations needed and formulate a detailed labor and equipment usage factor of production.

This work operation is performed within the limits of the foundation as defined by the project plans. It will be only the bottom surface area of the bearing points of the structure, and any form lines required for the construction of the foundation.

Application

This calculation includes the entire contact surface area of the bottom of the structure at the ground. The quantity reported from the takeoff is in a square foot summary of calculations.

Example: Operation time duration analysis

Operation. Final grading.

Area to be graded: 30'w * 50'l	= 1500 sf
Method of grading	= hand
Production type	= nonproductive
Constructed area type	= virgin excavation
Type of material	= compacted soil

Example: Production factoring

Operation. Final grading.

Crew members	= 2.0
Production unit	= square foot surface
Unit cycle time	= 2.5 minutes/sf
Productivity factor	= 55-minute hour
Cycles per hour: (55/2.5)	= 22.0 cycles/hour
Production per hour: 22.0 * 1	= 22.0 sf/hour
Shift duration	= 8 hours
Production per shift: 8 * 22.0	= 176 sf shift
Required shifts: 1500/176	= 8.52 shift
Total craft-hours: 8.52 * 8 * 2	= 136.32 craft-hours
Operation production factor: 136.32/1500	= 0.091 sf/craft-hour

The operation of final grading must be accounted for under most conditions of foundation preparation. The function allows for accurate grade control and assurance of constructing the structural component to the proper dimensions, thus eliminating a yield loss to incorporated permanent materials.

When preparing a detailed takeoff of work operations for a given item of work, all the necessary component items and work operations must be accounted for and incorporated in the final costing summary.

Dewatering and Subaquifer Preparation

Procedure

Dewatering and subaquifer preparation occur when the foundation area is encroached on or constructed within limits of water, such as river, stream, lake, or oceanic structures. This also is referred to as marine construction.

Simple dewatering practices can occur within a dry-type foundation that has simply been flooded by rain or surface runoff. Dewatering of this condition can be achieved by simple pumping. When conditions exist of uncontrolled surface water permeating the foundation area, a temporary or permanent diversion or a bypass of the water source may be required. In some cases, merely damming the water source is permissible and sufficient until the foundation is constructed. When a water flow is obstructed, temporary pumping may be required to prevent the backup of unwanted water in other areas.

Subaquifer foundation construction techniques are somewhat detailed and require great skill and experience since the foundation preparation is accomplished without dewatering the area. The work is required to be completed underwater.

Major foundation dewatering requires a perimeter seal, barrier, or dam to separate the foundation area from the water source. The perimeter separator is commonly referred to as a *cofferdam*. Depending upon the actual site, the dewatering process may be a one-time procedure or a continuous operation, varying upon the influx and influence of the water source and the ability to seal off the foundation area.

Simple dewatering usually consists of merely pumping the surface water from an excavated foundation area. This permits ongoing work to progress in a dry condition and also assures the density and integrity of the bearing surface. If the ponded water has altered the bearing surface by saturation, or if erosion has occurred, causing the bearing surface to become unsound, the unsuitable material is removed and a suitable backfill material constructed in its place. The foundation bearing surface must remain intact and to its designed profile and template.

The foundation preparation that is required to be performed underwater or in a subaquifer condition requires a detailed plan and careful consideration. This type of preparation normally occurs in a situation of marine-type construction where the bearing foundation cannot be dewatered in order to construct the proposed structure in a dry condition. This may also exist as a prestage type of construction, along with other techniques that would enable the structure to be constructed in a dry-type condition. In conjunction with this method, a concrete seal is required. A foundation of this type could be founded on bedrock or suitable soils.

A seal is a structural component that acts as a sealant or bottom liner to stop the flow of water from the underlying soils at the bottom of the cofferdam. The sequence of operation consists of a cofferdam, foundation preparation of the bearing surface, the foundation seal, and the dewatering procedure.

Once the cofferdam has been constructed, any required excavation must be completed. Owing to normal restraints of a cofferdam (tight conditions, reach and depth, and field logistics), a *clamshell* method is normally chosen. The clamshell, which is an attachment to a crane, gives much versatility to the site access. The limitations then become the reach and lifting capacity of the crane. The crane is usually located within close proximity to the excavation site, situated on stable ground, *crane mats,* or a barge within the water. Clamshell excavation allows for full vertical movement of the bucket and gives much more reach than conventional methods and a great variance of depth due to the length of cable on the crane. This removes a majority of the material from within the cofferdam.

When excavating by clamshell methods within the confines of a cofferdam, areas of material will still be inaccessible, primarily around the inside perimeter of the cell. To completely remove the remaining material along and adjacent to the perimeter, an *airlift* method is used. Care must be taken to remove all silt, loose rock, or remaining material from the foundation bearing lines, as shown by the plans, to ensure a proper profile is obtained.

Airlifting material is a technique by which excess or unwanted material is removed by way of vacuuming or suction. An airlift attachment is a pipe, suspended from the crane, which is sealed on the top end and has a discharge pipe attached to it. An inlet hose by which compressed air enters is fixed near the top of the unit. When the bottom of the airlift pipe is submerged in the water and compressed air is engaged, a suction is created. When this pipe is probed at the bottom of the cofferdam excavation, it removes the unwanted or loose material from within the cell.

Airlifting can only remove the material that is loose or that can be loosened by the suction power, and which can enter the mouth of the suction pipe. Other material may need to be removed by water jetting.

Rock excavation within the cofferdam must first be loosened and removed by some conventional method. The bottom foundation surface must be profiled within the allowable tolerances detailed by the project specifications. Normally, unless a specific elevation is specified, the excavation depth is taken to a sound rock elevation. The proposed elevation can be found within the soil boring log specified by the project. There will be a specific soil boring analysis for each critical foundation area of the structure. This is specified by a solid-type formation which will give suitable bearing to the foundation structure.

To obtain this elevation, some loose or unsuitable layers of the top surface of the rock layer may have to be removed. With this type of specification, the actual bottom elevation of the excavation may vary until a suitable bearing surface is achieved.

Under normal soil excavation within a cofferdam, the designed final bottom elevation is usually obtained. Regardless of the type of excavation, the final bearing surface of a marine-type structure, within a cofferdam cell, must be verified for a suitable bearing surface prior to any further construction.

Questions of estimate

The takeoff and quantifying of this type of foundation preparation require greater detail and research than the normal methods owing to the variability and somewhat unknown conditions. This section is defined by two types of preparation methods, normal dewatering and cofferdam or cell dewatering and preparation.

Normal dewatering:

1. Is the water in the foundation area ponded surface water or a continuous entry?
2. If a continuous entry, what methods must be done to control and stop the flow of water?
3. Will a temporary bypass channel or pumping be required?
4. What is the volume of water to be removed?
5. What is the distance of discharge?
6. Based on the volume of removal, what size pump is required to maintain an allowable time for the schedule?
7. What soil conditions will exist after the foundation area is dewatered?
8. Will any remedial work, excavation, or backfill be required for the final foundation bearing surface?
9. What measures can be implemented to prevent future water entry?
10. Is the work direct compensatory or indirect?

Cofferdam cell: dewatering and foundation preparation

1. Is the cell accessible from land, or will marine water equipment and methods be required?
2. Can the cell be dewatered?
3. What are the normal and high water depths?
4. What type of excavation exists within the cofferdam cell?
5. Will a concrete seal be required?
6. Will an airlift be required?
7. What is the volume of water to be removed?
8. What is the distance of discharge?
9. Based on the volume, what size pump will be required to maintain an allowable schedule?
10. What preparatory or remedial work will be required?
11. Will underwater divers be needed?
12. Can the water entry be controlled?
13. What final grading, if any, will be required?
14. What conditions will exist after dewatering?
15. Is the required work direct compensatory or indirect?

This listing of work activities will aid in the preparation and formulating of the construction estimate and give a clear picture to the variable conditions that exist and must be recognized with dewatering and subaquifer type work.

These work activities are confined within an excavated foundation area or a controlled condition of a cell cofferdam, whereas the cofferdam itself must be constructed in order to perform the foundation work under controlled conditions.

Foundation preparation must only be detailed and summarized to the work activity directly associated with the specified operation that is required to give a suitable and stable bearing surface to the foundation area.

Application: normal dewatering

The operation of dewatering is defined and associated by the direct proportionate relationship of flow of water, water volume, size of pump, and time required for control within the excavated foundation area. When a continuous flow or uncontrolled flow exists, the unwanted water flow must first be contained, diverted, or eliminated from the work area prior to the foundation dewatering. If surface runoff water is encountered, this procedure becomes just the function of its removal from the foundation area.

The controlling work activity is defined by the volume of water to remove and the necessary work required to prevent any future flow or encampment. In constructing a foundation area, steps should be taken to prevent water from entering the foundation area when possible, thus eliminating the need and cost of dewatering. In some cases, though, this is not possible or foreseen. A contingency within the estimate should be considered for this.

The calculation for simple dewatering of ponded water is the time required to remove the water volume within the foundation area. This also includes the necessary setup and proper disposal and control of the discharged runoff.

Example: Operation time duration analysis

Operation. Simple dewatering.

Volume cf	= 20′w * 40′l * 18/12″ d = 1200 cf
Volume gallons	= 7.48 gal/cf*1200 cf = 8976 gal
Pump capacity	= 90 gallons per minute
Entry flow	= none
Dewater time	= 8976 gal/90 gal/min = 99.73 min
	= 99.73 min/55-min hour = 1.81 hours
Soil saturation	= 3″
Soil removal	= 3/12″*20′w*40′l = 200 cf
Backfill required	= 3/12″*20′w*40′l = 200 cf

Example: Production factoring

Operation.   Simple dewatering.

Crew members	= 1.0
Production unit	= gallons per cubic foot
Unit cycle time: 90 gal/min	= 90 gal/min
Productivity factor	= 55-minute hour
Cycles per hour: (55 * 90)	= 4950
Production per hour: 4950 * 1	= 4950 gal per hour
Shift duration	= 8 hours
Production per shift: 8 * 4950	= 39,600 gal per shift
Required shifts: 8976.0/39,600	= 0.23 shift
Total craft-hours: 8 * 0.23 * 1	= 1.84 craft-hours
Operation production factor: 1.84/8976	= 0.0002 craft-hour/gal

The above example shows a dewatering condition with no threat of continuous water entry, so the condition that existed required minimal pumping and remedial soil work within the bottom surface.

Where the area of contour and slope drainage affects water control, it becomes necessary to divert or control the water flow to another area of the work site. Surface water from runoff or uncontrolled streams can be an aggravating source in some cases owing to the inconsistency of the amount and time of flow. This condition must be determined during the takeoff and analyzing of the project to assure the proper costing and time scheduling.

Example: Operation time duration analysis

Operation.   Water stoppage.

Source of flow	= low swale
Diversion method	= earth dike
Bypass required	= yes, temporary pumping
Pump size required	= 6 inch
Time required	= 10 days, complete foundation operation
Number of diversions	= 2.0 each

This example shows an uncontrolled water source obstructing the foundation area and not allowing the area to be properly dewatered. For this, a temporary diversion of the water source must be made and a time duration must be realized as a requirement for the uninterrupted completion of the foundation work. This time duration must account for not only the required time of the foundation preparation but the time duration needed to construct the structure foundation and any adjacent work related thereto. Also, any diversion of water and the relocation of it must at no time damage any areas of construction or cause flooding or erosion.

Example: Production factoring

Operation. Water stoppage, dike.

Crew members	= 2
Production unit	= dike, each
Unit cycle time	= 8 hours, 1 shift
Productivity factor	= 60-minute hour
Cycles per hour: ($1/8$)	= $1/8$ cycle per hour
Production per hour: ($1/8 * 1$)	= $1/8$ dike per hour
Shift duration	= 8 hours
Production per shift: ($1/8$ ea $* 8$)	= 1.0 ea per shift
Required shifts: (2.0 ea/1.0)	= 2.0 shifts
Total craft-hours: 2 shifts $* 8$ hours $* 2$	= 32 craft-hours
Operation production factor: 32 craft-hours/2 ea	= 16.0 craft-hours/ea

This example shows the condition of water stoppage by utilizing an earth constructed dike or berm. It requires two dikes to control the flow of water. The quantifying of the time duration for this activity is justified by determining the time required to construct each dike, maintain the stability of it, and remove it upon completion of the foundation. This portion is for the dike only; the pumping is a separate step of this operation. The unit for this activity is considered to be *lump sum* for each dike, or simply a unit of each for the total quantity required. It should be a consideration to view each dike independently for the required time duration, as separate dikes or diversions will require different time allocations for specific conditions.

The next step for this example is factoring the time required to pump the controlled or dammed water around the foundation area back to an area of natural flow.

Example: Production factoring

Operation. Water stoppage, pumping.

Crew members	= 1.0 ea
Production unit	= dike, each
Unit cycle time	= 8 hours, 1 shift
Productivity factor	= 60-minute hour
Cycles per hour: ($1/8$)	= $1/8$ cycle per hour
Production per hour: ($1/8 * 1$)	= $1/8$ each per hour
Shift duration	= 8 hours
Production per shift: ($8 * 1/8$)	= 1.0 ea per shift
Required days	= 10.0 ea
Required shifts: (3 ea $* 10$)	= 30.0 shifts
Shift hours per day: 8.0 hours $* 3.0$ shifts $* 1$	= 24.0 shift hours
Required craft-hours: 24.0 ea $* 10$ days	= 240.0 craft-hours
Operation production factor: 240	= 240 craft-hours/ea

This portion of the operation shows the pumping requirements for the diversion. This requires continuous pumping for water control. Since the pump will be in operation for the entire shift, 3 shifts per day, an attendant should be allocated to the pump to assure continuous and uninterrupted operation. As with the normal pumping operation, the sizing of the pump capacity should be calculated. Pumping of the diversion dike will be required for the entire duration that the foundation is under construction. This is the time from the initial excavation to the completed backfill, or unoccupied time of work activity to the foundation area, in other words, the time at which the water flow can be returned to its natural flow, or permanent relocated flow, without obstructing the foundation work area.

Summary: Dual activity

Operation. Water stoppage.

Dike	= 2.0 ea
Craft-hours: 2.0 * 16.0	= 32.0 craft-hours
Pumping	= 2.0 ea
Craft-hours: 2.0 * 240.0	= 480.00 craft-hours
Total craft-hours: 32.0 + 480.00	= 512.00
Craft-hours per unit: 512.0/2.0	= 256.0 ea
Craft-hour factor	= 256.0 per dike

The formulating of this approach gives the example of arriving at a total duration for a lump-sum activity type of work. This example shows a method of performing this in a detailed but simplified manner. The one rule of this procedure is that once the separate operations are factored, the final summary must be compiled by applying the total duration and craft-hours into the controlling unit of work, keeping units compared to similar units.

Application: Cofferdam Cell Dewatering

A complicated form of dewatering occurs with river or open water construction, where structure construction is within the water. When a supporting member of a structure spanning a waterway must be founded on suitable bearing, the foundation area must be accessible in order to perform the work required.

A cell or *cofferdam* must be constructed in the waterway to permit and contain the dewatering of the foundation area. This could consist of an earth dam if the conditions permit but usually is constructed of steel *sheet piling*. The piling forms a somewhat watertight and protective cell which enables the foundation area to be properly prepared and constructed under a term of controlled conditions. This cell is constructed at the area of the structure location in the water. It extends to the bottom of the water and into the underlying soil at the floor. This extension, or *toe,* must be driven securely and substantially into the soil or to the rock line to give support to the bottom of the cofferdam. A fair amount of *freeboard* must extend above the waterline at the surface to assure that no water will overflow into the cell. In areas of fluctuating water levels, high tides, and severe wave action this is an important consideration.

Cofferdam. (*Courtesy of Bethlehem Steel Corporation.*)

Within the cofferdam, a series of supports and bracing referred to as *walers, ringers,* and *struts* exist. These not only control the perimeter shape of the cell but keep the cell from collapsing and buckling under the water pressure.

A cofferdam will have equal pressure to the vertical walls until the cell is dewatered. After dewatering and during the foundation construction the perimeter of the cell must withstand the hydraulic pressure of the water against the outside walls. The deeper the cofferdam, along with the existing current, the greater the pressure per wall surface area. This entails a more sophisticated bracing system within the cell. One important note to detail is that the walers and bracing are permanent components of the cofferdam cell and must not obstruct the construction of the foundation or support structure.

A circular cell will withstand greater exterior pressures than will a square cell. In some conditions a double cell cofferdam may be required. This is merely two cells constructed around each other with some form of backfill material placed between them. This type would be commonly used under severe conditions.

The major dewatering of the cofferdam occurs upon the completion of cell construction. This is the time duration to primarily dewater the cell with the major pumping concern and have the cell in a controllable condition. If a concrete seal is required, dewatering would occur after the seal is constructed and the airlift excavation is complete.

Once the major dewatering has been completed, the normal pumping operation can be maintained to control minor seepage or surface water.

The major dewatering must be approached to remove the water from within the cell in a timely manner, but also in a manner that is not detrimental to the structural integrity of the cell. The exterior pressure must be monitored under a controllable situation.

Upon completion of the structure component construction and prior to the removal of the cofferdam, the cell should be refilled with water. This procedure equalizes the pressure from the interior to the exterior and provides a more controlled removal of the cofferdam.

Example: Operation time duration analysis

Operation. Major dewatering, marine construction.

Type construction	= river cell cofferdam
Underlying soil	= rock
Water depth	= 22′ normal mean (26′ high flood)
Current and water flow	= yes, medium
Cell size (waterline):	= 20′ * 20′ * 22′ deep
Cell water volume: 20′ * 20′ * 22′	= 8800 cf
Production type	= gallons
Volume of water: 8800 cf * 7.48 gal/cf	= 65,824 gal
Secondary pump required	= yes, 3″
Repump to fill	= yes
Pump size required: 6″, capacity 400 gal/minute	= use theoretical 350 gal/min
Operation unit of measure	= each per cell

Example: Production factoring

Operation. Major dewatering.

Crew members	= 2.0 ea
Production unit	= gallons
Unit cycle time	= 350 gal/min
Productivity factor	= 55-minute hour
Cycles per hour: (55 * 350)	= 19,250
Production per hour: 19,250 * 1	= 19,250.0 gal/hour
Shift duration	= 8 hours
Production per shift: 8 * 19,250	= 154,000.0 gal/shift
Required shifts: 65,824.0/154,000	= 0.43 shift
Total craft-hours: 8 * 0.43 * 2	= 6.88 craft-hours
Operation production factor: 6.88/65,824	= 0.0001 craft-hour/gal

Summary

Dewatering time: 65,824/350 gpm	= 188.07 minutes
Duration hours: 188.07 min/55-min hour	= 3.42 hours
Required craft-hours: 3.42 * 2	= 6.84 craft-hours

With regard to rounding this example comparison checks with the formula activity method. This comparison gives the example of the operations involved with a major dewatering operation. As with the other scenarios given, this quantifying method shows another way to format a quantity take-off and summary for differing conditions.

The example shows the requirements for determining the time duration for major dewatering of a cofferdam. The cell is calculated by volume in gallons for the amount of standing water to be removed. The time required for the dewatering is controlled by the type and size of pump used. This is somewhat governed by the conditions that exist. The industry standard for pump classification is gallons per minute the pump will discharge. To determine the pump size required, one must realize the logistics and economics of the operation. The cost of acquiring a pump to perform the work in a reasonable amount of time and the cost of the labor involved must be recognized.

The output capacity of the pump is governed by the engine horsepower, revolutions per minute of the impeller, the head, or the vertical distance the pump must lift the water, and the resistance of the horizontal distance of the discharge line.

The time required for the setup and demobilizing must be detailed along with the standby time that may be required.

Once the main cell is dewatered, a smaller secondary pump will be required to maintain the removal of seepage water and/or any surface water that may enter the foundation area.

The time required for the secondary pump is determined, usually by the amount of time the cell or cofferdam will be active. It is wise to assume this pump will be required and will be dedicated for that amount of time. The required labor for the operation of this pump is determined by logic for the amount of allowable water that may be in the cell at any given time. This normally should not be any appreciable amount. If the cell has difficulty in being sealed for water seepage, this would be a more involved operation. With some conditions in regard to controlling the amount of water buildup at the bottom of the cell, a *sump* may have to be constructed for an area in which the water can accumulate. This also gives a pool for the pump to be able to maintain its prime for pickup.

The work operation control and quantity detail will be determined by the craft-hours required to remove the water from the cell, and the associated work. The time duration needed will be a realistic estimate of the total and complete operation.

This dewatering operation, normally carried out from the waterway, will include any necessary scaffolding, temporary flotation equipment, and the time to transport the equipment and personnel from shore to the work site.

The time impact associated with this operation should be calculated for the total duration required to perform the task of dewatering the cell to the point of being able to perform the manual labor and work activities associated with the foundation construction in a somewhat dry condition. The duration, even though calculated by volume of water removed in gallons, should be converted back to a cost per cell for comparative summary incidental to the cofferdam.

Foundation Preparation Summary

As this section shows, the operation of foundation preparation can require a wide variety of detailed work involving a diversity of labor and equipment utilization. The preparing of the bearing foundation area can be a costly and detailed procedure. The estimator must carefully determine the work required and form a detailed and accurate takeoff of quantities to prepare an accurate and competitive cost analysis and also form a true schedule of the time duration required to complete the work.

These comparison cost examples detail only the requirements to perform the actual work activities at the work site. No allowance was accounted for nor were time duration craft-hours calculated for any setup or mobilization or demobilization. In formulating actual costs and required quantities, I prefer to account for the setup and mobilization differently because of the variance of the functions. This will be detailed under the appropriate section.

COMPONENT TAKEOFF

Foundation Preparation

The following outline contains the normal work operations that account for the performance and cost coding of foundation preparation construction components.

1. **Unclassified excavation**

   All work related to the performance of excavation and disposal of the unclassified category. Quantity should be reported in total cubic yards of excavated material along with neat line volume.

2. **Dry excavation**

   All work related to the performance of excavation and disposal of the dry category. Quantity should be reported in total cubic yards of excavated material along with neat line volume.

3. **Wet excavation**

   All work related to the performance of excavation and disposal of the wet category. Quantity should be reported in total cubic yards of excavated material along with neat line volume. Any dewatering required should be properly accounted for.

4. **Rock excavation**

   All work related to the performance of excavation and disposal of the rock category. Quantity should be reported in total cubic yards of excavated material along with the neat line volume. Involved with this is an accounting of drilling and shooting, if required.

5. **Unsuitable excavation**

   All work related to the performance of excavation and disposal of the unsuitable category. Quantity should be reported in total cubic yards of excavated material along with the neat line volume.

6. **Foundation preparation**

   All work related to the preparation of the excavated foundation as specifications require. This includes all cleaning, dentil work, compaction, removal of loose rock and material, and the verification of stability. Any removal of unsuitable material should be accounted for. Quantity should be reported in square feet of foundation bottom area with a detailed accounting of work performed.

7. **Extra depth excavation**

   All work related to performance of extra depth excavation and disposal for a specific category of work. Quantity should be reported and detailed in total cubic yards of material excavated along with the neat line volume for each specified classification of material.

8. **Rock drilling**

   All work related to the performance of the drilling of holes in a rock foundation. Quantity should be reported in actual lineal feet and diameter of hole drilled along with the number of holes.

9. **Foundation backfill**

   All work related to the backfill and compaction of a specified material for a foundation area. Quantity should be reported in cubic yards of material placed and compacted with a detail of specific type.

10. **Bedding material**

    All work related to the addition and compaction of foundation fill material for an area of undercut, unsuitable, or demarcation area. Quantity should be reported in type of material placed, square feet of bottom surface area, and total cubic yards of material.

11. **Dewatering**

    All work related to the performance of properly dewatering and maintaining a dry foundation. Quantity should be reported in craft-hours and equipment hours utilized along with the type of existing water condition. Hydraulic volume of cell calculation, along with pump volume, should be noted.

12. **Cleaning and dentil**

    All work related to the performance of cleaning and final preparation of the foundation area. Quantity should be reported in square feet of bottom surface area of the excavated foundation.

13. **Bracing and shoring**

    All work related to the temporary bracing and shoring of a foundation area, excluding steel sheeting. Quantity should be reported in lineal feet, pounds, etc., and individual material of items used.

14. **Airlift excavation**

    All work related to the performance of airlift excavation. Quantity should be reported in cubic yards of excavated material, square feet of bottom surface area, and hours of equipment used for this method.

15. **Other—excavation and foundation**

    This account will be used for other operation accounts not specified here. Quantity should be reported in a representative format relating to the work performed and the appropriate unit of measure.

Chapter 12

Foundation Piling

TECHNICAL SECTION

Introduction

The purpose of designing a pile-supported foundation is to transmit the loads from the structure through the soft earthen strata into more supportive soils or rock.

The two types of design theory used are friction bearing and point or tip bearing.

With a friction-type design, the pile develops bearing by skin friction. As the pile picks up friction on its perimeter, it transfers load triangularly downward and laterally into the adjacent soils, for the entire length of the pile. This is used when the tip of the pile cannot achieve bearing in specific soils.

The design of tip or point bearing relies on virtually the entire load of the pile's being transmitted to the bottom of the pile into a firm stratum below, preferably shale or rock.

Although these types can be independent of one another, generally a pile design can contain a combination of both.

Lateral forces are a concern for some structures. In this type of design, it becomes important for the piling to withstand lateral pressures, such as water currents, wind and horizontal soil pressures, plus the vertical load of the structure.

Soil conditions and the type of piling used will have a great effect on the pile-driving operation and the production achieved. Test piles can be helpful to a contractor and owner to determine exactly how a pile will react, but this is generally after the fact of design and the bidding stage. A dynamic formula will aid this prematurely but is not always an accurate method to establish a bearing capacity or hammer type. This calculation produces a formulated blow count to measure and determine bearing. This is derived from the average penetration per blow and the driving energy produced by the hammer.

In all types of piling designs, the bearing achieved and penetration requirements are the key factors.

The four major types of piling used today are concrete, timber, steel H pile, and steel pipe or shells.

The factors used for design and choice of piles are structural life expectancy, type of structure and its intended use, availability of materials, and foremost the soil conditions. Each type of pile has its own unique properties and usefulness with intention to produce specific designs. Any type of flat-bottomed pile is known as displacement pile, as the driving operation displaces the soils beneath it (concrete, pipe, and shell). These are usually considered for skin friction piles or nonrock landing piles. When displacement is either a problem or a concern, or the tip elevation is designed for rock, H pile will be considered.

Key factors that are discussed are test piles, permanent piles, and bearing piles. With regard to blow count, the refusal and hammer ratings control the size and type of hammer.

The absolute values of this section remain as follows: test piles will always be of the same size and type as the permanent design pile but not necessarily the same length. In driving of the test pile, the hammer must remain constant. No deviations of the hammer must be allowed. This is due to the performance activity of the chosen hammer to determine the end result bearing value and energy transmitted from the hammer. The bearing values are determined either one of two ways, by blow count or by refusal. A particular hammer is chosen by ratings based on the energy developed by the ratio calculations of a specific hammer. These ratios along with various types of hammers having their own specific properties vary widely depending on the types of soils and the type of piling being driven. A rule of thumb to initialize hammer sizing is that the weight of the hammer is equivalent to the pile weight plus the driving head.

Item: Piling, Permanent Foundation, Specifications

This work consists of furnishing all labor, material, and equipment to drive various types of *piling*. It includes any preboring, drilling pilot holes; furnish, placing, and removal of casings or any other means necessary to drive the piling to the proper plan *bearing* and *pile tip* as indicated on the plans and specifications. The contractor will be required to perform this work to conform reasonably to the details shown.

Usually, the contractor will be furnished, from the owner, a detailed list which will show the type, number, and estimated length of piles. In many cases, this will only be for informational purposes because piling does vary from project to project. Many variables will determine the actual length driven. The type of piling used will be a given based on the design of the structure and the *loading* involved. Piling is, however, used for supporting a structure or substructure, which could be earthen substance.

There are various types of piling for which their *properties* of design and *section modules* dictate certain uses and functions. Many techniques of installation are associated with these various types of pilings. The two main methods involve either driving or predrilling or augering.

In any case the contractor may drive or install *test piles*. This may be required by the owner or done at the contractor's option. This will be an absolute determination of the *permanent* pile lengths involved, based on the *substrata* and soil properties and the final bearing value achieved.

The final bearing along with the tip elevation will be derived from data by a *load test* applied to the test pile, and in some instances a load test applied to one or more of the permanent piles. This would be done to substantiate the determining values of the test pile.

In some instances bearing can be achieved at a higher elevation than required. The deeper *penetration* is taken into account because of the lateral stability needed for a particular structure. In this case *preboring* or *jetting* may be a requirement for the contractor to achieve the desired tip elevation. In other words, the piling cannot be damaged by excessive driving owing to the early encountered bearing and/or, one other determining factor, *refusal*. In this case the piling will have very little or no *skin friction bearing* but will have tip bearing. Regardless, the designer will have considered the type of bearing required and the appropriate type of piling needed for that design. All bearing piles are formulated by the minimum number of hammer blows per unit of pile penetration. This initial formula determines the desired and specified bearing value of the piling and is substantiated by a load test.

Types of Piling

Timber piles

Timber piles are usually of specified wood such as douglas fir, larch, southern yellow pine, norway pine, or red oak of a nominal dimensional size, which is predetermined. Depending on the use intended, they are treated or untreated. Treating can be of either a pressure method or creosote coated. Regardless, this will be specified by the designer.

Timber pilings are usually designed for use in areas of light loading since wood piles cannot generally support heavy loads. Main uses of wood pilings are in marine and water-type designs such as *dolphins, fender systems,* docks, and low-lying areas. Wood piling can also be used for temporary supports and falsework owing to their ease of removal or cutting off.

The design features and bearing values used in timber piling are tip and friction.

The type of driving equipment needed for timber piling can vary tremendously. One must first determine the bearing value required, the *driving conditions* that are to be encountered, and of course the equipment available. (See hammers and related equipment detailed later in this section.)

Timber piling can be driven by one of several methods of equipment type: *single-acting steam or air* hammers, *double-acting steam or air* hammers,

Wood pile. (*Courtesy of International Construction Equipment.*)

diesel hammers having an unrestricted rebound of *ram,* and diesel hammers having enclosed rams. The contractor must determine which type of hammer is best suited for the situation and specifications. Once a hammer is chosen, the contractor must submit this to the owner for approval. (See pages 220, 236, and 258 for pile hammer formulas to determine hammer size.) The owner will determine, based on a formula, if this hammer is properly sized for the desired application and if it will achieve the proper *energy* to set the pile to the designed bearing. This is an important and relative issue an estimator must take into account when bidding a project to determine equipment cost and crew size needed.

The contractors' requirements of timber piling can also vary greatly. One must take into account the soils in which the piling is to be driven. Soil conditions greatly impact the driving of piling. It must be determined whether *protective boots or shoes* are specified. Also, based on relative soil boring charts, estimators must determine, to the best of their ability, on information given, if any obstructions are going to be encountered in the driving operation. Such items could be boulders, tree stumps, etc., which can cause much havoc on a driving operation. Such data can sometimes be derived from local historical data.

Conditions such as this along with the appropriate size and method of determining the proper hammer must be taken into consideration regarding any type of pile-driving operation.

Concrete piles

Concrete piles are of a manufactured nature of *prestressed concrete.* They are made to a specific length which is obtained from test pile information. The concrete pile length has to be predetermined because of the design nature of the pile. The reinforcing is heavier at the ends owing to the tip force and the driving force. They are constructed of portland cement concrete usually having a *design strength* of a minimum of 5000 psi. The core of the pile has spiral reinforcing along with wire strands which are prestressed. This type of design gives the pile astonishing strength and resistance property.

A concrete pile must be manufactured to its final reasonably closely anticipated length or designated spliced lengths. A concrete pile can be *cut off* or *added on to* but only after it has been fully driven to bearing. The cutoff or add-on is usually only a few feet, but the governing specifications will control this.

Concrete piles can be manufactured in various lengths and sizes, depending on the final tip elevation and penetration desired and the amount of *skin friction* of the design. The greater the perimeter of the pile, the greater the skin friction and the overall load bearing of each individual pile.

The driving operation of concrete piles must be a thoroughly scrutinized determination of the estimator, who must take into account the type of hammer required, along with the handling of the pile.

Concrete piles can be of substantial weight. An estimator has to take this, the weight of the heaviest pile, along with the total hammer weight and *leads* into account along with the reach of the crane based on job conditions, when sizing and pricing the appropriate crane.

In all cases, handling of concrete piles is a specified procedure. Specific lifting points must be taken into account when handling the piles. The longer the pile, the more detailed the lifting system and pivot points.

In some cases a second rig may be needed for hoisting and handling. There is, however, only a fixed length that will be encountered by the estimator. This will be governed in some areas by weight laws of transportation agencies or just a length that is impossible to transport safely. In the case of multipiece piles, there is a procedure for mechanically *splicing* the piles. This is an added expense to the contractor, so the estimator must research it to its fullest extent: how many pieces need to be spliced, what kind of splices are approved, and additional equipment and labor that is required to do the splice.

Again, obstructions and/or tough, severe driving conditions must be researched by the estimator. Some types of soils, such as sand, can cause a compacted coning effect at the pile tip in a relatively short distance of driving and cause a refusal effect that continuous driving of the pile will cause severe damage to its structure integrity. In this case, an estimator must predetermine this at the bidding stage. A scenario such as this would require the *jetting method* to aid the pile to its required tip elevation. This is an added expense to the driving operation in both equipment and labor.

Another option to aid in driving would sometimes be augering or predrilling a starter hole to aid the driving operation and eliminate some of the strikes required. This option may also be used to get through a certain

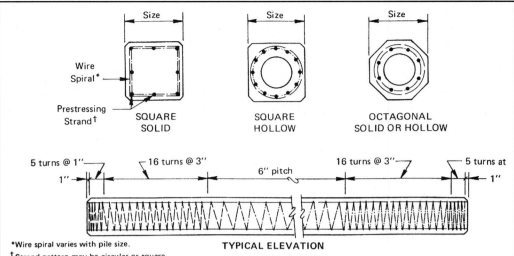

*Wire spiral varies with pile size.
†Strand pattern may be circular or square.

Size in.	Core Dia. in.	Area in.^{2}	Weight lb/ft	Moment of Inertia in.^{4}	Section Modulus in.^{3}	Radius of Gyration in.	Perimeter ft	Allowable Concentric Service Load, Tons[2] for f'_c of 5000 psi	6000 psi
SQUARE PILES									
10	Solid	100	104	833	167	2.89	3.33	73	89
12	Solid	144	150	1,728	288	3.46	4.00	105	129
14	Solid	196	204	3,201	457	4.04	4.67	143	175
16	Solid	256	267	5,461	683	4.62	5.33	187	229
18	Solid	324	338	8,748	972	5.20	6.00	236	290
20	Solid	400	417	13,333	1333	5.77	6.67	292	358
20	11	305	318	12,615	1262	6.43	6.67	222	273
24	Solid	576	600	27,648	2304	6.93	8.00	420	515
24	12	463	482	26,630	2219	7.58	8.00	338	414
24	14	422	439	25,762	2147	7.81	8.00	308	377
24	15	399	415	25,163	2097	7.94	8.00	291	357
OCTAGONAL PILES									
10	Solid	83	85	555	111	2.59	2.76	60	74
12	Solid	119	125	1,134	189	3.09	3.31	86	106
14	Solid	162	169	2,105	301	3.60	3.87	118	145
16	Solid	212	220	3,592	449	4.12	4.42	154	189
18	Solid	268	280	5,705	639	4.61	4.97	195	240
20	Solid	331	345	8,770	877	5.15	5.52	241	296
20	11	236	245	8,050	805	5.84	5.52	172	211
22	Solid	401	420	12,837	1167	5.66	6.08	292	359
22	13	268	280	11,440	1040	6.53	6.08	195	240
24	Solid	477	495	18,180	1515	6.17	6.63	348	427
24	15	300	315	15,696	1308	7.23	6.63	219	268

(1) Form dimensions may vary with producers, with corresponding variations in section properties.
(2) Allowable point bearing loads based on $N = A_c(0.33f'_c - 0.27f_{pe})$; f_{pe} = 700 psi. Check local producer for available concrete strengths. See Sec. 1.4.4 (E) of AASHTO Standard Specifications for Highway Bridges for ground capacities of piles unless subsoil investigations are conducted.

Section and sizes of concrete pile. (*Courtesy of Precast Concrete Association.*)

CONCRETE PILE PICK POINT CRITERIA (sample)

Points	Pick Up Points
1	0.292 L ― 0.708 L / L
2	0.207 L ― 0.586 L / L
3	0.145 L ― 0.355 L / L
4	0.107 L ― 0.262 L / L

Concrete Pile Pick-up Data

Pile Size	Approx. Weight per Lineal FT	Maximum Lengths for Pick-up Points			
		1-POINT "L"	2-POINT "L"	3-POINT "L"	4-POINT "L"
10"	104	47	66	95	129
12"	150	51	73	104	141
14"	204	55	78	112	152
16"	267	62	88	126	171
18"	338	64	90	129	175
20"	417	69	97	138	188
22"	504	69	98	140	190
24"	600	72	102	146	198

* This chart represents a sample illustration and shall not be depicted as a firm criteria for every condition, owner, or engineer.

Concrete pile lifting points.

type of soil stratum that may restrict the *resistance* of the hammer. Again the soil conditions and specifications pertaining to a specific project must be thoroughly researched by the estimator. Some specifications do not allow predrilling or jetting or will have limitations to these procedures.

The type of hammer required will be similar to those discussed, either diesel or air/steam powered, but they must be formulated to the specifications of the owner and be most economical to the contractor. A hammer must be relative to the soil, job conditions, and method used.

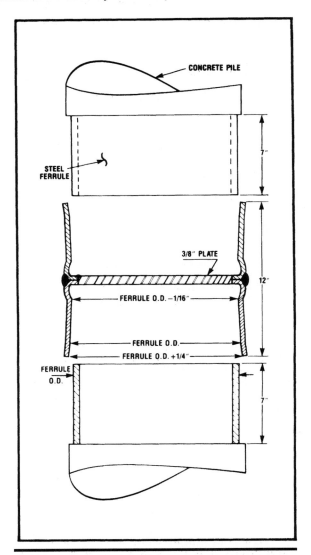

Concrete pile splices. (*Courtesy of International Construction Equipment.*)

One other important issue an estimator must consider is ground *displacement,* both subsurface and surface. The driving of piling displaces the earth at the pile tip.

The more lineal feet of piling that is driven in a confined close-spaced *pattern* or *cluster,* the greater the displacement becomes. Thus, in some cases, the tougher the driving operation. The estimator must be fully aware of this in researching the soil borings. As will be discussed later in this section, the key to production pile driving is units per hour. Pile driving is a unique operation because of the unknown subsurface data.

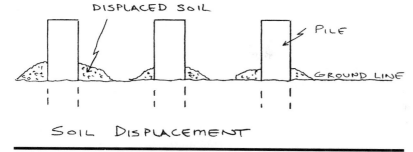

Concrete pile footer and ground displacement.

The second type of ground displacement is *upheaval*. This occurs during the driving of predominately flat-bottom piles. As the pile is being driven, the earth is uplifted around the perimeter of the pile.

This is something an estimator must take into consideration when pricing a pile-driving operation. A calculation must be made as to the amount and type of material that will be produced. Then a cost of removing that material from the already excavated pile area must be determined. All designs of foundations require a certain excavated elevation and template, and a specific dimension of the piling to be extended into the footer.

Any displacement of ground, whether it be over or under the template limits, must be corrected prior to the permanent foundation construction.

Cast-in-place concrete piling

Cast-in-place concrete piles can be achieved by several methods, normally chosen by the designer. The various types consist of concrete cast in place, either predrilled augered holes or steel shells drilled or driven to the required bearing depth. Both types must conform to the owner's specifications.

The first method, *drilled* or *augered* holes, is constructed to the minimum penetration detailed on the construction drawings or by the owner's specifications. In most cases the contractor must conform to a detailed plan of alignment and location as shown on the plans. Pilings must be plumb, vertical, and of uniform wall diameter for the entire length. In some instances, however, pilings are installed on a *batter* for lateral support. This will be discussed later in this section.

An important incidental factor an estimator must take into account with this type of operation is subsurface groundwater. All water encountered in the hole must be removed prior to the concrete's being placed. The standards and specifications normally allow some water even in what is considered a dry hole; this the estimator must verify based on specific conditions. Also, any loose material at the bottom of the hole must be removed to assure proper bearing.

In some instances, it is not possible for the configuration of the hole to remain intact during augering operations. This is due to unstable ground conditions. In this case a steel casing may be considered by the contractor or specified by the owner. Again, this must be acknowledged by the estimator in researching the soil borings and *differing site conditions* of the project.

Abnormal subsurface conditions usually can be overcome. If casing is required, it may be left in place or removed during the concrete placing operation. If left in place, normal specifications follow certain procedures as to the location of the bottom of the casing relative to the top of the concrete and a predetermined top elevation relative to the permanent foundation. An estimator must be aware of the exact procedures involved along with labor, material, and equipment needed to properly cost this operation.

One other type of uncontrollable situation is the impracticability of removing groundwater. In this scenario, specifications allow for the placing of the concrete either by first pouring *seal concrete* and then dewatering or by complete placing of the concrete by an approved, underwater method, using a *tremie*. In this case, the concrete must be placed without trapping water within the concrete.

A complete section on underwater concrete follows, but estimators must be fully aware of the encounters that lie before them in pricing a piling operation.

After the hole is deemed approved for concrete placement, normally the reinforcement cage, if required, is installed. This must be done immediately prior to the placement of the concrete to prevent any loose debris or material from contaminating the bottom of the hole. The cage is normally *pretied* and then installed as one unit (see section on reinforcing).

Always keep in mind that with any operation being performed in or around the excavated hole, the integrity of the entire unit must remain intact. The cage must be properly installed and supported to uphold this. The estimator must recognize appropriate methods of pricing this operation to conform to the specification. Every project has specific and concise specifications regarding these and all piling operations, and the estimator must thoroughly research this.

Augered holes are normally designed and used for an ease of penetration for the pile. The prebored hole allows the pile to be lowered to an elevation without causing damage to the pile. This application is usually performed in areas or situations of obstructions or differing conditions within a given elevation of the piling operation. Another use for preboring or augering is in the aid of a template, aiding the pile to be held in a vertical position.

A second method of constructing cast-in-place piling is by using thin-walled, spiral or longitudinal welded, corrugated coil *steel shells*. This method can be obtained by driving either with or without a *mandrel*. In either case, care must be taken not to damage the wall, destroy the configuration, or collapse or buckle the shell. This is a tapered pile based on the design, and usually the owner prepares specifications regarding the *modulus* design of the shells. Again, as in any piling operation, the piles must be driven to a proper and desired depth and bearing.

As discussed with augered piling, the inside of the shells must be kept clean of all loose material, debris, and water. If not, contractors will be required to perform this task at their own expense prior to concrete placement. Normally, an estimator will take a factor into account for a percentage of this work (see section on takeoff). This type of pile requires end closures. The contractor must maintain during the entire driving operation that the tip and the sidewalls remain intact and not damaged. It is also recommended that a protecting driving head or ring be used to protect the top of the pile from becoming damaged from the pile hammer's impact.

A mandrel system of driving does somewhat protect the shell. This is a telescopic solid tube that is inserted inside the shell. The driving is done with this instead of actual driving on the perimeter of the shell. It has a series of steps and tapers which align to the shape and configuration of the shell, thus keeping equal pressure on both sides of the shell to the required tip elevation.

Once the shell is driven, the mandrel is removed and the shell is ready for concrete.

An important specification is enforced with this type of operation, the fact that ground displacement is an item of concern. During the driving operation, especially on a close pattern, ground displacement may cause collapse to an adjacent pile.

The ground specifications address this issue and require the contractor to have multiple piles driven to resistance and inspected prior to the placement of any reinforcing or concrete. The same is true after the concrete is placed. The specifications limit the contractor from driving any new piles adjacent to and within a certain distance of any piles that have been poured with concrete until the concrete has obtained a specified strength. This is done to maintain the integrity of the design.

This type of piling is usually used for soil conditions that do not allow skin friction bearing but require more of a tip bearing.

Pipe piling

Pipe piling is of a thick-walled, spiral, or longitudinally welded pipe that is of a design to drive independently. Depending on the design, it varies in wall thickness and diameter. It too will be filled with concrete but normally not reinforcing steel. It has a flat bottom, closure plate, or a *conical point*.

As in cast-in-place piles, the interior of the pipe pile must be kept free of debris and water. Extensive driving or not having the proper size hammer can collapse or distort pipe piling. Again, the estimator must be fully knowledgeable of this in preparing an estimate and sizing of the hammers and driving techniques needed.

Ground displacement is still an issue with pipe piles, as the flat-bottom pipe will displace more soil as adjacent piles are driven in the foundation. Estimators must recognize this issue in preparing bids in regard to differing *production cycles*.

Pipe piling can be manufactured in various lengths but is usually governed by local and state laws of overlength loads. Again, this must be considered in the bid preparation. Long piling, as detailed on the plans, may require

Pipe pile installation. (*Courtesy of International Construction Equipment.*)

splices. If splices are needed, they must be of proper owner specifications and must be priced accordingly.

There are several methods of splicing pipe piling: *butt welds, mechanical spliced ring,* or a welded *chill ring.* The specifications will detail the appropriate one or give optional choices.

As discussed before, subsurface obstructions, damaged piles, and adverse driving conditions must be considered by the estimator. Once the pipe piles are driven to the proper tip elevation, bearing, and location, they can be prepared for the placement of concrete.

After the concrete is placed, as in the shell piles, no driving can be done within a specified distance until the concrete has obtained design strength.

The type of design that pipe piles would normally be used for is areas of soils that have virtually no upper-strata bearing value, such as swampy, peat, or unstable soft dry. In these areas, it is dependent on a lower-lying stratum for tip bearing or a definite landing point for the pile tip.

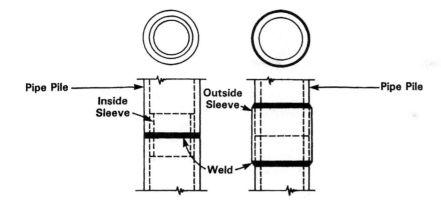

SPLICES FOR PIPE PILES

PATENTED SPLICER FOR PIPE PILE

Pipe pile splice. (*Courtesy of Associated Pile and Fitting Corp.*)

Pipe piles can be driven by conventional air/steam or diesel hammers of the appropriate size and *ram weight*.

Based on specific job conditions, preboring or jetting may be required to assist in the operation. This must be given careful consideration, as it may be a costly item. The owner's specifications will detail this. In the previously discussed types of piling, cast-in-place shell and pipe, the placing of the concrete is a critical procedure. Procedural care must be taken not to *segregate* the concrete. The concrete must be placed in a suitable manner to maintain the integrity of the mix. It must be placed with a tube or tremie, which is extracted as the pour progresses.

Normally, specifications require a *grout* mixture to be placed at the bottom of the pile first to act as a cushion for the normal mix of concrete specified. Usually, the top 10 feet of the pile is required to be consolidated by an acceptable vibratory method.

If placing concrete in an underwater condition, the tremie or fill pipe must maintain its tip within the concrete at all times to avoid trapping water within the pile (see underwater concrete section).

Steel H piles

Steel H piles are of a rolled steel H cross section referred to as a structural shape. The design of an H pile varies in weight per foot and size of section.

The project specifications will dictate what type and size are required and

Bethlehem H-Piles

Section Number	Weight per Foot	Area of Section	Depth of Section	Flange Width	Flange Thickness	Web Thickness	Axis X-X I_x	Axis X-X S_x	Axis X-X r_x	Axis Y-Y I_y	Axis Y-Y S_y	Axis Y-Y r_y	Surface Area
	lb	A in.^{2}	d in.	b_f in.	t_f in.	t_w in.	in.^{4}	in.^{3}	in.	in.^{4}	in.^{3}	in.	ft^2/ft
HP14 ×	117	34.4	14.21	14.885	.805	.805	1220	172	5.96	443	59.5	3.59	7.11
	102	30.0	14.01	14.785	.705	.705	1050	150	5.92	380	51.4	3.56	7.06
	89	26.1	13.83	14.695	.615	.615	904	131	5.88	326	44.3	3.53	7.02
	73	21.4	13.61	14.585	.505	.505	729	107	5.84	261	35.8	3.49	6.96
HP12 ×	74	21.8	12.13	12.215	.610	.605	569	93.8	5.11	186	30.4	2.92	5.91
	63	18.4	11.94	12.125	.515	.515	472	79.1	5.06	153	25.3	2.88	5.86
	53	15.5	11.78	12.045	.435	.435	393	66.8	5.03	127	21.1	2.86	5.82
HP10 ×	57	16.8	9.99	10.225	.565	.565	294	58.8	4.18	101	19.7	2.45	4.91
	42	12.4	9.70	10.075	.420	.415	210	43.4	4.13	71.7	14.2	2.41	4.83
HP8 ×	36	10.6	8.02	8.155	.445	.445	119	29.8	3.36	40.3	9.88	1.95	3.29

Steel H-pile details. (*Courtesy of Bethlehem Steel Corp.*)

H-pile point. (*Courtesy of Associated Pile and Fitting Corp.*)

specify the steel composite needed. They are driven as to the planned location, depth, and bearing, or in some cases, refusal. Refusal occurs when piling is designed to be driven into bedrock as a bearing or landing point, at which time the pile is refused any additional depth without damage to the pile. As in the other discussions on piling, any damaged or bent piles are replaced at the contractor's expense.

H piling has a high skin friction value, meaning that it picks up bearing on its sides as it is driven. Still, the final determination is usually made by a load test.

In the case of piling being driven to rock or refusal, pile points are normally specified. These protect the pile tip from damage and are designed for rock penetration.

These are affixed to the pile by means of welding. This, again, is a cost the estimator must recognize as, in most cases, these are not a direct item. In some soil conditions, as discussed, obstructions are encountered that prevent the accurate driving of the piles to the required tip or bearing. These include boulders or unsound rock strata, tree stumps, etc. In some cases, the soil borings of the project depict this and should be researched and accounted for by the estimator.

All piling must be kept clean and free from damage during all handling, transporting, storage, and unloading. A well-planned *laydown yard* at the job site is always an asset, but in the eyes of the estimator, it is an item that must be priced out.

H piles, like pipe piles, often require greater driven lengths than can be easily transported to the project site. In this case, again, splices are required. They can differ also from a butt-welded splice to a mechanical splice or a welded chill ring.

This type of piling is usually designed for areas where penetration into a rock stratum is desired or in bony soil where boulders may be encountered. In dense clay and sand areas, H pile can also be designed with skin friction bearing values. H pile is a widely used type of piling.

H-pile splice. (*Courtesy of Associated Pile and Fitting Corp.*)

The driving equipment normally used for H piles are conventional hammers such as air/steam or diesel. In some soil conditions, however, a *vibratory hammer* may be considered because of the coning effect or *resistance* problems. This will be outlined in your specific project requirements but must be given careful consideration as to the costing of the operation and the overall production achieved.

Pile cutoffs

Pile cutoffs are a normal occurrence in driving operations for virtually any type of piling and should be accounted for by the estimator. Depending on the type of pile, there are procedural methods in performing the cutoff. The cutoff is due to the fact that when piling is driven, the exact point of bearing and final tip elevation are somewhat unknown, at least down to the inches involved owing to the exact landing point of the pile. The top cutoff elevation is required based on dimensions of the piling in the footer or foundation designated by the designed plans.

The cutting procedures for any type of steel piles, whether they are H or pipe, are normally by the use of a cutting torch. However, a steel cutting saw could be used. The *cutoff elevation* is determined by the top of pile elevations shown on the plans and is marked on each of the piles. In the case of cast-in-place and pipe piles, the final cutoff elevation of the steel casing or steel pipe should be determined and made prior to the placement of the concrete.

The procedures for cutting prestressed concrete pile are by the use of a concrete cutting saw. The same procedure for establishing the final top of the pile elevation is followed as in steel piles. Once this is established, a single saw cut is made around the perimeter of the pile and through the prestress strands. It is imperative to make sure the cut continues through the strand.

Once this is completed, the pile cutoff piece can be snapped off and discarded. Some specifications detail the prestress strands of the concrete pile to extend into the foundation. In this case, the sawing operation would not cut through the strands because they must remain in place. In other words, the contractor would only make a saw cut into the perimeter of the pile and within approximately $\frac{1}{2}$ inch of the strands.

This would enable the contractor to have a clean-cut edge at the point of "top of concrete" and remove the excess of concrete, thus exposing the strand to become part of the foundation pour. This specification, however, must be realized at the time of bidding to compensate for the additional pile footage that would be required and wasted in order to retain the strands. (See p. 215.)

This may or may not be a compensated item but incidental to the contractor. In the case of prestressed concrete piles, the cutoff piece is no longer usable owing to design properties, but since normally the contractor is given the exact lengths of pile to cast, the owner's provisions usually compensate the contractor a specified percentage of the unit price.

In the case of steel piling, the cutoff could be spliced onto another pile and reused or salvaged (see splice section).

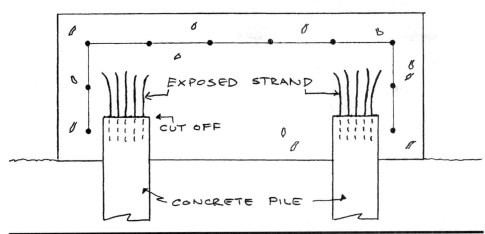

Concrete pile exposed strand.

Regardless of the type of pile or the method of cutoff, the pile must be landed to its final bearing and/or tip elevation prior to commencing the cutoff.

In some instances, the piling is driven to its proper cutoff elevation without performing a cutoff procedure. However, I would not recommend anticipating this occurrence 100 percent of the time.

Pile buildups

Pile *buildups* occur in prestressed concrete piles. They happen when a concrete pile is driven past the required cutoff elevation but achieves bearing within the exposed area of the foundation. There are specific specifications for an allowable length of a buildup and the method of performing the work.

The normal procedure is to chip away or remove a specified amount of concrete from the existing pile, exposing the reinforcing and strands. As in a cutoff method, the perimeter saw cut must be perpendicular to the axis of the pile. Reinforcing, as used in the pile or otherwise specified, is then securely fastened to the exposed steel. The top of the existing pile must then be treated and prepared with a bonding agent to adhere to the new placement. The buildup is then formed and poured to its proper and final elevation. In any case, once a build is attached, no redriving of the pile is permitted.

Splicing

The splicing of piling is an allowable standard to permanent lengths of piles longer than those which can be delivered or manufactured. There are many types and methods of splices for all types of piling. A splice is needed simply for when the ordered length of pile is insufficient to obtain the specified tip elevation or bearing value. In this case, an extension of the same cross section of the pile must be added, and a manner of splicing must be used.

Steel piles can be spliced by a variety of methods. A *butt weld* is the most common. In this case, each end of the pile is prepared in a manner suitable to the specifications and welded directly together. A *chill ring* is a collar adapter inserted between the two pieces of steel piling and then welded to each piece. A third type of splice is a *mechanical splice*. This type is an adaptive collar inserted between the two pieces, and is held in place by friction; no welding is needed. As in the latter two types of splices discussed, they are the same shape and cross section as the pile.

(1)

H pile (1) and pipe splice. (*Courtesy of Associated Pile and Fitting Corp.*)

These types of splices can be used for H pile, casing, pipe, and shell and are a practical way to continually add length to piles prior to achieving tip elevations or bearing.

Prestressed concrete piles can be spliced by dowelling. Steel dowels are drilled and epoxied into one of the piles.

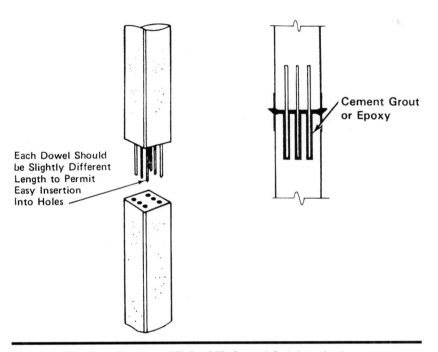

Concrete pile splice. (*Courtesy of Federal Highway Administration.*)

Then adjacent holes are drilled into the second pile. The two pieces are then epoxied together with a metal shell splice attached around the perimeter. This is the best proven method to splice these types of piles.

Timber piles can be spliced by means of *strapping*. The strapping must encircle the pile and be tensioned as tightly as possible. Another method is by using steel strips and bolting. This is similar to a mechanical splice.

The exact type of splice recommended and performed will be specified by the owner. The estimator must research this and understand it. With some types of splices, additional equipment or *templates* will be needed for support and handling. This, as in any operation, must be appropriately priced out.

Pile points and shoes

Points and shoes are adapters that fit onto the bottom of the piles. These are used for protection to the pile, enclosing the bottom of the pile, and for rock cutting in some cases.

218 Takeoff and Cost Analysis Techniques

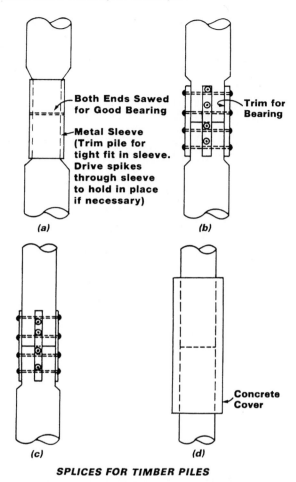

Timber pile splice. (*Courtesy of Federal Highway Administration.*)

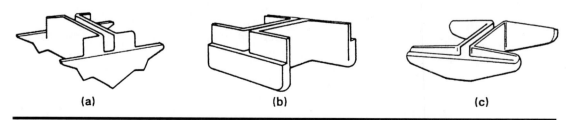

H pile and pipe points and shoes. (*Courtesy of Associated Pile and Fitting Corp.*)

Points and shoes are normally welded onto steel piles. In the case of rock points, they are designed to aid the pile in the landing or seating of the pile into the bedrock.

The specific points or shoes required along with the payment schedule will be detailed by the owner.

Bearing and tip elevation

Piling is used where the soil conditions of a specific job do not allow the safe bearing or load of a structure to be placed on it. In this case, piling is driven to a subsurface stratum capable of supporting the structure along with its designated loads. Two values are normally taken into account in the design of piling, a specified *bearing value* or a specific *tip elevation.*

A bearing value is the absolute minimum value of tons that a single pile will support. Based on prior soil boring data, the piling design is derived by the weight of the structure, as designed with live loading, against the number of piles needed to support it to the subsurface stratum. Bearing is derived by the tip of the pile landing itself on a specified material stratum or by the perimeter of the pile obtaining *skin friction,* or by a calculation of both. Again, this is based on actual soil data. The tip elevation method is used when the pile itself is designed to be driven to a specified depth, either for a certain material stratum for bearing or for *lateral stability.* In some soils, bearing can be obtained fairly quickly at a shallow elevation, but the pile has none or very little lateral stability. In this condition, the pile must be driven deeper to obtain this and may require jetting to assist the extra depth so as not to damage the pile from excessive driving force.

There are normally two methods of obtaining pile bearing information, by a test pile or by formula.

A *test pile* is a pile being of the same cross section as that of the permanent piles and being driven at a predetermined location on the project. This may be a requirement by the owner or an option of the contractor. A requirement in the driving of the test pile is not only that the pile be the same as the permanent, but that the specific *pile hammer* used to drive the test pile also be used for all the permanent piles related to the test pile. This is because of the information in calculating the hammer calibration and *blows per foot* that are derived from this operation. In some areas, a *wave equation* is also performed to conduct tests on the properties of the pile and the forces the hammer is applying. The test pile can be driven either to refusal or to a tip elevation specified. This will establish an absolute determination by the owner as to the reaction of the piling design in the specific soils.

A formula can be applied for the driving operation. This will assure that the actual bearing being produced is within the guidelines of the preestablished bearing forces rated for the hammer. This method is also used on permanent piles. To obtain a bearing value by formula, the following standard procedure is used:

Estimating bearing capacity

When load tests are not required, the bearing capacity of each pile shall be estimated by use of one of the following formulas, as appropriate:

1. For gravity hammers

$$R = \frac{2WH}{S + 1}$$

2. For single-acting steam or air hammers and for diesel hammers having unrestricted rebound of the ram

$$R = \frac{2WH}{S + 0.1}$$

3. For double-acting steam or air hammers and diesel hammers having enclosed rams

$$R = \frac{2H(W + AP)}{S + 0.1} \quad \text{or} \quad R = \frac{2E}{S + 0.1}$$

where R = safe bearing capacity, lb
W = weight, lb of striking parts of hammer
H = height of fall, ft
A = area of piston, in^2
P = pressure, psi, of steam, air, or other gas exerted on the hammer piston or ram
E = manufacturer's rating for foot-pounds of energy developed by double-acting steam or air hammers, or 90 percent of the average equivalent energy, in foot-pounds, developed by diesel hammers having enclosed rams as evaluated by gauge and chart readings
S = average penetration, inches per blow, for the last 5 to 10 blows of a gravity hammer or the last 10 to 20 blows for steam, air, or diesel-powered hammers

The above formulas are applicable only when:

a. The hammer has a free fall.
b. The head of the pile is not crushed.
c. The penetration is reasonably quick and uniform.
d. There is no sensible bounce after the blow.
e. A follower is not used.

Twice the height of the bounce shall be deducted from H to determine its value in the formula.

In case water jets are used in connection with the driving, the bearing power shall be determined by the above formulas from the results of driving after the jets have been removed.

Definition of hammers

Air/steam hammer. There are two types of external power-driven hammers, single-acting and double-acting. A single-acting air/steam hammer functions by the air pressure entering the hammer housing and raising the striking mechanism to a trip lever which releases it to fall by gravity and strike the pile. This type of hammer energy is rated by computing the sum of the weight of the ram or striking mechanism times the height the ram falls. These formulas vary on different sizes of hammers.

The double-acting air/steam hammer functions by using air/steam to raise the striking mechanism, using the same theory as the single-acting. But in place of a gravity fall, this type of hammer applies air/steam pressure to assist in the falling of the ram by applying the pressure to the top of the ram or piston. The method gives additional energy to the striking weight on the ram by adding downforce. This type of hammer energy is rated by computing the sum of the weight of the ram or striking parts, the height the ram falls, plus the amount of additional pressure being effectively added to the piston. These types of hammers are designed by specific blows per minute to develop their specified foot-pounds of delivered energy. Great care must be taken to follow the manufacturer's recommendations properly for compressor size, inlet hose size, cushion material, and other factors which will control the performance of the hammer to its utmost capacity and design.

In both types of hammers, the ram or piston impacts a driving head which then transfers the force to the pile in foot-pounds. A foot-pound is the energy produced by the hammer by the ram weight times the distance of the fall in feet.

In the driving of piles, the hammers produce *set* in the pile. This is what is used to determine the proper function of the hammer in relationship to the bearing values and proper penetration ratios by blow count and energy distributed. When driving heavy piles, such as concrete, one must take into account the additional weight of the pile. This should be used only when the weight of the pile exceeds the weight of the ram.

These methods will produce the required blow count per foot at the required bearing elevation.

Any other types of formulas or methods used will be determined by specific projects.

Another way to substantiate either the above formula or the actual bearing obtained is by a *load test*. A load test is a specified amount of weight applied to a pile. This test can be performed as either a test pile or a permanent pile. A load test is performed by a loading apparatus constructed at the site of the test pile.

The two most frequent methods are that in which material of weight greater than the tons required is stacked above the pile, or that in which ten-

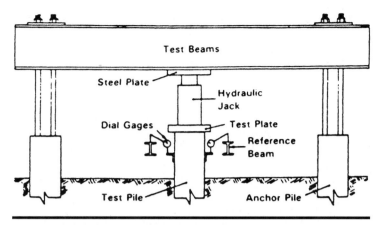

Load test stand. (*Courtesy of Federal Highway Administration.*)

sion or anchor piles are driven to withstand the uplift of the tons of downforce required by the test.

In both types of apparatus, a jack is placed between the jacking beam and the pile to be tested. Downforce is then applied to the pile in increments and procedures set forth by the owner. Any movement is then monitored in the pile as the force is applied, up to the required tons of testing force. This usually exceeds the designed bearing tons for a safety factor. Once the total force is reached, the pile is further monitored for a specified time period for movement. This eliminates any rebound effects that may occur.

These procedures are detailed in the specifications as for the costing procedures to be done by the estimator. Each type of piling has differing properties and is designed for different applications.

When safe loads or safe bearing by design is not found to be achieved by these methods, longer and/or additional piles may be required.

Procedures in the driving of permanent and/or test piles must be fully understood by the estimator. Conditions, methods, and specifications vary between projects, and procedures of installation differ tremendously in cost and duration.

Handling. The handling of piling is of utmost importance. It is a structural component and care must be taken with regard to damage. Piling is usually delivered in bulk loads. Thus unloading, stockpiling, and rehandling are common. The laydown and storage yard must be such that the piles can be stacked in an orderly manner and sized. It must be kept easily accessible for both the unloading equipment and the reloading equipment. The piling must be supported so that no deflection or cracking occurs. If the pile is *coated,* care must be taken to prevent damage to the coating, by both marring or, in some cases, sunlight. The piling is the contractor's responsibility until it is driven in the completed form of the specifications.

Driving methods. Piling, as shown on the drawings, must be driven at the location shown and to the designed bearing and tip elevation and the required specifications. Whether they are *vertical piles* or *battered piles,* they must adhere to the tolerances detailed. This is the requirement of the contractor, who must perform so as not to damage the pile.

Methods of driving may be assisted by means of *water jetting* and/or *preboring holes*. Water jetting is a method of displacing the soil material adjacent to the pile tip by high-pressure water forced through a single- or double-tipped nozzle. This eliminates the coning effect built up by hard driving and allows the pile to reach a desired tip elevation without damage. At a specific elevation the jetting operation must cease and the pile be driven by the hammer to bearing and final penetration. As the jets are removed from the hole, they are used to encase the pile with the disturbed material, thus allowing for a seating of skin friction. The water pressure must be applied at a constant rate so as to control the constant alignment of the pile.

Preboring is another method either used to assist the driving operation or required to remove specific material that would prohibit or hinder the design and driving. Preboring is usually done to a specified depth where driving would resume normally or to the point where the hammer would seat the pile.

Pile prebore with pile hammer and leads. (*Courtesy of International Construction Equipment.*)

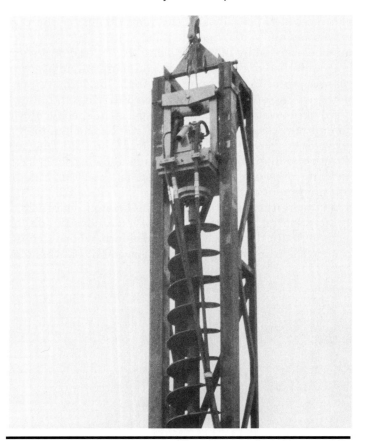

Pile augering. (*Courtesy of International Construction Equipment.*)

Preboring is done with an auger and eliminates the soil friction of the pile in that location. Some specifications require some kind of backfill to fill the void created.

Augering is a method which, in most cases, fully seats a pile to its final elevation such as some shell pile designs, *caissons,* and cast-in-place piles. The auger excavates to a predetermined tip or bearing elevation.

In all driving operations, the driven piling and equipment is supported by means of *leads, templates,* or both. The leads are supported from the crane. They can be either *fixed* or *swinging.* They are needed for alignment of the pile and for stability.

Templates are used for ground support and proper alignment of location at the point of pile entry. The pile must be constantly supported in line and in position while being driven. Guides and braces are also used for rigid lateral stability to ensure that no damage is done to the pile. These types of supports are also required for battered piles on which, during driving operations, much force is placed on the horizontal axis of the pile.

Pile leads. (*Courtesy of International Construction Equipment.*)

Another attachment or fixture that can aid in alignment and support is a bottom spotter attached to the leads. This is used to control the area at which the pile is penetrating the soil.

Again, these procedures must be researched and fully understood by the estimator.

Hammers. As discussed, a *pile hammer* is the tool used to produce the energy needed to force the pile into the ground. They are powered by air, steam, and diesel and in some cases are assisted by other means. Depending on the pile, size, conditions, and specifications, a wide variety of sizes of hammers are manufactured in regard to performance and *ram weight*.

An air/steam hammer has an external source delivering the air/steam to it. They are of a somewhat simple design in that the air/steam is used to lift the ram used to drive the pile. There is not a great deal of moving and complicated parts. Unlike diesel hammers, the air/steam is not dependent on *resistance* of the pile to fire the hammer. As long as the air/steam source is supplying

Pile template. (*Courtesy of Kiewit Construction Group.*)

the volume needed, the hammer will function. Different-sized hammers require specific volumes of pressure based on the manufacturer's recommendations. This is a specific requirement that controls the formulated bearing value of the hammer.

Diesel hammers. Diesel hammers are a self-sufficient unit that have their own power source. They work on the principle of internal combustion firing that raises the ram. Once manually started by a trip mechanism with the crane, these hammers require resistance from the pile to continue to operate. Both types of hammers can be operated from the ground.

The principle of achieving the desired bearing can be derived from either type of hammer; it is usually the preference of the contractor as to which type is used. Logistics, certain job conditions, and specifications do, however, sometimes dictate the type and should be researched by the estimator. Different hammers do control productivity.

Based on the formulas of bearing values and test pile results, a pile hammer must be calibrated and maintained in accordance with these specifications.

To size a hammer initially, it must be capable of developing the minimum energy rating required by the bearing formula value. Normally, for air/steam hammers, the weight of the ram, or striking parts, must be at least one-third the weight of the pile being driven, but some specifications vary and have

Bottom spotter. (*Courtesy of International Construction Equipment.*)

minimum and maximum requirements. Diesel hammers, owing to the variance of power supplied by the throttle, are size-based or energy distributed.

The information required by the owner for hammer approval usually is:

- The weight of the ram
- Diameter and length of the ram
- *Driving head* and *bonnet* or *anvil* weights
- *Capblock* and *cushion material* types and specifics
- Net weight of hammer
- Piston areas

This information is compiled and used to formulate an analysis of the hammer's functions and capabilities. The actual driving is then measured in blows per foot initially to blows per inch at bearing level, thus picking up a

ICE MODEL 40S DIESEL PILE HAMMER

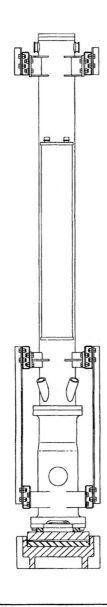

WORKING SPECIFICATIONS
- Rated energy .... 40,000 ft-lbs (5,530 kg-m)
- Minumum energy .... 16,000 ft-lbs (2,212 kg-m)
- Stroke at rated energy .... 10' (3.04 m)
- Maximum obtainable stroke .... 10'2" (3.1 m)
- Speed (blows per minute) 38-55
- Bearing based on EN formula .... 200 tons (181 tons)

WEIGHTS
- Bare hammer .... 7,500 lbs (3,402 kg)
- Ram .... 4,000 lbs (1,814 kg)
- Anvil .... 575 lbs (261 kg)
- Typical operating weight with cap .... 8,600 lbs (3,900 kg)

CAPACITIES (adequate for normal day)
- Diesel fuel tank .... 11 gal (42 l)
- Lube oil tank .... 5 gal (19 l)

DIMENSIONS OF HAMMER
- Width (side to side) .... 20" (508 mm)
- Depth .... 29" (737 mm)
- Centerline to front .... 13 3/4" (349 mm)
- Centerline to rear .... 15 1/4" (387 mm)
- Length (hammer only) .... 15'9" (4.80 m)
- Operating length (top of ram at max. stroke to pile) .... 27'2" (8.28 m)

DIMENSIONS OF LEADS
- Face width of guide rails .... 4-8" (100-200 mm)
- Distance between guide rails .... 20 1/2-26 1/2" (520-670 mm)

INTERNATIONAL CONSTRUCTION EQUIPMENT, INC.
Corporate offices: 301 Warehouse Drive, Matthews, NC 28105, USA
800 438-9281 & 704 821-8200   FAX 704 821-6448   Telex 572385 ICE INTL

Hammer specifications. (*Courtesy of International Construction Equipment.*)

Air/steam hammer. (*Courtesy of Vulcan Iron Works Inc.*)

high resistance. This must be arrived at without imposing damage or driving stress to the pile. Once a system is approved, it must be maintained throughout the operation.

One of the most important components of the hammer is the cushion material and capblocking. This has a great impact on the effectiveness of the hammer and driving results. These are items that take the impact of the striking parts of the hammer and protect the end of the pile from damage. Depending on the life (the number of piles that it lasts), an estimator must cost this item as a supplied material to the job. Based on driving conditions, it could be a costly item.

An air hammer provides constant energy, and a diesel provides variable energy based on its specifications. It is recommended that specific project requirements be evaluated and individual pile hammer manufacturer recommendations be viewed. It must be maintained that any pile damaged during the entire operation will be replaced at the contractor's expense.

230 Takeoff and Cost Analysis Techniques

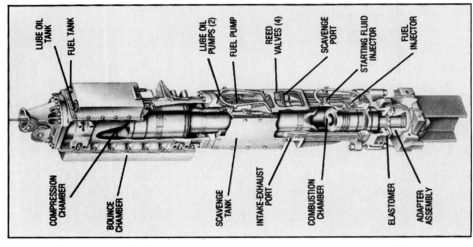

Diesel hammer. *(Courtesy of International Construction Equipment.)*

Diesel hammer. *(Courtesy of International Construction Equipment.)*

Vibratory driver/extractor. (*Courtesy of International Construction Equipment.*)

Item: Sheet Piling, Permanent and Temporary

Specifications

Sheet piling is of a structural shape, designed to contain earthen material or water, for either excavation or backfill purposes. It can be driven to refusal or to a specified tip elevation. It can be used to obstruct and retain water from a certain area. This method is used widely in *cofferdam* and *cell* construction in water-related operations. Sheet piling is also needed for areas of excavation when poor soil conditions exist. Used in this way, the sheets retain and support ground or support an adjacent structure and permit excavation to be performed in a tight and restricted area.

Sheet piling is designed and manufactured in a wide variety of shapes, sizes, and strengths or *modulus sections*. Depending on its desired use and purpose, a designer or user of sheet piling will choose the appropriate type.

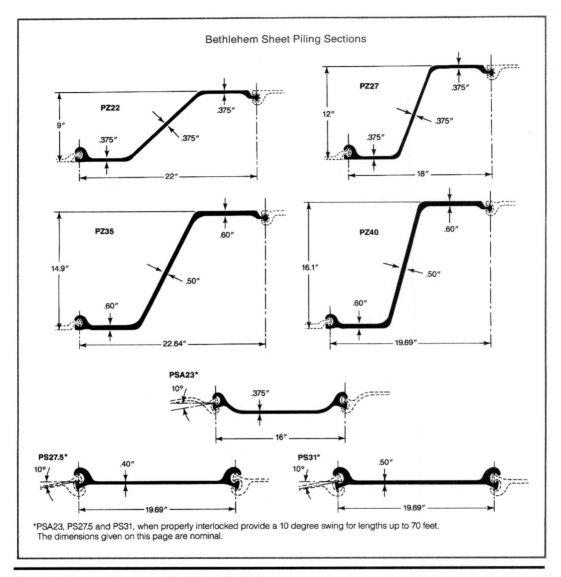

Sheet pile designs. (*Courtesy of Bethlehem Steel Corp.*)

    Sheet piling can be driven with either conventional hammers or *vibratory hammers*. Depending on the soil and project conditions, an appropriate hammer and method must be chosen. A vibratory hammer is usually the most common method. This type of hammer is designed with an eccentric motion to vibrate the piling into the ground. It is very useful in sand conditions. It is designed with a gripperlike feature that actually holds the pile. This can lift and place the sheets into position to drive.

Properties and Weights

Section Designation	Area sq in.	Nominal Width, in.	Weight in Pounds		Moment of Inertia, in.^{4}	Section Modulus, in.^{3}		Surface Area, sq ft per lin ft of bar	
			Per lin ft of bar	Per sq ft of wall		Single Section	Per lin ft of wall	Total Area	Nominal Coating Area**
PZ22	11.86	22	40.3	22.0	154.7	33.1	18.1	4.94	4.48
PZ27	11.91	18	40.5	27.0	276.3	45.3	30.2	4.94	4.48
PZ35	19.41	22.64	66.0	35.0	681.5	91.4	48.5	5.83	5.37
PZ40	19.30	19.69	65.6	40.0	805.4	99.6	60.7	5.83	5.37
PSA23	8.99	16	30.7	23.0	5.5	3.2	2.4	3.76	3.08
PS27.5	13.27	19.69	45.1	27.5	5.3	3.3	2.0	4.48	3.65
PS31	14.96	19.69	50.9	31.0	5.3	3.3	2.0	4.48	3.65

**Excludes socket interior and ball of interlock.

Note:
PS27.5 and PS31 interlock only with each other. All Bethlehem Z sections interlock with one another and with PSA23.

Interlock Strength:
PSA23, when properly interlocked, develops a minimum ultimate interlock strength of 12 kips per inch.

Excessive interlock tension results in web extension for section PSA23. Therefore, the interlock tension for this section should be limited to a maximum working load of 3 kips per in.
When properly interlocked, PS27.5 and PS31 in V-STAR 50, develop a minimum ultimate interlock strength of 20 kips per inch.

Sheet pile properties. (*Courtesy of Bethlehem Steel Corp.*)

A vibratory hammer will, in some soil conditions such as a gumbo-type clay, experience difficulty with clay absorbing the vibrations, preventing the hammer from functioning properly. In this case, a conventional hammer such as a diesel must be used. When using a conventional hammer, leads or guides must be used to control the hammer. One such guide is commonly referred to as a monkey stick.

Sheeting is designed in Z patterns for use in wall-type construction with varying widths and thicknesses and flat-type patterns, also of varying widths and thicknesses, for use primarily in circular or cell-type construction.

Regardless of the driving depth where sheeting is intended for structural support, a proper *toe* must be maintained. The toe is the length of sheeting that extends downward below the bottom of the proposed excavation. This is needed for bottom stability to withstand the horizontal movement of the retained earth and prevent a kick-out or collapse motion. This extra depth must be accounted for by the estimator, whether or not it will be a direct pay item.

Another important aspect of sheet-pile construction is *bracing,* i.e., struts, wales, and ringers. Bracing is required to support the cantilevered portions of the sheet wall upward from the toe support. The bracing must be designed with substantial structural integrity, as in the sheeting, to support the sheeted wall. Horizontal pressures, whether they are earthen or hydraulic, can be very misleading. The main intent of sheeting is to provide a safe work area that otherwise may not be feasible. Bracing can usually be a substantial, costly, and timely item and must be thoroughly understood by the estimator, in both structural and conditional methods. Bracing must be able to support the intended design but also and foremost provide a suitable and productive work area without obstructing it. In some instances, bracing can be a very intense item depending on the forces involved. Bracing is normally constructed of steel structural shapes such as H beams, angles, and wide-flange girders. It must be securely fastened to the sheeting to guard against slippage and also properly welded, bolted, or both. A certified drawing should be made and checked by a professional engineer as to the integrity and concepts involved for both pricing functions and safety reasons.

Sheeting can be used in a temporary situation or a permanent mode. A temporary use is when the sheeting is driven for construction purposes and then extracted for reuse at a later time. In this case, the project requires a temporary protection and/or access for a specific area. With cases of permanent sheeting, the project requirements dictate that sheeting be left in place and not removed. The pricing of sheeting in regard to these two conditions is then greatly affected. An estimator must understand the specifications exactly and consider the intended use.

Cutoff, waste, and overage. In working with sheet piling, as in foundation pile, cutoffs and waste are involved. The reasons are basically the same. Sheeting is driven to either tip elevation or bearing. In the area of permanent sheeting, it is usually specified that a top and final elevation be maintained, and after the driving operation, the piles would need trimming. With temporary sheeting, it must be realized that the constant use and reuse of the sheets cause damage to the bottom, depending on the driving conditions, and to the top from the hammer impact. Though care must be taken in driving, the hammer does mar the top. At the time of each extraction, the top few inches of the sheets are usually trimmed off.

Sheets are designed and manufactured with *locks*. This is the key principle in the design as to interlocking or holding the sheets together as one unit. Also, the locks provide somewhat of a watertight situation when used in cofferdams. Care must be taken to protect these locks for the integrity of the sheets. If the sheets become damaged or not trimmed properly, the locks could fail or prevent the wall from interlocking together. This also refers back to the need of proper bracing to prevent buckling. The estimator must realize this waste and *damage* involved in pricing sheet piling in both permanent and temporary reusable conditions and should also consider either the most effective lengths to purchase for the specified conditions or the best reusable length for economic reasons. As in foundation pile, the equipment needs and labor force differ in sheet-piling construction, based on the specific needs of the project.

Templates

Templates are devices used to keep the pile in its proper alignment during the driving operation. They are used for several purposes: to keep the actual plan location in tolerance, to control or assist in the proper batter or vertical tolerances, and to assist in abnormal driving conditions such as water or unsuitable soils where physical staking may be a problem and areas where the actual driving may move the pile by either subsurface conditions or lateral resistance from a tight cluster situation. Also, in a jetting situation, a template keeps the piles in line while awaiting driving owing to the normally larger displacement of soils by the jets. They keep the pile in a true and desired alignment.

There are many ways to construct templates depending on job physical conditions, types of piles to be driven, or the nature of performance of the work. Templates must be thought out so as not to interfere with the permanent pile location. They can range from a simple alignment beam to an extravagant structure used not only for alignment but for lateral stability. In these cases, often false piles or anchor piles must be driven to hold the template in place.

Materials used for templates are generally H-pile beams and other steel, but wood mats also work on a simple alignment scheme.

Leads

Leads are a steel structure used to support and guide the hammer during the driving operation. They also support the pile within the leads and guide the pile for proper location and alignment.

Leads can be of generally two types, either swinging or fixed. Swinging leads are the most common and widely used because of their simplicity in handling, efficiency, and cost. They are fixed to the crane by a hoisting cable and are free at the bottom. This allows the leads to be rotated and aligned by boom placement and control. Once in position, the leads are placed in the ground by means of points at the bottom, or anchored to the template to control proper driving location. At this point, the pile can be either plumbed or battered to its proper vertical alignment.

Fixed leads are attached to the crane by numerous methods. They can hang from the boom tip like swinging leads but are attached at the bottom by means of a mechanical guide or spotter, or can be fixed to the boom tip and supported at the bottom back to the crane. These methods involve more equipment but give full control to the positioning of the piles, not only for location but for batters, both simple and compound.

In any kind of pile-driving operation, certain principles of crane safety must always be adhered to not only for full load and line capabilities but for forward, aft, and side pressures put on the boom by the leads (including the hammer and pile weights involved).

Measurement and payment

Measurement and payment for foundation piles are normally made by the unit linear foot driven. Concrete piles, however, are generally paid for by the ordered length established by the owner for production. Normally, concrete pile cutoffs are compensated to the contractor.

The measurements for piles driven are established for the length of pile to the cutoff elevation.

Test piles are normally measured by one complete unit, as are the load tests.

Cast-in-place piles are measured to include the cost of the steel pipe or shell and the concrete.

Normally, any type of piling and preboring or jetting will not be measured or paid for unless it is specified by the owner.

Splices and points are normally a conditional item as to any direct payment.

Sheet piling is calculated, generally, by the square foot of direct contact area as needed and type specified by the owner and may not include any toe or additional top extensions.

As in any type of construction specifications and requirements, an estimator must thoroughly read and understand the specific project requirements as to direct and incidental materials and payments. The contract requirements will outline each item measurement and payment.

Pile hammer formulas (Pile Driving Guide by Floyd M. and Irving Cleveland)

Set

$$\frac{\text{Inches of penetration}}{\text{Blow count}} = \text{set}$$

Drop hammer formulas. To find the required set for specified bearing value (normal piles):

$$S = \frac{2WH}{R} - 1.0 \tag{1}$$

To find the bearing value having obtained a certain set:

$$R = \frac{2WH}{S + 1.0} \tag{2}$$

where S = set of the pile under the last few blows, in inches (at point of near refusal or tip elevation)
W = weight of the striking parts, in tons
H = fall of the striking parts, in feet
R = safe or designed bearing value, in tons
1.0 = constant

To find the required set for bearing value (heavy pile concrete):

$$S = \frac{2WrH}{R} \times \frac{Wr}{Wr + Wp} \tag{3}$$

To find the actual bearing value having obtained a certain set:

$$R = \frac{2WrH}{S} \times \frac{Wr}{Wr + Wp} \qquad (4)$$

where S = set of the pile under the last few blows, in inches (at point of near refusal or tip elevation)
Wr = weight of striking parts, in tons
Wp = weight of pile, in tons
H = fall of striking parts, in feet
R = safe or designed bearing value, in tons

Formula to obtain weight of a square concrete pile:

$$Wp = \frac{0.521 b^2 L}{1000} \qquad (4a)$$

where b = side dimension of pile, in inches, and L = length of pile, in feet.

Single-acting hammer formulas (air/steam). To find the required set for a specified bearing value:

$$S = \frac{2Wh}{R} - 0.1 \qquad (5)$$

To find the actual bearing value having obtained a certain set:

$$R = \frac{2Wh}{S + 0.1} \qquad (6)$$

where S = set of hammer under the last few blows, in inches (at point of near refusal or tip elevation)
W = weight of striking parts, in tons
H = fall of striking parts, in feet
R = safe or designed bearing value (in tons)

To find the required set for bearing value (heavy piles concrete):

$$S = \frac{2WrH}{R} - 0.1 \frac{Wp}{Wr} \qquad (7)$$

To find the actual bearing value having obtained a certain set (heavy pile concrete):

$$R = \frac{2WrH}{S + 0.1\,(Wp/Wr)} \qquad (8)$$

where S = set of pile under the last few blows, in inches (at point of near refusal or tip elevation)
Wr = weight of striking part, in tons
Wp = weight of pile, in tons
H = fall of striking part, in feet
R = safe or designed bearing value, in tons

Double-acting hammers (air/steam and diesel hammer formulas). To find the required set for a specified bearing value:

$$S = \frac{2E}{R} - 0.1 \tag{9}$$

When the weight of the pile is to be considered:

$$S = \frac{2E}{R} - (0.1 + 0.01 Wp) \tag{10}$$

To find the actual bearing value having obtained a certain set:

$$R = \frac{2E}{S + 0.1} \tag{11}$$

When the weight of the pile is to be considered:

$$R = \frac{2E}{S + (0.1 + 0.01 Wp)} \tag{12}$$

where S = set of pile under the last few feet, in inches
Wp = weight of pile, in tons
E^* = energy rating of hammer, in foot tons
R = safe or designed bearing value, in tons

---

*It is a policy to reduce the manufacturer's energy rating by 10 to 15% to allow for losses in the hammer and power unit for wear and usage.

TAKEOFF QUANTIFICATION

Definition

Foundation piles are the structural members that create an interim support between the foundation of a structure and suitable bearing on which the structure is to be seated. The need for foundation piling occurs when the existing conditions do not meet the design for the load of the structure or allow the support required by the lateral requirements of the structure's design. These conditions can occur with land- or water-type structures and differ greatly depending on the actual site conditions. The intended design, economics, and types of foundation piling suited for the area, with respect to structural integrity and capacity limitations of design, are conditions that determine the types and size of piling required.

The purpose of foundation piling is to extend the load of the structure to a suitable and appropriate bearing elevation to support the structure. Piling not only gives vertical support but can also achieve stability for horizontal movement to lateral forces.

The pile configuration within a structure can consist of one pile to give the desired support and bearing, or a *cluster*-type configuration in which the

Pile site.

desired load bearing is distributed over a group of piles. This distribution can consist of even point loads over equally spaced piles of equal size and length, or varying load distribution over differently sized piles of length and spacing.

Foundation piling is driven or augered into the ground to a specified underlying layer of strata that is structurally able to support the full *dead* load along with the *live* loading, as designed, of the bridge structure. These types of foundation piling are referred to as *bearing piles,* or piling that is designed for the *tip* of the pile to be landed in a stratum or formation capable of supporting the structure.

When conditions are present that do not offer a suitable bearing elevation by either economics or general conditions of the demographic area, a *friction*-type design is used for the foundation piling. The soil offers no substantial end bearing. The friction-type bearing is achieved by the perimeter surface, or *skin,* of the pile seizing against the soils as the pile is driven into the ground. The pile is designed to a certain elevation, with respect to the properties of the soils and the pile, which will give the pile a minimum specific load capability.

Friction piles rely directly on the natural friction that develops between the pile surface and the soil stratum that exists, to develop design load bearing. In wet conditions, the load-bearing friction capacity requires time to develop. At the time of driving, the vibration draws the water toward the pile, which offers little friction resistance. Over a given time period, the skin friction is regained as the soils absorb the water, and the resistance is developed.

Piles that are directly cast into their specified position are *caissons*. These consist of augered holes drilled to a specified bearing elevation. This type of pile relies almost fully on a tip elevation of bearing. Depending on varying soils and underlying conditions, steel sleeves may be required to line the perimeter of the hole to decrease the chance of hole collapse. The augered hole, of a specified diameter, is then filled with structural concrete and usually a cage of reinforcing steel. These types of foundation piles are considered to be of the *cast-in-place* type rather than the premade structural type of steel shape, prestressed precast concrete, and wood.

Foundation site showing foundation pile, sheeting, and bracing.

Pile designs are based on four criteria. The first is the load or actual weight of the structure and the second the intended loads that will be permitted to be implied onto the structure, with an anticipated safety factor applied. This gives the designer a full calculated bearing load to the foundation area. Given the footing parameters of the foundation area, this is averaged out to a given pounds per square foot of load to the soils.

This information is then applied to the soil analysis that is the third criterion of design. This determines if the soils within the foundation area of the structure can fully support the implied and intended loads, both vertically and laterally. If this analysis is negative, or the foundation area is nonresponsive to the loads, the fourth criterion is met, the need for foundation piling to distribute the load bearing of the structure to a suitable elevation.

Once the need for piling has been determined, the next step is to ascertain the most economical type of pile that will produce the required results. Within the foundation area, specifically each bearing area, there is a need to support a given load. This is then broken down to each piece of pile and a specific load per pile. By ascertaining this information, a designer then determines (1) how many pieces of pile can physically be put in the area; (2) what type will perform under the given conditions, both structurally and demographically; (3) the size of piling required; (4) the type of bearing required; (5) the length of pile required.

The normal types of foundation pile are driven members and drilled caissons. Driven piles can consist of a variety of types. The most common are rolled steel H shaped: *precast, prestressed* concrete; concrete-filled *steel tube*; *shell*; and *timber* piles.

The drilled type can be either concrete-filled earth augered or earth augered with a steel liner or jacket and concrete-filled *caissons. Augered piles* are those where a requirement of nonfriction exists or a condition in the underlying soils prohibits the accurate driving of the pile member. An example is rocks or boulders that may deflect the direction and alignment of the pile. With this condition, a prebored hole of a specified diameter is drilled to a specified depth for the pile member to be inserted, after which a driving operation may be required to further drive the pile to its tip elevation and firmly *seat* the pile. Each of the types of piling contains its own structural characteristics and properties, and is used with certain criterion design conditions.

Introduction

Estimating and takeoff of foundation piling must be performed meticulously and accurately. Many conditions and variables exist that must be determined for a justifiable summation. The core components of foundation piling are controlled by the type of pile required and the *driving conditions* that exist at the site.

Foundation piling is installed and constructed by a variety of methods and equipment. Normally the first piece, and one of the main units, of equipment needed is a *lifting crane*. Cranes are available in a wide range of types, sizes, and capacities. These vary greatly in cost, with respect to both ownership and operation. As discussed in the equipment section, the equipment needs of a project should always be determined by the best-suited piece that is available to perform the task required, both economically and performancewise.

The lifting crane categories that would normally suit foundation piling work are conventional crawler and truck cranes. One determination for a specific type within a project is the actual site conditions present. Would a truck carrier crane be usable or would a crawler crane be better suited for maneuverability and stability? The second requirement for crane choice is lifting capacity. The full loads that will be required to be hoisted along with the reach the crane will have to perform will determine this. When calculating lift capacity, all components of weight must be accounted for to be assured they are within the general capacities of the crane. Cranes are rated by the amount of weight to be hoisted within a given radius of reach for the amount or length of boom and the boom angle at the time of hoist. (See crane section for capacity determination.)

Pile operation. (*Courtesy of International Construction Equipment.*)

The total hoisting weight should be a combination of the weight of the longest piece of pile, the *pile hammer* and accessories, *leads,* and *auger,* if required, and other necessary components of the crane.

The second piece of controlling equipment is the pile hammer. This unit is the driving mechanism that "pounds" the pile into the ground. As with cranes, pile hammers are available in a wide variety of sizes and types. Pile hammers are operated by either air/steam or diesel firing. In some instances, the type of operation desired is a matter of choice, but conditions exist, both driving and logistics, that warrant one over the other.

Hammers are manufactured by driving capacity produced by the energy developed within the hammer combined with the operating weight of the hammer. Pile hammers are categorized by the energy developed; the weight of the hammer; the *stroke* of the driving mechanism; and the weight of the *ram,* or the cylinder providing the impact to the pile. The *driving head* is a variable component depending on the type of pile to be driven. This is the cap that is seated over the end of the pile and is located at the bottom of the hammer directly below the *anvil.* Within the head exists *cushion material* which protects the top of the pile from damage due to the impact of the driving force. This cushion material does, however, dissipate some of the developed energy to the piling component.

The last piece of necessary equipment normally required is the pile leads. The leads are a steel framework that guides the hammer and encompasses the piling member. They are fixed to the ground at the location point of the pile entry. They are required to keep the alignment and vertical control of the piling member to assure proper positioning. The steel leads can be fixed to the crane by a mere cable hookup, known as *swinging leads,* or pinned to the boom tip of the crane, known as *fixed leads.*

Augered piles and drilled caissons are performed by using a hydraulic powered drill, or auger, mounted on a *spotter* or *fixed leads.* This mechanism is suspended by the crane and is used in lieu of a pile hammer and driving the specific pile into the soil.

The alignment of the piling member is referred to as vertical or *batter.* This is the angle of descent where the pile enters the ground. The leads cradle the pile member and maintain the desired or specified batter.

A *template* of sorts is normally required at the point of entry to assure the exact location of the pile member is maintained. This is derived from the layout location of the foundation area. A template can be nothing more than wooden "stake" markers or can be an elaborate steel framework, dependent on conditions. Each pile must be maintained to its designed location and batter for proper load and transfer distribution to the structure.

The driving conditions are the actual site conditions that exist at the specific foundation area. Keep in mind they can differ within the job site, and even in some cases within the bridge structure itself. These conditions are a determining factor in the rate of production achieved in the driving operation.

The first and foremost condition to realize is access to the foundation site. Can the equipment, labor, and materials access the work area or do provisions such as haul road or secondary transport need to be addressed? Second, and probably the most controlling factor to pile-driving productivity, is the underlying soil conditions at the foundation site. The soil conditions were a determining factor during the design of the foundation. This determined the type of piling, the type of bearing required, and the *tip elevation* the pile must be driven to for the stability and load-bearing requirements of the structure.

Pile batter. (*Courtesy of International Construction Equipment.*)

Soils and underlying strata range through clay, sand, silt, muck, and rock. These differing conditions react to and impact the driving conditions. Each has its own characteristics with respect to *driving resistance* and *bearing values.*

The driving resistance is the amount of force required to drive the pile productively to the desired tip elevation, without damage to the pile. This is referred to as *blows per foot* of driven pile. The higher the *blow count* is per lineal foot of pile, the harder or denser the soil, the greater the resistance, the greater the skin friction or tip for bearing value.

The underlying soil conditions of a pile cluster driving site normally vary greatly in strata from the ground surface elevation to the tip elevation of the pile. The varying blow count is a determining factor in analyzing the pile-driving production. An example of blow count variances is one condition of the soil strata averaging from a blow count of 3 strikes per foot at the beginning, to a count of 40 blows at the tip elevation. This means three hammer strokes are required to move the pile 1 lineal foot into the ground at the start of the driving operation, and 40 hammer strokes to drive the pile 1 lineal foot at the tip elevation. This is the pile *refusal,* or resistance, that controls the production.

The differing soil conditions that reflect on varying production are also a determining factor in choosing the type of pile that will be used in the design, as some types of piling react differently, both structurally and productively, in different soil. Also the pile hammers must be accurately chosen, as they do react differently with piles and soil. Minimum and maximum requirements also exist with respect to the energy output of the hammer to certain specific pile and driving conditions. The pile must be driven into the ground without harm to the pile by the force of produced hammer energy but must also be within the performance expected by the production estimate. The choosing and sizing of the pile hammer will definitely be a controlling factor in the production and efficiency of a pile-driving operation.

A condition with that of the varying blow count of 3 to 40 means as the pile is driven into the ground, the driving will become harder, thus slowing the production. The blow count for the entire length of each pile should be averaged to obtain the best productivity, but keep in mind, one should analyze the soil borings to determine the blow count that best controls the specific condition.

An example is a pile that is 50 feet long being driven into a stratum with the blow count ranging from 3 to 40 blows per lineal foot. The controlling features of the blow count are that even though the initial count is 3 blows per foot, there exists a 10' layer stratum of 70 blow count material that must be penetrated. Once through this stratum, the blow count decreases to a varying condition of 10 to 20 count material, then progresses up to the 40 blow count stratum at the tip elevation.

The condition that exists is that there is not a progressive or uniform increase of the blow count resistance in the soil. A detailed summary must be formulated as to the depth of each of the controlling strata, the average blow count within each stratum, the total length of pile as it is affected by each stratum, and the blow count for the tip elevation. The need to realize the strata and blow count within the tip elevation area is normally that a pile may have little or virtually no resistance until it reaches the desired tip elevation area. If this is not realized, the average low blow count of the stratum suggests a quick or high production factor, but in reality, the last seating or driving time required for a few feet at the tip may absorb twice the time to drive as the majority of the pile did to that point.

Soil displacement during a cluster pile-driving condition must become a factor in the production analysis and blow count determination. The soil displacement may occur as the progression of the driving operation extends across the foundation area. As the piles are driven, the displacement of the underlying soil from the pile tip penetration is generally moved to the area of least resistance. This area is usually extended away from the pile-driving area, to the area next in line for driving to commence.

As the driving continues and more pieces are placed in the area of confinement, this displaced area becomes denser, thus increasing the blow count per foot. As the density increases, so does the resistance to drive the next piece, even though the soil borings reflect a relatively uniform and consistent blow count. This condition also reacts with upheaval limitations of the soils in displacing the bottom elevation of the foundation area. It may change the previ-

LOG OF BORING

PROJECT: International Paper Co. JOB NO.: E-363   ELEVATION: 110 Ft.
Secondary Dust Collector
BORING NO.: 1   TYPE BORING: ASTM D 1586, D 1587   DATUM: Plant
(4" Rotary Wash w/Bentonite Mud)
LOCATION: Moss Point, Miss.   DATE: 11/28/77   GR WATER: 3 Ft.

DEPTH IN FEET	LOG	DESCRIPTION	SAMPLE NO.	SPT N_i	SPT N_s	W C %	ATTERBERG LIMITS L.L.	ATTERBERG LIMITS P.I.	DRY UNIT WT pcf	% MINUS #200	(tsf) SHEAR STRENGTH	UNIFIED CLASS
0		Shell	1	21	39							
		Medium to loose brown	2	9	14	17						
		fine sand w/traces of	3	5	7							
		charcoal @ 10'	4	7	10	22				15.2		SM
10			5	5	7							
		Organic material (charcoal)	6	2	2							
		Medium white fine sand	7	20	24	23						
20		Dense to medium white fine sand w/thin clay layers	8	31	35							
			9	10	11	29				20.3		SM
30			10	3		62						
		Soft to very soft gray clay	T-1			68	66	43	59		C=0.40 ∅=0	CH
			11	0		58						
40		Stiff brown clay with thin sand layers	12	9		27	27	5				CL-ML
			T-2									
		Soft to stiff brown clay w/organic material	13	3		34						
			T-3			46	47	22	76		C=0.66 ∅=0	CL
50			14	8		25						
		Very dense brown and white fine to coarse sand w/a small amount of gravel	15	100+	100+	20				6.0		SP
60			16	100+	100+							
			17	100+	100+							
70		Stiff gray clay	18	8		47	72	44				CH
		B.T. @76.0'	19	10		50						

NOTE: Circulation was lost at 12 ft. 15 ft. of 4 inch casing set.

Soil boring analysis.

ously graded area that was to be the final preparation of the foundation bottom surface.

Differing soils along with different types of pile, with respect to bottom end surface, will alter this condition and greatly affect the end results. Tighter spacing within the pile cluster is also a controlling feature. Dense soils such as sand, large flat-bottom piles such as prestressed precast concrete, and a tight cluster spacing from pile centerline to adjacent pile centerline will affect the driving condition from the first pile driven to the last pile driven.

The project soil boring logs show the location of the test bore with relationship to the foundation area. They map the soil strata and characteristics to an elevation of at least bottom pile tip and suitable bearing location, thus showing the anticipated length of the proposed piling. This will be vital information to determine the productivity of the pile-driving operation.

The tip of the pile is a factor in the driving operation. In some conditions and pile designs, a blunt- or flat-end pile is used instead of a pointed or nonrestrictive end. Dependent on soil conditions, a blunt-ended or flat-bottom pile will produce a cone effect that will restrict the pile from being driven properly. This condition must be eliminated by means of either augering or *jetting*. This procedure will remove the soil buildup restriction at the tip, thus allowing the pile to penetrate to its proper and intended tip elevation. This condition would normally exist only with piles that are to achieve an actual tip bearing elevation rather than friction bearing.

Special *pile points* are used when the pile is designed to be landed in a rock stratum and the permanent pile tip may experience damage or unequal tip pressures. Rock points are designed to penetrate into the rock strata to give a substantial seat for the bearing or, with some conditions, penetrate through a secondary stratum as a cutting tool to advance to the primary strata. Pile points are normally used for a protective end and a closure with regard to pipe and shell piles. Regardless of the use, the pile point usually takes the blunt and tip impact of refusal during the driving operations.

Augered pile and drilled caissons are two types of piling designs and types that require *preboring* and *drilling*. Augered piling conditions occur when the underlying soil requires that the length of pile, or partial length, be installed in a prebored hole. When an augering operation is required, there will exist *spoil,* or excess material that is removed from the hole. This material must be accounted for and a budget of cost and craft-hours must be formulated.

Sometimes the design of foundation piles and the soil analysis do not fully assure the bearing value that is intended. This is a normal condition of friction pile, or bearing piles that may seat on a questionable shelf, or a pile design that has some tip bearing but not full structural integrity and the need for skin friction is used in combination.

To assure the intended and required load bearing is achieved, a *test pile* and/or a *load test* is required. A test pile is a structural member of the exact properties of the designed permanent pile. It is installed close to the actual foundation area of the permanent piles. The test pile may be driven to a deeper elevation than that of the permanent piles solely for further assurance of the stability and integrity of the intended tip elevation. The specific driving equipment intended for the permanent piles must be used on the test pile to assure the performance of the hammer in evaluating the bearing values.

The reaction of the hammer in regard to the blow count per lineal foot can be equated to foot-pounds of bearing applied to the pile member. To further assure this calculation, a load test may be required. This procedure entails the actual loading of specified weight being equally applied to the driven pile in a downward position, evaluating for movement of the member. A load test can also be applied to a permanent pile member.

The load test procedure is applied by a calibrated jack that asserts a specified force, usually two to three times the designed bearing load per pile, over a period of time, with the movement of the pile monitored.

Depending on the intended design, the load test will show little or no movement to the pile from the load applied. A framework or template of some sort must be constructed for the load test procedure. The requirement of this apparatus is to assure the load implied by the jack to the pile is fully asserted to the pile, without any uplift by the jacking forces. This will control the monitoring of the test pile movement for the full duration of the test. If a 600-ton test is required to be performed to the test pile, the jacking apparatus must be able to withstand the uplift of at least 600 tons so the entire applied load is transferred to the pile. Once this is maintained, the test pile can be monitored for assurance of the intended design and anticipated tip or friction bearing against the actual bearing being achieved at the site.

If the test fails, the designers will review the test results and make a determination as to the corrective actions required. A fix may be the addition of more pieces to the cluster to distribute the entire load of the structure over a higher number of piles, thus reducing the load capacity required for each pile; or extension of the length of the piles to project into a deeper stratum that will withstand the desired bearing by either tip bearing or added skin friction. Depending on the overdesign of the test pile requirements and the closeness of the actual results to the intended, the designer may elect to approve the results.

The type of pile specified will be a factor in the type of equipment chosen to handle and drive the pile. The type, length of each piece, and perimeter size will control the weight of each piece of pile. When determining this, the longest piece that is required to be handled and driven (without *splicing*) will be the controlling agent. Along with this weight, the weight of the pile hammer, leads, or auger must be added to this calculation for one of the factors that will determine the size of the lifting crane required.

The splicing of pile can be a requirement due to design length with regard to the physical transportation, logistical site conditions that inhibit the proposed length such as overhead restrictions, or design lengths that logistically prohibit long lengths or give limitations to equipment capacities in the actual driving operation. Pile splices can be either *mechanical* or conventional.

A mechanical splice is a separate type of coupling that is manufactured and installed between two pieces of pile that incorporate them structurally to one.

A conventional splice is one in which two pieces of pile are joined by means of welding or other methods not entailing a coupling device.

Some splices require a cure period for set. This condition requires a template or clamp to support the pile during this period so as not to confine and restrict the use of the hoisting and driving equipment. Most pile splices are made during the driving or augering procedure. The first piece of pile must be driven in its proper location and alignment, and to an elevation where a few feet of the top of the pile is above ground. At this point the splice is made to the second piece. Once the piles are spliced together, the driving operation can continue to the final tip elevation.

Steel piles are normally spliced by one of several procedures. A steel *chill ring* is attached between the two adjoining pieces and welded. This is a prefabricated piece that would be of the same configuration and section as the pile. Another option would be merely steel plates welded across the joint around the perimeter of the pile. The third method is a *mechanical coupling.* This too is a prefabricated piece that attaches to the two piles by friction. This coupling is fastened securely when the driving begins on the second pile. The mechanical coupling usually requires little or no welding.

Splices for concrete piles are performed by several methods also. One type of splice is a group of steel dowels drilled and epoxied into each end of the pile, and a saddle or clamp applied for set and cure. Another form of splice is a steel saddle attached around the perimeter of the piles at the joint.

Timber and wood piles are spliced by dowels and saddles. The splicing of all types of piles must be performed accurately and by specification. The added piece must maintain the exact alignment of the initial piece. Improper splicing can cause undue stress to each pile and distort the driving force of the hammer, causing inaccurate bearing calculations. Improper splices can also pose a failure to the splice during driving and/or the permanent position.

A firm template or cradle must be constructed to assure proper alignment and stability during the curing period, if applicable, and driving of certain piling, predominantly prestressed precast concrete. Precast piles are somewhat fragile in a horizontal position and when certain undue stress and point loading is applied. These piles particularly must be supported during a splicing procedure and until the cure period has elapsed for the epoxy compound. With many conditions, a template can be constructed at each bent, or driving location, which can act as both an alignment tool and a support apparatus.

In order to protect the pile end as it is being driven, a tip or bottom plate is used. These plates are attached to the bottom pile tip by either a friction fitting, welding, or a form of buildup during manufacturing. A form of tip attachment or protection and aid to driving is a *stinger.* This piece can be field attached or attached during the manufacturing process. It is most commonly used with precast concrete piles and is designed for use in hard driving conditions.

Piles are often driven to an elevation that is greater than that designed. When a pile is premanufactured to a specific predetermined length, such as precast concrete, it sometimes cannot be spliced owing to the project specifications. When this occurs, a procedure referred to as a *pile buildup* must be performed. A buildup also has length limitations and often does not permit driving or redriving of the permanent pile afterward. Usually, a pile buildup has a length range of under 10 lineal feet. A pile buildup is constructed by formwork, is of the same section as the permanent pile, and is then poured with a high-strength concrete.

When piling is not quite driven to its exact tip elevation, the condition of *pile cutoff* may exist. This is the procedure of trimming the top of the pile to the cutoff elevation shown, or the point which is the maximum or minimum allowed to extend into the foundation of the structure.

When piling is shown, other than piling with a predetermined and specified length such as precast concrete, an allowance should be factored for overrun, trimming, or merely a cutoff. This is normally a percentage per piece of pile based on experience, demographic location, driving conditions, or order length. The additional footage, along with the time and cost associated with this actual work, must be accounted for in the project planned footage as incidental work within the estimate.

Takeoff

Pile takeoff is a procedure that requires an understanding of numerous activities, equipment specifications and capacities, and a geotechnical understanding of soils, their properties and characteristics.

The pile-driving operation contains controlling activities that, when properly analyzed and combined, will produce an effective and productive solution that can be applied efficiently to the driving condition that exists.

The components that exist and must be detailed are:

1. The soil boring logs and analysis
2. The type of piling that is required
3. The site conditions that exist
4. The equipment best suited for the operation

The takeoff procedure must be formulated into a simple but effective format that will limit the possibility of error but will outline any required condition that may exist. This format may not always apply to every condition but will contain an outline that covers most activities and possibilities and serves as a checklist. The priority goal in the entire pile operation is to drive or auger the most pieces in a given time with the most effective utilization of labor and equipment. The entire budget, cost analysis, and activity schedule will be controlled by this operation.

Regardless of the type of pile design, the takeoff procedure will follow the same format. The first step is to determine the type and style of pile designed. Second, the project and specific site conditions must be detailed. The most controlling item to isolate and scrutinize, with great specifics, is the soil boring and log analysis. Regardless of the type of pile and the site conditions, this component will control the entire pile procedure. Once these steps, which are the most variable, have been completed, the core takeoff can continue.

The core takeoff will commence with summarizing the number of pieces; grouping by type and length; separating and quantifying by driving condition, driving type, and site area; determining splices and points; the requirement of templates; and identifying and calculating any secondary components incidental to the pile such as predrilling, concrete fill, and *redriving*.

These procedures will produce a detailed summary of the pile components for the project and will result in a professional and understandable takeoff with realistic production and costing.

The control of activity for producing a pile computation will be one of two determining methods. The format will be either driving piles to a specified bearing and tip elevation, using a variety of methods and functions; or augering piles to a specified bearing and tip elevation, using a variety of methods and functions.

Operation. Driving.

1. Conventional driving method
2. Jetting with final driving
3. Prebore with final driving

Operation. Augering.

1. Conventional augering
2. Augering with final driving

Procedure

When preparing a detailed takeoff and structuring an outline, the first item to recognize is the type and size of pile intended. This component will serve as the control for the entire estimate and the preparation of the cost summary and budget.

With this preparation, a list of questions must be formulated to construct the outline and prepare the procedure. This listing will serve under any condition with any type of pile used in today's market.

To prepare an estimate, the number of piles must be determined along with their length and weight. This must be formulated by individual sites first, starting with specific foundation areas. A category of separation by type and perimeter size is grouped. This detail then consists of the number of pieces within a foundation area, the assumed plan length of each piece, any splices that may be required, and any pile tips required. This is the time to list and detail the template requirement, as each individual driving site may dictate different templates.

The soil analysis pertaining to specific pile foundation sites must be determined, as they too may differ by specific site areas. The soil boring logs will show in detail the actual soil and rock strata that exist at the location of the pile foundation area. This will be to a depth of at least the tip elevation intended for the pile, and possibly farther to ensure a stable bearing elevation.

Along with this specific underlying condition, the analysis will give a driving condition that was prepared from an actual field test of driving a *spoon* with a specific weight and stroke. These blows per given increment will relate to blows per lineal foot for the pile operation. This information is used in formulating a theoretical analysis that can be used for determining a realistic pile-driving productivity factor.

The information is then enhanced with specific site condition details that may alter the actual driving productivity. Among these are access restriction for both equipment mobilization and actual delivery of the permanent materials; site preparation to enable a normal working condition; water-type conditions; or overhead and utility restriction.

A demographic analysis must next be detailed as to specific work environment, crew productivity, and labor relation craft contracts.

Questions of Estimate

1. What type of pile is designed?
2. Are there accurate, detailed soil borings and logs?
3. Is this a land- or water-type operation?
4. What are the site conditions?
5. What access is available (delivery and production)?
6. What are the specifications regarding pile hammers?
7. What is the demographic concern?
8. What labor contracts exist?
9. Are there equipment restrictions or concerns?
10. How many pieces of piling?
11. What is the average length and weight?
12. What is the longest and heaviest pile?
13. Are different types of pile designed?
14. Will pile points be required?
15. Will splices be required and how many?
16. Will augering or predrilling be required?
17. What is the quantity of displaced material?
18. Will dewatering be required?
19. What is the template requirement?
20. What is the anticipated blow count?
21. Is the design based on tip or friction bearing?
22. If tube piles are designated, what volume and type of concrete are required?
23. How much waste or added footage per pile?
24. What is the number of pile cutoffs required?
25. Are test piles required?
26. Are load tests required?

The above list of questions will aid in the determination and quantifying of the detailed pile estimate and component listing.

The drawing on page 253 is the example used in the application procedure and bid preparation.

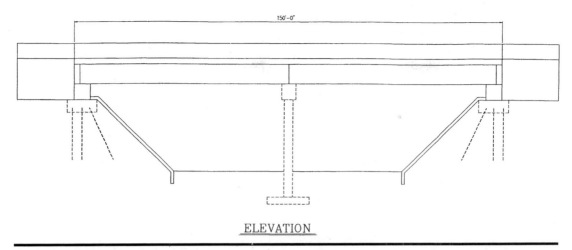

ELEVATION

Actual sketch for foundation pile, two sites (north and south abutment).

Application

The following example defines a conventional pile takeoff with a determination being resolved for a quantified summary and productivity factor.

There are two pile-driving locations, a north and south abutment. Each given foundation site requires a cluster of 15 steel H piles, having a section of 12"×12" and a weight of 53 pounds per lineal foot. The anticipated design length is 44 lineal feet. The spacing is symmetrical at 3 rows of 5 piles, 5 feet on centers. The front row of piles at each location are required to be battered at a 0.5':6.0' ratio; in other words, for every 6 vertical feet of pile, the batter is ½ foot. The top of the pile is designated to extend into the footing of the structure 1.0 lineal foot. The batter multiplier will be 1.025, or a 2½ percent loss time factor.

The pile tips are fitted with a rock point shoe for specified seating at the bearing point. The cutoff or trim waste is calculated at 1.5 lineal feet per pile, or 30 * 1.5 = 45 additional feet required for the project. This requires an order length per piece of 45.5 lineal feet, or a total requirement of 1365.0 lineal feet of 12"×53# steel H pile.

The soil boring logs show a series of fragmented shale layers and clay with a limestone formation at the bearing elevation of −43'.

The sample boring log shows a blow count of 75 blows per foot at bearing with a variance of 15 to 40 throughout the length of the pile. This was determined using a 2" 50-pound spoon with a stroke of 30 inches. The increment of measurement for the blow count is 1 foot.

The pile has a bearing design of 50 tons per pile with a hammer specification of a minimum of 12,000 foot-pounds of energy. The minimum blow count by specification is 70 blows per foot or refusal.

254 Takeoff and Cost Analysis Techniques

WORK SHEET

Project **BRIDGE # 17** Estimator **JDN** Item No. **BORE #13**
Type of Work **LOG FOR 12" H PILE** Date **10-13-99** Sheet No. **1**

SOIL LOG ANALYSIS

Elevation	Hammer Blows	Description
0.00	5, 8, 6	FIRM CLAY / ORGANIC ROOTS / GRAVEL
	10, 12, 15	FIRM CLAY
-10.00	15, 20, 22	WEATHERED FRAGMENTED LIMESTONE LAYERS
	27, 30, 32	
-20.00	28, 31, 33	FIRM CLAY SEAMS WITH LIMESTONE FRAGMENTS
	37, 40, 44	WEATHERED FRAGMENTED LIMESTONE LAYERS
-30.00	28, 50, 70	FIRM CLAY SEAMS WITH LIMESTONE FRAGMENTS
	49, 51, 55	FRACTURED FRAGMENTED LIMESTONE LAYERS
-40.00	50, 59, 60, 70	
-43.00	75, 80	PLANNED TIP ELEVATION
		SOLID LIMESTONE FORMATION
-50.00		

SPOON 2"
50 LB HAMMER
30" STROKE

PIER 1

Soil log.

Pile factoring formula: Nardon method

Function	Calculation
1. Type and size	Given
2. Total pieces required	Lines 12 + 13
3. Total lineal feet	Given
4. Average length per piece	Lines 3/2
5. Number of driving locations	Given
6. Average number of pieces per location	Lines 2/5
7. Craft-hours consumed per day	Lines $(8 * 9) + 1$ hr (foreman)
8. Work hours per shift	Given
9. Number of crew members	Given
10. Moving time to each location	Given
11. Productivity time per hour	Given
12. Number of straight pieces	Given
13. Number of battered pieces	Given
14. Ratio percentage of battered	Lines 13/2
a. Batter	Given
b. Batter multiplier	Given
15. Longest pile	Given
16. Required splices	Given
a. Percent of pieces spliced	Lines 16/2
17. Soil analysis	
a. Average blow count per pile length	Given
b. Increment per blow	Given
c. Hammer stroke	Given
d. Soil factor	Lines $(17a/17b)/17c$
e. Theoretical minutes per lf	Lines $17d/11$
f. Theoretical driving time per pile	Lines $4 * 17e$
g. Batter multiplier	Lines $(14b * 14) + 1$
18. Cycle time duration (in minutes)	
a. Set template	Given
b. Rig pile	Given
c. Hoist pile	Given
d. Set hammer	Given
e. Set batter $x = ?$ minutes	Line $14 * x$
f. Drive pile	Lines $17f * 17g$
g. Unhook pile	Given
h. Splice time $x = ?$ minutes	Line $16a * x$
i. Additional piece (spliced)	$\text{Sum}(18b...18e + 18g) * 16a$
j. Auger prebore shaft	Given
k. Other activity	Given
l. Total duration per average pile	Sum $18a...18k$
19. Driving time per pile	Line $18l$
20. Productivity factor per hour	Line 11
21. Required crew hours per pile	Lines 19/20
22. Average lineal feet per crew hour	Lines 4/21
23. Average lineal feet per shift	Lines $8 * 22$
24. Number of production shifts required	Lines 3/23
25. Location moving hours	Lines $10 * 5$
26. Required relocation days	Lines 25/8
27. Required operation days	Lines 26 + 24
28. Net production per day (lf)	Lines 3/27
29. Production factor craft-hours per lf	Lines 7/28
30. Lineal feet per craft-hour	Lines 28/7

Descriptive

Line 1: This is descriptive of the size and type of pile required.

Line 2: This is a calculation, sum of straight and battered pieces, lines 12 + 13 (actual in ground full pieces).

Line 3: This is required, given from the quantity takeoff procedure of actual determined footage of same type.

Line 4: This is a calculation, total lineal feet divided by number of pieces, lines 3/2.

Line 5: This is required, given from the quantity takeoff procedure of actual driving locations forcing a crew move.

Line 6: This is a calculation, total pieces divided by driving locations, lines 2/5.

Line 7: This is a calculation, multiplying the number of crew personnel times the hours worked per shift, lines 8*9, plus 1 additional hour for foreman.

Line 8: This is required, actual work hours per shift.

Line 9: This is required, actual number of personnel per crew.

Line 10: This is required, given duration in hours to move between driving locations.

Line 11: This is required, given an actual productivity time per work hour in minutes, i.e., 50 minutes productive per hour.

Line 12: This is required, given the actual number of straight pieces derived from the quantity takeoff procedure.

Line 13: This is required, given the actual number of battered pieces derived from the quantity takeoff procedure.

Line 14: This is a calculation, dividing the number of battered pieces by the total number of pieces, lines 13/2.

Line 14a: This is the batter of the battered pile (i.e., 1 to 3) given as 3.

Line 14b: This is a given. The ratio of loss time for the battered piles to a constant of 1 for straight. A 10% loss time factor is represented as 1.1.

Line 15: This is required, given the longest pile, including spliced lengths, to be handled.

Line 16: This is required, given the actual number of splices determined by the quantity takeoff procedure.

Line 16a: This is a calculation, dividing the number of splices by the actual number of piles, lines 16/2, shown as percent.

Line 17a: This is required, given the averaged blow count within the length of driven pile derived from the soil reports.

Line 17b: This is required, given the unit of measurement the soil report used as an increment per blow count, usually 0.5′ or 1.0′.

Line 17c: This is required, given the stroke of the hammer used in the soil boring report, in feet.

Line 17d: This is a calculation, dividing the maximum blow count by the increment per blow, then dividing by the hammer stroke, lines ($17a/17b$)/$17c$.

Line 17e: This is a calculation, dividing the soil factor by hourly productivity factor, lines $17d/11$.

Line 17f: This is a calculation, multiplying the average length per pile by the production per minute, lines $4 * 17e$.

Line 17g: This is a calculation, given the percent of loss productivity for battered piles, times the percent of battered piles, lines ($14b * 14$) + 1.

Line 18a: This is required, given the actual time in minutes to set the template for the location of the pile. This time must be distributed for the total pieces of pile.

Line 18b: This is required, given the actual time, in minutes, to rig and hook each piece for hoisting.

Line 18c: This is required, given the actual time, in minutes, to hoist each piece to the top of the leads for the hammer seating.

Line 18d: This is required, given the actual time, in minutes, to seat the hammer on the pile piece.

Line 18e: This is required with a calculation, given the x time to set the leads to the specified batter, then multiplying the designated time by the percent of job required battered pile, $x*$line 14.

Line 18f: This is fixed, given the calculation of lines $17f * 17g$

Line 18g: This is required, given the actual time, in minutes, to unrig the hammer from the driven piece and prepare for the next piece.

Line 18h: This is a calculation, given the actual time, in minutes, to attach the splice to both pieces required to be spliced, within the leads, multiplied by percent of pile to be spliced, $x * 16a$.

Line 18i: This is a calculation, a summation of the accumulated time from rigging the second piece pile through to the unrigging, then multiplying the accumulated time by the percent of spliced pile (sum lines $18b...18e + 18g$) $*$ line $16a$.

Line 18j: This is required, given the time, in minutes, to prebore the length of determined shaft.

Line 18k: This is an open line for any additional activity that may be required, to be given in minutes.

Line 18l: This is a calculation, a summation of section 18 inclusive, sum (lines $18a...18k$).

Line 19: This is fixed, given the value of line 18*l*.

Line 20: This is fixed, given the value of line 11.

Line 21: This is a calculation, dividing the total time duration by the hourly productivity factor, lines 19/20.

Line 22: This is a calculation, dividing the average lineal feet per piece by the crew hours per pile, lines 4/21.

Line 23: This is a calculation, multiplying the shift hours per day by the average production per crew hour, lines 8 * 22.

Line 24: This is a calculation, dividing the required project lineal footage by the average lineal feet produced per day, lines 3/23.

Line 25: This is a calculation, multiplying the relocating time between driving locations by the number of driving locations, lines 10 * 5.

Line 26: This is a calculation, dividing the required relocating hours by the shift hours per day, lines 25/8.

Line 27: This is a calculation, a summation of required relocating days and the required production days, lines 26 + 24.

Line 28: This is a calculation, dividing the required lineal footage by the required operation days, lines 3/27.

Line 29: This is a calculation, dividing the determined craft-hours per shift by the net lineal feet produced per day equaling the required production factor for this type and size of pile, lines 7/28.

Line 30:

n = lineal feet driven per craft-hour, lines (28/7)

The above formula will aid with the determination and calculating of driving time productivity. This was derived to be used with the formula of bearing analysis for determination of hammer size and hammer performance.

The following formula will be used as a guide to assure the performance of a given hammer and bearing value required. This formula should be used when piles are not required to be driven to absolute refusal but are required to have specific tip elevations and driving resistances.

Dynamic pile formula

$P = ((2WrH)/(S + 0.1)) * C$: for single-acting air and open-end diesel hammers

$P = ((2H(Wr + Ap))/(S + 0.1)) * C$: for double-acting or differential air hammers

$P = ((2E)/(S + 0.1)) * C$: for closed-end diesel hammers

where P = safe bearing value in pounds
S = average penetration rate, in inches per blow, for the last 10 to 20 blows
$C = (Wr + e^2 Wp)/(Wr + Wp)$
Wr = weight of the ram, in pounds
Wp = weight of the pile, driving head, anvil, mandrel, and follower, in pounds
e = coefficient of restitution for the capblock material:
    0.80 laminated micarta or plastic
    0.80 steel on steel (no capblock)
    0.50 hardwood
    0.25 softwood
    0.80 steel cable biscuit
H = stroke, in feet, for air/steam hammers, or minimum observed stroke for open-end diesel hammers at the time penetration rate readings are taken for the determination of bearing value. The maximum stroke for open-end diesel hammers will be determined by the test pile, or a maximum of 8.5 feet
E = energy of closed-end diesel hammers, in foot-pounds
A = area of piston, in square inches, for double-acting hammers
p = actual air/steam pressure measured by a needle gauge at the head of the hammer, in pounds per square inch

Pile hammer evaluation formula

This formula can be used to quickly evaluate a chosen pile hammer to determine the capacities and results.

$$\text{Pile bearing (in tons)} = 2E/(S + 0.01)/2000$$

where E = hammer energy, in foot-pounds
S = pile set, in inches per blow

The pile hammer chosen has the following specifications:

Rated energy: 30,000.0 foot-pounds maximum	
17,000.0 foot-pounds minimum	
Ram stroke	5'11"
Blows per minute	80–84
Power rating	76.3 hp
Bearing, formula based	150 tons
Hammer weight	13,400.0 pounds
Ram weight	5070.0 pounds
Anvil	1740.0 pounds
Typical operating weight	15,870.0 pounds

Given the information of this project;
  The pile bearing is 50 tons per pile.
  The energy rating of the chosen hammer is 17,000.0 to 30,000.0 ft-lb. Using a variable cycle hammer, we choose a 17,000.0 ft-lb rating.
  The inches per blow from the soil report is

$$12''/70 \text{ blows per foot} = 0.171 \text{ inch per blow}$$

Solving the formula:

$$X = 2(17,000.0)/(0.171 + 0.01)/2000$$

$$X = (34,000/0.181)/2000 = 93.92 \text{ tons}$$

$$\text{Project} \qquad\qquad = 50.0 \text{ tons}$$

According to this formula, the chosen hammer will exceed the minimum ton bearing required by project specification.

Given the above scenario and formulas, the following will detail the takeoff and production factoring method for driven piles (applicable for steel, concrete, tube, and timber).

The pile hammer anticipated for use is a double-acting diesel hammer with a ram weight of 5070 pounds and a rated stroke of 5'11".

Example: Operation analysis

Furnish and drive pile.

Type	= steel H 12"×53#
Spacing	= 3 rows @ 5' centers
Design length	= 44 lf
Quantity	= 30.0 each
Template requirement	= simple staking
Points	= 30.0 each
Splices	= not required
Battered pieces	= 10.0 each
Batter	= 0.5 on 6
Batter factor	= 1.025
Driving locations	= 2.0 each
Trim/cut off waste	= 1.5 lf each
Soil boring data:	
Material, fragmented shale and limestone	
Split spoon 2" @ 50 lb	
Hammer stroke 30" (2.5')	
Increment 1.0'	
Design resistance:	
Elevation −10'	= 15 blow count/lf
Elevation −18'	= 30 blow count/lf
Elevation −25'	= 40 blow count/lf
Elevation −32'	= 70 blow count/lf
Elevation −37'	= 55 blow count/lf
Elevation −43'	= 75 blow count/lf
Average blow count	= 47.5/lf

Example: Production factoring, time duration

Operation. Drive 12″×53# steel H pile.

Crew members	= 6.0
Production unit	= lineal feet
Productivity factor	= 50-minute hour
Shift duration	= 8 hours
Relocate time	= 4 hours

With the given information, the *pile factoring formula* can be implemented using the prescribed format.

1. Type and size	= 12″×54# steel H
2. Total pieces required	= 30.0 ea
3. Total lineal feet	= 1320.0 lf
4. Average length per piece	= 44.0 lf
5. Number of driving locations	= 2.0 ea
6. Average pieces per location	= 15.0 ea
7. Craft-hours consumed per day	= 49.0 craft-hours
8. Work hours per shift	= 8.0 hours
9. Number of crew	= 6.0 ea
10. Moving time to each location	= 4.0 hours
11. Productivity time per hour	= 50.0 minutes
12. Number of straight pieces	= 20.0 ea
13. Number of battered pieces	= 10.0 ea
14. Ratio percentage of battered	= 0.33%
a. Batter	= 0.5
b. Multiplier	= 1.025
15. Longest pile	= 44.0 lf
16. Required splices	= none
a. Percent of pieces spliced	= 0.0%
17. Soil analysis	
a. Average blow count	= 47.5 avg blow count
b. Increment per blow	= 1.0 ft
c. Hammer stroke	= 2.5 ft
d. Soil factor (47.5/1)/2.5	= 19.00 factor
e. Theoretical minutes per lf	= 0.38 minute
f. Theoretical driving time/pile	= 16.72 minutes
g. Batter multiplier	= 1.34
18. Cycle time duration (in minutes)	
a. Set template	= 0.00 minute
b. Rig pile	= 2.00 minutes
c. Hoist pile	= 1.50 minutes
d. Set hammer	= 2.50 minutes
e. Set batter: 2.50 minutes	= 0.83 minute
f. Drive pile	= 22.43 minutes
g. Unhook pile	= 2.00 minutes
h. Splice time	= 0.00 minute
i. Additional piece (splices)	= 0.00 minute
j. Auger prebore shaft	= 0.00 minute
k. Other activity	= 0.00 minute
l. Total duration per pile	= 31.27 minutes
19. Driving time per pile	= 31.27 minutes
20. Productivity factor per hour	= 50 minutes
21. Required crew hours per pile	= 0.63 crew hours

22. Average lineal feet per crew hour	= 70.36 lf
23. Average lineal feet per shift	= 562.91 lf
24. Number of production shifts required	= 2.34 shifts
25. Location moving hours	= 8.0 hours
26. Required relocate days	= 1.0 shift day
27. Required operation days	= 3.35 work days
28. Net production per day (lf)	= 394.63 lf shift
29. Production factor	= 0.1242 craft-hour/lf
Unit cycle time: 31.27 min/44.0 lf	= 0.711 minute per lf
Cycles per hour: (50/31.27)	= 1.599 cycles per hour
Production per hour: 1.599 * 44.0′	= 70.36 lf per hour
Production per shift: 8 * 70.36 lf	= 562.85 lf per shift
Required shifts: 1320.0′/562.85	= 2.345 shifts
Required relocate days	= 1.0 shift day
Total required operation days	= 3.35 shifts
Total craft-hours: 3.35 * 8 * 6 workers	= 160.8 craft-hours
System check: 1320 lf * 0.1242	= 163.9 craft-hours check
Operation production factor: 163.9/1320	= 0.1242 check

The pile point component will have to be addressed next to complete the operation. For this procedure, it has been elected to perform the task of welding the point to the pile at a staging area, prior to the driving operation. This will maintain a controlled environment for the pile point operation and not restrict the driving operation by the possibility that the driving crew may have to wait for the welding crew.

This is a simple operation that will require one person as a welder performing a quantity of one point per hour worked.

Operation. Attach pile points.

Crew members	= 1.0
Production unit	= each
Productivity factor	= 60-minute hour
Shift duration	= 8 hours
Production per shift	= 8.0 each

The next required operation to be performed after the driving is completed is the trimming to the specified cutoff or top elevation. This will require one person to perform this activity. The task is performed by defining the trim elevation on each pile and cutting the excess pile away.

Operation. Trim pile to cutoff elevation.

Crew members	= 1.0
Production unit	= each
Productivity factor	= 60-minute hour
Shift duration	= 8 hours
Production per shift	= 24.0 each

The operation duration shows a production of 24.0 cuts or a quantity of 24 piles to be trimmed within one shift. With the quantity of pile being 30 pieces, the time requirement for one person to complete this task is 1.25 shift durations.

Once these procedures have been completed, the task of applying a cost from the production operations, to the previously compiled labor and equipment crew, can be done.

For calculating purposes, the shift duration for the pile-driving operation is rounded to 3.35 required shifts.

Material Allocation

After the production has been determined and the labor and equipment crews have been allocated, the direct and incidental consumed materials for the operation must be summarized and the cost determined.

Operation.   Furnish 12"×53# steel H pile.

Direct material:	
▪ Required pieces	= 30.0 each
Design length	= 44.0 lf
Overrun and waste	= 1.5' each piece
▪ Order length	= 45.5 lf each
▪ Total feet required	= 1365.0 lf
▪ Splices required	= none
▪ Pile points required	= 30.0 each
Indirect material:	
Welding rod ⅜"	= 60.0 pounds
Cushion material (micarta)	= 6.0 pieces
Oxygen and acetylene	= 1.0 tank ea

The material list for this operation is comprised of direct materials that are accounted for and detailed by the project specification requirements as embedded items. The indirect materials are those which are needed to perform the operation, regardless of the specific direct materials.

The project requires 30 pieces of 12"×53# steel H pile to be 44.0 lineal feet long. This length is a permitted length to be transported to the project without special variances; thus the pieces are ordered in the full length required. This eliminates the need of splicing.

The need for trimming exists owing to any variance in actual driving conditions, the possibility of slight damage to the top of the pile from driving, and the specified top or cutoff elevation required in the footing foundation. With regard to these parameters, the determination to allow or add 1.5 lineal feet to the order length has been allocated. This makes the total order length the contractor must allow for 45.5 lineal feet per piece, which is an indirect cost or nonreimbursable cost relating to the additional footage (1.5 each) from that specified.

The specifications detail a pile point at each pile location; thus the quantity of points are one per piece or 30 points required.

The indirect material costs related to the pile-driving operation are the cushion material required within the pile hammer. The assumptions are made that each piece of cushion material will last for five pile drives, equating to the requirement of six pieces of cushion material to be ordered.

The pile points are required by specification to be welded to the tip of the pile. The calculation is made that each point will require 2 pounds of $\frac{3}{8}''$ welding rod to complete the operation per specification.

The cutoff or trimming is somewhat of a variable, but the assumption is made that 90 percent of the pile will need to be trimmed to the proper elevation after driving. This will require that one tank each of oxygen and acetylene be present at the site.

This material listing must be allocated for cost for the operation. The cost of all associated materials and applicable sales taxes, shipping charges, etc., will be directed toward the prime operation item and project required unit and measurement:

$$12'' \times 53\text{\# H pile, 30 pieces @ } 44.0' = 1320.0 \text{ lf}$$

The consumed material list associated with the direct and indirect materials should be detailed as required for the operation. This method provides a clear and detailed tabulation for review and distribution when the contractor has become the successful bidder.

Pile layout

The example discussed shows a need of simple staking for the pile location template. These are merely a wooden stake driven at the location of each pile. When more control is required, a pile location template can consist of a horizontal guide beam, set parallel with each row of piles, or a detailed grid system or framework which may even be supported from the ground. This type of system would also aid in vertical support to each pile as well as location alignment. An example of a sophisticated template of this nature may be required with large concrete pile to aid in productivity and splicing.

The template must be designed and constructed for an ease of adaptability to differing pile locations within the grid, between driving sites. The actual location of each pile varies; the template must be versatile, to a point, to be quickly adjusted to these variances. With some conditions, multiple templates are more economical.

The elevated alignment template required legs driven into the ground for lateral support. This type of system must be built for quick dismantling and movement from one driving location to another.

Operation. Construct alignment template.

1. Construct framework.
2. Place template.

3. Elevate system, if required.
   a. Drive support legs.
   b. Remove support legs.
4. Move system.

Test pile

The test pile summary is derived with a somewhat different approach. Normally, the test pile is done prior to the production piles for the data and bearing research.

The driving operation remains to have the same method of calculating the driving time, but the actual production factors will decrease because of driving only one pile. The setup, mobilization, and shutdown duration considerably increase. This is an operation which consists of a single pile drive, normally one per foundation location. The crew must then relocate to the next site and perform the function again.

The following duration components will be designated for a test pile-drive operation.

Operation. Drive test pile.

1. Move to driving site.
2. Prepare work area.
3. Set up equipment.
4. Drive pile.
5. Prepare to relocate.

The pile load test, whether being performed on a test pile or a permanent pile, will be conducted in the same manner of methodology and quantity summation. The complete operation for the performance of the load test, including the construction of a jacking frame or template, must be accounted for in this item.

The load test consists of the construction of a jacking frame to withstand the jacking forces, or a platform to load with the amount of weight required to withstand or apply the designed load for the test.

If a frame is to be constructed for uplift pressure, the operation will include driving for pilot piles around the perimeter of the pile to be tested. The piles or legs will act as the stand for the jacking framework to be attached. These legs must be designed for ample skin friction to withstand the uplift forces of the load test jack when the jack applies the required downforce to the test pile.

A jacking platform consists of a mass of weight, exceeding the amount of the required test load, suspended from a series of blocking which is stacked on the ground around the perimeter of the pile to be tested. From this platform, the downward jacking forces are applied to the pile.

Within the template itself, regardless of which type is used, a jacking beam must be installed directly over the test pile to distribute the jacking force equally to the framework from the jack. The jacking beam must be of a section to have zero deflection from the implied load.

A variable secondary operation exists within the category of performing the actual load test, that of constructing the jacking template. Conditions and specification may dictate the type of actual template needed, but the function and performance of the actual load test virtually remain the same.

Once the jacking template is in place, the load test can proceed. The actual test consists of the setup of the required and calibrated jack with the capacity to adequately load the pile with the required pressure. The load is then applied in progressing increments over a specified period of time. This allows a consistent increase of applied load to the pile without any structural damage. This method also compresses any crush or expansion within the jacking system and templates.

After the full load has been applied, the jack gauges are monitored, again for a specified period of time, for any settlement or downward movement to the pile. The monitoring is inspected by the gauge for loss of pressure and by an incremented measurement device which shows any possible vertical movement. A bit of caution: The jacking platform must also be monitored for any uplift movement which would be detrimental to the test and show a reading of failure.

A failing test requires the repeat of part or all of these procedures. An inspection must be made as to the reason of the failure, and corrective measures taken.

When the test is successfully completed, the jack force must be lowered in a systematic manner as it was applied. The platform or jack stand must then be removed.

Prime operation.   Load test pile.

1. Jack setup
2. Applying the incremented load
3. Jack monitoring
4. Jack removal

Secondary operation.   Construct jacking template.

1. Drive legs or build stand
2. Build and attach jacking frame
3. Install jacking beam
4. Remove and dismantle system

When tube or pipe pile is used, the listed procedures may apply with the addition of a third operation, the concrete fill within the tube.

Pipe piling by its own section fails to meet the required design criteria needed for the support of the foundation without collapse. The intent of the pipe is merely a driven casing for which a structural component can be introduced.

After the driving has been completed, the operation of the placement of the concrete must be summarized. The concrete can be placed by numerous methods but must maintain its structural value. With this scenario, the placement will be performed by a conventional crane using a bucket to make the transfer between the delivery vehicle and the foundation site. With proper conditions, the concrete truck may be permitted to discharge the concrete directly into the pipe pile, or a concrete pump may be used for placement.

Example: Operation analysis

Concrete placement, pipe pile.

Type	= 12 ¾″ pipe pile
Dimension, ID	= 12″ inside diameter
Spacing	= 4 rows @ 6.0′ center
Design length	= 50 lf
Quantity	= 40.0 each
Concrete type	= 3500 psi
Driving locations	= 2.0 each

This operation requires the placement of concrete within steel pipe piles of a 12 ¾″ outside diameter. The calculation and quantifying requires a volume of concrete to be determined for each pile, and a total requirement of concrete along with an overrun and waste factor.

This scenario also requires a crew location move between pile location sites. The calculations are based on an inside diameter dimension of 12″ and a length of 50 feet per pile, for 40 pieces of pile.

The required calculation for the volume of concrete will be pi times the radius squared (πr^2 in feet), times the length of each pile in feet, converted to cubic yards, times the number of required piles. Added to this will be an overrun/yield loss factor of 18%. This high loss is due to the somewhat small area of containment and the number of pieces and locations.

Calculation

$((((3.1416 * (0.5)^2) * 50.0)/27) * 40.0) * 1.18\%$
$(((3.1416 * 0.25 * 50.0)/27) * 40.0) * 1.18\%$
$(((0.7854 * 50.0)/27) * 40) * 1.18\%$
$((39.27/27) * 40) * 1.18\%$
$(1.4544 * 40) * 1.18\%$
$58.176 * 1.18\% = 68.65$ cubic yards of concrete required

The operation is based on 2000.0 lineal feet of pipe pile which will have 0.0343 cubic yard of concrete per lineal foot of pile.

Example: Production factoring, time duration

Operation. Concrete placement, pipe pile.

Crew members	= 5.0
Production unit	= cubic yards
Unit cycle time	= 10 minutes/cy
Productivity factor	= 50-minute hour
Cycles per hour: 50/10	= 5.0 cycles/hour
Production per hour: 5.0 * 1	= 5.0 cy/hour
Shift duration	= 8 hours
Production per shift: 8 * 5.0	= 40.0 cy/shift
Required shifts: 68.65/40	= 1.72 shifts
Relocate time: 2 hours/8	= 0.25 shift
Total required shifts	= 1.97 shifts
Total craft-hours: 1.97 * 5 * 8	= 78.80 craft-hours
Operation production factor: 78.8/68.65	= 1.148 craft-hours/cy

This item of additional component operations must be added to the arrived costs affiliated with driving the pile. This example of differing work, though an added operation, is directly associated with the required in-place item of pipe piling.

Section summary

The section examples show the outline setup of compiling a detailed takeoff summary for the operation of driving 12"×53# steel H pile in a given foundation site. It produces the end result production factor by analyzing the soil report and calculating the driving resistance of the pile as it penetrates the soil. From this information, the hammer size is also verified so it is functional and economically sized to perform the piling operation properly.

The entire operation must be researched and analyzed for the proper detailing and quantifying of the material and data available.

Each piling location within the site must be approached with an independent view, as each operation may have a complex variability different from others.

COMPONENT TAKEOFF

Foundation Piling

The following outline contains the normal work operations that account for the performance and cost coding of foundation piling construction components.

1. **Templates, build and install**

   All work related to the size and type of template needed, building and moving the template from each location to obtain proper line and grade. Quantity should be reported in lineal feet placed and moved.

2. **Steel H piling**

   All work related to driving H pile to the proper line and grade of type specified. Quantity should be reported in driven length (lineal feet) after cutoff and number of pieces.

3. **Steel tube-pipe piling**

   All work related to driving tube and pipe pile to the proper line and grade of type specified. Quantity should be reported in driven length (lineal feet) after cutoff and number of pieces.

4. **Prestressed precast concrete piling**

   All work related to driving concrete pile to the proper line and grade. Quantity should be reported in actual pile length of type specified and number of pieces.

5. **Timber piling**

   All work related to driving timber pile to the proper line and grade of type specified. Quantity should be reported in actual lineal feet driven to the proper cutoff elevation, and number of pieces.

6. **Sheet piling driven**

   All work related to driving sheet pile to the proper line and grade of the type specified. For permanent sheeting, quantity should be reported in actual square feet driven after cutoff but including toe and number of pieces. For temporary sheeting, quantity should be in square feet prior to cutoff trimming but for an average grade line established and number of pieces.

7. **Sheet piling extracted**

   All work related to extracting sheet pile of type specified. Quantity should be reported in square feet removed prior to cutoff trimming and number of pieces.

8. **Splices**

   All work related to the installation of pile splices of type specified. Quantity should be reported in number of each installed as per type of pile.

9. **Concrete fill**

   All work related to the installation of structural poured concrete of the type specified to fill tube-pipe pile. Quantity should be reported in cubic yards placed including waste, overrun, number of pieces, and lineal feet of pile placed.

10. **Pile cutoff**

    All work related to the cutting off of a permanent pile to its proper elevation. Quantity should be reported in actual lineal feet cutoff and number of pieces with respect to steel and concrete piles.

11. **Trimming**

    All work related to the cutting off or trimming of steel sheet piles. Quantity should be reported in square feet and number of pieces trimmed.

12. **Drilling and augering**

    All work related to the predrilling and augering of holes for permanent piles at a specified diameter. Quantity should be reported in lineal feet and number of holes drilled of a specified diameter. If poured concrete and/or reinforcing steel is required, quantity should be in cubic yards placed and pounds placed, including overrun and waste, of the type installed respectively.

13. **Pile extensions**

    All work related to the adding onto a pile previously driven. Quantity should be reported in lineal feet added and number of pieces of a specified type. For precast concrete piles, additionally, quantities should be reported for square feet of formwork, cubic yards of concrete, and pounds of reinforcing steel including waste and overrun, and square feet of rubbing and patching. Quantities of number of top preparation should be reported.

14. **Pile stripping**

    All work related to the stripping, removal of concrete, and exposing of prestressing strand of precast prestressed concrete piles. Quantity should be reported in number of pieces and lineal feet exposed, by pile size.

15. **Encasing piles**

    All work related to the concrete encasing of steel H piles. Quantity should be reported in square feet of formwork, cubic yards of concrete, and pounds of reinforcing steel including waste and overrun, square feet of rubbing and patching, and lineal feet and number of pieces encased of a specified size and type.

16. **Test pile**

    All work related to the driving and installation (and, if necessary, the removal) of the test pile of a specified size and type. Quantity should be reported in lineal feet driven of the predetermined length and lineal feet removed.

17. Load test

All work related to the performance of the load test at either a test pile and/or a permanent pile. Quantity should be reported in each of the load test performances completed per size and type of pile specified.

18. Load test support mechanism

All work related to the construction and removal of the framework and templates necessary to withstand the load test forces applied to the test pile. Quantities should be reported as follows:

- Legs: Lineal feet and type of pile legs and number of each driven to obtain the proper bearing for the load test.
- Upper frame: Lineal feet, type, and pounds of structural steel used to support the implied loads of the test.
- Dunnage: Number of pieces and pounds of live load material applied to framework for the performance of the test.
- Jack beam: Lineal feet, size, and pounds of structural steel used and installed for, the jacking forces applied, during the performance of the test.
- Removal: Detailed account of the removal of the above detailed system.

19. Pile points

All work related to the installation of pile points of the size and type specified. Quantity should be reported in each of the points installed.

20. Piling (other)

This account will be used for other operation accounts not specified here. Quantity should be reported in a representative format relating to the work performed and the appropriate unit measure.

Chapter 13

Substructure Unit

TECHNICAL SECTION

Introduction

This section includes the technical specifications for constructing the substructure unit of the bridge structure. Detailed in perspective are specified procedures, materials, and equipment, and an outline of various components of the substructure unit. This describes the work consisting of constructing the portland cement concrete portions of the substructure to the required dimensions with or without reinforcing steel.

The equipment requirements shall be of sufficiently rated capacity and condition to place the concrete at an able pour rate to maintain a monolithic, uniform unit.

The forming systems shall be of sufficient strength and capacity to properly contain the concrete within the specified lines and grade. The falsework shall be of such designed strength and capacity to support any formwork and the imposed loads of concrete, steel, equipment, and labor force.

The permanent materials shall be of approved sources and be of the designed specifications.

Definition

The *substructure* is the intermediate portion of a bridge which bears the load of, and supports, the *superstructure*. It is a very integral part of the entire bridge design which distributes the weight and *load* to the foundations.

Substructure units comprise various unique components which give it many interesting and intriguing characteristics. When estimating a substructure unit, one must recognize and acknowledge these components as to the task of work involved.

Components

The basic components of a substructure unit are the *abutments* and the *piers*. These can be formulated into a wide variety of designs, but the end purpose remains the same. They transfer the load of the superstructure to the foundations and act as lateral supports to the stability of the entire structure.

A substructure, as is true of most of the bridge structure, consists of concrete and reinforcing steel. These two vital components give a bridge its integrity. Both consist of various designs which determine their independent and individual strengths.

Concrete is classified by its *compressive* strength, or pounds of pressure per square inch it can withstand. Reinforcing steel is manufactured by grade, which gives it its *tensile* strength. Separately, concrete and reinforcing steel have their own characteristics and minimal strengths; but when used together, they have a tremendous amount of structural uses.

Common classes of concrete designs, or *mix designs,* for normal substructures are 3000 and 3500 psi. These represent the pounds per square inch of compression or force it can withstand. In many cases, however, special and independent designs, depending on the end result and use in the structure, can vary from these two most common types.

The two most commonly used types of reinforcing steel are grades 40 and 60. These two, depending on their individual components, vary in designed strength.

These can be further referenced and detailed in the technical section of Chap. 17, "Reinforcing Steel."

Abutments. An abutment consists of various basic components but is a major part of the structural unit. The abutments, or *end bents,* are located at the ends of the bridge unit. They serve multiple purposes for the design of the unit. The beams and superstructure bear on the abutments at the ends of the bridge. They serve as bearing for the approach slabs and a tie from the bridge to the earthen fill. They also act as a containment structure for the earthen fill and the approach to the bridge. In essence, the abutments are the beginning and ending of the bridge unit.

Abutment footers. The footer units can bear on a prepared earth or rock foundation or a pile supported foundation. Footers can be of direct load-bearing design to the piling or of a spread bearing design which distributes the load over a wider, more uniform area of the foundation.

Abutment walls. The next component of an abutment structure is the *abutment wall*. This is the unit in which the beam structures are seated. Located on the wall are the *beam seats*. These are the actual bearing locations for the individual beams. The beam seats may or may not contain *position dowels* for the beams, depending on whether the design at this point in the bridge calls for the beams to be *fixed* or *expansive*.

Backwalls. The adjacent component to the abutment wall is the *backwall*. This part of the abutment structure separates the earthen fill on the

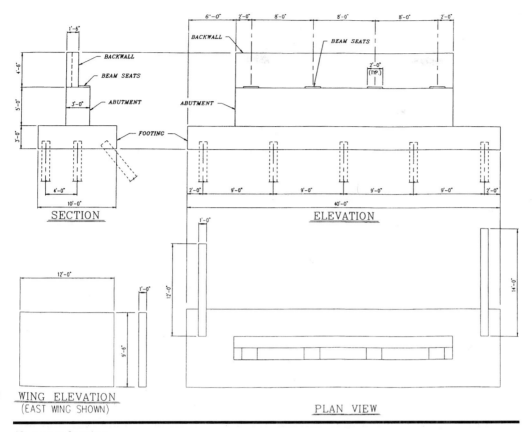

Abutment drawing.

approach side from the beam structure on the bridge side. It serves as an integral part of the structure. The backwall serves as the bearing point of the deck of the superstructure as well as the seat on which the approach slab rests. It becomes a uniform point linking these bearing points to the foundation as the pieces of the bridge are built. The backwall can be constructed as an independent wall to the beam structure or can be poured, as a dependent wall, monolithically encompassing the beam structure and the deck. This, again, depending on the design, is one of the many unique features an estimator must be aware of in a bridge.

Wing walls. The *wing walls* are also a type of backwall which bears on the footer foundations. They serve as support for fill and the approach to the bridge structure. The wing walls flair off of the backwall in a direction to support and retain the fill as the design calls for. They, too, can contain a seat for the approach slab to bear on.

Numerous types of construction methods and techniques are used in bridge construction and certain design methods which will be discussed later in this chapter. These methods lend themselves to both abutments and pier construction as well as superstructure construction.

276 **Takeoff and Cost Analysis Techniques**

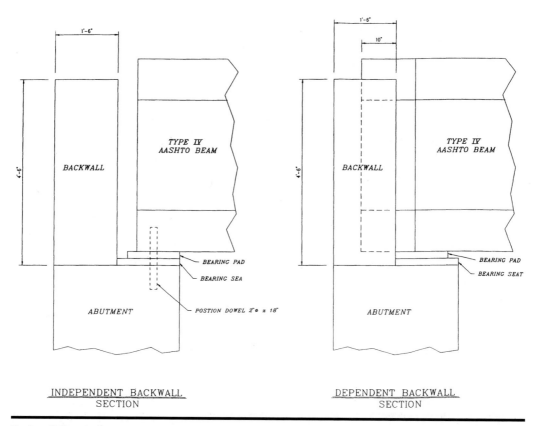

Backwall drawing.

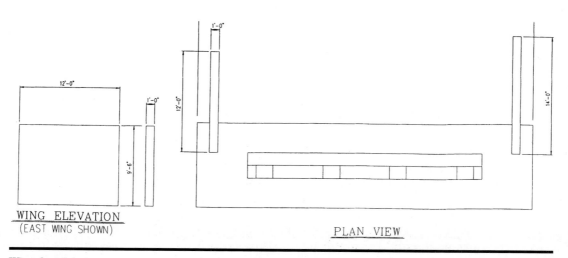

Wing drawing.

Piers

Piers are the intermediate supports of the bridge structure. They, too, are needed for support of the beam and superstructure units, but only for a *multispan* bridge. On *simple single-span* bridges, the beam and superstructure units are supported and bear on the abutments only. These vertical members can consist of various shapes and sizes.

The piers, or intermediate bents, also consist of specific components but can vary somewhat depending on the intended design. The piers, like abutments, have to bear on a suitable foundation. The footers can be of pile supported or of spread design on a prepared earth or rock foundation. Conventional piers are comprised of a *footer, columns,* and *caps*. Occasionally, within columns, there is a need for *struts, web walls,* and *crash walls*. The struts and web walls serve as intermediate supports to tie multiple columns together, giving them added structural stability. This design is often used for tall columns to aid in lateral support. Crash walls act as protection to the base of the pier structure and also for the columns together to act as one unit rather than multiple independent towers.

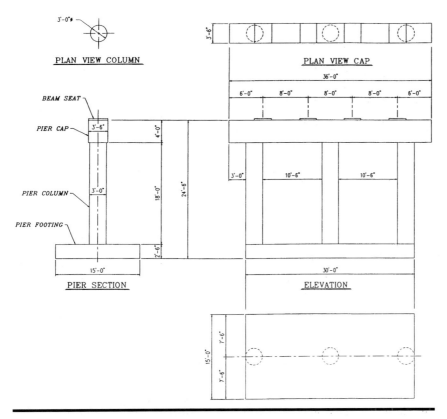

Pier drawing.

Pier footers. The footer of a pier structure is a poured unit constructed to support the columns and caps. Another type of method used in foundations is *drilled caissons*. As discussed in the piling sections, these are used in place of conventional piling. In some designs, the pier columns bear independently on these caissons, thus replacing the footer. A *grade beam* is sometimes used in this design, which acts somewhat as a strut, serving as a structural tie between the columns.

Columns. The columns are the next integral part of the pier construction. They are the vertical members which support the cap and transfer the load bearing of the beam and superstructure to the foundation. The designs of columns range from a single column with a *hammerhead* cap design to multiple columns with a conventional cap design. In section they can be square, round, octagonal, etc. Much of this is individual owner requirements.

Pier cap. The pier cap is the next component in the pier structure. The cap is a horizontal member of the pier which directly bears the load of the beam and superstructure units.

Within the cap, as in abutment walls, are the beam seats. These are the specific areas of the bearing points of the beams and also control and adjust the final elevations for the superstructure. In a *superelevated* section where the top of the cap is not symmetrical or where each beam has to be a different elevation, the beam seats become *riser pads* to adjust for this. This is also true in the abutment sections.

As in the abutment section, these, too, may or may not contain position dowels or *anchor bolts* for the beams. The number and location of pier units within a bridge structure are determined by numerous factors. Ground terrain, existing roads and transportation facilities, rivers, and obstructions are a few deterrents.

The design deterrent is the length of the span. But with today's technology, greater span lengths have become more common as the challenge of engineering minds has somewhat overcome this barrier. Conventional span lengths are determined by the material composite of the beams, whether concrete or steel, both of which also have various factors. These, along with modern technology, are discussed in detail in other sections.

Demographic areas along with financial budgets are the two other key factors in the design of bridge structures.

Joints. Within the various components of the substructure lie pertinent methods of construction techniques.

Construction and *control joints* are important items and must be fully understood by the estimator. There are fixed areas of joints the designers choose, and there are optional joints.

Optional joints are set up by the designers to aid the contractor in construction methods and ways of constructing a specific unit.

Keyway joints are used along with construction joints. These are of important use to tie or lock the joints together in either a horizontal or vertical

TYPICAL CONSTRUCTION JOINT

TYPICAL DOWELLED CONSTRUCTION JOINT

TYPICAL KEYWAY CONSTRUCTION JOINT

TYPICAL EXPANSION JOINT

Joint drawing.

plane. These are most commonly used in conjunction with rebar for tension in construction joints.

Control joints are usually designed with no rebar continuation through them, but they sometimes have a keyway joint for lateral control. A control joint's purpose is for controlled movement, either *compressive* or *expansive,* within a component.

Vertical control joints will be continuous through every adjacent component.

The measurement and payment of substructure units are determined by the units of direct materials, which are controlled by a variety of incidental materials and construction methods. Concrete is measured by the *cubic yard* of *neat line* placement, and reinforcing steel is measured commonly by the pound, obtained from a *bar list*.

The equipment needs, labor forces, and *permanent* and *indirect* materials vary widely with uses and methods in the construction of substructure units. The estimator must be well versed in construction techniques and takeoff procedures to fully understand these. The following sections outline this in detail as to standard practices and conditioning one to rhythmic procedures of adaptability.

Forms and formwork. Substructures are most commonly constructed using *removable forms* to contain the concrete mix. Forms can be of metal, wood, or a combination of both and can be lined with plastic, fiberglass, or a rubberized *form liner*.

Metal forms are constructed of a fabricated structural unit. They can be of normal standard sizes or of a custom design to fit a specific need and purpose. Standard metal forms are ganged together from multiple sections to make one unitized member or are used individually.

Combination forms are of metal and wood composite. Usually the structural frame is of metal angle or tube and the facing is of wood skin. These, too, can be either ganged together or used individually but are normally ganged. These forms are commonly referred to as *hand-set* owing to their being lightweight and easy to handle.

Wood forms are probably the most common and oldest method of forming a concrete structure. Even though the current technology and need for speed and efficiency have led us to metal forms, there is still a much-needed market and use for wood forms, both independent use and in conjunction with metal. Special uses, irregular sizes, closures, cost, and user preference are some advantages and/or requirements for wood forming systems. Wood forms are also built up of a structural banding of wood and have a wood face or skin.

For a detailed discussion on these types of forms and methods, refer to Chap. 3, "Formwork and Falsework."

Forms for concrete structures will be *mortartight* and adhere to the lines and dimensions of the formed area. The major concern is that they be of sufficient strength to prevent appreciable deflection and support the *live load* during the placement of concrete. Control must be taken in the placing of concrete within the limits of the forms to control the deflection and not cause undue stress to the form designed specifications. An even rate of placement must be maintained, controlled by delivery, slump, and ambient temperature.

Concrete placement. Prior to the placement of concrete, all *embedded* materials required within the section, both permanent and indirect, must be in place. This includes reinforcing steel, mechanical form inserts, and miscellaneous embedded items. All work associated with the construction and securing of the formwork must be completed, including temporary *bracing, shoring,* falsework, and all *fillets* and *chamfered corners*.

A thorough inspection of the system and unit must be performed. The removal of all foreign material such as dirt, sawdust, debris, and water must be completed. The interior of the forms must be cleaned of all dirt, mortar, and foreign material. The removable forms must be well oiled and/or thoroughly coated with an approved form oil to aid in the release of the formwork. All *contact* areas will be lightly moistened with water immediately prior to concrete placement.

Forming systems must follow the manufacturer's recommendations and specifications when applicable.

Standard engineering practices must be adhered to. Normal specifications for deflection will not exceed $\frac{1}{360}$ of the span or unit under full load. The design load of formwork will consist of the sum of the *dead* and live vertical load with an assumed horizontal load.

When concrete containing a retarding admixture of fly ash or other *pozzolan* replacement is used, the formwork will be designed for a lateral pressure equal to that exerted by a fluid weighing 150 lb per cubic foot, unless a special mix design or manufacturer's recommendations supersede this.

The removal of formwork will not take place until the concrete has reached sufficient design strength or percentage of strength required for form stripping.

Once strength is obtained, all forms must be removed from the finished structure, and the exposed surface area *rubbed and patched.*

Care should be taken during the formwork system removal not to damage the structural concrete unit or the form system. The forms should be well cleaned of *laitance* and foreign materials, oiled, and stored in such a manner to preserve the integrity of the form panel.

An approved *curing compound* should be adequately applied to the structural concrete unit to continue the appropriate curing time for the designed strength and uniform appearance.

TAKEOFF QUANTIFICATION

Introduction

This section defines the components and technique required to compile a detailed takeoff and identify the variables of a substructure unit.

A substructure unit, though it is compiled as a singular unit, is comprised of numerous individual components. These components, which are constructed separately, are internally connected to form the entire structural unit. Each structure has its own design as to shape, intended use, and structural integrity, but the basic components are required in some format. The variance of circumstances and requirements to the use of the structure may tend to add or alter these components, but the desired function remains the same.

The takeoff of the substructure units will be comprised of the format of square feet and cubic yard units of measurement, the concrete surface contact

Minimum Required Percentage of 28-Day Strength ($f'c$)

	Standard element	Percentage of 28-day strength
1.	Columns, walls, footers, abutments, mass piers (not yet supporting loads)	50
2.	Pier caps, struts, caps, cross beams (not continuously supported)	80
3.	Pier caps (continuously supported)	60

area, in square feet, for forming and stripping, and a cubic yard measurement for the volume of required concrete. These are the two controlling operations of the methodology used.

The secondary items such as preparation and curing will be relative to the operation that it is applicable for, normally square feet and cubic yard unit measurements. Falsework and scaffolding, heat and housing, will be treated as specific units, and incidental embedded items will be identified as each independent unit. These units, however, should be treated as incidental to the controlling unit of measure, which is cubic yards of concrete. At most, the incidental operation units may be accounted for and realized within a secondary operation such as forming or preparation and then summarized within the controlling item component.

Following an outlined, procedural format will aid with the preparing of an accurate and detailed takeoff.

Definition

The substructure area of the bridge structure, with regard to quantity takeoff, will be defined as the structural components extending from the limits of the foundation preparation or piling area to the beam structure portion. It will consist of the concrete portions with the work operations of preparation, falsework and scaffolding, forming, heating and housing, placing embeds, concrete placement, stripping, and curing of said components.

Takeoff

The takeoff of substructure units must be completed in a systematic manner with regard to the construction techniques to be utilized, and the order of construction to be performed on the unit and specific components. Normally, the substructure units are constructed from the foundation unit upward and outward. This manner, a stacking method, plays an important part in the takeoff procedure.

Categorizing components

Each component must be identified as to the required form surface contact area and the component contact area. This will determine the surface requirements of the component. Once the surface areas have been defined, the volume area of each component will be identified. The functional difference dictates the takeoff and quantifying of component volumes and surface areas. The formwork contact area is the requirement needed to construct the formwork to the image and shape of the component, and to withstand and contain the volume of concrete. It is the surface that contacts with the concrete.

The component contact area is the part or surface area of each component that comes in direct contact with another structural component and does not require any formwork. This area can be either a horizontal or vertical surface. Areas of blockouts, bulkheads, and recesses are quantified by additional and separate takeoff procedures.

By categorizing various components, and work operations within each component, one will be able to structure a library of specific or average craft-hour production factors. These factors will then be used in the formulating of the labor requirements for each component of a project. The work operations should be maintained as a constant for summarization.

Every component within a project will be calculated for a labor usage and performance by the factoring method, applying a craft-hour factor to the quantity to calculate the required craft-hours. By categorizing like or similar work operations within similar components, a firm and assured production factor can be maintained.

The procedures outlined in this section make the assumption, with regard to formwork, that the actual fabrication and construction of the wood formwork is previously done prior to the discussed operation. The method of unit craft-hour production factoring is primarily the function of erecting the formwork system with minor construction and/or alterations to the forms themselves.

The approach taken by this book is to attach the cost of actual form construction to the formwork as a direct cost. (See Chap. 3 on form type and construction.)

Within these parameters of work items, there exist certain details of work performance that are requirements of performing the actual work. The formwork operation, of which the prime function is the setting of the formwork, will include any minor modifications required, the construction of minor formwork, and the application of form oil or form release compound which aids in the removal of the formwork.

The concrete placement operation, of which the prime function is the pouring of the concrete, will include the final cleaning and preparation of the formed area prior to placement, and the covering, curing, and protection of the freshly placed concrete until the end of the required cure period.

The formwork removal operation, of which the prime function is stripping the formwork from the surface area, will include the cleaning of the formwork, and minor required repairs, reoiling the formwork for skin or surface protection, and properly stacking or storing readying for removal from the immediate site.

Rubbing, patching, and surface treatment will be of the function to fill and smooth formwork and hardware holes, and voids and crevasses within the concrete surface, and uniformly and cosmetically apply a paste to the concrete surface. This operation is not to be associated with surface texturing, which is a direct item of work independent of this. All surface areas of a component may not be required or have the need to be rubbed or patched. With some specifications, a different form of rubbing or patching may be required for specific areas of the component. This must be accounted for within the takeoff procedure.

Blockouts, beam seats, and bulkheads

Blockouts are to be calculated by the actual surface area required for the blockout or recessed area and should be quantified in a square foot item summary. This includes the direct contact surface area, plus the perimeter thickness of dimension area required by the plans.

Beam seats are a component of an abutment or pier, which are the bearing surface areas which the beam structure components set on. These are categorized in two areas of performance, surface beam seats and riser blocks. The beam seats are located on the top of the abutment wall or pier cap and are spaced equally and symmetrically to the beams or girders of the beam structure.

Bulkheads within the formwork sections are formed areas which are intended for the containment of the concrete at the end of a component section, or midpoint of the section. This is merely an end closure section or an intermediate separator which becomes a formed construction joint. The construction joint, or a defined joint between two placement sections, falls within a formed section of multiple designated pours. The classification of bulkheads is end or intermediate, and is accounted for in a unit of square feet.

The construction of falsework, scaffolding, and shoring will consist of only the direct requirement to support the formwork. This item is not intended for major structural falsework and scaffolding. These items are addressed elsewhere.

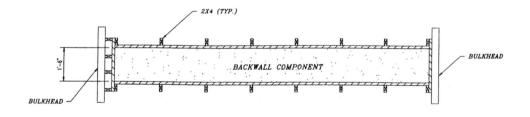

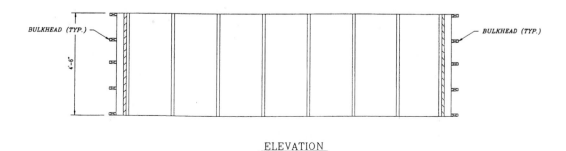

Vertical bulkhead construction joint.

Heating and housing constitute all required work necessary to build, maintain, and remove the enclosure needed to adequately protect a given component from the elements of raw weather.

The requirement of accounting for incidental embedded items will be accumulated within the formwork summary, for which the need of embeds is initiated. The embeds are normally anchors that permit and aid in the attachment of future formwork, falsework, or scaffolding. These are summarized in a quantity of each and distributed over the required formwork section.

In the performance of takeoff measurements, the occurrence of varying dimensions within the component arises. The solution to this, for ease of calculation, is averaging. This is simply taking the difference between two given points as the governing dimension. An example: a wall that is 20 lineal feet long, having vertical dimensions of 10' at one end and 6' at the opposing end, decreasing at a constant rate. Taking the averaging solution for calculating, the vertical dimension is 8', thus calculating the wall as 8' by 20'.

The performance of a substructure takeoff requires the identifying of defined measurement areas and actual structural components defined by the project plans.

The outline of the takeoff begins at the footing or lowest level of the substructure unit and progresses upward to each individual component.

The main units of the substructure are abutments, piers or intermediate bents, and retaining walls.

Abutment Unit

The substructure abutment units are located at each end of the bridge structure where the normal roadway portion meets the bridge portion. Within the substructure abutment unit, the normal components are defined as:

1. Abutment and wing footings
2. Abutment walls
3. Backwalls
4. Wing walls
5. Beam seats

Within these structural components exist the following operations of work:

1. Preparation of the work area
2. Falsework, scaffolding, shoring construction
3. Formwork installation
4. Bulkhead installation
5. Incidental embedded item installation
6. Placement and curing of concrete
7. Formwork removal
8. Rub and patch concrete surface

286   **Takeoff and Cost Analysis Techniques**

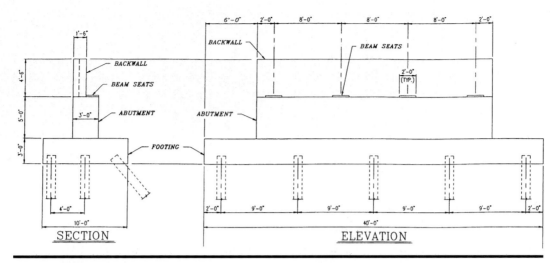

Abutment footer component.

The abutment structures are the units which begin and end the bridge unit. The preparation for the construction of the substructure unit begins with the verification of the foundation unit. The foundation unit, i.e., foundation piling, excavation and/or backfill, and foundation preparation will be completed and identified prior to the component.

Abutment footer

The footer component is the first order of activity. The footer unit normally rests and is founded directly on the foundation area; therefore, there is no requirement of formwork for the bottom surface area. This is referred to as the component contact area. The grade and elevation of the foundation contact area are verified as to defined tolerances for control of the footer construction. These checks should include elevation and dimension verification.

The perimeter of the footer component is then calculated for formwork contact surface area. This dimension calculation requires the continued perimeter of the footer measurement multiplied by the height measurement from the bottom elevation to the top elevation of the footer. This computation is then quantified in a unit of square feet of required formwork, summarized as: formwork, abutment footer.

The bulkhead takeoff requirement is summarized at this time. This calculation consists of the length of the bulkhead multiplied by the height, multiplied by the number of bulkheads required by plans, and summarized as: required bulkhead, abutment footer.

The embedded item quantifying would be defining any incidental embeds required within the abutment wall. These would be summarized in a unit of each as: abutment footer, embeds.

The quantity of formwork stripping, or removal, will be identical for the required formwork placement quantity, summarized as: formwork stripping, abutment footer.

The concrete requirements are calculated by a simple volume determination, the width, multiplied by the depth, multiplied by the length, of the total area of the footer to be filled by the concrete. When there are quadrants of variable volumes and sizes, each rectangular area is calculated independently, then summarized together. This quantity, of cubic feet, is then translated to a volume measurement of required cubic yards: concrete placement, abutment footer.

Rubbing and patching is calculated by the surface area, in square feet, of the component that is required to be completed. The quantity is then summarized as: abutment footer, rub and patch.

Abutment wall

The abutment wall component rests on the footer component. With this known, there will exist no formwork requirement for the bottom horizontal surface of the wall section, or component contact area.

The wall itself is comprised of a front side, back side, and two end areas. The formwork requirement is calculated by multiplying the length of the wall by the "average" height, to determine the square foot surface area. This calculation is performed for the front face as well as the back face. The average height in elevation is determined because the wall normally has variable vertical dimensions. The component unit is summarized as: required square feet of formwork, abutment wall.

If an intermediate construction joint exists within the wall, it requires an internal bulkhead forming system. This is summarized as a required square foot: abutment wall, internal bulkhead.

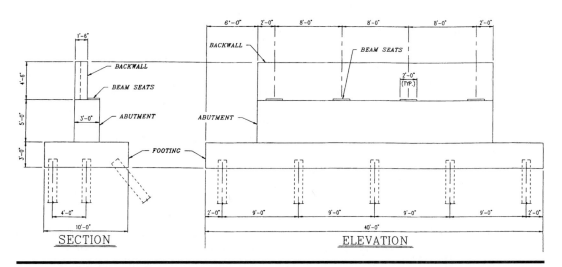

Abutment component.

The vertical ends of the abutment are classified as end bulkheads. The calculation of this operation is the width of the end multiplied by the height of the end, by the number of ends required. This is completed and summarized in square feet as: abutment wall, end bulkhead. The requirement of an internal bulkhead is calculated in the same manner and summarized in square feet as: abutment wall, intermediate bulkhead.

The embedded item quantifying would be defining any incidental embeds required within the abutment wall. These would be summarized in a unit of each as: abutment wall, embeds.

The quantity of formwork stripping, or removal, is identical for the required formwork placement quantity, summarized as: formwork stripping, abutment wall.

The concrete requirements are calculated by simple volume methods, multiplying the wall length, by the wall height, by the depth or thickness, averaging dimensions when required. This cubic foot volume is then converted to a unit measurement of cubic yards, summarized as: concrete placement, abutment wall.

Rubbing and patching is calculated by the surface area, in square feet, of the component that is required to be completed. The quantity is then summarized as: abutment wall, rub and patch.

The surface beam seats are a calculation of direct surface area in square feet with no volume dimension. That would be the width by the length, by the number of required beam seats per abutment. The quantity summary would be the required square foot: abutment wall, surface beam seats.

The riser blocks are calculated in square feet by the perimeter of the riser multiplied by the height of each riser, by the number of required riser blocks for each abutment. The quantity summary would be the required square feet: abutment wall, riser block.

The rubbing and patching operation for the beam seats would be a calculation of required surface area to be treated. The summation would be in square feet of: abutment beam seat, rub and patch.

Abutment backwall

The abutment backwall is the structural component of the abutment structure which is located above the abutment wall and directly behind the beam structure unit. The abutment backwall is constructed and rests directly on the abutment wall. The wall is comprised of a front side, back side, and two ends. There may be an intermediate construction joint within the abutment backwall, which would require an internal bulkhead.

The abutment backwall component may be constructed independent of the beam unit or be dependent to the beam unit, thus being identified as a dependent or independent backwall unit.

The backwall component also acts as a retaining structure for the earthen backfill behind the bridge unit. This is the structural support for the roadway approach slab to the structure. The top design of a backwall component may entail a bearing seat for the approach slab section to rest on. This will assure

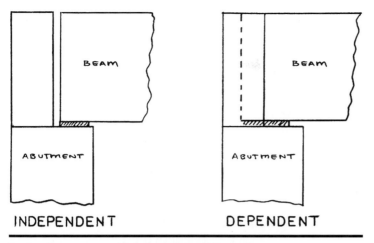

Backwall component. (*a*) Independent. (*b* Dependent.

a firm foundation and a secure seat for structural integrity. When the wall requires a bearing seat for a pavement slab, it is suggested to maintain this type of wall as a separate component for factoring. This is due to the need of additional formwork requirements that will differ in the production from a normal flat-surfaced wall.

An independent backwall is constructed as a separate wall component that is not connected or encompasses the beam structure unit. This wall is normally constructed prior to the placement of the beam structure. The independent backwall normally is the location of the expansion movement area of the beam and superstructure unit.

A dependent backwall is constructed after the beam structure is in place. The dependent backwall encompasses the ends of the beams and fixes them to the abutment unit. This normally is the fixed or stationary portion of the beam and superstructure unit.

The takeoff of the backwall component is similar to the abutment wall. The bottom surface of the backwall bears on the abutment wall, thus not requiring the need of formwork for this horizontal area, component contact area.

The backwall may also have a variance to elevation which would require an averaged height calculation. The formwork requirement would be the surface contact area of the front and back faces of the backwall component.

Independent backwalls. The formed surface area for a front face of an independent backwall would calculate as the length, from the left side to the right of the wall, multiplied by the average height, giving square feet of wall area. The backside of the backwall would be calculated in a similar manner with the exception of an approach slab seat, if required. This dimensional area would be added to the surface calculation for the formed area. The quantity summation would be:

1. The requirement of square feet: independent abutment backwall, formwork
2. The requirement of square feet: independent abutment backwall and roadway seat, formwork

Dependent backwalls. A dependent backwall section requires a somewhat different calculation with respect to the area around the beam unit. The front side of the backwall must be formed within the perimeter shape of the corresponding beam. This is referred to as scribing. The formwork shaping, or scribing, will vary in production with respect to different beam types, and the number of beams within the dependent backwall. An additional factor must be applied to the craft-hour production factor for this operation.

The illustration shows a beam section with four beams, which is the view of the front face of a dependent backwall section. The contact surface formwork requirement for the backwall is the closure area between the beams, or interior bays, and the end area adjacent to the outside beams. (The interior bay view is a similar occurrence for diaphragm construction with a beam unit.) The formed interior panels required are three, since there are four beams. The requirement is one less interior bay than the number of required beams, plus normally two exterior panels.

The calculation requirement to arrive at the surface contact area for the formwork is done by determining the horizontal dimension between the beam centerlines, and the vertical height or elevation difference from the top of the abutment wall to the top of the proposed backwall (averaging the height if required). This will be the dimensions used for the interior bay calculation for

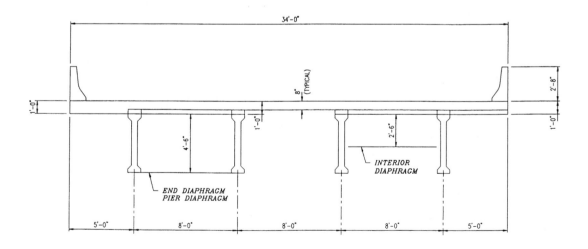

SUPERSTRUCTURE CROSS SECTION

Four-beam unit section component.

the backwall. The width of each interior bay will be calculated because the beam spacing may vary.

This format uses a theoretical vertical line from the center to center of beam for measurement purposes. It aids in the ease of takeoff, even though the volume or shape of the beam is not reduced from the measurement. This method accounts for the additional labor required to scribe the beam shape and the added formwork usage in the formwork production factor for a dependent backwall, front face.

With precast concrete beams, for the additional labor consumption of scribing, it is suggested to use a factor of 0.01*, multiplied by the end section surface area of the beam in square inches, multiplied by the number of beams within the dependent backwall component. The factor is categorized as: crafthour per square inch factor. This is the reasoning behind the need for categorizing structural components.

Steel girder scribing is factored using a different method. The end section total perimeter dimension, in feet, is used rather than the end section surface area, in inches. This dimension is multiplied by a 0.35* production factor for the additional labor requirement, used for the formwork shaping, or scribing.

The calculation for the front face of the dependent backwall is the summation of the multiplication of the width by the average height of each bay. This is the surface contact area of the interior bay areas of the backwall.

The end closure panels are, again, derived by the centerline method using the exterior beams. The square foot measurement required is the width from the beam centerline to the outside edge of the backwall, multiplied by the average height of the wall. For a given section, there normally will be two closure panels, one at each end of the abutment unit.

The summation of the front face of the dependent backwall will be the surface square feet of each of the interior bays, plus the surface square feet of each of the exterior closure panels, plus the scribing hours required with regard to total square feet.

The next calculation required is the surface contact area of the backface of the dependent backwall component. This is derived using the same method as with the independent backwall backface, the full surface wall calculated as the length of the wall multiplied by the average height, giving the required dimensions in square feet. The quantity summation would be:

1. The requirement of square feet: dependent abutment backwall, formwork
2. The requirement of square feet: dependent abutment backwall and roadway seat, formwork
3. The factor application for scribing beam area for a dependent abutment backwall (0.01 * beam end area surface square inches or 0.35 * beam end section lf, by the number of beams, for the additional labor usage)

The end sections of the backwall component would be classified as end or

---

*NOTE: These factors vary based on specific experience.

closure bulkheads. These exist with both conditions of backwalls. The surface contact area calculation required for the formwork will be the average thickness of the backwall section multiplied by the height of each end section, in square feet. When an approach slab seat is present, this end area must be added to the calculation in required square feet. The summation would be categorized as: end bulkhead square feet, abutment backwall.

The embedded item quantifying would be defining any incidental embeds required within the abutment backwall sections. These would be summarized in a unit of each as: (1) independent abutment backwall, embeds, or (2) dependent abutment backwall, embeds.

The requirement for the removal of the formwork system will be the identical square feet as required with the placement of the formwork. The quantity summation would be classified as:

1. The requirement of square feet: independent abutment backwall, formwork removal
   a. The requirement of square feet: independent abutment backwall and roadway seat, formwork removal
2. The requirement of square feet: dependent abutment backwall, formwork removal
   a. The requirement of square feet: dependent abutment backwall and roadway seat, formwork removal

Concrete volume requirements for placement calculations are compiled using the volumetric methods.

An independent backwall is calculated for concrete volume by simply multiplying the averaged width, by the height, by the thickness, then converting the resulting cubic feet to cubic yards. The volume requirement of a roadway bearing seat is computed using the same method, then added to the volume required for the backwall.

The dependent backwall calculation requirements for the concrete entail a somewhat more complicated method. To ease this, again, it is suggested to use the methodology used with the square foot takeoff. Since the dependent backwall encompasses the beam units, it is necessary to reduce the volume of concrete that is displaced by the beams.

The primary concrete volume calculation is completed by the same method as done with the independent backwall. The next procedure required is to reduce the displaced volume consumed by the end area of the beam section. This calculation is simply done by multiplying the end area square inches of a beam, by the dimension the beam extends into the backwall, then multiplying the volume cubic inches by the number of required beams. This is then converted to cubic yards. The total displaced cubic yards for the beam section is then subtracted from the primary volume, giving the net volume of required cubic yards.

The quantity summation volume requirement would be classified as: (1) concrete placement volume, cubic yards, independent abutment backwall, and (2) concrete placement volume, cubic yards, dependent abutment backwall

A calculation for surface area of required rubbing and patching will be the surface area required for this procedure. This will be reported in a square foot dimension of:

1. Independent abutment backwall, rub and patch
2. Dependent abutment backwall, rub and patch

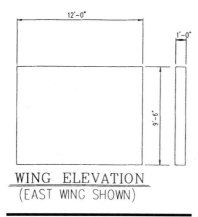

Wing component.

Abutment wing wall

The abutment wing wall components of an abutment unit are intended for the containment of fill and side support for the approach roadway area. They are located at the exterior perimeter of the abutment section, and extend either back toward the roadway or laterally from the abutment wall.

An abutment wing wall rests directly on the abutment footer and usually abuts the abutment wall and abutment backwall. There usually exist two abutment wing walls, one on each side of the abutment section, which are comprised of a front face, a back face, and one end.

The quantity takeoff requirements are similar to the other components of the abutment unit, forming, stripping, concrete placement, and rubbing and patching.

The front face formwork is calculated by multiplying the width of the wing wall by the height, averaging dimensions when required, to determine the square feet. This calculation is then repeated for the next wing wall, and so on. The width of the wall is determined by locating the point at which the wing wall abuts the abutment and backwall section, to the free end, or the transition end of the wing wall.

The back face of the wing wall is calculated using the same method. The quantity summation for the forming requirements is in square feet as: abutment wing wall, formwork.

The embedded item quantifying would be defining any incidental embeds required within the abutment wing wall sections. These would be summarized in a unit of each as: abutment wing wall embeds.

The requirement for the removal of the formwork system will be the identical square feet as required with the placement of the formwork within the abutment wing walls. The quantity summation would be classified as: The requirement of square feet: abutment wing wall, formwork removal

The end bulkhead requirement will occur at one end of the abutment wing wall, or the transition end. This closure bulkhead will be the end surface area at each abutment wing wall. The formwork calculation will be the height of the end section multiplied by the average width. The quantity summation will be in required square feet as: abutment wing wall, end bulkhead formwork.

The removal of the bulkhead will be the identical square feet summarized in the formwork and listed as: abutment wing wall, end bulkhead, formwork removal.

Concrete volume calculations are performed using the volume method. The area of the abutment wing wall is calculated by multiplying the average dimensions of width, by the height, by the thickness, converting the arrived-at cubic feet into cubic yards. This procedure is then repeated for subsequent wing walls, and added together. The quantity summation for concrete placement volume in cubic yards is: concrete placement volume, abutment wing wall.

The rubbing and patching requirements would be calculated for the surface area of the abutment wing wall, requiring the treatment. The quantity summation would be in required square feet of: abutment wing wall, rub and patch.

Pier Units

The pier units, or interior bents, are the intermediate supports that continue the bridge structure span throughout the entire structural unit. These intermediate units support the beam and superstructure units of the bridge and are required because of span limitations.

The pier units and intermediate bent structures consist of these normal components and are defined as:

1. Pier footings
2. Columns
3. Struts
4. Pier caps
5. Beam seats

Within these structural components, the following categories of work operations exist:

1. Preparation of the work area
2. Falsework, scaffolding, shoring construction

3. Formwork installation
4. Bulkhead installation
5. Incidental embedded item installation
6. Placement and curing of concrete
7. Formwork removal
8. Rub and patch concrete surface

Pier footer

The pier footer components are, in theory and function, identical to the abutment footers. They act as the foundation support tied to the foundation structure unit, such as the foundation piling or the foundation area.

The takeoff procedures for the pier footer are performed in the same manner as for the abutment footer. The footer rests directly on the foundation area, thus requiring no formwork operation for the bottom horizontal area, component contact area.

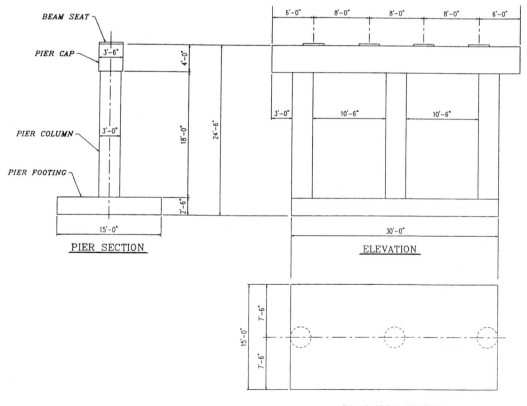

Pier footer component.

The square foot forming requirements for the pier footer are derived by summing the perimeter of the footer dimensions and multiplying by the designed thickness of the footer. This function is repeated for all the required pier footers. The takeoff quantity summation is calculated in required square feet for: pier footer, formwork.

The embedded item quantifying would be defining any incidental embeds required within the pier footer sections. These would be summarized in a unit of each as: pier footer, embeds.

The requirement for the removal of the formwork system is the identical square feet as required with the placement of the formwork within the pier footer.

The quantity summation would be classified as the requirement of square feet: pier footer, formwork removal.

Concrete placement volume calculations are performed using the volume method. The pier footer concrete volume requirements are calculated by multiplying the footer width, by the length, by the thickness or depth, then converting the cubic feet to cubic yards. This procedure is followed for each individual pier footer. The quantity summation required would be compiled as: pier footer, cubic yard placement.

The rubbing and patching requirements would be calculated for the surface area of the pier footer, requiring the treatment. The quantity summation would be in required square feet of: pier footer, rub and patch.

Pier column

The pier columns are the intermediate support component that connects the pier footer to the pier cap. The pier columns vary in section, number, and design, individually from square, to round, to a solid rectangular form. The height or elevation also is a variance due to surface conditions and requirements at the bridge site, and the number of columns required for a given pier vary. This is due to the width of the bridge at each pier, the total load applied to the pier, and logistics of design and structure location.

With some conditions of pier columns having a great height in elevation, there exists a need of a component referred to as an intermediate strut. This strut is a horizontal, midpoint structural member spanning between each independent column to add lateral stability to the columns.

Formwork takeoff. The formwork takeoff of pier columns first requires a determination of the type and shape of the column. The bottom horizontal surface of the pier column rests directly on the pier footer, thus requiring no formwork, this portion being the component surface contact area.

Square-shaped columns require a calculation of the summation of the perimeter of a column, multiplied by the height of the pier column. The procedure is repeated for each column; then each column is added together, giving a quantity per pier. The summation of square foot requirements is listed as: pier column, square, formwork.

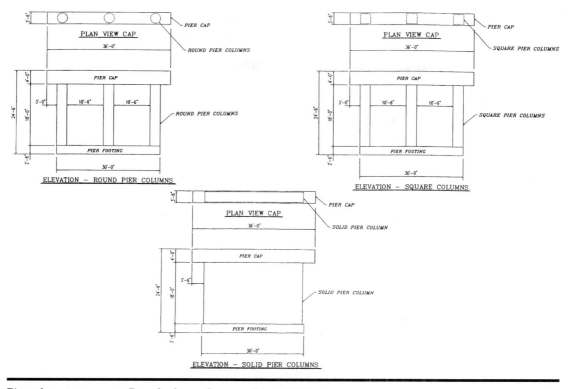

Pier column component. Round column. Square column. Solid column.

Round columns are calculated in a different manner. As with a square column, the perimeter must be calculated. Since these are round, the formula required is *pi* multiplied by the diameter of the column in feet, multiplied by the height of the column. This will give the square feet per column. The procedure is repeated for each column; then the columns are summarized, giving a total square feet accumulation for each pier. The quantity summation for square feet required is: pier column, round, formwork.

Rectangular or solid pier columns are calculated by a similar method used in square columns, with the exception that one exists per pier. The required square feet of formwork is calculated by summing the perimeter dimensions of the column, then multiplying by the average height. The quantity summation for square feet required is: pier column, solid, formwork.

The embedded item quantifying would be defining any incidental embeds required within the pier column sections. These would be summarized in a unit of each as: pier column, embeds.

The requirement for the removal of the formwork system will be the identical square feet as required with the placement of the formwork within the pier column. The quantity summation would be classified as: the requirement of square feet:

1. Pier column, square, formwork removal
2. Pier column, round, formwork removal
3. Pier column, solid, formwork removal

Concrete takeoff. Concrete placement volume calculations are performed using the volume method. For square pier columns, the calculation will be multiplying the width, by the depth, by the height of the column, then converting the required cubic feet summation into cubic yards. This procedure is repeated for each column; then each column is summarized giving a required quantity for the entire pier.

The concrete placement volume calculation requirement for a round column is computed by multiplying *pi* by the radius of the column, squared (in feet). The result of this formula is then multiplied by the height of the column, producing a cubic foot volume. This quantity is then converted to required cubic yards.

This procedure is repeated for each column; then each column is accumulated, giving a total required concrete placement quantity for the pier.

When calculating the required concrete placement volume for a solid pier, the formula used is, multiply the width, by the depth, by the average height. The result, in cubic feet, is then converted to required cubic yards per pier.

The takeoff quantity summation for required concrete placement is defined in cubic yards as:

1. Pier column, square, concrete placement
2. Pier column, round, concrete placement
3. Pier column, solid, concrete placement

The rubbing and patching requirements would be calculated for the surface area of the pier column, requiring the treatment. The quantity summation would be in required square feet of: pier column, rub and patch.

Intermediate strut

The takeoff procedure of the intermediate strut requires the surface accumulation of the surface contact area of the component member. The strut is being supported between the columns, at a point in elevation midway between the pier footer and pier cap. With the strut being built after the column construction, there is no horizontal component contact area as found with the other members. At the point of contact at the pier columns, a vertical component contact surface area does exist at each column.

If the columns are constructed in a two-tier manner, as a lower column unit and an upper column unit, the strut can be built during this operation. When this occurs, there will exist a horizontal component contact area, only that surface point at which the bottom surface of the strut contacts the top of the lower column units.

Strut formwork takeoff. The strut formwork takeoff requires a calculation of the bottom, or soffit, of the member, plus the side surface areas. If the columns are constructed in a two-tier method, end closure formwork requirements to the strut will exist. With this procedure, the strut is constructed over the columns, full width.

The strut must have a form of structural support to aid in the formwork, placement, and removal procedure due to the lack of a horizontal component contact surface. The strut also requires a platform for the personnel to work from. This can be accomplished by a scaffolding system and falsework or shoring, or a self-supported form system anchored to the previously constructed lower columns.

The formwork takeoff procedure requires a calculation of surface contact area of the sides and bottom soffit. When the strut is constructed between the columns, the quantity calculation is the surface area of the height of the sides multiplied by the length dimension between each column, for both sides. The

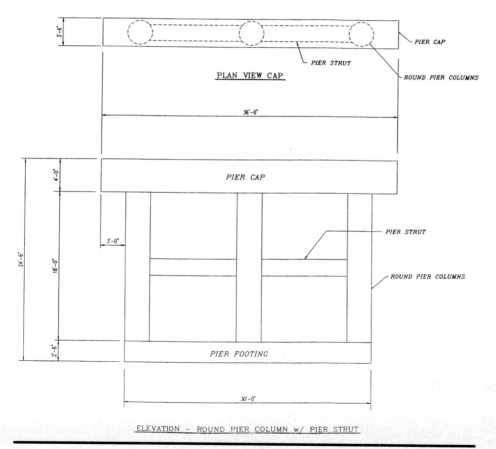

Pier strut component.

next calculation is the area of the width of the bottom multiplied by the length dimension between each column.

When the strut is constructed over the columns, the quantity calculation is the surface area of the height of the sides multiplied by the length dimension from the outside of the left farthest column, to the outside of the right farthest column, for both sides. The next calculation is the area of the width of the bottom multiplied by the length dimension between each column. The quantity summation required in square feet is listed as:

1. Pier strut, interior, formwork (interior struts)
2. Pier strut, exterior, formwork (full width struts)

End bulkhead requirement. The end bulkhead requirement will occur only when an exterior strut is constructed. This closure bulkhead will be the end surface area at each exterior column and will be the same configuration as the columns. The formwork calculation will be the height of the end section multiplied by the width, or ½ the perimeter diameter multiplied by *pi* (if circular columns) for each end. The quantity summation will be in required square feet as: pier strut, end bulkhead formwork.

The removal of the bulkhead will be the identical square feet summarized in the formwork and listed as: pier strut, end bulkhead, formwork removal.

The requirement for the removal of the formwork system will be the identical square feet as required with the placement of the formwork within the pier strut. The quantity summation would be classified as the requirement of square feet:

1. Pier strut, interior, formwork removal
2. Pier strut, exterior, formwork removal

Other items. The embedded item quantifying would define any incidental embeds required within the pier strut sections. These would be summarized in a unit of each as: pier strut, embeds.

The scaffolding or shoring requirements would be quantified as each unit required per pier strut.

Concrete placement volume calculations are performed using the volume method. For an interior strut, the calculation will be the width of the strut, multiplied by the length, multiplied by the height or depth, between each column; then each span is summarized and the cubic feet is converted to cubic yards.

For an exterior strut, the calculation will be the width of the strut, multiplied by the length, multiplied by the height or depth; then the cubic feet is converted to cubic yards. The quantity summation for the strut concrete volume requirement is listed in cubic yards as:

1. Pier strut, interior, concrete placement volume
2. Pier strut, exterior, concrete placement volume

The rubbing and patching requirements would be calculated for the surface area of the pier strut, requiring the treatment. The quantity summation would be in required square feet of: pier strut, rub and patch.

Pier cap

The pier cap is the structural component that is located at the top of the pier, on which the beam structure unit bears. The pier cap, owing to the width requirement of fully encompassing the limits of the beam structure, requires portions of the cap to be not resting on or cantilevered from the column section. Therefore, there exists the multiple bottom horizontal condition; part of the cap is required to be formed, and the part which actually bears on the columns or the component contact area.

The quantity takeoff requirements for a pier cap start with the underside, or soffit, portion of the cap. The continuing requirements are the side surface area and the end or closure surface areas. Keep in mind, the pier cap must have a method of structural support to aid in the formwork, placement, and removal procedure owing to the lack of a horizontal component contact surface. The only portion of the pier cap that has a component contact area is the horizontal surface portion directly over the pier columns. The pier cap also

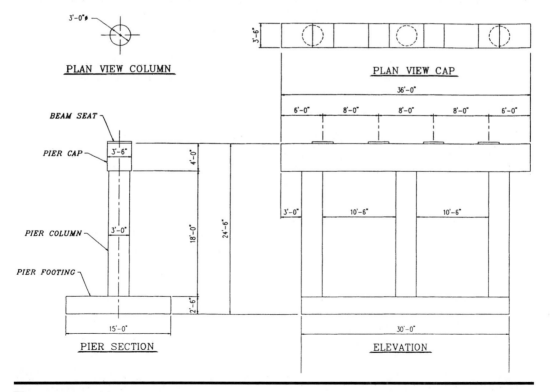

Pier cap component.

requires a platform for the personnel to work from. This can be accomplished by a scaffolding system and falsework or shoring, or a self-supported form system anchored to the previously constructed lower columns.

The formwork requirements for the pier cap will be calculated similar to the pier strut. The bottom horizontal surface, or soffit, is computed by multiplying the width of the soffit by the length of each soffit panel between each column, then summing each quantity. The side panels are calculated by multiplying the average height of each side in elevation by the length of the side, for each side. These side panel quantities are then added together for total side panel square feet, then summarized with the bottom soffit quantity.

The contact surface area quantity for required square feet would be summarized as: Pier cap, formwork

The requirement for the removal of the formwork system is the identical square feet as required with the placement of the formwork within the pier cap.

The quantity summation would be classified as the requirement of square feet: Pier cap, formwork removal

If an intermediate construction joint exists within the pier cap, it would require an internal bulkhead forming system.

The vertical ends of the pier cap are classified as end bulkheads. The calculation of this operation is the width of the end multiplied by the height of the end, by the number of ends required. This is completed and summarized in square feet as: pier cap, end bulkhead.

The requirement of an internal bulkhead would be calculated in the same manner and summarized in square feet as: pier cap, intermediate bulkhead.

The embedded item quantifying would be defining any incidental embeds required within the pier cap sections. These would be summarized in a unit of each as: pier cap, embeds.

The scaffolding or shoring requirements would be quantified as each unit required per pier cap.

Concrete placement volume calculations are performed using the volume method. For a pier cap, the calculation is the width of the cap, multiplied by the length, multiplied by the height or depth, and the cubic feet is converted to cubic yards.

The quantity summation for the pier cap concrete volume requirement is listed as: Pier cap, concrete placement volume

The rubbing and patching requirements are calculated for the surface area of the pier cap, requiring the treatment. The quantity summation will be in required square feet of: pier cap, rub and patch.

Pier cap beam seats

The surface beam seats, as with the abutment wall, are a calculation of direct surface area in square feet with no volume dimension. That would be the width by the length, by the number of required beam seats per pier cap. The quantity summary would be the required square feet: pier cap, surface beam seats.

The riser blocks are calculated in square feet by the perimeter of the riser multiplied by the height of each riser, by the number of required riser blocks for each pier cap. The quantity summary would be the required square feet: pier cap, riser block.

Retaining Wall Unit

A retaining wall is a structural unit designed to retain or support earthen fill, roadways, or other units of a varying elevation to the adjacent area.

Retaining wall and subsequent units contain these normal components and are defined as:

1. Wall footings
2. Wall units
3. Counterforts and struts
4. Parapets

Within these structural components, the following categories of work operations exist:

1. Preparation of the work area
2. Falsework, scaffolding, shoring construction
3. Formwork installation
4. Incidental embedded item installation
5. Placement and curing of concrete
6. Formwork removal
7. Rub and patch concrete surface

Retaining wall footer

The retaining wall footer unit normally rests and is founded directly on the foundation area; therefore, there is no requirement of formwork for the bottom surface area. This is the component contact area. As with the abutment footer, the grade and elevation of the foundation contact area are verified as to defined tolerances for control of the footer construction. These checks should include elevation and dimension verification.

Retaining walls are normally constructed in continuing sections of varying height elevations. With this in mind, the quantity takeoff procedure should be performed for each component within each section, and summarized accordingly. These sections are referred to as panels, which will consist of each identified component.

The first footer section is identified as the beginning unit, and will be considered to have four sides, with each subsequent section being an abutting section, and will have three sides, and so on. This will be assumed for each component, walls, parapets, etc., unless a *skip pour* method is used.

The perimeter of the first footer component is then calculated for formwork contact surface area. This dimension calculation requires the continued perimeter of the footer measurement multiplied by the height measurement from the bottom elevation to the top elevation of the footer. The next subse-

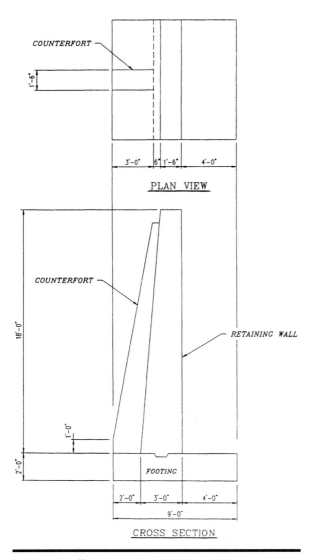

Retaining wall component.

quent footer sections are calculated by summing the two sides and the open end, then multiplying by the height of each footer section. These computations are then quantified in a unit of square feet of required formwork, summarized in square feet as: retaining wall footer, formwork.

When the total perimeter of all the footers is chosen to be formed in a continuous system, the need of intermediate bulkheads will exist at each of the construction joints. The bulkhead takeoff requirement would be summarized at this time. This calculation would consist of the length of the bulkhead multiplied by the height, multiplied by the number of bulkheads required by plans, and summarized in square feet as: required bulkhead, retaining wall footer.

The embedded item quantifying would be defining any incidental embeds required within the wall footer. These would be summarized in a unit of each as: retaining wall footer, embeds.

The quantity of formwork stripping, or removal, will be identical for the required formwork placement quantity, summarized as: formwork stripping, retaining wall footer.

The quantity of bulkhead stripping, or removal, is identical for the required bulkhead placement quantity, summarized as: formwork stripping, retaining wall footer bulkhead.

The concrete requirements are calculated by a simple volume determination, the width, multiplied by the depth, multiplied by the length, of the total area of the footer to be filled by the concrete. When there are quadrants of variable volumes and sizes, each rectangular area is calculated independently, then summarized together. This quantity, of cubic feet, is then translated to a volume measurement of required cubic yards: concrete placement, retaining wall footer.

Rubbing and patching is calculated by the surface area, in square feet, of the component that is required to be completed. The quantity is then summarized as: retaining wall footer, rub and patch.

Retaining wall panels

The first wall section is identified as the beginning wall unit and will be considered to have two sides, a front face and a back face and two end bulkheads. Each subsequent wall section, being an abutting section, will have two sides, and so on and one end bulkhead. This will be assumed for each wall panel throughout the wall unit. The bottom horizontal surface area that will rest on the footer component will be considered as the component contact area, thus requiring no formwork.

The front face of the wall panel component is calculated for formwork contact surface area. This dimension calculation requires the width of the wall panel multiplied by the average height measurement from the bottom elevation to the top elevation of the wall panel. The back face of the first panel is calculated in the same manner, and the two square foot quantities are summarized. The next subsequent wall sections are calculated in the same manner. These computations are then quantified in a unit of square feet of required formwork, summarized as: retaining wall panels, formwork.

A note to remember: there exist conditions when the backwall panels of retaining walls are battered for structural integrity, and the bottom of the wall panel is thicker in section than the top. When this condition exists, the thickness of the panel must be averaged for dimension and volume calculations. Also with this condition, the actual height of the back face, for the formwork requirement, will be longer than that of the front face owing to the slope.

When the total, continuous wall panels are chosen to be formed in a continuous system, the need of intermediate bulkheads will exist at each of the construction joints. The bulkhead takeoff requirement would be summarized at this time. This calculation would consist of the average width of the bulkhead multiplied by the height, multiplied by the number of bulkheads required by plans, and summarized in square feet as: required bulkhead, retaining wall panel. The same calculation is applied for end bulkheads.

The embedded item quantifying would be defining any incidental embeds required within the wall panel. These would be summarized in a unit of each as: retaining wall panel, embeds.

The quantity of formwork stripping, or removal, will be identical for the required formwork placement quantity, summarized as: formwork stripping, retaining wall, panel.

The quantity of bulkhead stripping, or removal, will be identical for the required bulkhead placement quantity, summarized as: formwork removal, retaining wall panel, end or intermediate.

The concrete requirements are calculated by a simple volume determination, the width, multiplied by the average thickness, multiplied by the height, of the total area of each wall panel to be filled by the concrete. Each wall panel will be calculated independently, then summarized together. This quantity, of cubic feet, will then be translated to a volume measurement of required cubic yards: concrete placement, retaining wall panel.

Rubbing and patching is calculated by the surface area, in square feet, of the component that is required to be completed. The quantity is then summarized as: retaining wall panel, rub and patch.

Retaining wall parapet and subsequent lift wall panel

The retaining wall parapets are a top wall section intended for a barrier or defined separator, not necessarily a structural support or retention component. An additional lift of second tier wall panels are those which are stacked on the lower wall construction. This method is normally chosen for structural design requirements, constructability due to height, or of differing conditions with respect to wall panel usage.

The quantity takeoff procedures for retaining wall parapets or additional lift pours are performed identical to those of a normal retaining wall, first lift. The first parapet or wall section is identified as the beginning wall unit and will be considered to have two sides, a front face and a back face, and two ends. Each subsequent wall section, being an abutting section, will have two sides and one end, and so on. This will be assumed for each parapet or wall panel throughout the wall unit. The bottom horizontal surface area that will rest on the lower wall component will be considered as the component contact area, thus requiring no formwork.

Subsequent or additional lift wall construction and takeoff will require one more work activity than the lower wall unit. The stacked lift method requires the formwork to be supported on the lower wall section, as well as a scaffolding requirement for the workforce. The higher or more additional lifts that are required, the more distressing or slowing of the production operation. The normal work requirements associated with this method will be factored within this work operation, thus the need of the additional and separate work operation classifications, by lift are required.

The front face of the parapet and wall panel component is calculated for formwork contact surface area. This dimension calculation requires the width of the wall panel multiplied by the average height measurement from the bottom elevation to the top elevation of the wall panel. The back face of the first panel is calculated in the same manner, and the two square foot quantities are summarized. The next subsequent parapet and wall sections are calculated in the same manner. These computations are then quantified in a unit of square feet of required formwork, summarized as:

1. Retaining wall parapet panels, formwork
2. Retaining wall additional lift #_____, formwork

A note to remember: as with the lower wall units, there may exist conditions when the back wall panels of retaining walls are battered for structural integrity, and the bottom of the wall panel is thicker in section than the top. When this condition exists, the thickness of the panel must be averaged for dimension and volume calculations. Also with this condition, the actual height of the back face, for the formwork requirement, will be longer than that of the front face owing to the slope.

When the total, continuous parapet and wall panels are chosen to be formed in a continuous system, the need of intermediate bulkheads will exist at each of the construction joints. The bulkhead takeoff requirement would be summarized at this time. This calculation would consist of the average width of the bulkhead multiplied by the height, multiplied by the number of bulkheads required by plans, and summarized in square feet as: required bulkhead, end or intermediate, retaining parapet and wall panel.

The embedded item quantifying would be defining any incidental embeds required within the parapet and wall panel. These would be summarized in a unit of each as: retaining parapet and wall panel, embeds.

The quantity of formwork stripping, or removal, will be identical for the required formwork placement quantity, summarized as:

1. Formwork removal, retaining wall parapet panel
2. Formwork removal, retaining wall additional lift #_____

The quantity of bulkhead stripping, or removal, will be identical for the required bulkhead placement quantity, summarized as: formwork stripping, retaining wall panel bulkhead.

The concrete requirements are calculated by a simple volume determination, the width, multiplied by the average thickness, multiplied by the height, of the total area of each wall panel to be filled by the concrete. Each parapet and wall panel will be calculated independently, then summarized together. This quantity, of cubic feet, will then be translated to a volume measurement of required cubic yards for:

1. Concrete placement, retaining wall parapet panel
2. Concrete placement, retaining wall, additional lift #_____

Rubbing and patching is calculated by the surface area, in square feet, of the component that is required to be completed. The quantity is then summarized as: retaining parapet and wall panel, rub and patch.

Counterforts

A counterfort is a structural support which is located on the back face of a retaining wall unit. It serves as a vertical strut for counteractive movement, or to prevent a forward movement or rollover effect to the wall section. A counterfort normally is designed as a tapered component, from the bottom to the top.

The takeoff of a counterfort is done in accordance with the normal procedures outlined. The counterfort can be constructed during and with the wall component as an integral section, or separately after the construction of the wall panel. The order of operation method should be addressed at takeoff time for proper allocation of labor usage.

The formwork contact surface will consist of the two vertical sides along with the vertical back face. The front side of the counterfort (the surface against the wall panel) and the horizontal bottom surface are designated as component contact areas.

The formwork requirement for a side of the counterfort is calculated by multiplying the average width by the height of the counterfort. The adjacent side is calculated in the same manner, then the square foot quantities of the two sides are summarized. The back face of the counterfort is calculated, multiplying the width by the vertical or sloping height. These quantities are summarized for the required square feet as: retaining wall counterfort, formwork, for each counterfort.

The embedded item quantifying would be defining any incidental embeds required within the counterfort. These would be summarized in a unit of each as: retaining wall counterfort, embeds.

The quantity of formwork stripping, or removal, is identical for the required formwork placement quantity, summarized as: formwork removal, retaining wall counterfort.

The concrete requirements are calculated by a simple volume determination, the width, multiplied by the average thickness, multiplied by the height, of the total area of each counterfort to be filled by the concrete. Each counter-

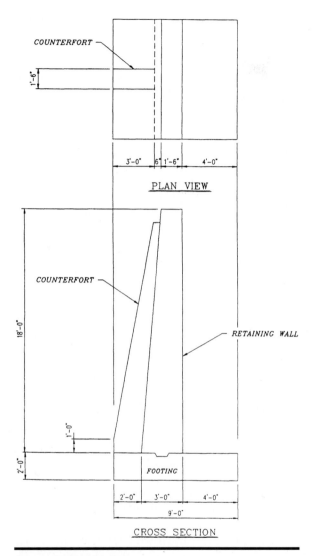

Counterfort component.

fort will be calculated independently, then summarized together. This quantity, of cubic feet, will then be translated to a volume measurement of required cubic yards for: Concrete placement, retaining wall counterfort

Rubbing and patching is calculated by the surface area, in square feet, of the component that is required to be completed. The quantity is then summarized as: retaining wall counterfort, rub and patch.

Procedure

The preparation of a detailed and accurate takeoff for the components of substructure work operations requires a thorough understanding and scope of the project, site conditions, logistics of the area, and construction techniques. The work operation will demand the most economical and practical approach to the forming and placement requirements of the project, to be competitive and productive.

With the variable methods available to perform and utilize formwork systems, the estimator must give great thought to the performance of the product. The form system must be versatile and labor "friendly." In other words, the form system chosen must provide the most economic solution to both ownership and labor consumption. The end result of this will aid in a more productive function for the entire component operation, forming, embeds, placement, and form removal.

The choices of systems to evaluate, as shown in the formwork section, are conventional wood, metal and wood handset or lightweight, or metal high-production gang systems. Different substructure components can provide a method of choice for formwork, but some, feasibly, cannot. Subsequently, the method chosen will dictate and control the production factors and indirect forming costs.

Initially, the determination must be made, based on the conditions, which system types, or portions or multiples of, to use. When a repetitious condition exists and the facilities permit, a metal gang formwork system could be used, whereas if a custom, or one-use-type requirement exists, a wood form system would be most economical. A condition which would require a variable, the combination of both systems in unison, may be chosen.

In some cases, the choice of form systems will somewhat control the concrete placement procedures, but regardless, the form system must withstand and support the load and force of the concrete. A lighter-duty form system will require a slower method of placement and will add to the cost by reducing the placement production per hour, or hourly pour rate. A heavy-duty form system will allow a more aggressive placement procedure but also may require larger equipment for the handling of the form system. The cost and time associated with the embedded anchors and hardware for the formwork requirements will be affected also, dependent on the choice of systems. With these scenarios, a well-proved, efficient, and all-round procedure must be chosen and planned at the time of quantity takeoff.

The choice of form system will also affect the amount of rubbing and patching requirements for form holes caused from the hardware used.

The takeoff of formwork systems requires the need to most effectively form, or cover, the surface contact area with the least overrun of formwork, in other words, the sizing of the surface area of the form to the area to be formed. This must be first analyzed by recognizing the best use of a common-sized larger form, then filling around it with smaller-sized forms. This will produce the best and most effective form system, efficiently utilize the required labor for both erection and removal, and provide for a realistic concrete placement procedure.

Substructure Unit 311

The placement of concrete is a variable, based on economics and ease, even once the formwork system has been chosen. Concrete can be placed by conventional methods, crane and bucket or truck, or by pumping, the cost of which and labor usage must be determined at the bidding stage. The pumping method may be a higher cost for a small placement, but for a large volume placement may be the most economical and timesaving operation.

The conditions of the concrete placement will also affect the overrun, or yield, of the volume required.

Questions of Estimate

When analyzing the substructure takeoff, these questions formatted in this method will best outline the work operations required for the given components:

1. What site conditions exist?
2. What access is available for equipment?
3. Is staging or part width construction required?
4. What is the type of structure?
5. How many spans are shown?
6. What type of construction methods are required?
7. What are the substructure units?
8. What are the required unit components?
9. What is the method of formwork?
10. What is the method of concrete placement?
11. What will the concrete yield be?
12. Is there a need of falsework, shoring, or scaffolding?
13. What specialty equipment is required?
14. Are there traffic conditions within the work area?

Application

The following example details a given substructure quantity takeoff. This shows the production requirements, producing a craft-hour production factor, along with a detailed summary listing of required formwork, concrete volumes, material needs both permanent and incidental, and equipment usage.

A given multispan structure has the following substructure units:

South abutment unit

North abutment unit

Center pier unit

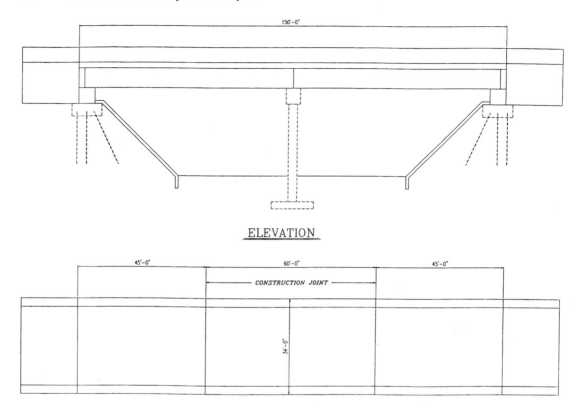

Elevation view, two-span structure.

Within each of the symmetrical abutment units there are the components of:

Abutment footer

Abutment wall

Abutment backwall (1 independent, 1 dependent)

East abutment wing wall

West abutment wing wall

Abutment beam seats, 4 each

Within the center pier unit there are the components of:

Pier footer

Pier columns, round, 3 each

Pier cap

Pier beam seats, 4 each

For these components, the required work operations will be:

Formwork and formwork removal

Installation of form hardware, embeds

Determination of beam type, end section area

Concrete placement

Rub and patch concrete surface

The following is a listing of required component dimensions as per the project plans:

Abutment unit (north and south symmetrical)			
Abutment footer:	3′ thick	10′ wide	40′ long
Abutment wall:	3′ wide	5′ high	36′ long
Abutment backwall:	1.5′ thick	4.5′ high	36′ long
Beam type: precast concrete type IV			
End area: 789 square inches			
East wing wall:	1′ thick	9.5′ high	12′ long
West wing wall:	1′ thick	9.5′ high	14′ long
Beam seats, ea:	1.5′ × 2′		
Pier unit:			
Pier footer:	2.5′ thick	15′ wide	30′ long
Pier columns:	3′ diameter	18′ high	3 ea
Pier cap:	3.5′ wide	4′ high	36′ long
Beam seats, ea:	3.5′ × 2′		

Example: Operation Analysis, Abutments and Center Pier

North abutment footer

Operation. Form and strip, abutment footer. Unit: square feet.

The operation of forming and stripping of the abutment footer will be calculated in the following manner for the required surface area:

The perimeter multiplied by the thickness:

$$(10' + 40' + 10' + 40') * 3' = 300.0 \text{ sf form}$$

$$= 300.0 \text{ sf strip}$$

Operation. Form and strip, end and intermediate bulkheads. Unit: square feet.

The operation of forming and stripping of the bulkheads will be calculated in the following manner for the required surface area, the width multiplied by the height by the number required:

$$\text{None required} = 0 \text{ sf form}$$
$$= 0 \text{ sf strip}$$

Operation. Install embeds, abutment footer. Unit: each.

The operation of installing embeds for the formwork of the abutment footer will be calculated in the following manner:

$$\text{None required} = 0 \text{ each}$$

Operation. Concrete placement, abutment footer. Unit: cubic yard.

The operation of concrete placement required within the abutment footer will be calculated in the following manner for the required volume; the yield of the placement will be calculated and added during this time, for a total volume requirement.

$$\text{Required yield: } 10\%$$

The multiplying of the width, by the length, by the thickness, converted to cubic yards, plus the yield.

$$((10' * 40' * 3')/27) * 1.10 = 48.88 \text{ cy concrete gross}$$
$$((10' * 40' * 3')/27) = 44.44 \text{ cy concrete neat}$$

Operation. Rub and patch, abutment footer. Unit: square feet.

The operation of rubbing and patching for the abutment footer will be calculated for the area of treatment in the following manner for the required surface area.

$$\text{Required area: none}$$

The width multiplied by the height:

$$\text{None required} = 0 \text{ sf rub and patch}$$

North abutment wall

Operation. Form and strip, abutment wall. Unit: square feet.

The operation of forming and stripping of the abutment wall will be calculated in the following manner for the required surface area, the length multiplied by the height, front face:

$$5' * 36' = 180.0 \text{ sf form}$$
$$= 180.0 \text{ sf strip}$$

The length multiplied by the height, back face:

$$5' * 36' = 180.0 \text{ sf form}$$
$$= 180.0 \text{ sf strip}$$

Operation. Form and strip, end and intermediate bulkheads. Unit: square feet.

The operation of forming and stripping of the bulkheads will be calculated in the following manner for the required surface area, the width multiplied by the height, by the number required:

$$(3' * 5') * 2 = 30.0 \text{ sf form}$$
$$= 30.0 \text{ sf strip}$$

Operation. Install embeds, abutment wall. Unit: each.

The operation of installing embeds for the formwork of the abutment wall will be calculated in the following manner:

$$20 \text{ required} = 20.0 \text{ each}$$

Operation. Concrete placement, abutment wall. Unit: cubic yard.

The operation of concrete placement required within the abutment wall will be calculated in the following manner for the required volume; the yield of the placement will be calculated and added during this time, for a total volume requirement.

$$\text{Required yield: } 8\%$$

The multiplying of the width, by the length, by the height, converted to cubic yards, plus the yield.

$$((3' * 36' * 5')/27) * 1.08 = 21.6 \text{ cy concrete gross}$$
$$((3' * 36' * 5')/27) = 20.0 \text{ cy concrete neat}$$

Operation. Rub and patch, abutment wall. Unit: square feet.

The operation of rubbing and patching for the abutment wall will be calculated for the area of treatment in the following manner for the required surface area.

$$\text{Required area: front face}$$
$$\text{both ends}$$

The width multiplied by the height:

$$(5' * 36') + 2(3' * 5') = 210.0 \text{ sf rub and patch}$$

North abutment backwall, independent

Operation. Form and strip, abutment backwall. Unit: square feet.

The operation of forming and stripping of the abutment backwall, independent, will be calculated in the following manner for the required surface area, the length multiplied by the height, front face:

$$4.5' * 36' = 162.0 \text{ sf form}$$
$$= 162.0 \text{ sf strip}$$

The length multiplied by the height, back face:

$$4.5' * 36' = 162.0 \text{ sf form}$$
$$= 162.0 \text{ sf strip}$$

Operation. Form and strip, end and intermediate bulkheads. Unit: square feet.

The operation of forming and stripping of the bulkheads will be calculated in the following manner for the required surface area, the width multiplied by the height, by the number required:

$$(1.5' * 4.5') * 2 = 13.5 \text{ sf form}$$
$$= 13.5 \text{ sf strip}$$

Operation. Install embeds, abutment backwall. Unit: each.

The operation of installing embeds for the formwork of the abutment backwall will be calculated in the following manner:

$$15 \text{ required} = 15.0 \text{ each}$$

Operation. Concrete placement, abutment backwall. Unit: cubic yard.

The operation of concrete placement required within the abutment backwall, independent, will be calculated in the following manner for the required volume; the yield of the placement will be calculated and added during this time, for a total volume requirement.

Required yield: 8%

The multiplying of the width, by the length, by the height, converted to cubic yards, plus the yield.

$$((1.5' * 36' * 4.5')/27) * 1.08 = 9.72 \text{ cy concrete gross}$$
$$((1.5' * 36' * 4.5')/27) = 9.00 \text{ cy concrete neat}$$

Operation. Rub and patch, abutment backwall. Unit: square feet.

The operation of rubbing and patching for the abutment backwall will be calculated for the area of treatment in the following manner for the required surface area.

Required area: front face

both ends

The width multiplied by the height:

$$(4.5' * 36') + 2(1.5' * 4.5') = 175.5 \text{ sf rub and patch}$$

North abutment east wing wall

Operation. Form and strip, abutment wing wall. Unit: square feet.

The operation of forming and stripping of the abutment wing wall will be calculated in the following manner for the required surface area, the length multiplied by the height, front face:

$$9.5' * 12' = 114.0 \text{ sf form}$$
$$= 114.0 \text{ sf strip}$$

The length multiplied by the height, back face:

$$9.5' * 12' = 114.0 \text{ sf form}$$
$$= 114.0 \text{ sf strip}$$

Operation. Form and strip, end and intermediate bulkheads. Unit: square feet.

The operation of forming and stripping of the bulkheads will be calculated in the following manner for the required surface area, the width multiplied by the height, by the number required:

$$(1' * 9.5') * 1 = 9.5 \text{ sf form}$$
$$= 9.5 \text{ sf strip}$$

Operation. Install embeds, abutment wing wall. Unit: each.

The operation of installing embeds for the formwork of the abutment wing wall will be calculated in the following manner:

$$5 \text{ required} = 5.0 \text{ each}$$

Operation. Concrete placement, abutment wing wall. Unit: cubic yard.

The operation of concrete placement required within the abutment wing wall will be calculated in the following manner for the required volume; the yield of the placement will be calculated and added during this time, for a total volume requirement.

Required yield: 9%

The multiplying of the width, by the length, by the height, converted to cubic yards, plus the yield.

$$((1' * 12' * 9.5')/27) * 1.09 = 4.60 \text{ cy concrete gross}$$

$$((1' * 12' * 9.5')/27) = 4.22 \text{ cy concrete neat}$$

Operation. Rub and patch, abutment wing wall. Unit: square feet.

The operation of rubbing and patching for the abutment wing wall will be calculated for the area of treatment in the following manner for the required surface area.

Required area: front face

one end

The width multiplied by the height:

$$(9.5' * 12') + (1' * 9.5') = 123.5 \text{ sf rub and patch}$$

North abutment west wing wall

Operation. Form and strip, abutment wing wall. Unit: square feet.

The operation of forming and stripping of the abutment wing wall will be calculated in the following manner for the required surface area, the length multiplied by the height, front face:

$$9.5' * 14' = 133.0 \text{ sf form}$$

$$= 133.0 \text{ sf strip}$$

The length multiplied by the height, back face:

$$9.5' * 14' = 133.0 \text{ sf form}$$

$$= 133.0 \text{ sf strip}$$

Operation. Form and strip, end and intermediate bulkheads. Unit: square feet.

The operation of forming and stripping of the bulkheads will be calculated in the following manner for the required surface area, the width multiplied by the height, by the number required:

$$(1' * 9.5') * 1 = 9.5 \text{ sf form}$$

$$= 9.5 \text{ sf strip}$$

Operation. Install embeds, abutment wing wall. Unit: each.

The operation of installing embeds for the formwork of the abutment wing wall will be calculated in the following manner:

$$6 \text{ required} = 6.0 \text{ each}$$

Operation. Concrete placement, abutment wing wall. Unit: cubic yard.

The operation of concrete placement required within the abutment wing wall will be calculated in the following manner for the required volume; the yield of the placement will be calculated and added during this time, for a total volume requirement.

$$\text{Required yield: } 9\%$$

The multiplying of the width, by the length, by the height, converted to cubic yards, plus the yield.

$$((1' * 14' * 9.5')/27) * 1.09 = 5.37 \text{ cy concrete gross}$$

$$((1' * 14' * 9.5')/27) = 4.93 \text{ cy concrete neat}$$

Operation. Rub and patch, abutment wing wall. Unit: square feet.

The operation of rubbing and patching for the abutment wing wall will be calculated for the area of treatment in the following manner for the required surface area.

$$\text{Required area: front face}$$
$$\text{one end}$$

The width multiplied by the height:

$$(9.5' * 14') + (1' * 9.5') = 142.5 \text{ sf rub and patch}$$

North abutment beam seat

Operation. Prepare surface, abutment beam seat. Unit: square feet.

The operation of preparing of the abutment beam seats will be calculated in the following manner for the required surface area, the length multiplied by the width by the number of beam seats:

$$(1.5' * 2') * 4 = 12.0 \text{ sf surface}$$

South abutment footer

Operation. Form and strip, abutment footer. Unit: square feet.

The operation of forming and stripping of the abutment footer will be calculated in the following manner for the required surface area, the perimeter multiplied by the thickness:

$$(10' + 40' + 10' + 40') * 3' = 300.0 \text{ sf form}$$

$$= 300.0 \text{ sf strip}$$

Operation. Form and strip, end and intermediate bulkheads. Unit: square feet.

The operation of forming and stripping of the bulkheads will be calculated in the following manner for the required surface area, the width multiplied by the height by the number required:

$$\text{None required} = 0 \text{ sf form}$$

$$= 0 \text{ sf strip}$$

Operation. Install embeds, abutment footer. Unit: each.

The operation of installing embeds for the formwork of the abutment footer will be calculated in the following manner:

$$\text{None required} = 0 \text{ each}$$

Operation. Concrete placement, abutment footer. Unit: cubic yard.

The operation of concrete placement required within the abutment footer will be calculated in the following manner for the required volume; the yield of the placement will be calculated and added during this time, for a total volume requirement.

$$\text{Required yield: } 10\%$$

The multiplying of the width, by the length, by the thickness, converted to cubic yards, plus the yield.

$$((10' * 40' * 3')/27) * 1.10 = 48.88 \text{ cy concrete gross}$$

$$((10' * 40' * 3')/27) = 44.44 \text{ cy concrete neat}$$

Operation. Rub and patch, abutment footer. Unit: square feet.

The operation of rubbing and patching for the abutment footer will be calculated for the area of treatment in the following manner for the required surface area.

$$\text{Required area: none}$$

The width multiplied by the height:

$$\text{None required} = 0 \text{ sf rub and patch}$$

South abutment wall

Operation. Form and strip, abutment wall. Unit: square feet.

The operation of forming and stripping of the abutment wall will be calculated in the following manner for the required surface area, the length multiplied by the height, front face:

$$5' * 36' = 180.0 \text{ sf form}$$

$$= 180.0 \text{ sf strip}$$

The length multiplied by the height, back face:

$$5' * 36' = 180.0 \text{ sf form}$$

$$= 180.0 \text{ sf strip}$$

Operation. Form and strip, end and intermediate bulkheads. Unit: square feet.

The operation of forming and stripping of the bulkheads will be calculated in the following manner for the required surface area, the width multiplied by the height, by the number required:

$$(3' * 5') * 2 = 30.0 \text{ sf form}$$

$$= 30.0 \text{ sf strip}$$

Operation. Install embeds, abutment wall. Unit: each.

The operation of installing embeds for the formwork of the abutment wall will be calculated in the following manner:

$$20 \text{ required} = 20.0 \text{ each}$$

Operation. Concrete placement, abutment wall. Unit: cubic yard.

The operation of concrete placement required within the abutment wall will be calculated in the following manner for the required volume; the yield of the placement will be calculated and added during this time, for a total volume requirement.

$$\text{Required yield: } 8\%$$

The multiplying of the width, by the length, by the height, converted to cubic yards, plus the yield.

$$((3' * 36' * 5')/27) * 1.08 = 21.6 \text{ cy concrete gross}$$

$$((3' * 36' * 5')/27) = 20.0 \text{ cy concrete neat}$$

Operation. Rub and patch, abutment wall. Unit: square feet.

The operation of rubbing and patching for the abutment wall will be calculated for the area of treatment in the following manner for the required surface area.

Required area: front face

both ends

The width multiplied by the height:

$(5' * 36') + 2(3' * 5') = 210.0$ sf rub and patch

South abutment backwall, dependent

Operation. Form and strip, abutment backwall. Unit: square feet.

The operation of forming and stripping of the abutment dependent backwall will be calculated in the following manner for the required surface area.

Requirement: beam centerline dimensions = 8.0'

exterior beam to wall end = 6.0'

beam type: PCB IV AASHTO

end area: 789 sq in

The length, from the center of beam to the center of beam, multiplied by the height, multiplied by the number of interior bays, plus the length, from the center of the exterior beam to the end of the backwall, multiplied by the height, multiplied by each end, front face: The end area of each beam (square inches), multiplied by the number of beams, multiplied by a factor of 0.01.

$((8' * 4.5') * 3) + ((6' * 4.5') * 2) = 162.0$ sf form dependent

$= 162.0$ sf strip dependent

$(789'' * 4) * 0.01 = 31.56$ craft-hours beam

The length multiplied by the height, back face:

$4.5' * 36' = 162.0$ sf form

$= 162.0$ sf strip

Operation. Form and strip, end and intermediate bulkheads. Unit: square feet.

The operation of forming and stripping of the bulkheads will be calculated in the following manner for the required surface area, the width multiplied by the height, by the number required:

$(1.5' * 4.5') * 2 = 13.5$ sf form

$= 13.5$ sf strip

Operation. Install embeds, abutment backwall. Unit: each.

The operation of installing embeds for the formwork of the abutment backwall will be calculated in the following manner:

$$15 \text{ required} = 15.0 \text{ each}$$

Operation. Concrete placement, abutment backwall. Unit: cubic yard.

The operation of concrete placement required within the dependent abutment backwall will be calculated in the following manner for the required volume; the yield of the placement will be calculated and added during this time, for a total volume requirement.

$$\text{Required yield: } 11\%$$

The multiplying of the width, by the length, by the height; then deducting the volume of the intrusion of each beam, converted to cubic yards, plus the yield.

$$\text{Beam requirement, type} = \text{IV concrete AASHTO}$$

$$\text{quantity} = 4 \text{ each}$$

$$\text{end area} = 789 \text{ sq inches}$$

$$\text{intrusion} = 10''$$

NOTE: (1.0 cubic yard has 46,656 cubic inches)

$$(((1.5' * 36' * 4.5')/27) - ((789 * 10 * 5)/46{,}656)) * 1.11 = 9.05 \text{ cy gross}$$

$$(((1.5' * 36' * 4.5')/27) - ((789 * 10 * 5)/46{,}656)) = 8.15 \text{ cy neat}$$

Operation. Rub and patch, abutment backwall. Unit: square feet.

The operation of rubbing and patching for the abutment backwall will be calculated for the area of treatment in the following manner for the required surface area.

$$\text{Required area: front face}$$

$$\text{both ends}$$

The width multiplied by the height:

$$(8' * 4.5' * 3) + (6' * 4.5' * 2) + 2(1.5' * 4.5') = 175.5 \text{ sf rub and patch}$$

South abutment east wing wall

Operation. Form and strip, abutment wing wall. Unit: square feet.

The operation of forming and stripping of the abutment wing wall will be calculated in the following manner for the required surface area, the length multiplied by the height, front face:

$$9.5' * 12' = 114.0 \text{ sf form}$$
$$= 114.0 \text{ sf strip}$$

The length multiplied by the height, back face:

$$9.5' * 12' = 114.0 \text{ sf form}$$
$$= 114.0 \text{ sf strip}$$

Operation. Form and strip, end and intermediate bulkheads. Unit: square feet.

The operation of forming and stripping of the bulkheads will be calculated in the following manner for the required surface area, the width multiplied by the height, by the number required:

$$(1' * 9.5') * 1 = 9.5 \text{ sf form}$$
$$= 9.5 \text{ sf strip}$$

Operation. Install embeds, abutment wing wall. Unit: each.

The operation of installing embeds for the formwork of the abutment wing wall will be calculated in the following manner:

$$5 \text{ required} = 5.0 \text{ each}$$

Operation. Concrete placement, abutment wing wall. Unit: cubic yard.

The operation of concrete placement required within the abutment wing wall will be calculated in the following manner for the required volume; the yield of the placement will be calculated and added during this time, for a total volume requirement.

Required yield: 9%

The multiplying of the width, by the length, by the height, converted to cubic yards, plus the yield.

$$((1' * 12' * 9.5')/27) * 1.09 = 4.60 \text{ cy concrete gross}$$
$$((1' * 12' * 9.5')/27) = 4.22 \text{ cy concrete neat}$$

Operation. Rub and patch, abutment wing wall. Unit: square feet.

The operation of rubbing and patching for the abutment wing wall will be calculated for the area of treatment in the following manner for the required surface area.

Required area: front face

one end

The width multiplied by the height:

$$(9.5' * 12') + (1' * 9.5') = 123.5 \text{ sf rub and patch}$$

South abutment west wing wall

Operation. Form and strip, abutment wing wall. Unit: square feet.
The operation of forming and stripping of the abutment wing wall will be calculated in the following manner for the required surface area, the length multiplied by the height, front face:

$$9.5' * 14' = 133.0 \text{ sf form}$$
$$= 133.0 \text{ sf strip}$$

The length multiplied by the height, back face:

$$9.5' * 14' = 133.0 \text{ sf form}$$
$$= 133.0 \text{ sf strip}$$

Operation. Form and strip, end and intermediate bulkheads. Unit: square feet.
The operation of forming and stripping of the bulkheads will be calculated in the following manner for the required surface area, the width multiplied by the height, by the number required:

$$(1' * 9.5') * 1 = 9.5 \text{ sf form}$$
$$= 9.5 \text{ sf strip}$$

Operation. Install embeds, abutment wing wall. Unit: each.
The operation of installing embeds for the formwork of the abutment wing wall will be calculated in the following manner:

$$6 \text{ required} = 6.0 \text{ each}$$

Operation. Concrete placement, abutment wing wall. Unit: cubic yard.
The operation of concrete placement required within the abutment wing wall will be calculated in the following manner for the required volume; the yield of the placement will be calculated and added during this time, for a total volume requirement.

$$\text{Required yield: } 9\%$$

The multiplying of the width, by the length, by the height, converted to cubic yards, plus the yield.

$$((1' * 14' * 9.5')/27) * 1.09 = 5.37 \text{ cy concrete gross}$$
$$((1' * 14' * 9.5')/27) = 4.93 \text{ cy concrete neat}$$

Operation. Rub and patch, abutment wing wall. Unit: square feet.

The operation of rubbing and patching for the abutment wing wall will be calculated for the area of treatment in the following manner for the required surface area.

Required area: front face

one end

The width multiplied by the height:

$$(9.5' * 14') + (1' * 9.5') = 142.5 \text{ sf rub and patch}$$

South abutment beam seat

Operation. Prepare surface, abutment beam seat. Unit: square feet.

The operation of preparing the abutment beam seats will be calculated in the following manner for the required surface area, the length multiplied by the width by the number of beam seats:

$$(1.5' * 2') * 4 = 12.0 \text{ sf surface}$$

Abutment quantity summary

	Unit	North	South
Abutment footer			
Formwork place	sf	300.0	300.0
Formwork removal	sf	300.0	300.0
Bulkheads place	sf	0.0	0.0
Bulkheads remove	sf	0.0	0.0
Hardware and embeds	ea	0.0	0.0
Concrete placement	cy	48.88	48.88
Rub and patch	sf	0.0	0.0
Abutment wall			
Formwork place	sf	360.0	360.0
Formwork removal	sf	360.0	360.0
Bulkheads place	sf	30.0	30.0
Bulkheads remove	sf	30.0	30.0
Hardware and embeds	ea	20.0	20.0
Concrete placement	cy	21.6	21.6
Rub and patch	sf	210.0	210.0
Abutment backwall (independent)			
Formwork place	sf	324.0	0.0
Formwork removal	sf	324.0	0.0
Bulkheads place	sf	13.5	0.0
Bulkheads remove	sf	13.5	0.0
Hardware and embeds	ea	15.0	0.0
Concrete placement	cy	9.72	0.0
Rub and patch	sf	175.5	0.0
Abutment backwall (dependent)			
Formwork place	sf	0.0	324.0
Formwork removal	sf	0.0	324.0

	Unit	North	South
Abutment backwall (dependent) (*Cont.*)			
Bulkheads place	sf	0.0	13.5
Bulkheads remove	sf	0.0	13.5
Hardware and embeds	ea	0.0	15.0
Concrete placement	cy	0.0	9.05
Rub and patch	sf	0.0	175.5
Dependent beam factor	craft-hour		31.56
Abutment wing wall			
Formwork place	sf	494.0	494.0
Formwork removal	sf	494.0	494.0
Bulkheads place	sf	19.0	19.0
Bulkheads remove	sf	19.0	19.0
Hardware and embeds	ea	11.0	11.0
Concrete placement	cy	9.97	9.97
Rub and patch	sf	266.0	266.0
Abutment beam seats			
Surface beam seats	sf	12.0	12.0

Center pier footer

Operation. Form and strip, pier footer. Unit: square feet.

The operation of forming and stripping of the pier footer will be calculated in the following manner for the required surface area, the perimeter multiplied by the thickness:

$$(15' + 30' + 15' + 30') * 2.5' = 225.0 \text{ sf form}$$

$$= 225.0 \text{ sf strip}$$

Operation. Form and strip, end and intermediate bulkheads. Unit: square feet.

The operation of forming and stripping of the bulkheads will be calculated in the following manner for the required surface area, the width multiplied by the height by the number required:

$$\text{None required} = 0 \text{ sf form}$$

$$= 0 \text{ sf strip}$$

Operation. Install embeds, pier footer. Unit: each.

The operation of installing embeds for the formwork of the pier footer will be calculated in the following manner:

$$\text{None required} = 0 \text{ each}$$

Operation. Concrete placement, pier footer. Unit: cubic yard.

The operation of concrete placement required within the pier footer will be calculated in the following manner for the required volume; the yield of the placement will be calculated and added during this time, for a total volume requirement.

Required yield: 10%

The multiplying of the width, by the length, by the thickness, converted to cubic yards, plus the yield.

$$((15' * 30' * 2.5')/27) * 1.10 = 45.84 \text{ cy concrete gross}$$

$$((15' * 30' * 2.5')/27) = 41.67 \text{ cy concrete neat}$$

Operation. Rub and patch, pier footer. Unit: square feet.

The operation of rubbing and patching for the pier footer will be calculated for the area of treatment in the following manner for the required surface area.

Required area: none

The width multiplied by the height:

$$\text{None required} = 0 \text{ sf rub and patch}$$

Center pier column, round

Operation. Form and strip, pier column. Unit: square feet.

The operation of forming and stripping of the pier column will be calculated in the following manner for the required surface area.

Required diameter: 3'

3 each columns

18' each (height)

The diameter of each column, multiplied by *pi* (3.142), multiplied by the height, summarizing each column:

$$((3' * 3.142) * 18') * 3 = 508.94 \text{ sf form}$$

$$= 508.94 \text{ sf strip}$$

Operation. Install embeds, pier column. Unit: each.

The operation of installing embeds for the formwork of each of the pier columns will be calculated in the following manner:

$$3.0 \text{ each column required} = 9.0 \text{ each}$$

Operation. Concrete placement, pier column. Unit: cubic yard.

The operation of concrete placement required within the pier columns will be calculated in the following manner for the required volume; the yield of the placement will be calculated and added during this time, for a total volume requirement.

Required yield: 7%

The radius squared, multiplied by *pi* (3.142), multiplied by the height of each column, summarizing each column, converting to cubic yards, plus the yield.

$$(((((3 * 3') * 3.142) * 18) * 3)/27) * 1.07 = 60.51 \text{ cy concrete gross}$$

$$(((((3 * 3') * 3.142) * 18) * 3)/27) = 56.55 \text{ cy concrete neat}$$

Operation. Rub and patch, pier column. Unit: square feet.

The operation of rubbing and patching for the pier column will be calculated for the area of treatment in the following manner for the required surface area.

Required area: complete column

The diameter of each column, multiplied by *pi* (3.142), multiplied by the height, summarizing each column:

$$((3' * 3.142) * 18') * 3 = 508.94 \text{ sf rub and patch}$$

Center pier cap

Operation. Form and strip, pier cap. Unit: square feet.

The operation of forming and stripping of the pier cap will be calculated in the following manner for the required surface area, the summarizing of the length of each soffit panel, between each column multiplied by the width, plus the length of each end soffit panel multiplied by the width, summarizing, plus multiplying the length of each side by the height and summarizing, for the required surface area.

Required: dimension between columns 10.5'

dimension of end distance 3.0'

support, column embeds

$$(((10.5' * 3.5') * 2) + (3 * 3.5') * 2) + ((36' * 4') * 2) = 382.5 \text{ sf form}$$

$$= 382.5 \text{ sf strip}$$

Operation. Form and strip, end and intermediate bulkheads. Unit: square feet.

The operation of forming and stripping of the bulkheads will be calculated in the following manner for the required surface area, the width multiplied by the height, by the number required:

$$(3.5' * 4') * 2 = 28.0 \text{ sf form}$$

$$= 28.0 \text{ sf strip}$$

Operation. Install embeds, pier cap. Unit: each.

The operation of installing embeds for the formwork of the pier cap will be calculated in the following manner:

$$6 \text{ required} = 6.0 \text{ each}$$

Operation. Concrete placement, pier cap. Unit: cubic yard.

The operation of concrete placement required within the pier cap will be calculated in the following manner for the required volume; the yield of the placement will be calculated and added during this time, for a total volume requirement.

Required yield: 8%

The multiplying of the width, by the length, by the height, converting to cubic yards, plus the yield.

$$((3.5' * 36' * 4')/27) * 1.08 = 20.16 \text{ cy concrete gross}$$

$$((3.5' * 36' * 4')/27) = 18.67 \text{ cy concrete neat}$$

Operation. Rub and patch, pier cap. Unit: square feet.

The operation of rubbing and patching for the pier cap will be calculated for the area of treatment in the following manner for the required surface area.

Required area: soffit

both sides

both ends

The summarizing of the length of each soffit panel, between each column multiplied by the width, plus the length of each end soffit panel multiplied by the width, summarizing, plus multiplying the length of each side by the height and summarizing, plus the summation of multiplying the width by the height of each end, for the required surface area.

Required: dimension between columns 10.5'

dimension of end distance 3.0'

$$(((10.5' * 3.5') * 2) + ((3 * 3.5') * 2) + ((36' * 4') * 2) + (3.5' * 4') * 2) = 410.5 \text{ sf rub and patch}$$

Center pier cap beam seat

Operation. Prepare surface, pier cap beam seat. Unit: square feet.

The operation of preparing of the pier cap beam seats will be calculated in the following manner for the required surface area, the length multiplied by the width by the number of beam seats:

$$(3.5' * 2') * 4 = 28.0 \text{ sf surface}$$

Center pier quantity summary

	Unit	Center
Pier footer		
Formwork place	sf	225.0
Formwork removal	sf	225.0
Bulkheads place	sf	0.0
Bulkheads remove	sf	0.0
Hardware and embeds	ea	0.0
Concrete placement	cy	45.84
Rub and patch	sf	0.0
Pier column		
Formwork place	sf	508.94
Formwork removal	sf	508.94
Hardware and embeds	ea	9.0
Concrete placement	cy	60.51
Rub and patch	sf	508.94
Pier cap		
Formwork place	sf	382.5
Formwork removal	sf	382.5
Bulkheads place	sf	28.0
Bulkheads remove	sf	28.0
Hardware and embeds	ea	6.0
Concrete placement	cy	20.16
Rub and patch	sf	410.5
Pier cap beam seats		
Surface beam seats	sf	28.0

From the quantity summary list, the time duration and production factoring will be compiled.

Example: Production Factoring, Time Duration

The production factoring will be performed by using a defined crew to complete each of the associated work operations. Most of the work operation for the substructure units and components within is accomplished using the same defined crew. This is usually due to overlapping work operations, but each component production factor will still be based on the actual work operation and the craft-hours consumed. The smaller, more detailed components will require a different, smaller allocated crew owing to limited ability to perform for the operation.

For each operation, a crew must be formulated to determine the number and class of personnel required, along with a shift duration of time per workday. This will determine the total craft-hour requirements per shift, for the given operation. From this the production cycle of units produced per hour, for that given crew, the productivity per craft-hour unit will be determined.

The factoring procedure will be based on an average per project, equaling the project pay item schedule. The individual components will need to be factored because of complexity but should be averaged into the complete project quantity for a given component. A summation of each similar work operation for each structure, or work area, will be done and categorized by the operation performance.

As with all components, a productivity factor as to performance per hour will be used. This will be a normal contingency applied for all work operations.

This example details the production factoring procedure to compile the requirements for an abutment wall.

Abutment wall component

Operation. Formwork, place, abutment wall.

Production unit: = square foot, surface contact area	
Project requirements	= 720.0 sq ft
Crew members	= 7.0 ea
Formwork type	= wood
Forming condition	= average, 5
Shift duration	= 8.0 hours
Unit cycle time	= 0.85 minute/sf
Productivity factor	= 50-minute hour
Cycles per hour: (50/0.85)	= 58.82 cycles/hour
Production per hour: 58.82 * 1	= 58.82 sf/hour
Production per shift: 8 * 58.82	= 470.56 sf/shift
Required shifts: 720.0/470.56	= 1.53 shifts
Total craft-hours: 1.53 * 8 * 7	= 85.68 craft-hours
Operation production factor: 85.68/720	= 0.119 craft-hour/sf
Place formwork	= 0.119 craft-hour/sf

This factoring and allocation time should include all associated work related to performing the complete operation.

Example: Material Allocation

The material costs associated with each operation will be the direct permanent and indirect incidental materials and supplies required per specification. The cost of the formwork itself will be allocated as formwork, which is treated as an asset to the company. The formwork will be treated as equipment and will have an ownership rate applied to it for its use. This allocation, detailed in the formwork section of the book, deals with the cost associated with the formwork purchase, or construction, and reuse disbursing.

The consumable formwork required, that will not be reused, will be considered direct to the job as an incidental supply.

If there is a requirement for formwork that is not in inventory or is classified as specialty, the allocation of price or rental should be factored in the operation over any direct allocation.

The material listed will be summarized by operation and allocated to the direct pay item, which is the volume of neat line cubic yards of concrete.

Abutment wall

Operation. Construct, abutment wall.

Direct material:	
*Concrete, 3500 psi	= 43.2 cubic yards
Indirect material:	
Embeds, form hardware	= 40.0 each
Form oil (2 oz per sf)	
((720 + 120)*2)/128	= 13.13 gal
Formwork	
Walls: 2×6 wood panel forms	= 720.0 sf
End bulkhead: 2×4 wood panel forms	= 60.0 sf
Consumed lumber	= 40.0 lf
Incidental hardware, snap ties	
1 each per 5 sf = 840 sf/5	= 168.0 ea
Nails: 2.5 lb per 100 sf = (840/100) * 2.5	= 21.0 lb

*Denotes controlling item.

The material list is comprised of direct and indirect materials required to perform the operations. The direct material will be outlined, and required by specification, and the indirect will be required by the operation to be performed. All materials will be priced with all applicable sales and use taxes included.

The formwork is detailed separately owing to the different structural size of the formwork and the intended reuse, or life, of the panels.

The ratio of surface contact area to the control cubic yards will give a factor for a relationship of cost, both for the current job and for future reference. This will be taken from the total form area of the operation against the neat line cubic yards for the operation. The same can be done with the cost ratio.

Surface ratio, sf per cy:

$$840.0 \text{ sf}/40.0 \text{ cy} = 21.0 \text{ square feet per cubic yard}$$

Section Summary

The substructure section shows the basic skills and functions needed to perform an accurate and detailed takeoff and cost summary. These basic skills and formulations can be used to perform any quantity takeoff and cost summary for any given component. The requirement of detailing and a step-by-step procedure will prove to give the accuracy needed for a formatted outline.

COMPONENT TAKEOFF

Substructure Unit

The following outline contains the normal work operation accounts for the performance and cost coding of substructure construction components:

1. **Footer, form and strip**

    All work related to the forming of the footer to its proper line, grade, and dimension, and the stripping of same.

    Quantity should be reported in square feet of the concrete surface contact area for both forming and stripping of individual footers.

2. **Footer, embed item**

    All work related to the installation of the embed item components to their proper line, grade, and dimension, for the support of the form system. Quantity should be reported in each for the size and type of specific embed.

3. **Footer, pour**

    All work related to the placement of concrete of the specified type. Quantity should be reported in cubic yards including overrun and waste, and net amount of each individual placement.

4. **Footer, rub and patch surface**

    All work related to the required rubbing and patching of the concrete surface. Quantity should be reported in square feet of the concrete surface contact area required for the application.

5. **Pier columns, form and strip**

    All work related to the forming of the columns to their proper line, grade, and dimension, and the stripping of same. Quantity should be reported in square feet of the concrete surface contact area for both forming and stripping of individual columns.

6. **Pier columns, embed item**

    All work related to the installation of the embed item components to their proper line, grade, and dimension, for the support of the form system. Quantity should be reported in each for the size and type of specific embed.

7. **Pier columns, pour**

    All work related to the placement of concrete of the specified type. Quantity should be reported in cubic yards including overrun and waste, and net amount of each individual placement.

8. **Pier columns, rub and patch surface**

    All work related to the required rubbing and patching of the concrete surface. Quantity should be reported in square feet of the concrete surface contact area required for the application.

Substructure Unit

9. **Pier caps, form and strip**

    All work related to the forming of the pier cap to its proper line, grade, and dimension, and the stripping of same. Quantity should be reported in square feet of the concrete surface contact area for both forming and stripping of individual pier caps.

10. **Pier caps, embed item**

    All work related to the installation of the embed item components to their proper line, grade, and dimension, for the support of the form system. Quantity should be reported in each for the size and type of specific embed.

11. **Pier cap, pour**

    All work related to the placement of concrete of the specified type. Quantity should be reported in cubic yards including overrun and waste, and net amount of each individual placement. This item would include the volume of the seats and risers.

12. **Pier caps, rub and patch surface**

    All work related to the required rubbing and patching of the concrete surface. Quantity should be reported in square feet of the concrete surface contact area required for the application.

13. **Abutment, form and strip**

    All work related to the forming of the abutment to its proper line, grade, and dimension, and the stripping of same. Quantity should be reported in square feet of the concrete surface contact area for both forming and stripping of individual abutment walls.

14. **Abutment, embed item**

    All work related to the installation of the embed item components to their proper line, grade, and dimension, for the support of the form system. Quantity should be reported in each for the size and type of specific embed.

15. **Abutment, pour**

    All work related to the placement of concrete of the specified type. Quantity should be reported in cubic yards including overrun and waste, and net amount of each individual placement. This item would include the volume of the beam seats and/or riser pads.

16. **Abutment, rub and patch surface**

    All work related to the required rubbing and patching of the concrete surface. Quantity should be reported in square feet of the concrete surface contact area required for the application.

17. **Backwall, form and strip**

    All work related to the forming of the backwall section and type, of the substructure to its proper line, grade, and dimension, and the stripping of same. Quantity should be reported in square feet of the concrete surface contact area for both forming and stripping of individual backwalls, dependent and independent.

18. **Backwall, embed item**

    All work related to the installation of the embed item components to their proper line, grade, and dimension, for the support of the form system. Quantity should be reported in each for the size and type of specific embed.

19. **Backwall, pour**

    All work related to the placement of concrete of the specified type. Quantity should be reported in cubic yards including overrun and waste and net amount of each individual placement.

20. **Backwall, rub and patch surface**

    All work related to the required rubbing and patching of the concrete surface. Quantity should be reported in square feet of the concrete surface contact area required for the application.

21. **Wing wall, form and strip**

    All work related to the forming of the wing section to its proper line, grade, and dimension, and the stripping of same. Quantity should be reported in square feet of the concrete surface contact area for both forming and stripping of individual wings.

22. **Wing wall, embed item**

    All work related to the installation of the embed item components to their proper line, grade, and dimension, for the support of the form system. Quantity should be reported in each for the size and type of specific embed.

23. **Wing wall, pour**

    All work related to the placement of concrete of the specified type. Quantity should be reported in cubic yards including overrun and waste, and net amount of each individual placement.

24. **Wing wall, rub and patch surface**

    All work related to the required rubbing and patching of the concrete surface. Quantity should be reported in square feet of the concrete surface contact area required for the application.

25. **Intermediate struts, form and strip**

    All work related to the forming of the pier strut to its proper line, grade, and dimension, and the stripping of same. Quantity should be reported in square feet of concrete surface contact area for both forming and stripping of individual pier struts.

26. **Intermediate struts, embed item**

    All work related to the installation of the embed item components to their proper line, grade, and dimension, for the support of the form system. Quantity should be reported in each for the size and type of specific embed.

27. **Intermediate struts, pour**

    All work related to the placement of concrete of the specified type. Quantity should be reported in cubic yards including overrun and waste, and net amount of each individual placement.

28. **Intermediate struts, rub and patch surface**

    All work related to the required rubbing and patching of the concrete surface. Quantity should be reported in square feet of the concrete surface contact area required for the application.

29. **Beam seats and riser pads, form and strip**

    All work related to the forming of beam seats and beam riser pads to their proper line, grade, and dimension, and the stripping of same. Quantity should be reported in square feet of concrete surface area or concrete surface contact area for both forming and stripping of individual beam seats and/or riser pads, respectively.

30. **Beam seats and riser pads, embed item**

    All work related to the installation of the embed item components to their proper line, grade, and dimension, for the support of the form system. Quantity should be reported in each for the size and type of specific embed.

31. **Beam seats and riser pads, preparatory**

    All work related to the preparation of seats and risers for the placement of beams and bearings. This will include any necessary grinding or grouting to facilitate a flat surface. Quantity should be reported in square feet of flat surface area for the individual beam seats and/or riser pads.

32. **Beam seats and riser pads, pour**

    All work related to the placement of concrete of the specified type. Quantity should be reported in cubic yards including overrun and waste, and net amount of each individual placement. Use only if poured separately from substructure component.

33. Beam seats and riser pads, rub and patch surface

All work related to the required rubbing and patching of the concrete surface. Quantity should be reported in square feet of the concrete surface contact area required for the application.

34. Bulkheads, form and strip

All work related to the forming of a bulkhead required for a construction joint, and the stripping of same. Quantity should be reported in square feet of concrete surface contact area for both forming and stripping of individual bulkheads, along with the size and lineal feet of any keyway required.

35. Miscellaneous, rub and patch surface

All work related to the rubbing and patching of concrete surfaces. This will include all through tube holes, snap tie holes, other embedded and hardware holes, any irregular form deviations, small honeycombs, and any work necessary to produce a pleasing and true surface effect of the entire substructure unit. Quantity should be reported in square feet of the entire surface area of each individual substructure component.

36. Apply texture finish

All work related to the mechanical application of a surface treatment of specified type. Quantity should be reported in square feet of applied texture surface of each individual substructure component.

37. Other, form and strip

This account will be used for other operation accounts not specified here, for both forming and stripping. Quantity should be reported in square feet of concrete surface contact area for both forming and stripping of individual components or a representative format relating to the work performed and the appropriate unit of measure.

38. Other pour

This account will be used for other operation accounts for the placement of a specified concrete mixture not specified here. Quantity should be reported in cubic yards including overrun and waste, the net amount required, or a representative format relating to the work performed and the appropriate unit measure for each individual component.

Chapter 14

Beam Structure

TECHNICAL SECTION

Introduction

This section details the techniques and specifications needed to perform the construction and placing of the beam structure unit of a bridge. Detailed are the specified procedures, materials, and equipment along with an outline of the various types of beams and the necessary components and appurtenances. This describes the work consisting of erecting prestressed precast concrete beams, and rolled and fabricated structural steel girders. Included are normal standards and specifications and discussions as to span length with respect to beam designs and capacities.

The equipment requirements are of sufficient rating and capacity to properly erect the beam structure according to ground conditions, lifting capacities, and maneuverability.

Erection procedures outline proper techniques for prestressed precast concrete beams, the bearing structure, and the diaphragm units. Structural steel girders outline the main girders, cross frames and diaphragms, bolting and welding, and the bearing units.

The permanent materials are of approved sources and fabricated to the proper line, grade length, and camber as detailed in the design specifications.

Definition

The beam structure of a bridge is the intermediate structural support unit that bears on the substructure which creates the span of the superstructure. It is the independent self-supporting unit which carries the load of the superstructure and is the horizontal skeletal frame of the bridge.

The beam structure consists of individual structural components consisting of either concrete or steel. They are normally a prefabricated unit.

Concrete Beams

Concrete beams consist of a *prestressed, precast* member in a configuration of an *I beam,* a *bulb tee beam,* or a *box beam.*

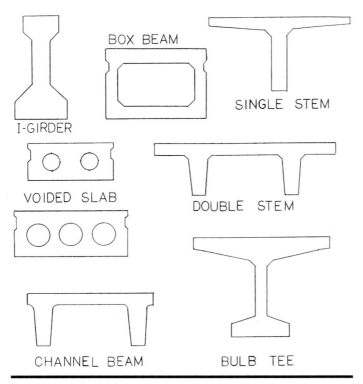

Concrete beam sections. (*Courtesy of Precast/Prestressed Concrete Institute.*)

These beams range in perimeter size as well as length, depending on the characteristics imposed by specific design standards of a particular project. Concrete beams, of *AASHTO Specifications,* have their structural limitations in span length capabilities.

The design properties of concrete beams within their interior structure vary tremendously, while the exterior size and shapes sometimes remain the same. A beam of a particular exterior size can vary differently in load and spanning capabilities within the design limitations of that particular category based on the interior design standards. This is controlled in concrete beams by *strand pattern,* which is the location of the *prestressing cables,* the size and number of strand used, the *drape pattern,* which determines the uplift or load capabilities and *deflections* of the girder, and the amount of reinforcing used.

Within these interior controlling items, AASHTO has set standard sizes of prestressed, precast concrete beams.

I-beams

In the I-beam category, the beam sizes are referred to as type 1, 2, 3, 4, 4 modified, 5, 5 modified, 6, and 6 modified.

I-Girders - PCI Standards

Thousands of bridges have been built utilizing the former standard AASHO-PCI I-Girders shown here. Many states have developed additional I-girder sections. Producers of prestressed concrete will be glad to furnish you with the dimensions and properties of the sections made locally. A cast-in-place deck provides composite action with the girders. Stay-in-place precast, prestressed concrete deck panels, which span between girders, are available for use with I-girders. They serve both as formwork for the cast-in-place slab concrete and as transverse positive moment reinforcement. Full depth precast, prestressed deck panels are also available.

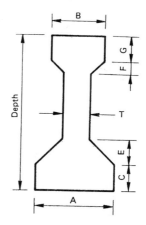

SECTION DIMENSIONS (INCHES)

Type	Depth	A	B	C	E	F	G	T
II	36	18	12	6	6	3	6	6
III	45	22	16	7	7½	4½	7	7
IV	54	26	20	8	9	6	8	8

GIRDER SECTION PROPERTIES

Type	Depth in.	Weight lb/ft	Area in.^{2}	I_x in.^{4}	y_b in.	S_b in.^{3}	S_t in.^{3}
II	36	384	369	50,980	15.83	3220	2528
III	45	583	560	125,390	20.27	6186	5070
IV	54	822	789	260,730	24.73	10543	8908

Concrete I-beam properties. (*Courtesy of Precast/Prestressed Concrete Institute.*)

Box beams

The box beams range from 36"×27" to 48"×42" in section.

Box beams

The box beams shown are the former AASHO-PCI standard sections. They can be used either as adjacent units with or without an added wearing surface or spaced apart in which case the deck slab is cast-in-place using integral precast, prestressed deck panels. Box beams for railway loadings have been standardized by AREA.

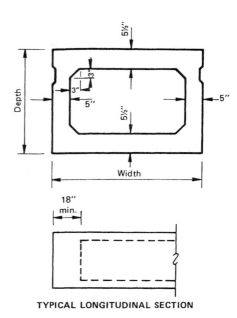

TYPICAL LONGITUDINAL SECTION

TYPICAL SECTION PROPERTIES

Type	Width ft	Depth in.	Weight lb/ft	Net Area in.^{2}	I_x in.^{4}	y_b in.	S_b in.^{3}	S_t in.^{3}
B I-36	3	27	584	561	50,334	13.35	3770	3687
B II-36	3	33	647	621	85,153	16.29	5227	5096
B III-36	3	39	709	681	131,145	19.25	6813	6640
B IV-36	3	42	740	711	158,644	20.73	7653	7459
B I-48	4	27	722	693	65,941	13.37	4932	4838
B II-48	4	33	784	753	110,499	16.33	6767	6629
B III-48	4	39	847	813	168,367	19.29	8728	8542
B IV-48	4	42	878	843	203,088	20.78	9773	9571

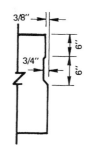

TYPICAL KEYWAY DETAIL

Concrete box beam properties. (*Courtesy of Precast/Prestressed Concrete Institute.*)

Bulb tees

The bulb-tee beams are sized as BT-54, BT-63, and BT-72.

Bulb tees

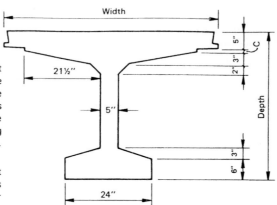

Integral deck bulb tee bridge sections are efficient and economical for spans of 60 ft or more. The sections shown here are available only in some parts of the United States and Canada. Designers wishing to use bulb tee sections should determine if they are available within an economical hauling range. In some areas, depths of 53, 65, and 77 in. are available for spans up to about 180 ft.

Because of the high section modulus to weight ratio (particularly when lightweight concrete is used for the top flange) the use of bulb tees for bridge construction is gaining wide acceptance. Thin-flange bulb tees are also highly efficient for use with cast-in-place concrete decks.

TYPICAL SECTION PROPERTIES*

Width ft	Depth in.	C in.	Weight lb/ft	Area in.^{2}	I_x in.^{4}	y_b in.	S_b in.^{3}	S_t in.^{3}
4	34	0	627	602	88,310	20.38	4340	6,480
5	29	1	708	680	64,110	18.48	3470	6,093
5	34	0	690	662	95,180	21.39	4450	7,548
5	41	1	771	740	157,840	26.18	6029	10,652
6	29	1	771	740	67,790	19.13	3544	6,866
6	34	0	752	722	100,600	22.25	4520	8,550
6	41	1	833	800	166,390	27.11	6139	11,975
7	29	1	833	800	70,930	19.68	3604	7,611
7	41	1	896	860	173,760	27.90	6228	13,265

* These sections are sometimes made with normal weight concrete web and bottom flange and lightweight concrete deck, in which case the weight and section properties differ from those shown.

Concrete bulb-tee beam properties. (*Courtesy Precast/Prestressed Concrete Institute.*)

Voided slabs

With regard to voided prestressed slab beams, they are categorized similar to box beams, from 36"×15" to 48"×21" in section.

Voided slabs

Voided slabs are similar to solid slabs except that they are cast with cylindrical voids to reduce dead load. The sections tabulated below are the former AASHO-PCI standards which can span up to 50 ft for HS20 loadings. Sections with widths and depths other than those tabulated are available from some precasting plants.

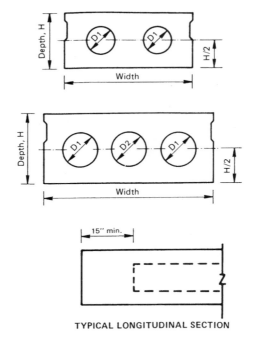

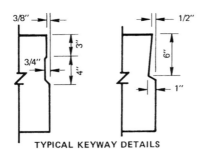

TYPICAL KEYWAY DETAILS

TYPICAL LONGITUDINAL SECTION

TYPICAL SECTION PROPERTIES

Width ft	Depth in.	No. of Voids	Void Dia. in.		Weight lb/ft	Net Area in.^{2}	I_x in.^{4}	S in.^{3}
			D1	D2				
3	15	2	8	—	457	439	9,725	1296
3	18	2	10	—	511	491	16,514	1835
3	21	2	12	—	552	530	25,747	2452
4	15	3	8	8	593	569	12,897	1720
4	18	3	10	10	654	628	21,855	2428
4	21	3	12	10	733	703	34,517	3287

Concrete void beam properties. (*Courtesy of Precast/Prestressed Concrete Institute.*)

Single tee

The single-tee-type on single stemmed beams range from 4′ to 6′ in width.

Single stemmed bridge sections

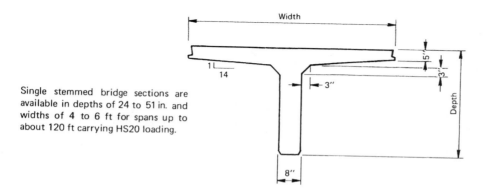

Single stemmed bridge sections are available in depths of 24 to 51 in. and widths of 4 to 6 ft for spans up to about 120 ft carrying HS20 loading.

TYPICAL SECTION PROPERTIES

Width ft	Depth in.	Weight lb/ft	Area in.^{2}	I_x in.^{4}	y_b in.	S_b in.^{3}	S_t in.^{3}
4	24	448	430	18,555	16.96	1094	2634
4	36	548	526	61,058	24.76	2466	5434
4	48	648	622	139,038	32.01	4344	8695
5	24	531	510	19,788	17.53	1129	3059
5	36	631	606	65,673	25.80	2545	6440
5	48	731	702	150,643	33.45	4504	10,353
6	24	620	595	20,782	17.94	1158	3431
6	36	720	691	69,323	26.62	2604	7388
6	48	820	787	160,146	34.64	4623	11,987

Concrete single-stem beam properties. (*Courtesy of Precast/Prestressed Concrete Institute.*)

Channel sections

The channel type ranges from 40″ to 66″.

Channel sections

Many precasting plants manufacture prestressed concrete channel sections for use in 20 to 60-ft span bridges. These plants will furnish you with information on the sections they produce.

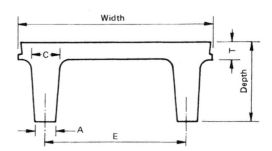

TYPICAL SECTION PROPERTIES

SECTION	Width in.	Depth in.	Slab T in.	Stems A in.	Stems C in.	Stems E in.	Weight lb/ft	Area in.^{2}	I_x in.^{4}	y_b in.	S_b in.^{3}	S_t in.^{3}
"LIGHT"	40	21	5	3.25	6	30	362	348	11,495	14.37	800	1734
	42	25	5	3.5	6	30	417	400	20,105	16.98	1184	2507
	46	23	5	4.62	6	36	438	421	18,507	15.46	1197	2453
	60	27	5	3.75	5.75	48	530	509	28,886	19.27	1499	3739
"MEDIUM"	36	20	4	5.5	7	29	358	344	12,283	12.37	993	1610
	48	20	5	6	7.5	40.5	461	443	14,663	13.05	1123	2110
	48	27	5	4.5	8	36	536	515	31,216	17.85	1748	3410
	60	35	5	4	8	48	688	660	67,648	23.86	2835	6074
"HEAVY"	36	24	4	7	9.25	27	488	469	24,241	14.00	1731	2425
	66	21	5	7.75	9.75	48	640	614	22,051	13.90	1586	3106
	66	27	5	7	9.75	48	730	701	43,738	17.82	2454	4764
	66	35	5	6	9.75	48	844	810	87,469	22.96	3810	7265

Concrete channel beam properties. (*Courtesy of Precast/Prestressed Concrete Institute.*)

Double tee

The double tee or double-stemmed type ranges from 5' to 8.'

Double stemmed bridge sections

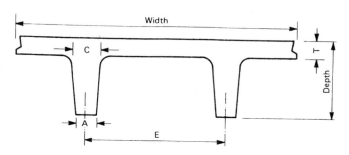

Integral deck double stemmed bridge sections are available in depths of 18 to 36 inches and widths of 5 to 8 ft. "Heavy" sections can span 60 ft and more with HS20 loading while lighter sections can be used for shorter spans.

TYPICAL SECTION PROPERTIES

SECTION	Width ft	Depth in.	Slab T in.	Stems A in.	Stems C in.	Stems E in.	Weight lb/ft	Area in.^{2}	I_x in.^{4}	y_b in.	S_b in.^{3}	S_t in.^{3}
"LIGHT"	5	27	5	4.50	8	36	599	575	33,740	18.60	1812	4020
	6	23	5	4.50	6.50	36	582	558	21,366	16.61	1286	3345
	6	27	5	4.50	8.00	36	662	635	35,758	19.15	1866	4560
	8	27	5	3.75	5.75	48	718	689	32,888	20.64	1593	5171
	8	35	5	3.75	6.50	48	820	787	72,421	26.20	2764	8230
"HEAVY"	5	36	6	6	8	30	812	780	90,286	23.69	3811	7334
	6	35	5	6	9.75	48	876	840	90,164	23.30	3870	7706
	7	35	5	6	9.75	48	938	900	95,028	23.91	3974	8569
	8	35	5	6	9.75	48	1001	960	99,299	24.45	4061	9412
	6	27	5	7	9.75	48	761	731	45,084	18.09	2492	5060
	7	27	5	7	9.75	48	824	791	47,486	18.58	2556	5640
	8	27	5	7	9.75	48	886	851	49,566	19.00	2609	6196
	6	21	5	7.75	9.75	48	671	644	22,720	14.11	1610	3298
	7	21	5	7.75	9.75	48	733	704	23,903	14.48	1651	3666
	8	21	5	7.75	9.75	48	796	764	24,920	14.80	1684	4019

Concrete double-stem beam properties. (*Courtesy of Precast/Prestressed Concrete Institute.*)

In today's technology and engineering practices, other modifications to these standards sometimes apply and are constantly being tested to increase length and usability of concrete beams.

Span length

I beams. With respect to the category sizes of I beams, as they increase in cross section, their span length capabilities increase.

Type I girders have a height of 2'4", a cross-section area of 276 square inches, and a length range from 30' to 45'.

Type II girders have a height of 3'0", a cross-section area of 369 square inches, and a length range of 40' to 60'.

Type III girders have a height of 3'9", a cross-section area of 560 square inches, and a length range of 55' to 80'.

Type IV girders have a height of 4'6", a cross-section area of 789 square inches, and a length range of 70' to 100'.

Type IV-S girders have a height of 4'6", a cross-section area of 681 square inches, and a range comparable to type IV.

Type IV-modified girders have a height of 5'0", a cross-section area of 909 square inches, and a range comparable to type IV.

Type V-modified girders have a height of 5'10", a cross-section area of 826 square inches, and a range of 100' to 130'.

Type V girders have a height of 5'3", a cross-section area of 1013 square inches, and a length range of 90' to 120'.

Type VI girders have a height of 6'0", a cross-section area of 1085 square inches, and a length range of 110' to 140'.

Type VI-modified girders have a height of 6'6", a cross-section area of 1133 square inches, and a range of 125' to 150'.

Box beams. The category of box beams increases their capabilities as the width and height section increases.

B I-36 has a dimension of 3' width and a 27" depth with a 561 square inch section. The span limits are 74'.

B II-36 has a dimension of 3' width and 33" depth with a 621 square inch section. The span limits are 86'.

B III-36 has a dimension of 3' width and 39" depth with a 681 square inch section. The span limits are 97'.

B IV-36 has a dimension of 3' width and 42" depth with a 711 square inch section. The span limits are 103'.

B I-48 has a dimension of 4' width and 27" depth with a 693 square inch section. The span limits are 73'.

B II-48 has a dimension of 4′ width and 33″ depth with a 753 square inch section. The span limits are 86′.

B III-48 has a dimension of 4′ width and 39″ depth with an 813 square inch section. The span limits are 96′.

B IV-48 has a dimension of 4′ width and 42″ depth with an 843 square inch section. The span limits are 103′.

Bulb tees. The bulb-tee girders are designed for and practical for spans over 60′ with a maximum range of 180′ with practical limits at the 150′ range.

BT-54 girders have a height of 54″, a cross-section area of 659 square inches, and a length range from 70′ to 125′.

BT-63 girders have a height of 63″, a cross-section area of 713 square inches, and a length range from 80′ to 135′.

BT-72 girders have a height of 72″, a cross-section area of 767 square inches, and a length range from 90′ to 150′.

Many design factors, such as number of and strand spacing, loading, number of girders, and geographical logistics, are determined when anticipating these types of girders.

Voided prestressed slab beams. Voided prestressed slab beams are similar to box beams in sectional shape but have circular cylinder void sections in lieu of the rectangular hollow section of the box beams.

SI-36 has a dimension of 3′ width and a 12″ depth with a 432 square inch section. The span range is 20′ to 29′. NOTE: This beam is of a solid section.

SII-36 has a dimension of 3′ width and a 15″ depth with a 432.5 square inch section. The span range is 28′ to 36′.

SIII-36 has a dimension of 3′ width and an 18″ depth with a 491 square inch section. The span range is 37′ to 46′.

SIV-36 has a dimension of 3′ width and a 21″ depth with a 530 square inch section. The span range is 45′ to 54′.

SI-48 has a dimension of 4′ width and 12″ depth with a 576 square inch section. The span range is 20′ to 29′. NOTE: This beam is of a solid section.

SII-48 has a dimension of 4′ width and a 15″ depth with a 569.2 square inch section. The span range is 28′ to 36′.

SIII-48 has a dimension of 4′ width and 18″ depth with a 628 square inch section. The span range is 37′ to 46′.

SIV-48 has a dimension of 4′ width and 21″ depth with a 703 square inch section. The span range is 45′ to 55′.

Double-tee beams. Channel girders and double-tee beams have a very economical short-span use.

C-I has a height of 17″, a width of 2′8″, and a section of 244 square inches. The span range is 15′ to 30′.

C-II has a height of 19.5″, a width of 2′9″, and a section of 288.25 square inches. The span range is 20′ to 35′.

C-III has a height of 21″, a width of 3′4″, and a section of 348 square inches. The span range is 25′ to 40′.

C-IV has a height of 25″, a width of 3′6″, and a section of 400 square inches. The span range is 30′ to 50′.

In addition to these common types of sections, many other sections are available. All weights of girders are calculated assuming the concrete weight of 150 lb per cubic foot.

Precast, prestressed concrete girders. The concept of precast, prestressed concrete girders has provided a technique for a minimal height design in the superstructure for clearance and approach limits. This technique utilizes a depth-span ratio as low as 1:32 in most types of concrete girders.

Concrete beams are of simple design and can normally be erected quickly with minimum traffic disruptions. They are widely used with a low maintenance factor.

Prestressed concrete beams have proved to be low in first cost and maintenance and have a high durability to weather and fire resistance.

Along with their integral deck components, the prestressing technique has offered a unique concept for bridge design and construction.

Prestressed, precast concrete beams, however, do have their capacities, limitations, and preference. Along with these and other technical designs and longer span capabilities, structural steel beams will always have their need and use.

Bridge structures can be designed utilizing one type or multiple types of prestressed concrete girders within the beam structure unit. Again, this is determined by the length of the span, the imposed loads, number of beams needed, and the economics involved within the design. In some cases, the end result can be achieved more economically by adding more beams of a smaller cross section or by using fewer beams of a larger cross section, provided the length of the span falls within the structural design properties of the intended member.

Manufacture. Prestressed, precast beams are manufactured in a controlled environment at a manufacturing facility, then transported to the project site. The means of transportation and/or logistics is sometimes a controlling item for the use of precast concrete beams.

Concrete beams are cast in steel *beds* of the same cross section as the intended design member. The length is controlled by the use of *bulkheads* within the beds. The strand cable is inserted in the beds. The *drape pattern* and location of the strand cables are controlled by positioning *templates, hold-downs,* and *chairs.* The cables are then prestressed to their proper and designed tension prior to concrete placement. The insertion of the reinforcing steel stirrups usually occurs during the placement of concrete. Other embedded items such as *bearing plates* and *lifting cables* must be properly placed prior to the concrete placement.

Specifications. Material specifications of prestressed, precast concrete members require a minimum of a 5000 psi concrete mix design. The embedded item requirements are the reinforcing steel grades of 40 or 60 and the prestressing strand is of grade 250. Special project requirements may alter this, but all materials must meet ASTM Standards pertaining to their individual categories.

Prestressed, precast concrete beams are designed and constructed with *camber*. This is a slight vertical arch in the beam from end to end. This is achieved by the amount of stress and the drape pattern of strand cables. The beam is cast in a flat plan; and during the detensioning process, the girder obtains its camber. The camber of a beam is a design feature that, when loaded, depending on the span length and imposed load of the structure, makes the girder flatten out to its theoretical designed location.

Technology has permitted, along with the precise design standards, the method of posttensioning multiple precast girders together longitudinally to form one longer structural member, thus allowing longer span capabilities. This is accomplished with the addition of conduit-type void ducts placed within the girder at the precast facilities.

The precast member is cast as a normal prestressed, precast unit. The beam is then transported to the project site and erected. These types of beams, which would be multiple individual members within a span, would need temporary support or some kind of falsework. This allows the proper positioning of the girders and prepares them for posttensioning.

Posttensioning. The posttensioning process is similar to prestressing except it is done and completed after casting. The strand cable is inserted through ductwork previously cast in the concrete members. Through a series of calculated procedures, the cables are stressed by hydraulic methods to their proper tension. After this procedure, the ductwork is pressure grouted to fill the remaining voids in the girders. This entire operation enables the multiple concrete girders to become one structure member within the span of the beam unit. (See illustration page 352.)

Diaphragms. Diaphragms are integral members of the beam structure which tie the individual beams together transversely. They are designed to transfer the specific load of a certain area of the deck more uniformly over the skeletal framework of the beam structure. The diaphragms are designed to make the beam structure a rigid framework.

With prestressed, precast concrete beams, diaphragms can consist of either cast-in-place reinforced concrete or steel. With the beam structure, the types of diaphragms needed, depending on the intended design, are end diaphragms, pier diaphragms, and intermediate-span diaphragms.

Posttension activity. (*Courtesy of Dywidag Systems.*)

The end diaphragms tie the ends of the beams of an individual span together transversely. The pier diaphragms tie the ends of the beams of two individual spans together transversely, making one continuous unit. The intermediate diaphragms tie the beams together at the midportion of a span. This could be with multiple points of the span, i.e., half, thirds, etc.

The cast-in-place diaphragms are formed of either wood or metal. The formwork must be well supported and follow the specifications detailed in the Formwork Section.

Bearing and diaphragm details

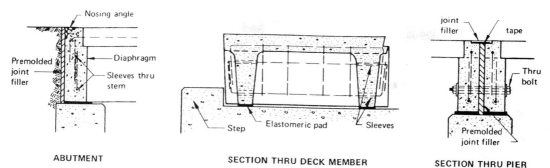

This bearing employs only elastomeric pads. The deck members are retained transversely by steps cast at the ends of the abutments and piers. The end diaphragm shown may be either cast onto the deck member in the precasting plant or field poured.

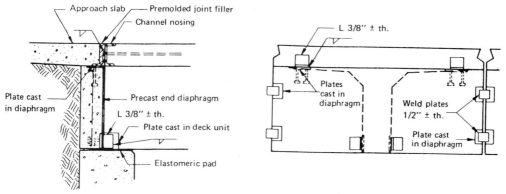

Precast concrete end diaphragms are attached through weld plates. Diaphragms can be attached to each bridge member at the precasting plant or at the bridge site.

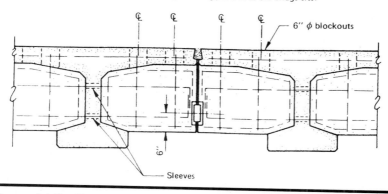

Concrete diaphragm details. (*Courtesy of Precast/Prestressed Concrete Institute.*)

Continuity details

Continuity is an effective means of eliminating joints and increasing load carrying capabilities of precast, prestressed beams of all types. These suggested diaphragm details provide continuity for carrying superimposed dead loads and live loads. The use of diaphragms at other locations is not recommended. When applied, this concept will result in more economical structures.

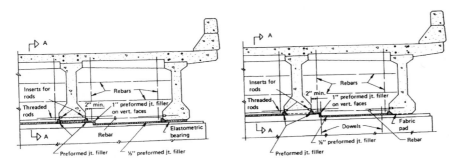

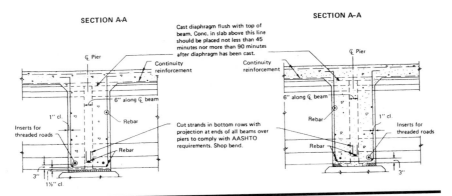

Concrete diaphragm details. (*Courtesy of Precast/Prestressed Concrete Institute.*)

Embedded items. Embedded items such as *threaded rebar inserts* or continuous *through bolts* must be properly installed prior to concrete placement. The threaded rebar inserts are screwed into the beam at designated locations. This serves as an embedded tie between the beam and the diaphragm. A continuous through bolt is embedded in the diaphragm and runs continuously from the exterior girder across the interior beam structure to the other exterior girder through holes provided for in the beams. After the installation of these embedded items, the forms are prepared and installed as specified, and the reinforcing steel and concrete are placed.

Normally, the type of concrete used is that which is specified for the superstructure deck.

Beam seats, bearing plates, and bearing pads. The beams are a designed structural member which bears its load and the load of the superstructure deck onto the substructure, either piers or abutments. The beams rest on a specific area known as *beam seats*. These seats are a prepared area of the substructure.

On the bottom of the precast beams at this bearing area, there is a *bearing plate*. This is a steel device which transfers and distributes the load of the girders to the bearing area without damaging the girder. Between the bearing plate and the concrete seat of the substructure is the *bearing pad*. This individual structural component is a rubber compound embedded with steel. This serves as a cushion between the beam and the beam seat.

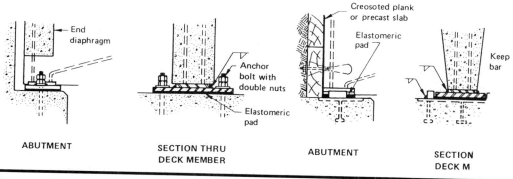

Elastomeric bearing details. (*Courtesy of Precast/Prestressed Concrete Institute.*)

Bearing pads vary in size, thickness, and internal compound depending on the design. Care must be taken when preparing the beam seat and the actual setting of the beam. Full and consistent bearing must be obtained in this area. An approved grout pad may be needed to assure a flat and true bearing surface.

Within this same area of the beam structure lie the *anchor* and *position dowels*.

The anchor area is usually the *fixed end* or permanently seated end of the precast beam. This area is not intended for movement or expansion. The beam is anchored or bolted through the bearing area to the substructure at this point.

At an expansion point in the structure, the beams are designed for controlled movement. This, the *expansion end* of the beams, is seated on the bearing area by means of *positioning dowels* or a sliding anchored base plate within the bearing area. This mechanism keeps the beams in the proper location but assures the required movement due to expansion. They, too, are a self-supporting structural member.

Individual prestressed, precast concrete beams of a bridge unit appear to be the same but have many different and unique characteristics. The estimator must be fully aware of these in order to obtain a proper bid in regard to erection sequence of the entire beam structure.

The top of a precast beam contains reinforcing bar *stirrups*. These are loop bars cast in the beam to form a monolithic tie between the beam structure and the superstructure known as a *composite* deck. All the structures' controlling points are maintained through the entire bridge, expansion, and control joints.

Measurement and payment of prestressed, precast concrete structural members are calculated by the actual linear foot of the specified beam. This includes all material, equipment, and labor necessary to fabricate and erect the unit, complete in place as per plans and specifications.

Structural steel girders

Structural steel girders have many of the same designed characteristics and serve the same purpose as their counterparts, concrete girders. They form the intermediate skeletal frame of the bridge which supports the superstructure. Structural steel beams can be used in simple and multispan bridges.

Steel beams consist of either *rolled* or *fabricated* steel composite. Rolled beams are of a standard uniform size; I-beam shapes consist of a known and consistent thickness, width, and height in cross section.

Rolled beams. This beam is rolled and formed at a steel mill processing facility from liquid molten ore containing specific material properties from which the steel is graded. These girders are classified by the width of the web and the weight per lineal foot of section. (See illustration page 358.)

Depending on the intended use, the designer determines which appropriate cross section is needed. A structural steel manufacturing facility prepares the girders by detailing them as per specifications.

The *interior stiffeners, end treatment for length, shear connectors, bearing treatment, anchors at the bearing points, splice details,* and areas of *cross frame* and *diaphragm* attachments are areas of attention to the stock beam.

Fabricated beams. These are constructed at a manufacturing facility from steel plate of specified thickness and composition. They are of a variable cross section and could vary in thickness, width, and height. This type of I-beam girder is normally intended for use in long-spanned structures, structures

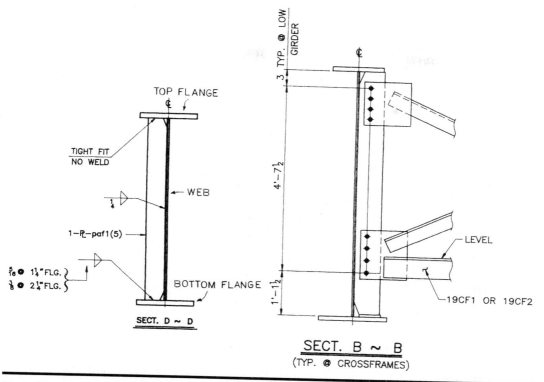

Steel girder details. (*Courtesy of Hartwig Manufacturing Corporation.*)

carrying an abnormal amount of load, and multispan composite bridges. A designer can formulate a girder based on unique and custom features rather than a stock prefabricated girder of common features. They are built up and constructed of steel plate, angle, and other steel shapes of welded connections. The design of these girders allows specific loading and deflection at certain areas of the girder and distributes the imposed loads to the designed bearing areas. They usually contain a thicker and heavier section at specified areas and transition downward to the ending or lesser load-bearing or distributing area.

The weight per linear foot of this type of beam is not consistent and varies greatly depending on the designed section. The individual girders can only be averaged as to weight per foot of section.

Specific beam areas. Within the total beam structure of a fabricated girder are similar areas of a rolled design, stiffeners, bearing area, connecting points of cross frames and diaphragms, splice details, and shear connectors. Specifications of these specific areas can be of welding, bolted, or a combination of both.

A steel girder consists of a *web* and *top* and *bottom flanges*.

WIDE FLANGE SHAPES

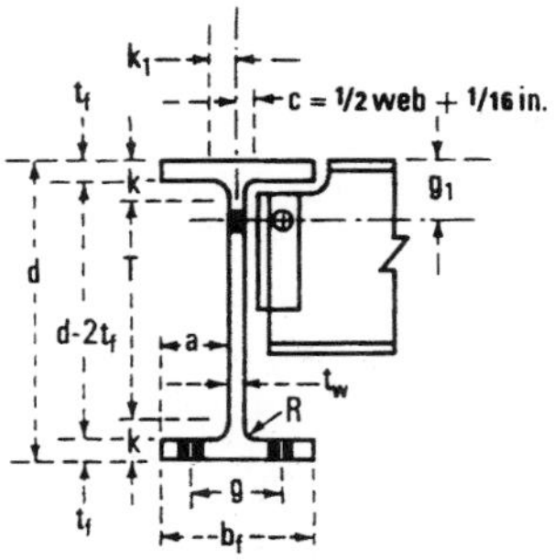

Approximate Dimensions for Detailing

Section Number and Nominal Size	Weight per Foot	Depth of Section	Flange		Web Thickness	d-2t_f	a	T	k	k_1	g_1	c	Usual Flange Gage
			Width	Thickness									
		d	b_f	t_f	t_w								g
in.	lb	in.	in.	in.	in.	in.	in.	in.	in.	in.	in.	in.	in.
W36 x	**300**	36¾	16⅝	1¹¹⁄₁₆	¹⁵⁄₁₆	33⅜	7⅞	31⅛	2¹³⁄₁₆	1½	3¾	⁹⁄₁₆	5½
B36A	**280**	36½	16⅝	1⁹⁄₁₆	⅞	33⅜	7⅞	31⅛	2¹¹⁄₁₆	1½	3¾	½	5½
36 x 16½	**260**	36¼	16½	1⁷⁄₁₆	¹³⁄₁₆	33⅜	7⅞	31⅛	2⁹⁄₁₆	1½	3½	½	5½
*R = .95	**245**	36	16½	1⅜	¹³⁄₁₆	33⅜	7⅞	31⅛	2⁷⁄₁₆	1⁷⁄₁₆	3½	⁷⁄₁₆	5½
	230	35⅞	16½	1¼	¾	33⅜	7⅞	31⅛	2⅜	1⁷⁄₁₆	3½	⁷⁄₁₆	5½
W36 x	**194**	36½	12⅛	1¼	¾	33¹⁵⁄₁₆	5⅝	32⅛	2³⁄₁₆	1³⁄₁₆	3½	⁷⁄₁₆	5½
B36	**182**	36⅜	12⅛	1³⁄₁₆	¾	33¹⁵⁄₁₆	5⅝	32⅛	2⅛	1³⁄₁₆	3¼	⁷⁄₁₆	5½
36 x 12	**170**	36⅛	12	1⅛	¹¹⁄₁₆	33¹⁵⁄₁₆	5⅝	32⅛	2	1³⁄₁₆	3¼	⅜	5½
*R = .75	**160**	36	12	1	⅝	33¹⁵⁄₁₆	5⅝	32⅛	1¹⁵⁄₁₆	1³⁄₁₆	3¼	⅜	5½
	150	35⅞	12	¹⁵⁄₁₆	⅝	33¹⁵⁄₁₆	5⅝	32⅛	1⅞	1⅛	3	⅜	5½
	135	35½	12	¹³⁄₁₆	⅝	33¹⁵⁄₁₆	5⅝	32⅛	1¹¹⁄₁₆	1⅛	3	⅜	5½
W33 x	**240**	33½	15⅞	1⅜	¹³⁄₁₆	30¹¹⁄₁₆	7½	28⅝	2⁷⁄₁₆	1⅜	3½	½	5½
B33A	**220**	33¼	15¾	1¼	¾	30¹¹⁄₁₆	7½	28⅝	2⁵⁄₁₆	1⅜	3½	⁷⁄₁₆	5½
33 x 15¾	**200**	33	15¾	1⅛	¹¹⁄₁₆	30¹¹⁄₁₆	7½	28⅝	2³⁄₁₆	1⅜	3¼	⁷⁄₁₆	5½
*R = .90													
W33 x	**152**	33½	11⅝	1¹⁄₁₆	⅝	31⅜	5½	29¾	1⅞	1⅛	3¼	⅜	5½
B33	**141**	33¼	11½	¹⁵⁄₁₆	⅝	31⅜	5½	29¾	1¾	1¹⁄₁₆	3	⅜	5½
33 x 11½	**130**	33⅛	11½	⅞	⁹⁄₁₆	31⅜	5½	29¾	1¹¹⁄₁₆	1¹⁄₁₆	3	⅜	5½
*R = .70	**118**	32⅞	11½	¾	⁹⁄₁₆	31⅜	5½	29¾	1⁹⁄₁₆	1¹⁄₁₆	2¾	⁵⁄₁₆	5½
W30 x	**210**	30⅜	15⅛	1⁵⁄₁₆	¾	27¾	7⅛	25¾	2⁵⁄₁₆	1⁵⁄₁₆	3½	⁷⁄₁₆	5½
B30A	**190**	30⅛	15	1³⁄₁₆	¹¹⁄₁₆	27¾	7⅛	25¾	2³⁄₁₆	1⁵⁄₁₆	3¼	⁷⁄₁₆	5½
30 x 15	**172**	29⅞	15	1¹⁄₁₆	⅝	27¾	7⅛	25¾	2¹⁄₁₆	1¼	3¼	⅜	5½
*R = .85													
W30 x	**132**	30¼	10½	1	⅝	28⁵⁄₁₆	5	26¾	1¾	1¹⁄₁₆	3	⅜	5½
B30	**124**	30⅛	10½	¹⁵⁄₁₆	⁹⁄₁₆	28⁵⁄₁₆	5	26¾	1¹¹⁄₁₆	1	3	⅜	5½
30 x 10½	**116**	30	10½	⅞	⁹⁄₁₆	28⁵⁄₁₆	5	26¾	1⅝	1	3	⅜	5½
*R = .65	**108**	29⅞	10½	¾	⁹⁄₁₆	28⁵⁄₁₆	5	26¾	1⁹⁄₁₆	1	3	⁵⁄₁₆	5½
	99	29⅝	10½	¹¹⁄₁₆	½	28⁵⁄₁₆	5	26¾	1⁷⁄₁₆	1	2¾	⁵⁄₁₆	5½
W27 x	**177**	27¼	14⅛	1³⁄₁₆	¾	24¹⁵⁄₁₆	6⅝	23	2⅛	1¼	3¼	⁷⁄₁₆	5½
B27A	**160**	27⅛	14	1¹⁄₁₆	¹¹⁄₁₆	24¹⁵⁄₁₆	6⅝	23	2¹⁄₁₆	1¼	3¼	⅜	5½
27 x 14	**145**	26⅞	14	1	⅝	24¹⁵⁄₁₆	6⅝	23	1¹⁵⁄₁₆	1³⁄₁₆	3	⅜	5½
*R = .80													

Steel girder sections. (*Courtesy of Bethlehem Steel Corporation.*)

Interior stiffeners. Interior stiffeners are vertical members fastened perpendicular to the web. This forms a reinforced support to the web for imposed downforce and vertical support to the cross section of the beam.

End treatment. End treatment is the preparation and detailing of the individual ends of beams for either an exposed end or a spliced connection. This is the exact detailing for the proper length of the girder.

Splice detail. The splice detail is the point of connection of two beams at the ends. This permits a structural lengthening of beams for longer spans. This consists of either a bolted or welded connection point to the web and flanges. (See illus. p. 364)

Cross frames and diaphragms. Cross frames and diaphragms are methods of transversely connecting the girders together to form a rigid beam structure. Cross frames are a fabricated unit, usually of angle and plate, and are connected to the web section of the individual beam. The diaphragms are of a stock rolled section, usually channel, which is connected to the web section. Both cross frames and diaphragms can be connected by either a bolted or a welded section, depending on individual project specifications.

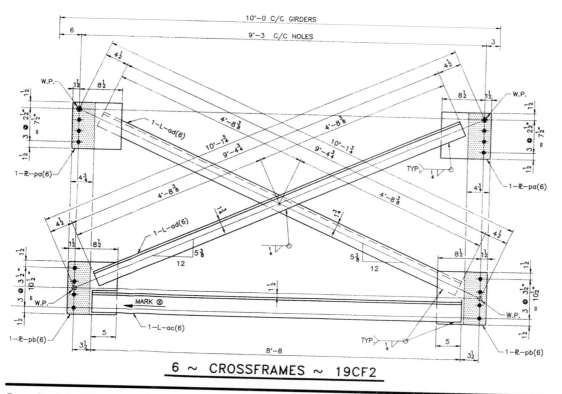

Cross frame and diaphragm details. (*Courtesy of Hartwig Manufacturing Corporation.*)

Bearing treatment. The bearing areas of a steel beam consist of numerous types, *flat bearing, rocker bearing, pot bearing,* and *fixed bearing* points.

The bearing section consists of a bearing plate attached to the bottom of the individual beam, an area of support or cushion between the bearing plate and the substructure, and the beam seat of the substructure. As with the concrete beams, the beam seat must be constructed flat and true to grade to assure a proper load and bearing displacement.

A flat bearing or slider bearing can be either fixed or expansive. This is a simple type of bearing structure. On the fixed bearing assembly or the slider assembly, anchor bolts are embedded in the substructure to assure the proper location and fastening of the girder.

A rocker assembly is a detailed form of movable mechanism at the expansion point of the structure. This permits the beam structure to have a horizontal movement but maintain its rigidity and location within itself. In most cases, steel composite beams have a greater expansion movement than concrete.

A pot bearing assembly allows a flat horizontal expansion movement and is usually designed for use in large girder structures such as *tub girders*.

The fixed bearing point of the beam unit is that of a permanently nonmovable anchored position to the substructure. As with the flat bearing area, this can be cushioned with an elastomeric bearing pad but can also be designed with flat steel plating, depending on the movement and design intended.

The anchor bolt assembly to the bearing locations must be accurately placed to assure proper alignment and function to the entire beam structure. As specifications allow, the actual anchor bolts can be either cast directly in the substructure unit or grouted in previously formed holes in the substructure unit.

The shear connectors or studs are the structural element used to connect the beam unit to the superstructure unit. These are calculated in the design of the movement with respect to contraction and expansion to the concrete deck and steel structure. (See illustration page 362.)

The designed area of attachment is known as the composite deck section, and the nonattached or free-movement area is the noncomposite sector. The shear connectors are attached to the steel girders by means of a welded connection.

ASTM and AASHTO specifications. Material specifications and requirements for structural steel girders follow ASTM standard specifications for properties. Because of the wide variety of design potentials and uses of steel girders, these standards are often used in conjunction with each other to give the fabricated steel many unique and individual physical characteristics with respect to loading and spanning capabilities.

Structural steel is produced by grade, which controls its physical properties. A common grade is 50. Classifications of steel from that grade range with respect to additive properties for hardness, flexibility, and other design uses. Normal ASTM classifications are A36, A325, A572, and A588. Design and structural uses of these types range from the main girders to the diaphragm

Beam Structure   361

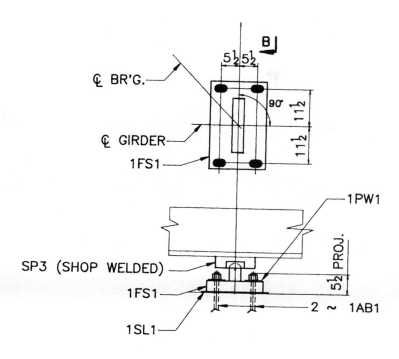

BRG. ELEV. @ PIER # 5

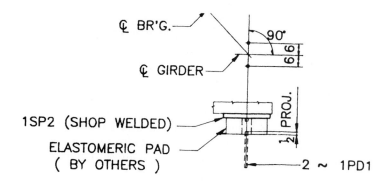

BRG. ELEV. @ PIER # 4

Bearing details. (*Courtesy of Hartwig Manufacturing Corporation.*)

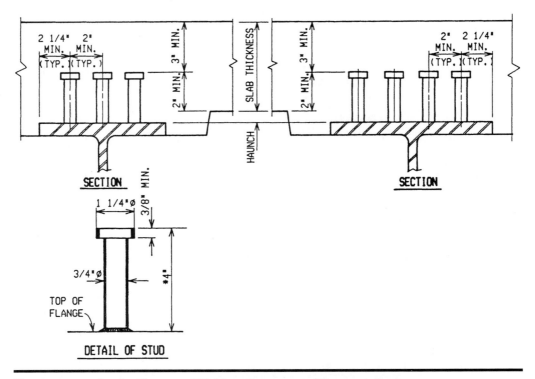

Shear connector details. (*Courtesy of Michigan Department of Transportation.*)

and cross frames, to stiffener plates, bearing assemblies, and bolts. Bolts are normally specified of a high tensile strength characteristic.

Span limitations along with load deflections of the section size of the girder are controlling items of the design and use of structural steel.

The standards of practice for use of structural steel follow AASHTO recommendations and procedures.

Erection procedures must take into account the necessary lifting equipment needed, along with temporary supports and falsework. With spliced beams, they must be braced, supported, and securely held in their permanently erected position during the bolting procedure. During assembly, all spliced parts must be properly aligned prior to the connection procedure, either welding or bolting. Normally, mechanical forces will not be permitted to force align a connection point. Any reaming, misaligned steel, or chipping must be an approved method prior to correcting. Welded connections as to size and passes and bolt tension as to the amount and type of method used will be explicitly outlined in the individual project specifications.

Used in the alignment of bolted steel splice connections are *drift pins, barrel pins,* and *temporary bolts* in which the percentage and number are outlined in the specifications.

The drift pins are used for quick alignment of the holes when landing a girder. These are tapered pins. Once aligned, barrel pins are inserted. Normally, the hole size of the bolted connections is $\frac{1}{16}''$ larger than the bolts. The barrel pins are a steel-shouldered pin that properly gauges this variance to equal amounts around the circumference of the hole. This fits the template of the girder. These variances assure that all the girders are properly aligned during final bolting. These barrel pins normally cover 25 to 33% but could be up to 50% of the hole pattern.

In conjunction with the barrel pins, temporary bolts are specified. These are used to firmly bolt the spliced girders together in the proper position. The temporary bolts cannot be reused for permanent bolts, as the tightening of them during tension causes stress, thread stretch, and elongation to the bolt. Thus, when the calibrated wrench or *turn of nut* method is used for rotation, a true reading cannot be obtained. The amount of temporary bolts needed will be specified in the project documents. This is an area that the estimator must thoroughly research to prepare a bid. (See illustration page 364.)

Once the girders have been secured in their proper position, the permanent bolts can be installed and tightened to their specified tension and in sequence with the removal of the temporary bolts and barrel pins.

The painting of structural steel must adhere to the project and manufacturer's specifications. Painting is a detailed and precise process of the structural item. The final paint coat or partial coating is normally done after all erection procedures are completed. For initial protection, the steel girders are usually prime-coated at the manufacturing facility. The cleaning, wash primer, and undercoating or final paint coat are items that need addressing in preparing a detailed estimate. These necessary procedures must be detailed and acknowledged. The painting of structural steel is controlled by moisture, temperature, and humidity and has specified limitations.

Painting. The painted surface must be free from dirt, oils, laitance, and moisture. The surface must be prepared and cleaned according to the project specifications. The primer and paint systems must be of an approved source and adhere to specific specifications. The equipment used must be approved and be able to disperse the paint properly to assure a proper and even distribution.

Measurement and payment. Measurement and payment of structural steel shapes and fabricated structural steel members are determined by the actual designed pounds of steel specified in the completed girder unit.

Normal assumed weights of materials:

Steel:	0.2833 lb per cubic inch
Cast iron:	0.2600 lb per cubic inch
Bronze:	0.3150 lb per cubic inch
Lead:	0.4110 lb per cubic inch

This normally includes all components fabricated from metal compounds listed for the completed unit, such as castings, rolled shapes, alloy steels,

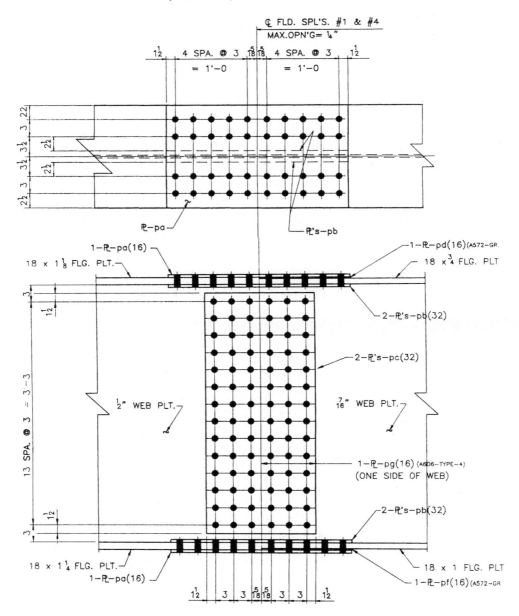

Splice details. (*Courtesy of Hartwig Manufacturing Corporation.*)

steel plates, anchor bolts and nuts, shoes, rockers, rollers, pins and nuts, expansion joints, roadway drains and *scuppers,* weld metal, embedded items, cradles and brackets, and other structural metals. Specific components must be verified.

The composition is determined based on final quantities from the fabrication. Delivery, erection, and final painting of the designed beam unit must be in accordance with project plans and specifications.

The equipment needed for erection of a structural beam unit, whether precast or steel, will be of sufficient size and capacity to safely handle the beam unit and necessary accessories. Any falsework, scaffolding, or bracing will be of sufficient design and capacity to support the intended loads implied.

Estimators must be fully aware of the specific project and payment specifications in regard to the beam structure, referencing incidental items such as bearing assemblies, anchors, tie-rods, plates, and the need of any temporary shoring or falsework. With structural steel, they should note the amount of temporary bolting required and incidental welding and optional field fabrication not performed by the fabricator, also the paint system or finish end coating, respectively, applied to structural and precast beams and the site conditions controlling delivery and erection procedures involved along with the capacities of needed equipment. These items are thoroughly outlined in the takeoff section.

TAKEOFF QUANTIFICATION

Introduction

The beam structure is an integral part of the bridge. This structure contains various components depending on the selected design. The two types of beam structure units are (1) prestressed, precast concrete beams and (2) structural steel girders.

The prestressed, precast concrete beams are designed with various shapes, lengths, internal features, and cross sections, each giving the beam its own characteristics. These beams, within the beam structure unit, have supporting components that aid in their ability to structurally support the bridge superstructure. These components are diaphragms, bearing units, and tension rods.

For structural bonding to the superstructure, or a composite section, steel stirrups are cast into the top of each beam.

The steel girders are also designed with varying features. The two types of steel girder designs are rolled beams and fabricated girders. As discussed in the technical section, the rolled beams are of a predetermined cross section, produced in a steel mill by hot rolling. Each rolled beam has identical properties.

The fabricated girders are of an independent design, suited specifically for a required purpose. Each girder has its own design characteristics and properties for the span distance and implied loads of the bridge. The fabricated girders allow for more specialty applications, loads, and greater span distances.

366 Takeoff and Cost Analysis Techniques

As with the concrete beams, the steel girders also have support components which aid in the structural integrity of the beam unit. These are defined as cross frames, diaphragms, splice plates, bearing units, stiffeners, and bolts.

The bonding to the superstructure deck, for composite tension, is done with components defined as shear connectors. These are attached to the top surface of the top flange of the steel girder.

Definition

The beam structure unit of a bridge is the structural unit that spans the substructure units, linking them together. It is the skeletal framework that supports the superstructure unit and its loads.

Takeoff

The quantity takeoff of a beam structure is compiled by identifying the primary components first, then defining the secondary components. The primary components are the beams or girders themselves. The secondary components are those associated with the type of required beam design, such as concrete diaphragms with precast concrete beams, or metal cross frames with structural steel girders.

The quantity takeoff provides and associates the components with the required craft-hours to produce the function, giving a productivity required for each operation. This is a summarized and detailed account of all required operations.

The erection operation for the beam structure members is calculated in a systematic format for each given erection method. The requirement of information shows the length of each member, and the weight of each member. The reach and the height limits of the hoisted member must be determined with respect to the crane(s) capacity and position. The crane that is available for the hoist must be analyzed to its maximum rated capacity in pounds, and for the rated capacity in pounds for the reach that is required, and the length of boom needed for the hoist height and reach ratio. Therefore, the chosen crane is capable to lift a defined member of weight, in pounds, to the maximum reach point required. If the crane is not logistically feasible, a larger crane or a multiple of dual cranes should be analyzed.

Prestressed, Precast Concrete Beams

The first item of the beam structure unit discussed is the concrete beam design. This section highlights the construction requirements and operations associated with the takeoff and quantifying of these components. It also identifies the span designs as to simple or continuous. A simple-span beam unit spans directly from one substructure unit to the next adjacent substructure unit. The continuous girder unit encompasses multiple substructure units.

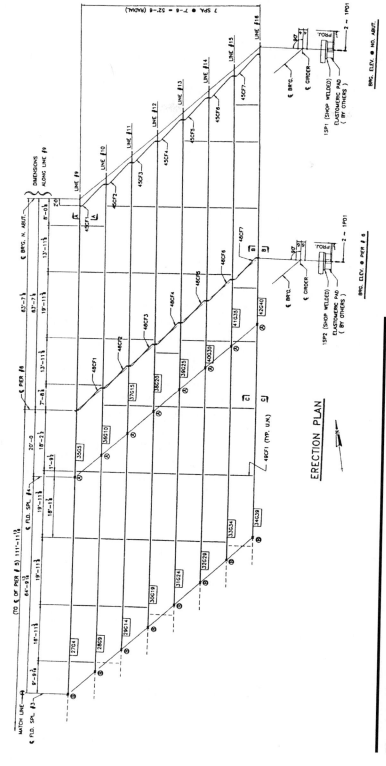

Plan view of beam structure.

The components that may be required for the prestressed, precast concrete beam operation are identified as:

1. Site and pad preparation
2. The concrete beams
3. Bearing assemblies
4. Position dowels
5. Concrete diaphragms
6. Tie-rod assemblies
7. Temporary bracing
8. Stockpile and rehandle
9. Relocate crew

Once the components have been identified, the site conditions, erection procedures, and construction methods must be outlined. The site conditions will control mobility and access of both delivery of materials and equipment needs and requirements. There can be restrictive conditions that will alter or affect the erection procedure. The erection procedures will outline the requirements and specifications for the actual setting of the beams or girders. The construction methods will detail the constructability requirements, the type of work operation required for the required beam unit and components, and any shoring, scaffolding, and bracing requirements.

The quantity takeoff procedure for a prestressed, precast concrete beam unit is accomplished by first identifying the type of beams required. Second, the associated components must be detailed. The third requirement is to define the site conditions, erection procedures, and construction methods required for the specific project.

Site and pad preparation

The requirement of this operation will be the preparation of a level pad for the hoisting equipment to stage or for the beams to be stockpiled. If wood crane mats are required for foundation stability, they should be accounted for under this operation.

The defined quantity unit for this operation will be summarized as required craft-hours to perform the work.

The concrete beam

The quantity takeoff of concrete beams is performed by accounting for the number of required beams and the lineal feet required, by type, by span. The procedure is done independently by bridge structure.

The outline requirement for the takeoff of the precast concrete beam operation for quantifying the components will be listed as follows:

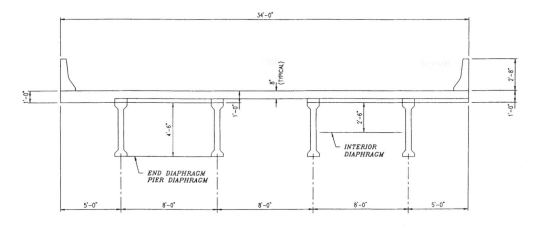

SUPERSTRUCTURE CROSS SECTION

Cross section of beam structure.

Precast concrete beam:

1. Beam type
2. Required quantity per span
3. Beam length
4. Number of spans
5. Total required beams
6. Total length required of type ―――――
7. The longest beam
8. The heaviest beam

Starting with the first span, identifying the type of beam required will determine the next steps. This will be the control of the takeoff.

For each span, the beams will be defined as the composite material type, which will be concrete. The type of beam as to their AASHTO designation (i.e., type IV) will need to be defined. The number of required beams defined in quantity of each will be listed next. The next procedure will be listing the design length for each beam, per span. From this listing, the weight of each beam will need to be determined. This information is required to establish the hoisting requirement. Establishing the longest and heaviest required beam is necessary to determine the maximum hoist per location. The crane power, in both capacity and quantity, will be controlled by this beam.

The properties of a given beam can be obtained from the AASHTO specification chart. This will provide the weight per foot of standard concrete beams. When a modified beam is designed, the design specifications will provide the section properties.

This outlined procedure must be performed for each span within the bridge structure, then summarized by beam type per bridge. The summary must include the required quantity of each type of beam; the required lineal feet per beam type; and the longest and heaviest beam, per type, within the bridge structure. Once this has been completed, the total length required is to be divided by the number of beams required to obtain an average length per beam type.

The delivery of the beams will affect the erection procedure. The number of beams that can be delivered per crew hour will be the basis for calculating the lineal feet erected per hour, including time loss for relocation or rehandling. Also, the time to brace, scaffold, or shore will be calculated as idled time to the erection crew, if it affects the hoisting duration.

The required work for this operation will be all work related to the beam erection, which includes site staging, rigging, and unloading. The moving time between spans and stockpiling to double handle are factored under separate operations. The defined control unit of measure for quantifying concrete beams is listed as:

Prestressed concrete beam erected:

Type _____, for the lineal feet required

Type _____, weight per beam

Elastomeric bearing pads

The next item of takeoff associated with the beam structure is identifying and quantifying the required bearing assemblies designed for the beams. This is accomplished by a listing per beam, per span, and categorizing by beam type.

These bearing assemblies are defined as elastomeric bearing pads, which are placed between the beam and the substructure beam seat. This rubber-type pad provides the beam a cushion and movement section, between the two opposing concrete surfaces. The bearing pad is a structure component that is designed to withstand the load imposed, including the beam weight, the load of the superstructure, and the live loads implied to the surface of the deck.

Manufactured within the rubber composite are steel shim plates. The design of the pad hardness, referred to as durometer, and the amount of and size of steel plates, differ within each pad based on the intended design. This will vary the unit square foot cost of the bearing pads.

This rubber pad is a part of the designed bearing assembly of the structure. A steel plate or bearing shoe is cast embedded into the concrete beam during the manufacturing process. The component section bears directly onto the bearing pad. There are varying designs and conditions of bearing section areas. The beam may rest directly on the pad with no other location or positioning assistance, or the section design may require positioning dowels to aid with the location assurance.

The bearing assemblies will be outlined in the following format:

1. Type of required pad
2. Number required per beam
3. Total number per span
4. Square feet required per pad and total
5. Total of each pad(s) required per beam type

The quantity takeoff procedure for this item is compiled as the number of each type of elastomeric bearing pads required, per span, per beam type. This will provide the quantity for craft-hour factoring. For material costing, it usually requires a quantity of required square feet. The listing of square feet per bearing pad, and an accumulation of bearing pad type, will accompany this list.

From this listing, a summary will be compiled. The defined control unit of measure for quantifying the bearing pads will be:

1. Elastomeric bearing pad, beam type _____, each
2. Elastomeric bearing pad, beam type _____, square feet

Position dowels

Within the bearing area there may exist a component defined as a position dowel. This component is embedded within the substructure beam seat and provides location control for the beam. The position dowel extends from the beam seat, through the bearing assembly, into the bottom surface of the concrete beam.

A position dowel is intended to maintain the location, both longitudinally and transversely, of the beam. The two types of positioning dowels are a fixed end and a sliding end.

A fixed-end position dowel is designed for a nonmovement bearing section area. The dowel will fit snugly into the bottom beam surface. A sliding-end design has an elongated slot or hole cast into the bottom beam surface, allowing the beam to travel longitudinally over the position dowel during expansion and contraction movement.

The most efficient method of installing this dowel is to drill and epoxy in place. This is simply drilling the required size hole into the substructure beam seat at the proper location, after the pier cap and beam seat is constructed. If permitted by specification, this procedure will ensure better control and quality for the positioning of the dowel. Otherwise, the position dowel will be monolithically cast into the beam seat during the construction of that substructure operation defined as hand setting position dowels.

One other alternative would be to cast plastic inserts in the cap beam seat.

This eliminates the need of drilling but requires the removal of the insert and the epoxying of the dowel in place.

Once the proper size hole is drilled, the position dowel will be epoxied in place, within the beam seat.

The outline required for the quantity takeoff of position dowels will be:

1. Type of dowel insertion (drill and epoxy, hand set)
2. The size of dowel required
3. The penetration of dowel (within the beam seat)
4. The quantity of dowels per beam
5. The number required per span
6. The total required per bridge
7. The number of required holes to be drilled
8. The number of required dowels to epoxy
9. The required volume of epoxy

The quantity factoring for these items is determined based on the operation to be performed. The drilling of dowel holes is quantified as the number of like-sized holes drilled, categorizing by diameter and depth. The epoxy application is quantified as the number, by size, of dowels required to be placed. The application of epoxy within the hole is calculated by the time required per hole and the volume of epoxy material required per hole. This is formulated by the cubic inches of open hole minus the cubic inches of area displaced by the dowel.

The operation of installing position dowels is calculated under two functions. The first operation is the performance time for drilling the required dowel holes. The second operation is the performance of the epoxy application and the insertion of the position dowels.

The drilling operation is calculated for the time, per dowel, to drill the hole and move to the next. The size and depth of the required hole will govern the production.

The epoxy application is calculated for the time, per dowel, to inject the required volume of epoxy and insert the dowel.

From this listing, a summary is compiled. The defined control unit of measure for quantifying the position dowels will be each required, by type of installation. The categorizing will be listed as:

1. Dowels drilled, each, by size, by beam type _____
2. Dowels placed, each, by size, by beam type _____
3. Dowels epoxied, each, by size, by beam type _____

Concrete diaphragms

Once the concrete beams are set into their final position, the diaphragm components are the next operation of construction. The diaphragms are structural components that bridge or transversely link each beam together and give a complete structural uniformity to the beam unit. The purpose is to take the independent structural capacity of each beam and unite them into a dependent system, distributing the implied loads evenly over the area section rather than over one beam. The diaphragms are constructed in a row, transversely across the beam section, between the exterior beams.

The concrete diaphragms are cast in place between the beams, at their designated location. The spacing of the beams may vary by design, and therefore the length of the diaphragm may vary per bay.

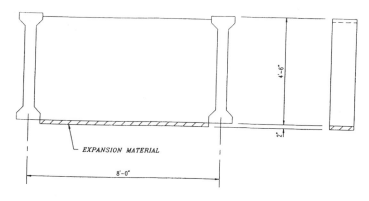

END DIAPHRAGM / PIER DIAPHRAGM

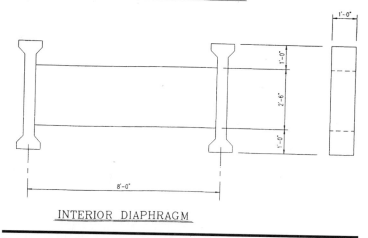

INTERIOR DIAPHRAGM

Diaphragm details. End. Midspan and interior. Composite pier.

Types. The category of types of diaphragms is defined as: end diaphragms, interior or midspan diaphragms, and pier composite diaphragms. The end type of diaphragm is located at the end of a beam span and is dependent to that span. The end diaphragm can be located at the abutment or pier location but does not connect adjacent spans. The end diaphragms can be suspended from the beams with no horizontal component contact area or can rest directly on the substructure component surface area beneath the diaphragm.

The interior diaphragms are located at the midpoint, longitudinally, of a span. They are a suspended component with no horizontal component contact area. There will be a required formed surface on two sides, plus the bottom soffit area.

The pier or composite diaphragms are those which connect two adjacent spans, creating a continuous beam unit of multiple spans. The composite diaphragms can be suspended from the beams with no horizontal component contact area or can rest directly on the substructure component surface area beneath the diaphragm. With this type of diaphragm, there may be a formed surface area at the exterior beam end area, the longitudinal spacing of the beam end dimension between the spans.

Work operations. All diaphragms will contain a vertical component contact area that abuts the beam surface. The diaphragm components consist of numerous work operations:

1. Falsework
2. Formwork placement
3. Embed item installation
4. Expansion paper
5. Concrete placement
6. Formwork removal
7. Rub and patch concrete
8. Falsework removal

These work operations are independent functions required to complete the component of concrete diaphragms.

The falsework requirement is needed for the logistics of constructing the diaphragm members. The diaphragms are transversely suspended between each beam and require a work platform for the construction operations. The falsework is suspended from the top or bottom flange of each adjacent beam, on both sides of each diaphragm, spanning the horizontal distance between each beam. The falsework not only serves as a platform for the work personnel but is the support for the diaphragm component itself during the construction and curing duration of the concrete.

The falsework operation includes all work necessary for the installation of the system at each diaphragm location. The work pertains to the beam supports, the hang-down rods or vertical members, and the horizontal planking or decking under the diaphragm component. The unit of measure description will be defined and quantified as the time duration for labor and equipment, and required materials for each required diaphragm.

The quantity summation will be listed as: diaphragm falsework, erection, each.

The removal of the falsework system will be the identical quantity as defined with the placement. The quantity summary will be listed as: diaphragm falsework, removal, each.

The formwork requirements are the quantity of concrete surface contact area. The required unit calculation of square feet will be each side area, plus the bottom soffit area, if required. As with the dependent abutment backwall, there will be scribing or shaping of the formwork to the configuration of the vertical side of the beam. This factor will be taken into account as with the backwall, referred to as the diaphragm beam factor. This formwork quantity is summarized for each type of diaphragm.

A concrete beam diaphragm requires the compound takeoff to the area around the beam unit. Each side of the diaphragm must be formed within the perimeter shape of the corresponding beam. This is referred to as scribing. The formwork shaping, or scribing, will vary in production with respect to different beam types. An additional factor must be applied to the craft-hour production factor for this operation.

Illustrated (p. 369) is a beam section with four beams, which is the view of the face of a diaphragm section. The contact surface formwork requirement for the individual diaphragm is the closure area between the beams or interior bays. The formed diaphragm section required will have three bays, since there are four beams. The requirement will be one less bay than the number of required beams.

The calculation requirement to arrive at the surface contact area for the formwork is done by determining the horizontal dimension between the beam centerlines, and the vertical height or elevation difference from the bottom horizontal elevation of the diaphragm, to the top of the proposed diaphragm (averaging the height if required). These will be the dimensions used for each bay calculation of the diaphragms, for each type of diaphragm. The width of each interior bay will be calculated because the beam spacing may vary.

This format uses a theoretical vertical line from the center to center of beam for measurement purposes. This method of format will aid in the ease of takeoff, even though the volume or shape of the beam is not reduced from the measurement. This method will account for the additional labor to scribe the beam shape, and the added formwork craft-hour usage in the formwork production factor for a concrete diaphragm.

With precast concrete beams, for the additional craft-hour consumption of scribing, it is suggested to use a factor of 0.0005,* multiplied by the end section surface area of the beam in square inches, multiplied by the number of scribed ends within the diaphragm section component, for each side panel (normally 4). The factor will be categorized as a craft-hour per square inch factor. This is the reasoning behind the need for categorizing structural components.

---

*0.0005 will vary depending on specific experience.

The quantity summary for the formwork placement requirements of concrete diaphragms will be categorized in square feet as:

1. Concrete diaphragm, end, formwork placement
   a. The factor application for scribing beam area for an end diaphragm (0.0005* beam end area surface, square inches, by the number of scribed ends, for the additional craft-hour usage)
2. Concrete diaphragm, interior, formwork placement
   a. The factor application for scribing beam area for an interior diaphragm (0.0005* beam end area surface, square inches, by the number of scribed ends, for the additional craft-hour usage)
3. Concrete diaphragm, composite, formwork placement
   a. The factor application for scribing beam area for a composite diaphragm (0.0005* beam end area surface, square inches, by the number of scribed ends, for the additional craft-hour usage)

The removal of the formwork system will be identical to the quantity of square feet placed and will be summarized as:

1. Concrete diaphragm, end, formwork removal
2. Concrete diaphragm, interior, formwork removal
3. Concrete diaphragm, composite, formwork removal

The embed or form hardware will be for the installation of each required, for each type of diaphragm. The quantity summation for the hardware embeds for concrete diaphragms will be listed as: (1) concrete diaphragm, embed installation, end; (2) concrete diaphragm, embed installation, interior; (3) concrete diaphragm, embed installation, composite.

When the end or composite diaphragms require a horizontal component contact surface, the design may designate expansion material to be placed between the diaphragm and substructure components. This quantity will be a calculation of the horizontal surface area requiring the material, in square feet designation, for each type of diaphragm. The summary will be identified as:

1. Concrete diaphragm, expansion material, end
2. Concrete diaphragm, expansion material, composite

The concrete volume requirement will be the calculation of cubic yards of concrete needed for each type of diaphragm, summarized by diaphragm type.

Concrete volume requirements for placement calculations will be compiled using the volumetric methods.

A concrete diaphragm component is calculated for concrete volume by sim-

---

*0.0005 will vary depending on specific experience.

ply multiplying the averaged width, by the height, by the thickness, then converting the resulted cubic feet to cubic yards, using the following format.

The concrete diaphragm calculation requirements for the concrete entail a somewhat more complicated method, as with the dependent backwall. To ease this, again, it is suggested to implement the methodology used with the square foot takeoff. Since the diaphragm dimensions encompass the beam units, it is necessary to reduce the volume of concrete that is displaced by the beams.

The primary concrete volume calculation is completed by the same method as done with the dependent backwall. The next procedure required is to reduce the displaced volume consumed by the end area of the beam section. This calculation is simply done by multiplying the end area square inches of a beam by the dimension the beam extends within the diaphragm row, then multiplying the volume cubic inches by ½ the number of affected beams for each diaphragm. The two beams that are affected are ½ of each beam dimension, owing to the centerline method of calculation. This is then converted to cubic yards. The total displaced cubic yards for the beam section is then subtracted from the primary volume, giving the net volume of required cubic yards, per each type of diaphragm. The composite diaphragm volume calculation must add the open volume dimensions between the longitudinal end section of the beams. This calculation would be the average height of the diaphragm row, multiplied by the dimensional thickness between the ends of the beams, multiplied by the length dimension from the outside of one exterior girder to the outside of the other exterior girder, then converted to a cubic yard requirement for a composite diaphragm row.

The quantity summation volume requirement would be classified as:

1. Concrete placement volume, cubic yards, end diaphragm
2. Concrete placement volume, cubic yards, interior diaphragm
3. Concrete placement volume, cubic yards, composite diaphragm

The operation of rubbing and patching will be the quantity of required surface area, in square feet, to be treated. This area, usually identical to the required formwork surface area, will be summarized as:

1. Concrete diaphragm, end, rub and patch
2. Concrete diaphragm, interior, rub and patch
3. Concrete diaphragm, composite, rub and patch

Tie-rod assembly

The tie-rod assembly is a steel component that extends through the diaphragm section, which clamps the beam unit together transversely. This connecting rod is quantified in a unit measure of lineal feet, required for each diaphragm section, and is categorized as:

1. Number of required rod assemblies
2. Diameter of rod, and length per piece
3. Total lineal feet required by beam type

The operation will be quantified for the work required to install the tie-rod assembly as specified. The tie-rod assembly will be summarized, by the type of beam that they are required for, as: tie-rod assembly, beam type ─────, lineal feet.

Temporary bracing

Temporary beam bracing is a secondary operation to the beam erection item. For interim protection until the completion of the diaphragms, this item may be necessary owing to the unstable condition that exists with certain types of beams. The design section of certain beams causes them to be top-heavy, or flexurally unstable. Concrete beams can sway or tip over because of their weight, length, and/or weather conditions.

The bracing operation will be the installation of temporary bracing and shoring to keep each beam stable and secure in position. The takeoff outline will be as listed.

1. The type of bracing required
2. The type of beam
3. The lineal feet of X bracing or top bracing per bay
4. The number of bays requiring bracing

The operation includes the full duration to install, maintain, and remove the required bracing. The bracing is quantified and summarized by the unit of each, per bay, per type of beam. The defined category is listed as: bracing, beam type ─────, each.

Stockpiling and rehandling

Owing to certain conditions that may exist, the concrete beams may need to be stockpiled or double handled at the site. This operation may be required because of delivery restrictions or needs, or logistic and handling requirements at the erection site. A site condition or erection condition that may cause this would be access by delivery vehicles, reach capacity of the crane with regard to point of hook to point of release, or staging and conflicting obstructions at the erection site.

This operation will include the duration for unloading, bracing, and securing, or reloading and/or retransport, to the time of the operation of final erection. The final erecting will be the operation accounted for in normal beam procedures, thus not accounting for that operation within stockpiling and rehandling.

The quantity outline for this operation will be as listed:

1. Equipment staging
2. Unload beams
3. Secure beams
4. Reload and transport (if applicable)

This operation will be quantified by type of required beam, for the total lineal feet required to be stockpiled or rehandled. The defined categories will be:

1. Stockpile, beam type _____, lineal feet
2. Rehandle, beam type _____, lineal feet

Structural Steel Girders

The second item of the beam structure unit discussed is the structural steel girder design. This section highlights the construction requirements and operations associated with the takeoff and quantifying of these components. It also defines the span designs as to simple or continuous. A simple span beam unit, similar to the concrete beam unit, spans directly from one substructure unit to the next adjacent substructure unit. The continuous girder unit encompasses multiple substructure units.

The components that may be required for the structural steel girder operations will be identified as:

1. Site and pad preparation
2. The primary steel girder erection
3. Secondary or stringer beam erection
4. Bearing assemblies
5. Anchor bolts
6. Steel diaphragms
7. Steel cross frame assemblies
8. Splicing
9. Shear connectors
10. Temporary bracing
11. Temporary erection scaffolding
12. Stockpile and rehandle
13. Painting
14. Relocate crew

380 Takeoff and Cost Analysis Techniques

Once the components have been identified, the site conditions, erection procedures, and construction methods must be outlined. As with the concrete beams, the site conditions will control mobility and access of both delivery of materials and equipment needs and requirements. There can be restrictive conditions that will alter or affect the erection procedure. The erection procedures will outline the requirements and specifications for the actual erection of the beams or girders. The construction methods will detail the constructability requirements, the type of work operation required for the required beam unit and components, and any shoring, scaffolding, and bracing requirements.

The quantity takeoff procedure for a structural steel girder unit is accomplished by first identifying the type of girders required, rolled or fabricated. Second, the associated secondary components must be detailed. The third requirement is to define the site conditions, erection procedures, and construction methods required for the specific project.

Site and pad preparation

The requirement of this operation will be the preparation of a level pad for the hoisting equipment to stage or for the girders and secondary components to be stockpiled. If wood crane mats are required for foundation stability, they should be accounted for under this operation.

The defined quantity unit for this operation will be summarized as: required craft-hours to perform the work.

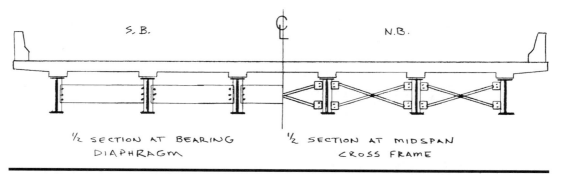

End section view of steel girder unit.

The structural steel beam unit, as with the concrete beam unit, consists of many components that comprise the designed section. The design of a steel beam unit has varying features depending on the intended use, with regard to span length and imposed load capacities.

Structural steel beam units normally have greater span limitations from the concrete beam unit. The greater the span, the heavier the section design, the heavier the girder. The rolled steel beams have a consistent section and weight; therefore, when a rolled beam is used, the total weight of the beam is easily calculated, the weight per lineal foot multiplied by the length of the beam.

A fabricated structural steel girder is constructed of multiple-thickness cross sections, of various grades of steel, and can have a variable height. This type of beam will normally have a differing cross section throughout its length, making it difficult to obtain a known weight per foot.

This type of girder will have to be calculated for the cubic inches of steel, by grade, within the design section of the girder. Once this has been accomplished and a total weight of the girder has been calculated, an average weight per lineal foot can be established.

Steel girders

The steel girders, either rolled or fabricated, are the primary components of the steel girder beam unit. The girders span the substructure units to form a continuous girder section between bearing points. Even though the design section shows a continuous girder, it may not be logistically possible to construct or ship a continuous girder, requiring it to have a field splice(s). This may be due to length, weight, or constructability limitations and compound deflections.

The quantity takeoff for the structural steel girders must include all necessary work required to complete the erection procedure, rigging, hoisting, and unloading. At this time, this would include the determination of the number of field splices and quantify the individual girders that comprise the entire unit. There are, however, two conditions that may exist: the required number of field splices to be performed in the air or erected position, or the number of field splices that can be performed on the ground.

The erected splice procedure requires the two beams adjacent to the splice to be handled and hoisted separately, with some instances of falsework or scaffolding requirements. The on-ground field splice procedure can be performed by attaching the two girders together with minimal hoisting requirements, then hoisting and erecting the longer combined section into place. These are variable conditions that need to be analyzed for each job occurrence.

The quantity takeoff will define the span type, required splices, and erection procedure, and commence with identifying the number of individual girders that need to be handled, advancing to the secondary components.

The quantity outline for the components of this operation will be as listed:

1. Type of span
2. Type of girder
3. The size of the section and height of girder
4. The number of girders
5. The length of the girders
6. The weight of each girder, and total

For a simple span girder unit, the quantifying of individual girders will be outlined using the following format. The first step is the number of required individual girders, to be handled, per span. This is simply the number of girders within the span unit, defined by each. They should be listed by girder type as to section weight or size by height.

Second, a listing of length for each girder will be composed, with a unit of lineal feet.

The third item, following the outlined procedure for identifying the weight per lineal foot, the total weight of each girder must be listed. This will determine the hoisting power required and the logistics of the erection procedure. This procedure will be continued for each span of the bridge unit, with the identifying listed as follows:

1. Number of girders per span, as each, summarized in total
2. The section or size of the girder design
3. The weight of each girder, in pounds, identifying the heaviest
4. The total weight of all girders, per bridge unit, in pounds

This format will identify the required information for the main girders of a simple span bridge unit.

For a continuous span unit, the quantity takeoff procedure will be altered as follows. The required number of individual girders that need to be handled, taking into account the required splices, will be defined, as each. The number of individual girders, required in length, to construct the continuous girder unit, an example being a designed continuous girder of 150', with one splice at the center midpoint. This would determine each individual girder component to be 75' in length, requiring the handling of two girders. Then the listed format, as detailed with the simple span, will prevail. When falsework is required for temporary support during erection, it will be defined and quantified under a separate operation.

The delivery of the girders will affect the erection procedure. The number of girders that can be delivered per crew hour will be the basis for calculating the lineal feet per hour, including time loss for relocation or rehandling. Also, the time to brace, scaffold, or shore will be calculated as idled time to the erection crew, if it affects the hoisting duration.

The outlined format of a quantity takeoff for the continuous girder unit will be defined as:

1. Number of individual girders per continuous unit, as each, summarized in total
2. The section or size of the girder design
3. The weight of each girder, in pounds, identifying the heaviest
4. The total weight of all girders, per bridge unit, in pounds

This format will identify the required information for the main girders of a continuous span bridge unit.

If girder splices are required, this would be the next procedure in the quantity takeoff.

Girder splices

The girder splice consists of a web splice, a top flange splice, and a bottom flange splice. The procedure of splicing can be accomplished by bolting or welding.

A bolted splice consists of the girder section being sandwiched together with two steel plates, one on either side of the web and flange, connected by high-strength bolts.

The quantity outline for the components of this operation will be as listed:

1. Type and method of splice
2. Number of splices (normally one web and two flanges per splice)
3. Quantity of temporary attachments

The quantity takeoff procedure for girder splices is summarized by splice type, either a web splice or flange splice, and by the method of performance, either bolting or welding. The work activity will consist of the necessary labor and equipment to perform the splice, along with any scaffolding or staging required to support the personnel to perform this operation. Any required temporary falsework to support the main girders, during erection, will be identified and quantified under a separate category.

The accumulation of the required bolted web splices will be done first. This is considered as a unit, per splice. The web splice will consist of two each steel plates. The procedure of installing both required plates and the temporary bolts to attach them will be considered as the required operation.

The quantity summation for bolted web splices will be listed for each bridge unit as: web splice, bolted, each.

The accumulation of the bolted flange splices will be done next. This is considered as a unit, per splice, for the top flange and the bottom flange. The flange splices will consist of three steel plates each, for the top and bottom flange. The procedure of installing the six required plates and the temporary bolts to attach them will be considered as the required operation.

The quantity summation for bolted flange splices, for each bridge unit, will be listed as: flange splice, bolted, each.

The accumulation of the welded girder splices will be done second, if required. This is considered as a unit, per splice. The welded web splice will consist of temporarily butt welding the girder sections together, including both the web and flanges. The procedure of the required tack welding and incidental work to attach them will be considered as the required operation.

The quantity summation for welded web splices will be listed for each bridge unit as: girder splice, welded, each.

Girder splice, bolts

The operation of installing the necessary permanent bolts for the splice section will be accounted for as each bolt, per girder splice unit, installed as the specifications detail.

The quantity outline for the components of this operation will be as listed:

1. Number of required permanent bolts, by size

This operation will be an accumulation of both the web splice and the flange splice bolts, per bridge unit. The quantity summation for this operation, per bridge structure, will be listed as: girder splice, bolting, each bolt, by size.

Girder splice, butt weld

The takeoff of required quantity for the permanent welding procedure, the girder splice will be performed by accounting for the lineal feet of welding required per pass. The welding of a girder butt splice is performed by a series of different-sized-diameter weld passes, building up the specified penetration.

The quantity outline for the components of this operation will be as listed:

1. Lineal feet of required weld, by size

The quantity summation will be listed as the total lineal feet of required weld, by size, for each bridge structure. It will be categorized by: girder splice, butt weld, bead size _____, lf.

Structural steel diaphragms

The structural steel diaphragms, as with the concrete diaphragms, are structural components that tie the girders together as one unit. They are connected to the web of the main girders, usually at the stiffener location. The takeoff quantity is compiled in each, per bridge structure. As with the splices, there exist two methods of performance, bolting in place and welding in place. This operation is for the placement and temporary attachment of each diaphragm, along with any required scaffolding needed to perform this activity. The temporary attaching would consist of temporary bolts, or tack welds.

The quantity outline for the components of this operation will be as listed:

1. Size of diaphragm
2. Number of diaphragms

The quantity summation, for each bridge structure, will be listed as: (1) girder diaphragm, bolted, each; or (2) girder diaphragm, welded, each.

Girder diaphragm, bolts

The operation of installing the necessary permanent bolts, for the steel diaphragm component, will be accounted for as each required bolt, per diaphragm unit, installed as the specifications detail. This operation will be an accumulation of bolts required, per diaphragm, per bridge unit.

The quantity outline for the components of this operation will be as listed:

1. Number of required permanent bolts, by size

The quantity summation for this operation, per bridge structure, will be listed as: girder diaphragm, bolting, each bolt.

Girder diaphragm, weld

The takeoff of required quantity for the welding procedure for the girder diaphragm will be performed by accounting for the lineal feet of welding required per pass. The welding of a girder diaphragm is performed by a series of different-sized-diameter weld passes, building up the specified penetration.

The quantity outline for the components of this operation will be as listed:

1. Lineal feet required weld, by size

The quantity summation will be listed as the total lineal feet of required weld, by size, for each diaphragm, summarized per bridge structure. It will be categorized by: girder diaphragm, weld, bead size _____, lf.

Structural steel cross frames

The structural steel cross frames, as with the structural steel diaphragms, are structural components that tie the girders together transversely as one unit. They are connected to the web of the main girders, usually at the stiffener location. The takeoff quantity is compiled in each, per bridge structure. As with the splices, there exist two methods of performance, bolting in place and welding in place. This operation is for the placement and temporary attachment of each cross frame, along with any required scaffolding needed to perform this operation. The temporary attaching would consist of temporary bolts or tack welds.

The quantity outline for the components of this operation will be as listed:

1. Size of cross frame
2. Number of cross frames

The quantity summation, for each bridge structure, will be listed as: (1) girder cross frame, bolted, each; or (2) girder cross frame, welded, each.

Girder cross frame, bolts

The operation of installing the necessary permanent bolts, for the steel cross frame component, will be accounted for as each bolt required, per cross frame unit, installed as the specifications detail. This operation will be an accumulation of bolts required, per cross frame, per bridge unit.

The quantity outline for the components of this operation will be as listed:

1. Number of required permanent bolts, by size

The quantity summation for this operation, per bridge structure, will be listed as: girder cross frame, bolting, each bolt.

Girder cross frame, weld

The takeoff of required quantity for the welding procedure of the girder cross frame will be performed by accounting for the lineal feet of welding required per pass. The welding of a girder cross frame is performed by a series of different-sized-diameter weld passes, building up the specified penetration.

The quantity outline for the components of this operation will be as listed:

1. Lineal feet required weld, by size

The quantity summation will be listed as the total lineal feet of required weld, by size, for each cross frame unit, summarized per bridge structure. It will be categorized by: girder cross frame, weld, bead size _____, lf.

Structural steel stringer beam

The steel stringer beams, or floor beams, usually rolled, are the secondary girder components of the steel girder beam unit. The stringer beams span transversely between the main girder sections to provide lateral support and floor sections.

The quantity takeoff for the structural steel stringer beams must include all necessary work required to complete the erection procedure. At this time, this would include the quantity of the individual girders that comprise the entire stringer unit.

The quantity takeoff will define the span, the required number of stringer beams, and the erection procedure.

The quantity outline for the components of this operation will be as listed:

1. Type of girder
2. The size of the section or height of girder
3. The number of girders
4. The length of the girders
5. The weight of each girder, and total

For a steel stringer beam girder unit, the quantifying of individual girders will be outlined using the following format. The first step is the number of required individual girders, to be handled, per span. This is simply the number of stringer girders within the span unit, defined by each. They should be listed by girder type as to section weight or size by height.

Second, a listing of length for each stringer girder will be composed, with a unit of lineal feet.

The third item, following the outlined procedure for identifying the weight per lineal foot, the total weight of each girder must be listed. This will determine the hoisting power required and the logistics of the erection procedure.

This procedure will be continued for each span of the bridge unit, with the identifying listed as follows:

1. Number of stringer girders per span, as each, summarized in total
2. The section or size of the stringer girder design
3. The weight of each stringer girder, in pounds, and heaviest girder
4. The total weight of all the stringer girders, per bridge unit, in pounds

Girder stringer beam, bolts

The operation of installing the necessary permanent bolts, for the steel stringer beam component, will be accounted for as each bolt required, per stringer unit, installed as the specifications detail. This operation will be an accumulation of bolts required, per stringer beam, per bridge unit.

The quantity outline for the components of this operation will be as listed:

1. Number of required permanent bolts, by size

The quantity summation for this operation, per bridge structure, will be listed as: girder stringer beam, bolting, each bolt.

Girder stringer beam, weld

The takeoff of required quantity for the welding procedure for the girder stringer beam will be performed by accounting for the lineal feet of welding required per pass. The welding of a girder stringer beam is performed by a series of different-sized-diameter weld passes, building up the specified penetration.

The quantity outline for the components of this operation will be as listed:

1. Lineal feet required weld, by size

The quantity summation will be listed as the total lineal feet of required weld, by size, for each stringer unit, summarized per bridge structure. It will be categorized by: girder stringer beam, weld, bead size _____, lf.

Anchor bolts

The anchor bolts are the components that fasten the bearing assemblies, or bearing plates, to the substructure unit. Within the bearing area there may exist a component defined as an anchor bolt. As with the position dowel, this component is embedded within the substructure beam seat and provides location control and fastening of the bearing assembly. The anchor bolt extends from the beam seat through the bearing assembly.

An anchor bolt is intended to maintain and position the bearing assembly, and directly, the girder from any undesired movement, both longitudinally and transversely.

The most efficient method of installing this anchor bolt is to drill and epoxy in place. This is simply drilling the required size hole into the substructure beam seat at the proper location, after the pier cap and beam seat is constructed. If permitted by specification, this procedure will ensure better control and quality for the positioning of the anchor bolt. Otherwise, the anchor bolt will be monolithically cast into the beam seat during the construction of that operation, defined as hand setting.

One other alternative would be to cast plastic inserts in the cap beam seat. This would eliminate the need of drilling but would require the removal of the insert and the epoxying of the dowel.

Some specifications may not permit the drill and epoxy method owing to the uplift movement of the bearing area.

Once the proper size hole is drilled, the anchor bolt will be epoxied in place, within the beam seat.

The outline required for the quantity takeoff of a structural anchor bolt will be as follows:

1. Type of bolt insertion (drill and epoxy, hand set)
2. The size of anchor bolt required
3. The penetration of bolt (within the beam seat)
4. The quantity of anchor bolts per bearing area
5. The number required per span
6. The total required per bridge
7. The number of required holes to be drilled
8. The number of required bolts to epoxy
9. The required volume of epoxy

The quantity factoring for these items would be determined based on the operation to be performed. The drilling of anchor holes will be quantified as the number of like-sized holes drilled, categorizing by diameter and depth. The epoxy application will be quantified as the number, by size, of dowels required to be placed. The application of epoxy within the hole will be calculated by the time required per hole and the volume of epoxy material required per hole. This will be formulated by the cubic inches of open hole minus the cubic inches of area displaced by the dowel.

The operation of installing anchor bolts will be calculated under two functions. The first operation will be the performance time for drilling the required dowel holes. The second operation will be the performance of the epoxy application, and the insertion of the position dowels.

The drilling operation will be calculated for the time, per dowel, to drill the hole and move to the next. The size and depth of required hole will govern the production.

The epoxy application will be calculated for the time, per dowel, to inject the required volume of epoxy and insert the dowel.

From this listing, a summary will be complied. The defined control unit of measure for quantifying the anchor bolts will be each required, by type of installation. The categorizing will be listed as:

1. Structural anchor bolts drilled, each, by size
2. Structural anchor bolts placed, each, by size
3. Structural anchor bolts epoxied, each by size

Structural steel bearing assemblies

The bearing assemblies for a structural steel unit will be quantified as each for the type required. The operation will include the necessary work to install each bearing assembly at its proper location.

There are multiple types of structural steel bearing assemblies: the expansion type, or rocker, which allows a movement of horizontal rotation to the beam unit; the slide bearing, which is designed to give horizontal shear movement to the beam structure; and a fixed bearing assembly, which gives no movement to the beam structure.

The anchoring method varies by design for bearing assemblies. They can be bolted to the substructure bearing seat and/or welded. The welding may occur between the bearing assembly and the base or shoe of the structural girder.

The outline required for the quantity takeoff of a structural bearing assembly will be as follows:

1. The type of bearing assembly required
2. The number of each type of bearing assembly required
3. The number of bolts in the bearing area
4. The lineal feet of welding required

The quantity summation for this operation, for each bridge structure, will be defined as:

1. Structural bearing assembly, rocker, each
2. Structural bearing assembly, slider, each
3. Structural bearing assembly, fixed, each

Elastomeric bearing pads

The next item of takeoff associated with the structural bearing assembly is identifying and quantifying the required bearing pads designed for the bearing area. This is accomplished by a listing per beam, per span, and categorizing by beam type.

These bearing pads are defined as elastomeric bearing pads, which are placed between the beam bearing and the substructure beam seat. This rubber-type pad provides the beam a cushion and movement section, between the two opposing surfaces. The bearing pad is a structure component that is designed to withstand the load imposed, including the beam weight, the load of the superstructure, and the live loads that are implied to the surface of the deck.

Manufactured within the rubber composite are steel shim plates. The design of the pad hardness, referred to as durometer, and the amount of and size of steel plates, differ within each pad based on the intended design. This will vary the unit square foot cost of the bearing pads.

This rubber pad is a part of the designed bearing assembly of the structure. A steel plate or bearing shoe is attached to the girder during the manufacturing process. The component section bears directly onto the bearing pad. There are varying designs and conditions of bearing section areas. The beam may rest directly on the pad with no other location or positioning assistance, or the section design may require anchor bolts to aid with the location assurance.

The bearing assemblies will be outlined in the following format:

1. Type of required pad
2. Number required per beam
3. Total number per span
4. Square feet required per pad and total
5. Total of each pad(s) required per beam type

The quantity takeoff procedure for this item will be compiled as the number of each type of elastomeric bearing pads required, per span, per beam type. This will provide the quantity for craft-hour factoring. For material costing, it usually requires a quantity of required square feet. The listing of square feet per bearing pad, and an accumulation of bearing pad type, will accompany this list.

From this listing, a summary will be compiled. The defined control unit of measure for quantifying the bearing pads will be:

1. Elastomeric bearing pad, each
2. Elastomeric bearing pad, square feet

Shear connectors

The shear connectors provide a connection point from the top girder flange to the concrete superstructure deck, and prevent shear movement between the girder and deck. The shear connectors, or studs, are attached to the top flange of the structural girder by a welding procedure. Dependent on the intended design, the connectors vary in length, diameter, number of rows, and longitudinal spacing on the girder.

The quantity outline for the components of this operation will be as listed:

1. The size and diameter of the shear connector
2. The number of connectors, by size

The quantity takeoff of the shear connectors is summarized by each, by size as: structural shear connectors, size _____, each.

Temporary falsework

The need of temporary falsework is a requirement when the structural steel girders need to be temporarily supported during the erection procedure. They would include scaffolding towers, jack stands, or piling supports, and the associated work required to perform the operation, from placement to removal.

This is a wide-range operation owing to actual project conditions. It is suggested to thoroughly determine the requirement of each condition and formulate this function as each, required.

The quantity outline for the components of this operation will be as listed:

1. The number of towers required
2. The foundation support and load required per tower
3. The height of each tower
4. The material requirements per tower
5. The time duration per tower
6. The removal and dismantle time

The quantity takeoff summation of this item will be defined, per bridge location and girder type, by each requirement: falsework and shoring, girder type _____, each.

Temporary bracing

Temporary beam bracing is a secondary operation to the girder erection item. For interim protection until the completion of the diaphragms or cross frames, this item may be necessary because of the unstable condition that exists with certain types of girders. The design section of certain girders causes them to be top-heavy, or flexurally unstable. Structural steel girders can sway or tip over owing to their weight, length, and/or weather conditions.

The bracing operation will be the installation of temporary bracing and shoring to keep each beam stable and secure in position. The takeoff outline will be as listed.

1. The type of bracing required
2. The type of girder
3. The lineal feet of X bracing or top bracing per bay
4. The number of bays requiring bracing

The operation will include the full duration to install, maintain, and remove the required bracing. The bracing will be quantified and summarized by the unit of each, per bay, per girder type. The defined category will be listed as: temporary bracing, each.

Stockpiling and rehandling

Because of certain conditions that may exist, the structural steel girders may need to be stockpiled or double handled at the site. This operation may be required owing to delivery restrictions or needs, logistic and handling requirements at the erection site, or splicing requirements. A site condition or erection condition that may cause this would be access by delivery vehicles, reach capacity of the crane with regard to point of hook to point of release, or staging and conflicting obstructions at the erection site.

This operation will include the duration for unloading, bracing and securing, or reloading and/or retransport, to the time of the operation of final erection. The final erecting will be the operation accounted for in normal girder procedures, thus not accounting for that operation within stockpiling and rehandling.

The quantity outline for this operation will be as listed:

1. Equipment staging
2. Unload girders
3. Secure girders
4. Reload and transport (if applicable)

This operation is quantified by type of required steel girder, for the total lineal feet and pounds required to be stockpiled or rehandled. The defined categories are:

1. Stockpile, girder type _____, lineal feet/pounds
2. Rehandle, girder type _____, lineal feet/pounds

Painting

Painting of structural steel girders and components requires a defined takeoff procedure, based on specification requirements. The method of final field painting of girders may require sandblasting, primer coat, wash coat or midcoat, and the final color coat of paint.

The painting operation includes all scaffolding and rigging necessary to access the work area along with any protective measures required to prevent environment damage, and the paint procedures themselves.

The quantity takeoff of painting structural steel components is composed of the required square foot of surface area designated to be painted. This surface area would apply to all coats of paint material and the required sandblasting.

To perform the takeoff, each component that is required to be painted must be identified as to the item and surface area.

The quantity outline for the components of the painting operation will be:

Sandblast surface area (square feet):

1. Structural main girders (including splices)
2. Stringer beams
3. Diaphragms
4. Cross frames
5. Bearing assemblies
6. Incidental components

Prime coat surface (square feet):

1. Structural main girders (including splices)
2. Stringer beams
3. Diaphragms
4. Cross frames
5. Bearing assemblies
6. Incidental components

Wash coat surface (square feet):

1. Structural main girders (including splices)
2. Stringer beams
3. Diaphragms
4. Cross frames
5. Bearing assemblies
6. Incidental components

Final paint coat (square feet):

1. Structural main girders (including splices)
2. Stringer beams
3. Diaphragms
4. Cross frames
5. Bearing assemblies
6. Incidental components

Field touch-up (square feet):

1. Structural main girders (including splices)
2. Stringer beams
3. Diaphragms
4. Cross frames
5. Bearing assemblies
6. Incidental components

The surface area calculations for the girder units and floor beams are made by computing the perimeter of each girder requiring treatment in square feet, and multiplying by the length of the girder. (NOTE: The top flange surface of the deck girder will not require painting owing to the concrete deck surface.) This calculation will then be summarized in square feet for each girder.

The diaphragm and cross frames will be calculated for the surface area of each requiring treatment, in square feet, and summarized together by diaphragm or cross frame.

The other components of the beam structure will be calculated using the same format of surface area. Each component will be categorized by type, by total surface area.

The quantity summation will be categorized by the controlling component item, and compiled by quantity of surface area square feet by the operation function:

1. Structural steel beam unit, sandblasting, square feet required
2. Structural steel beam unit, prime coat, square feet required
3. Structural steel beam unit, wash coat, square feet required
4. Structural steel beam unit, final coat, square feet required
5. Structural steel beam unit, touch-up, square feet required

These listed procedures will aid in the formulating of an accurate and detailed quantity takeoff. They will provide the fundamentals required to understand the component operation and formulate production factors.

Procedure

Preparing a detailed and accurate takeoff for the components of a beam structure and work operations requires a thorough understanding and scope of the project, site conditions, logistics of the area, and construction techniques. Each work operation will demand the most economical and practical approach to the erection system chosen, equipment requirements, and work techniques within the project, to be competitive and productive.

With the variable methods available to perform and utilize the related beam operations, the estimator must give great thought to the selected procedures. The end result of this will aid in a more productive function for the entire component operation of the various beam and girder units.

The erection of structural members requires a level work and staging area for the operation to function smoothly. The required secondary components must be accounted for, and ready, along with the preparation of the substructure units.

The primary categorizing units of measure will be, normally, lineal feet for the prestressed, precast concrete beam members, and pounds for the structural steel members. The secondary components, even with the differing unit control, will be summarized into these primary units for total cost distribution.

The site preparation and staging requirements will be summarized by the consumption of time, in craft-hours, needed to perform the operation. When it is required to relocate the hoisting equipment and crew from one span to another, this time duration will be summarized as craft-hours consumed to relocate. A relocation to a separate bridge site will be summarized under a different category, mobilization.

As with any detailed quantity takeoff procedure, every component of each operation must be accounted for. This will include all incidental work that is directly related to the operation as common practice and procedure. One must realize these items may not be directly referred to in the project specifications.

Questions of Estimate

The necessary and required questions for a properly prepared outline of a beam structure unit, as listed, will provide the general scope and nature of the work operations:

Prestressed, precast concrete beam unit:

1. What are the type, size, and section of the beam?
2. What are the site conditions?
3. How many crew relocations are required (number of spans)?
4. What is the longest and heaviest beam?
5. What is the cross section?
6. What are the hoisting requirements?
7. Will stockpiling and double handling be required?
8. What erection procedure will be implemented?
9. Is the erection staged?
10. Is bracing, falsework, or scaffolding required?
11. What are the secondary components?
12. Is specialty equipment required?
13. Are there traffic conditions within the work area?

396   Takeoff and Cost Analysis Techniques

Structural steel girder unit:

1. What are the type, size, and section of the girder?
2. What are the site conditions?
3. How many crew relocations are required (number of spans)?
4. What is the longest and heaviest girder?
5. Is splicing required and what type?
6. What is the cross section?
7. What are the hoisting requirements?
8. Will stockpiling and double handling be required?
9. What erection procedure will be implemented?
10. Is the erection staged?
11. Is bracing, falsework, or scaffolding required?
12. What are the secondary components?
13. What are the paint requirements?
14. Is specialty equipment required?
15. Are there traffic conditions within the work area?

Application

The application of the quantity takeoff procedure shows a clear understanding of the required format. The example listed is a general type of structure and is designed to show the methods that are described in this section. The actual procedures and outlined format in the example can be implemented for any given operation.

The given bridge example details the production requirements and produces a craft-hour production factor along with a detailed summary listing of required structural members, components, and material needs, both permanent and incidental.

The quantity takeoff of the bridge structure is completed in an orderly system of constructive priority. The beams are the control; therefore, the beams will be the first component of takeoff. From the beams, the position dowels and bearing assemblies should be the next components, followed by the diaphragm area.

The first application details a dual, simple span, prestressed, precast concrete beam unit, having a dependent backwall design.

A given multispan structure has the following beam structure units:

Span 1, simple span prestressed, precast concrete beam

Span 2, simple span prestressed, precast concrete beam

Each span is symmetrical in length and width, having dimensions of 74' longitudinally from center of bearing point to center of bearing point, and a

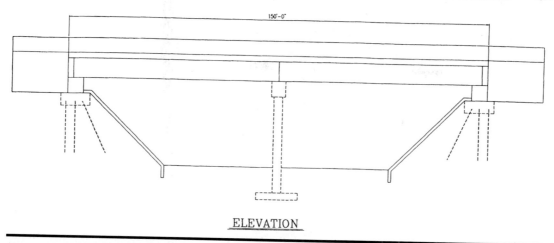

ELEVATION

Plan view of two-span structure.

beam structure unit width of 24′ transversely from center of outside beam to center of outside beam. The designated diaphragm locations that are designed will be as follows: span 1, one set of interior diaphragms, two sets of end diaphragms; span 2, one set of interior diaphragms and one set of end diaphragms (a dependent backwall is located in span 2).

Within each of the symmetrical beam structure units there exist the required components of:

    Precast concrete beams: AASHTO type IV I beam

    Position dowels

    Elastomeric bearing pads

    Concrete diaphragms:
      End diaphragms
      Interior diaphragms

    Tie-rod assemblies

For these components, the required work operations will be:

Precast concrete beams: AASHTO type IV I beam:

    Site staging and preparation
    Beam erection
    Location repositioning

  Position dowels
    Drill holes
    Epoxy dowels

Elastomeric bearing pads
　Place bearing pads

Concrete diaphragms, end and interior
　Place expansion material
　Scaffold and staging erection and removal
　Formwork placement and removal
　Installation of form hardware, embeds
　Determination of beam type, end section area
　Concrete placement: yield and overrun percentage
　Rub and patch concrete surface

Tie-rod assemblies
　Install tie-rods

The following are the required component dimensions per the project plans and sections.

Precast beam unit:

Precast concrete beam	
Type IV, 54″	= 75 lf ea
Number of beams per span	= 4.0 ea
Number of spans	= 2.0 ea
Position dowels	= 2″ × 18″ ea
Elastomeric bearings	= 2″ × 18″ × 24″ ea
Beam spacing	= 8′ centers
Concrete diaphragm	
End	= 1.0′ thick × 54″ high
Interior	= 1.0′ thick × 30″ high
Tie-rod assemblies	= $1\frac{1}{2}$″ diameter, lf

Example: Operation Analysis

Operation. Site preparation. Unit: Lump sum per location.

The operation of preparing the site area for the crane pads and beam erection is itemized with this step.

The duration required will be calculated by craft-hour consumption for the surface area square feet, in the following manner:

$$\text{Required shifts} = 1.25 \text{ ea}$$

The multiplying of the number of crew, by the shift duration, by the required shifts.

$$\text{Required shifts @ } 1.25 = 1.25 \text{ shifts per location}$$

Beam Structure

Operation. Beam erection. Unit: Lineal feet.

The operation of beam erection to determine the required lineal feet of beams, per size, is calculated in the following format: the summary of the lineal feet of each beam, for the number of beams per span, for each span of similar-type beams, plus the number of beams to handle and the average length per beam.

$$(75' * 4) * 2 \quad = 600.0 \text{ lf}$$
$$(4 * 2) \quad = 8.0 \text{ ea}$$
$$(600'/8 \text{ ea}) \quad = 75 \text{ avg/lf}$$

Operation. Relocation. Unit: Craft-hours.

The operation of relocating the erection crew is calculated as the required time to move from one span to the next erection area within the bridge site. (This is *not* the move in or move out at the site.) The calculation will be done for the consumed shift hours as:

Required: 4 hours for each relocation

The required moves multiplied by the shift duration per move.

$$(4 \text{ hours} * 1.0 \text{ ea}) \quad = 4.0 \text{ shift hours}$$

Operation. Position dowels. Unit: Each.

The operation of quantifying position dowels is compiled by a summation of the actual number of dowels required, per beam, per span.

Required: Span 1; 2 dowels per beam

Span 2; 1 dowel per beam (dependent backwall)

The number of dowels per beam, multiplied by the number of beams per span.

$$(2 * 4) + (1 * 4) \quad = 12.0 \text{ ea drill}$$
$$= 12.0 \text{ ea epoxy}$$
$$\text{Hole size} \quad = 2\tfrac{1}{8}'' \times 9''$$

Operation. Elastomeric bearing pads. Unit: Each, square foot.

The operation of quantifying the required number of bearing pads is calculated as follows for the required units, each and total square feet.

Required: 2 each, pads per beam

The number of bearing pads per beam, multiplied by the number of beams per span, summarizing each span.

$$(2 * 4) * 2 = 16.0 \text{ each}$$

$$((18" * 24")/144) * 16 = 48.0 \text{ sq ft}$$

Operation. Concrete end diaphragms, form and strip. Unit: Square feet.

The operation of forming and stripping of the concrete diaphragms is calculated in the following manner for the required surface area for all dia-phragms.

Requirement: beam centerline dimensions = 8.0′

number of end diaphragms = 9.0 ea

beam type, PCB AASHTO IV

beam end area, 789 sq in

The calculations that are required for the formwork surface area are:

The length, from the center of beam to the center of beam, multiplied by the side height plus the opposing side height plus the width of the bottom (if applicable), multiplied by the number of interior bays of similar width.

For the beam factor calculation, multiply the square inch end area of the beam by the number of diaphragm ends adjacent to the beams, by a factor of 0.0005, as described.

$$(8.0' * (4.5' + 4.5' + 0.0')) * 9.0 \text{ ea} = 648.0 \text{ sf form}$$

$$= 648.0 \text{ sf strip}$$

$$(9.0 \text{ ea} * (789 * 0.0005)) * 4.0 \text{ ends} = 14.20 \text{ craft-hours}$$

Operation. Tie-rod assemblies, end diaphragms. Unit: Lineal feet (center to center of exterior girders).

The operation of installing the diaphragm tie-rod assemblies will be quantified as lineal feet, as described, in the format listed:

Required: Centerline dimension 24′

The lineal feet calculation is performed by the number of diaphragm rows requiring tie-rods by the length of tie-rod.

$$3 * 24' = 72.0 \text{ lf}$$

Operation. Concrete end diaphragms, embeds. Unit: Each.

The operation of installing embeds for the formwork of the concrete diaphragms is calculated in the following manner:

None required = 0.0 each

Operation. Diaphragm expansion material. Unit: Square feet.

The operation of installing expansion material for the concrete diaphragms is calculated by square feet, in the following manner:

$$\text{Requirement: beam centerline dimensions} = 8.0'$$

$$\text{number of end diaphragms} = 9.0 \text{ ea}$$

The calculations that are required for the expansion surface area are:
The length, from the center of beam to the center of beam, multiplied by the width of the bottom of the contact diaphragm, multiplied by the number of interior bays.

$$(8.0' * 1.0') * 9.0 \text{ ea} = 72.0 \text{ sf expansion}$$

Operation. Concrete end diaphragm, concrete placement. Unit: Cubic yard.

The operation of concrete placement for concrete diaphragms is calculated for the cubic yard volume required using the following method; the yield loss of the placement will be calculated at this time:

$$\text{Required yield} = 14\%$$

$$\text{Beam type} = \text{PCB AASHTO IV}$$

$$\text{Beam quantity} = 4.0 \text{ per row}$$

$$\text{End area, beam} = 789 \text{ sq in}$$

$$\text{Intrusion} = 12''$$

$$\text{Dimension width} = 24.0'$$

The multiplying of the width of the diaphragm row, center of exterior beam to center of exterior beam, by the diaphragm height, by the diaphragm width, by the yield loss, converting to cubic yards. Then deducting the displacement of concrete by the volume of beam area within the diaphragm row. (NOTE: The two exterior beams are displaced by ½ the end area of each beam.) Then, summarizing the required rows,

$$((24.0' * 4.5' * 1.0') * 1.14)/27 = 4.56 \text{ cy row 1}$$

$$((24.0' * 4.5' * 1.0') * 1.14)/27 = 4.56 \text{ cy row 2}$$

$$((24.0' * 4.5' * 1.0') * 1.14)/27 = 4.56 \text{ cy row 3}$$

$$-((((2.0 + 0.5 + 0.5) * 789) * 12'') * 3)/46{,}656 = -1.83 \text{ cy total}$$

$$\text{Gross required} = 11.85 \text{ cy}$$

$$\text{Net required} = 10.17 \text{ cy}$$

NOTE: 1.0 cubic yard has 46,656 cubic inches.

Operation. Concrete end diaphragm, rub and patch. Unit: Square feet.

The operation of rubbing and patching for the concrete diaphragms is calculated for the area of treatment in the following manner for the required surface:

Required: both sides

The calculations that are required for the treatment surface area are:
The length, from the center of beam to the center of beam, multiplied by the side height plus the opposing side height plus the width of the bottom (if applicable), multiplied by the number of interior bays.

$$(8.0' * (4.5' + 4.5' + 0.0')) * 9.0 \text{ ea} = 648.0 \text{ sf rub and patch}$$

Operation. Concrete interior diaphragms, form and strip. Unit: Square feet.

The operation of forming and stripping of the concrete diaphragms is calculated as follows for the required surface area for all diaphragms.

Requirement: beam centerline dimensions = 8.0'

number of interior diaphragms = 6.0 ea

beam type, PCB AASHTO IV

beam end area, 789 sq in

The calculations that are required for the formwork surface area are:
The length, from the center of beam to the center of beam, multiplied by the side height plus the opposing side height plus the width of the bottom, multiplied by the number of interior bays.

For the beam factor calculation, multiply the square inch end area of the beam, by the number of diaphragm ends adjacent to the beams, by a factor of 0.0005,* as described.

$$(8.0' * (2.5' + 2.5' + 1.0')) * 6.0 \text{ ea} = 288.0 \text{ sf form}$$

$$= 288.0 \text{ sf strip}$$

$$(6.0 \text{ ea} * (789 * 0.0005)) * 4.0 \text{ ends} = 9.47 \text{ craft-hours}$$

Operation. Tie-rod assemblies, interior diaphragms. Unit: Lineal feet (center to center of exterior girders).

The operation of installing the diaphragm tie-rod assemblies is quantified as lineal feet, as described, in the format listed:

Required: Centerline dimension 24'

The lineal feet calculation will be performed by the number of diaphragm rows requiring tie rods by the length of tie-rod.

$$2 * 24' = 48.0 \text{ lf}$$

---

*0.0005 will vary depending on specific experiences.

Operation. Concrete interior diaphragms, embeds. Unit: Each.

The operation of installing embeds for the formwork of the concrete diaphragms is calculated in the following manner:

$$\text{None required} = 0.0 \text{ each}$$

Operation. Concrete interior diaphragm, concrete placement. Unit: Cubic yard.

The operation of concrete placement for concrete diaphragms is calculated for the cubic yard volume required using the following method; the yield loss of the placement is calculated at this time:

$$\text{Required yield} = 14\%$$

$$\text{Beam type} = \text{PCB AASHTO IV}$$

$$\text{Beam quantity} = 4.0 \text{ per row}$$

$$\text{End area, beam} = 789 \text{ sq in}$$

$$\text{Intrusion} = 12''$$

$$\text{Dimension width} = 24.0'$$

The multiplying of the width of the diaphragm row, center of exterior beam to center of exterior beam, by the diaphragm height, by the diaphragm width, by the yield loss, converting to cubic yards. Then deducting the displacement of concrete by the volume of beam area within the diaphragm row. (NOTE: The two exterior beams are displaced by ½ the end area of each beam.) Then, summarizing the required rows,

$$((24.0' * 2.5' * 1.0') * 1.14)/27 = 2.53 \text{ cy row 1}$$

$$((24.0' * 2.5' * 1.0') * 1.14)/27 = 2.53 \text{ cy row 2}$$

$$-((((2.0 + 0.5 + 0.5) * 789) * 12'') * 2)/46{,}656 = -1.22 \text{ cy total}$$

$$\text{Gross required} = 3.84 \text{ cy}$$

$$\text{Net required} = 3.22 \text{ cy}$$

NOTE: 1.0 cubic yard has 46,656 cubic inches.

Operation. Concrete interior diaphragm, rub and patch. Unit: Square feet.

The operation of rubbing and patching for the concrete diaphragms is calculated for the area of treatment in the following manner for the required surface:

$$\text{Required: both sides}$$

$$\text{bottom}$$

Takeoff and Cost Analysis Techniques

The calculations that are required for the treatment surface area are:
The length, from the center of beam to the center of beam, multiplied by the side height plus the opposing side height plus the width of the bottom (if applicable), multiplied by the number of interior bays.

$$(8.0' * (2.5' + 2.5' + 1.0')) * 6.0 \text{ ea} = 288.0 \text{ sf rub and patch}$$

Unit quantity summary: Precast beam

Concrete beam unit		Span 1	Span 2
Site preparation	ls	0.5	0.5
Beam erection	lf	300.0	300.0
Position dowels	ea	8.0	4.0
Elastomeric bearing	ea	8.0	8.0
Bracing	ea	0.0	0.0
Epoxy bonding compound	oz	48.0	24.0
Concrete diaphragms		Span 1	Span 2
Formwork place	sf	576.0	360.0
Formwork removal	sf	576.0	360.0
Beam factor	craft-hour	14.20	9.47
Bulkheads place	sf	0.0	0.0
Bulkheads remove	sf	0.0	0.0
Hardware and embeds	ea	0.0	0.0
Expansion material	sf	48.0	24.0
Concrete placement	cy (gross)	9.82	5.87
Rub and patch	sf	576.0	360.0
Tie-rod assemblies	lf	72.0	48.0

From the quantity summary list, the time duration and production factoring are compiled. The quantity summary list is formed by grouping the operations of similar components.

Example: Production Factoring, Time Duration

Operation. Site preparation.

Crew members	= 2.0 ea
Shift duration	= 8.0 hours
Production unit	= lump sum/area
Project quantity requirement	= 1.0
Unit cycle time	= 10.0 hours
Productivity factor	= 1.0
Cycles per hour (lump sum)	= 10.0 cycles
Production per hour	= 1.0 ea
Production per shift: 8/1	= 8.0 ea
Required shifts: 10.0/8.0	= 1.25
Total craft-hours: 2 * 8 * 1.25	= 20.0
Operation production factor: 20/1	= 20.0 lump sum
Operation factor	= 20.0 craft-hours lump sum

Beam Structure

Operation. Precast concrete beam erection. Required:

Beam delivery	= 1.25 hours/beam
Beam length	= 75.0 lf
Weight, 850 lb/lf	= 63,750.0 lb
Maximum reach radius	= 45'
Boom length	= 90'
Crane capacity	= 100,000.0 pounds
Lift capacity	= 45,000 lb @ 45'
Unit calculation: 63,750/45,000	= 1.42
Crane units, 1.42 units	= 2.0 ea

The first calculation required is the delivery production per hour, for the number of beams. Second, the beams per hour are translated to a lineal foot per hour.

Delivery: 75'/1.25 hours	= 60.0 lf hour
Crew members	= 9.0 ea
Shift duration	= 8.0 hours
Production unit	= lineal feet
Project quantity requirement	= 600.0 lf
Unit cycle time: 60 min/60'	= 1.0 lf min
Productivity factor	= 55-min hour
Cycles per hour: 55/1.0	= 55.0 cycles/hour
Production per hour: 55.0' * 1	= 55.0 lf hour
Production per shift: 8 * 55.0'	= 440.0 lf shift
Production shifts: 600.0'/440.0	= 1.36 shifts
Relocation duration per shift 4/8	= 0.50 shift
Required shifts	= 1.86
Total craft-hours: 1.86 * 8 hours * 9 crew	= 133.92 craft-hours
Operation production factor: 133.92/600	= 0.223
Operation factor	= 0.223 craft-hour/lf

The related diaphragm work, for this structure, consists of the same productive unit for both the end and interior; therefore the quantities are combined.

Operation. Concrete diaphragms, formwork placement.

Crew members	= 3.0 ea
Shift duration	= 8.0 hours
Production unit	= square feet
Project quantity requirement	= 936.0 sf
Formwork type	= wood
Forming condition	= severe, 3
Unit cycle time	= 2 min/sf
Productivity factor	= 50 minutes
Cycles per hour: 50/2	= 25.0 cycles
Production per hour: 25.0 * 1	= 25.0 sf/hour
Production per shift: 25 sf * 8	= 200.0 sf/shift
Required shifts: 936.0 sf/200.0 sf	= 4.68 shifts
Total craft-hours: 4.68 * 8 * 3	= 112.32 craft-hours
Beam factor craft-hours	= 23.67 craft-hours
Operation production factor: 135.99/936.0	= 0.145
Activity factor	= 0.145 craft-hour/sf

The listed formulation of factoring can be applied to any operation with a determination of realistic production, based on the given situation. This determination must be a realistic output of craft-hour consumption to provide the required task. The function is the full cycle of work required to perform that operation, staying focused only on that operation.

Example: Material Allocation

The material costs associated with each operation are the direct permanent and indirect or incidental materials and supplies required per specification. The cost of the formwork itself is allocated as formwork, which is treated as an asset to the company. The formwork is treated as equipment and has an ownership rate applied to it for its use. This allocation is detailed in the formwork section and deals with the cost associated with the formwork purchase or construction and reuse disbursing.

The consumable formwork required, that will not be reused, is considered direct to the job as an incidental supply.

If there is a requirement for formwork that is not in inventory or is classified as specialty, the allocation of price or rental should be factored in the operation over any direct allocation.

The material listed is summarized by operation and allocated to the direct pay item, which is lineal feet of concrete beams required, for the beam-related items, and the cubic yards of neat line concrete required for the concrete diaphragms.

The material allocation is performed for each operation, for each type of material required, permanent direct and indirect incidental supplies. The material cost is disbursed against the control operation item quantity, not the work item quantity.

Operation. Prestressed, precast concrete beam, type IV.

```
Direct material:
*Precast concrete beam type IV  = 600.0 lineal ft
Indirect material:
   Position dowels: 2" × 18"      = 12.0 each
   Elastomeric bearing pads       = 16.0 each
Incidental material:
   Epoxy bonding compound, 6 oz/hole
      (6 * 12)                    = 72 ounces
```

This material list is comprised of the direct permanent and indirect and incidental material required to perform the operations within the controlling item. The direct and indirect material are outlined and required by specifica-

---

*Asterisk denotes control items for operation.

tion, and the incidental supplies are required by the operation to be performed. All materials are priced with all applicable sales and use taxes included.

Any required formwork is detailed separately owing to the different required forming systems and the intended reuse, or life of the panels.

Operation. Concrete diaphragm.

Direct material:	
*Concrete 4000 psi	= 15.69 cy
Indirect material:	
Tie-rod assemblies	= 120.0 lf
Incidental material:	
Form oil (2 oz/sf)	
(936 * 2)/128	= 14.63 gal
Bagged cement (rub and patch)	= 5.0 ea
Consumed lumber	= 120.0 lf
Incidental hardware, snap ties	
1 each per 3 sf: 936/3	= 312.0 ea
Nails: 2.5 lb per 100 sf: (936/100) * 2.5	= 23.4 lb
Formwork:	
Wall panels, 2 × 4 wood forms	= 936.0 sf
All thread 3/4" (20' ea diaphragm)	
(20.0 lf * 15.0 ea)	= 300.0 lf

This material list is comprised of the direct permanent and indirect and incidental material required to perform the operations within the controlling item. The direct and indirect material are outlined, and required by specification, and the incidental supplies are required by the operation to be performed. All materials are priced with all applicable sales and use taxes included.

Any required formwork is detailed separately owing to the different required forming systems and the intended reuse, or life of the panels.

The ratio of surface contact area to the control cubic yards gives a perspective for a relationship of cost, both for the current job and for future reference. This is taken from the total form surface area of the operation against the neat line (or plan) cubic yards for the operation. The same can be done with the cost ratio.

Surface ratio, sf per cy

936 sf/13.39 cy = 69.90 square feet per cubic yard

Section Summary

The beam structure unit of a bridge structure contains many components and operations. This section details the techniques needed to perform an accurate quantity takeoff. These basic skills and formulations can be applied to any given takeoff and cost summary by following the order of construction and identifying each required component. This step-by-step procedure provides the accuracy needed for a formatted outline.

COMPONENT TAKEOFF

Beam Structure Unit

The following outline contains the normal work operation accounts for the performance and cost coding of beam structure construction components:

1. **Prestressed, precast concrete beams**

   All work related to the erection of concrete beams of the specified type. Included in this account is the necessary temporary bracing needed to support the girders. Quantity reported should be actual lineal feet of girders placed, number of each per span, number of individual spans, and type and size of beam.

2. **Tie-rods**

   All work related to the installation of the transverse tie-rods at diaphragm locations. Quantity should be reported in actual lineal feet of tie-rods required and number of locations installed.

3. **Position dowels and anchor assembles**

   All work related to the installation of location devices. Quantity should be reported in number of each required at individual bearing location.

4. **Elastomeric bearing**

   All work related to the installation of elastomeric bearings for precast girders or structural steel girders. Quantity should be reported in number of each required at bearing location.

5. **Diaphragms, form and strip**

   All work related to the forming of portland cement concrete diaphragms for the beam structure to their proper line, grade, and dimension, and the stripping of same. This will include all embedded hardware items and necessary supports needed to construct the form system. Quantity should be reported in square feet of the concrete surface contact area, including the bottom for a suspended diaphragm, for both forming and stripping of individual diaphragms. This should be calculated from the centerline of girder transversely to centerline of girder.

6. **Diaphragm, pour**

   All work related to the placement of concrete of the specified type. Quantity should be reported in cubic yards including overrun and waste, and net amount of each individual placement.

Beam Structure

7. **Diaphragms, rub and patch surface**

    All work related to the rubbing and patching of portland cement concrete diaphragm surfaces for the beam structure. Quantity should be reported in square feet of the concrete surface contact area, including the bottom for a suspended diaphragm. This should be calculated from the centerline of girder transversely to centerline of girder.

8. **Structural steel girders (main)**

    All work related to the erection of structural steel rolled or fabricated main girders. Included in this account is the necessary temporary bracing needed to support the girders. Quantity should be reported in total pounds, actual lineal feet of girders, number of each per span, number of individual spans, and type and size of girder.

9. **Structural steel girders (floor and miscellaneous)**

    All work related to the erection of structural steel or fabricated steel floor beams and miscellaneous girders. Quantity should be reported in total pounds, actual lineal feet of girders, number of each per span, and type and size of girders.

10. **Cross frames**

    All work related to the erection of fabricated cross frames onto the permanent girders, excluding permanent bolts or welding. Quantity should be reported in number of each required for individual spans.

11. **Diaphragms (structural steel)**

    All work related to the erection of fabricated diaphragms onto the permanent girders, excluding permanent bolts or welding. Quantity should be reported in number of each required for individual spans.

12. **Splices**

    All work related to the assembly of field splices for both web and flange, excluding permanent bolts or welding. Quantity should be reported in number of each individually completed splices required for both web and flange.

13. **Temporary bolts**

    All work related to the installation and removal of temporary bolts required for the individual area. Quantity should be reported in each individual bolt and specified area of installation.

14. **Permanent bolts**

    All work related to the installation of permanent bolts required for the individual area. Quantity should be reported in each individual bolt and specified area.

15. **Welding**

    All work related to the installation of permanent welds required for the individual area. Quantity should be reported by specified area, section size of weld, total lineal feet per area, number of passes per section size, and number of areas.

16. **Rockers and bearings**

    All work related to the installation of location and bearing devices. Quantity should be reported in number of each required at bearing area, type of device, and number of anchor bolts.

17. **Miscellaneous steel**

    All work related to incidental structural steel not specified here. Quantity should be reported in total pounds and type of unit.

18. **Falsework and scaffolding**

    All work related to the assembly, erection, and removal of temporary falsework and scaffolding needed for erection of precast or structural steel girders. Quantity should be reported in a detailed and specific unit of the work performed along with the type of system used.

19. **Texture surface finish**

    All work related to the application of a preapproved material compound to the surface of precast concrete beams. Quantity should be reported in actual square feet of surface application and lineal feet of beam applied.

20. **Painting, structural steel**

    All work related to the painting of a structural steel beam unit. Included is all preparation, sandblasting, primer, and final paint applications. Quantity should be reported in total pounds of structural steel painted, individual units of material used for each process and application, and square feet of surface.

21. **Other**

    This account will be used for other operation accounts not specified here. Quantity should be reported in a representative format relating to the work performed and the appropriate unit of measure.

Chapter 15

Superstructure

TECHNICAL SECTION

Introduction: Conventional Superstructure

The superstructure section describes the necessary technical specification for construction of the superstructure unit. Included are specified procedures, materials, and equipment, and an outline of the various components of the superstructure unit. This details the necessary work consisting of constructing the portland cement concrete portions of the superstructure to the required dimensions with or without reinforcing steel.

The placing and finishing equipment will be of sufficient rated capacity and condition to properly place and finish the concrete at an able-pour rate to maintain a monolithic, unitized unit.

The forming systems will be of sufficient strength and capacity to properly contain the concrete within the specified lines and grades. The falsework will be of such designed strength and capacity to support the formwork and the imposed loads of concrete, steel, equipment, and labor force.

The permanent materials will be of approved sources and be of the designed specifications.

Definition

The *superstructure,* or *deck,* is the portion of the bridge which is above either the *beam structure* or bearings of a *cast-in-place* deck. It is the riding surface and live load distribution system of the bridge structure.

The deck is the horizontal structural member of the bridge unit which completes the span.

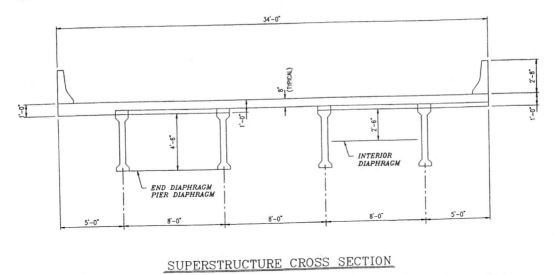

SUPERSTRUCTURE CROSS SECTION

Conventional deck section.

A *conventional* superstructure deck is constructed onto the beam structure which spans and distributes the weight to the bearing points of the substructure. This section excludes the beam structure from the superstructure because of the complexity of each of the items. The beam structure portion is discussed in detail in Chap. 14.

A cast-in-place superstructure deck is that of which the beam and deck portion are of monolithic design which comprises the entire section of the deck and spans itself to the substructure bearing points.

The superstructure comprises various individual and unique components which, as in the substructure, gives it many intriguing design characteristics. An estimator must constantly and fully understand these components and address them properly and entirely when compiling a *takeoff*.

This section entails the work necessary from the top of the beam structure, or bearings on a cast-in-place deck, upward to the completed riding surface, including the *barrier railing*. In other words, starting at the point of the *haunch* construction and moving upward, this is the area that rests on either the beam structure or the bearing point, as previously discussed. This is the connecting point, or *composite section* (positive movement), or *noncomposite section* (negative movement) of the superstructure unit. The base components of the conventional superstructure unit are the deck and barrier sections.

Deck

The deck unit is comprised of numerous components which are individual and unique to themselves but require exact detail by the estimator.

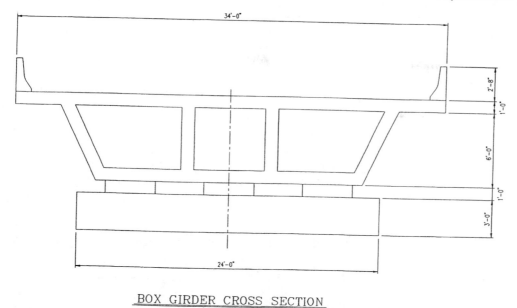

BOX GIRDER CROSS SECTION

Cast-in-place deck section.

Haunch. The haunch is usually a thickened area of the deck in which the structural bearing point of the deck to the beams is obtained.

Composite deck. A composite deck is tied to the beams. In the case of prestressed, precast girders, this is accomplished by the use of reinforcing steel *stirrups*.

In the case of structural steel, this is done with the use of *shear connectors*.

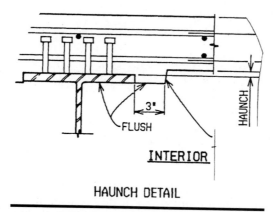

HAUNCH DETAIL

Haunch section. (*Courtesy of Michigan Department of Transportation.*)

Noncomposite deck. A noncomposite deck, or portion of the deck, is not tied to the beam structure. This design is used for tension, contraction, or expansion of the deck design independently from the beam structure.

A haunch or fillet section is the built-up section which is to be poured monolithically with the deck. The haunch section is designed for additional load bearing points to the beam for weight distribution. It is normally a thickened area of design.

Camber. The haunch, or fillet, is also used for adjustment of the theoretical *camber* of the beams to the actual as-built camber. The haunch dimensions are from the flange of the top of the beam to the bottom of the deck. The as-designed slab thickness of the deck is then from this point upward to the top of the finished deck.

If a beam has a theoretical camber design of 2 inches prior to loading and obtains an actual camber of 2.5 inches after fabrication, the haunch dimension must be adjusted for the ½ inch of deviation. This is accomplished by profiling the individual beams after erection but prior to loading with formwork, reinforcing, or concrete. Once these dimensions have been established, either plus or minus, the deck can be formed. This step is essential to obtain a proper finished deck elevation.

Deflection. This same dimension is needed later in the deck construction when establishing the *deflection* of the beam structure to obtain the proper slab thickness and proper finished elevation.

The top or finish elevation of the deck slab remains constant, whereas the sublying components, such as beam seat elevations and actual camber of the beams, may vary. Establishing this varying dimension for the deck grades is

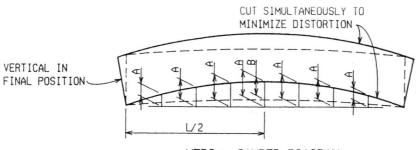

WEBS - CAMBER DIAGRAM

CUT WEB PLATES TO CAMBER INDICATED BY DIAGRAM.
A = TOTAL CAMBER REQUIRED WITH GIRDER LAYING ON ITS SIDE. (SHOW AT CENTER OF SPAN AND AT 1/4, 1/6, 1/8, OR 1/10 POINTS, AS APPROPRIATE WITH MAXIMUM SPACING OF APPROXIMATELY 25'.)
B = DEFLECTION OF ℄ GIRDER DUE TO ITS OWN WEIGHT, CENTER ONLY.

X dimension and camber. (*Courtesy of Michigan Department of Transportation.*)

known as the *X dimension*, or unknown. This must be field acknowledged and then referenced back to the theoretical grades from the design.

From the actual girder profile, the dimensions between the actual top of beam and designed top of the slab are referenced against each other. This dimension is checked against the theoretical *deflection* of the beam plus the adjustment of the as-built camber. This X dimension will assure the contractor of the proper deck thickness and proper slab elevation at the time of concrete placement.

The deck formwork and overhangs are then placed using these dimensions from a point referencing fills (added dimensions) or cuts (subtracted dimensions) of the centerline of each beam.

The next component is the deck itself, which is broken down into two areas, the interior portion and the overhang portion.

Interior section forming. The interior section is the portion of the deck from the center of the outside girder to the center of the outside girder, encompassing all the interior girders, and extends the length of the entire bridge unit (abutment to abutment). This section can normally be formed in one of three ways.

Removable forms. The first method is by the use of removable forms, either wood or metal. This is commonly referred to as conventional deck forming. With conventional deck forming, the interior portion of the deck unit is formed in such a fashion that upon completion of the deck structure, the forming system is removed, allowing the bottom concrete surface to be exposed.

The wood system is a series of dimensional lumber and plywood constructed to temporarily support the construction of the deck. This system is the most commonly used. The dimensional lumber *walers* are supported from embedded metal hangers and other assisted hardware. The wood deck and haunch template is then constructed.

A metal forming system consists of prefabricated steel forms which are designed for speed and consistency. This gang system is usually handled with some form of mechanical *travelers* or picking system. Both of these systems are supported off of the beam structure.

Stay-in-place forms. A second method of interior deck forming is by the use of metal *stay-in-place* forms or pans. This uses a corrugated rigid metal pan section which is designed to span the interior beam section and be self-supportive. The metal pans are supported on the beam structure by the use of metal angles. These angles are anchored to the individual beams by various methods: on precast girders, by embedded clips and welding; on steel girders, by either welding or by supportive straps spanning the top flanges. The haunch section template of the deck is also constructed at this time by the placement of the angles. The metal stay-in-place pan section is then anchored to the angles.

Interior Hangers

Interior Hanger Spacing Charts

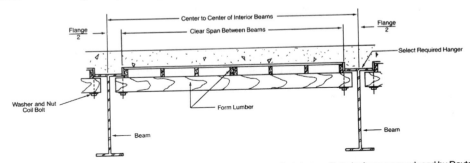

The following charts list the maximum safe hanger spacing for the various types of interior hangers produced by Dayton Superior. When the clear span on a project is not an even foot, the next larger clear span, from the chart, should be used in determining the maximum hanger spacing.

In many cases, the form lumber may control the maximum spacing between hangers, therefore the lumber must always be checked before a hanger spacing is determined for actual use.

These charts are based on the following formula:

$$\text{Maximum Hanger Spacing (Limited at 8'-0'' Maximum Centers)} = \frac{\text{S.W.L. per Side of Hanger}}{\text{Design Load, PSF} \times \left(\frac{\text{Clear Span, Feet}}{2}\right)}$$

		2,000 lbs. per Side Hanger Safe Working Load									
		Clear Span Between Beams									
Design Load PSF	Slab Thickness	3'-0"	4'-0"	5'-0"	6'-0"	7'-0"	8'-0"	9'-0"	10'-0"	11'-0"	12'-0"
		Maximum Interior Hanger Spacing									
125	6"	8'-0"	8'-0"	6'-4"	5'-4"	4'-6"	4'-0"	3'-6"	3'-2"	2'-10"	2'-8"
150	8"	8'-0"	6'-8"	5'-4"	4'-5"	3'-9"	3'-4"	2'-11"	2'-8"	2'-5"	2'-2"
175	10"	7'-7"	5'-8"	4'-6"	3'-9"	3'-3"	2'-10"	2'-6"	2'-3"	2'-0"	1'-10"
200	12"	6'-8"	5'-0"	4'-0"	3'-4"	2'-10"	2'-6"	2'-1"	2'-0"	1'-9"	1'-8"

		2,375 lbs. per Side Hanger Safe Working Load									
		Clear Span Between Beams									
Design Load PSF	Slab Thickness	3'-0"	4'-0"	5'-0"	6'-0"	7'-0"	8'-0"	9'-0"	10'-0"	11'-0"	12'-0"
		Maximum Interior Hanger Spacing									
125	6"	8'-0"	8'-0"	7'-7"	6'-4"	5'-5"	4'-9"	4'-2"	3'-9"	3'-5"	3'-2"
150	8"	8'-0"	7'-11"	6'-4"	5'-3"	4'-6"	3'-11"	3'-6"	3'-2"	2'-10"	2'-7"
175	10"	8'-0"	6'-9"	5'-5"	4'-6"	3'-10"	3'-4"	3'-0"	2'-8"	2'-5"	2'-3"
200	12"	7'-11"	5'-11"	4'-9"	3'-11"	3'-4"	2'-11"	2'-7"	2'-4"	2'-1"	1'-11"

		2,500 lbs. per Side Hanger Safe Working Load									
		Clear Span Between Beams									
Design Load PSF	Slab Thickness	3'-0"	4'-0"	5'-0"	6'-0"	7'-0"	8'-0"	9'-0"	10'-0"	11'-0"	12'-0"
		Maximum Interior Hanger Spacing									
125	6"	8'-0"	8'-0"	8'-0"	6'-8"	5'-8"	5'-0"	4'-5"	4'-0"	3'-7"	3'-4"
150	8"	8'-0"	8'-0"	6'-7"	5'-6"	4'-9"	4'-2"	3'-8"	3'-4"	3'-0"	2'-9"
175	10"	8'-0"	7'-1"	5'-8"	4'-9"	4'-1"	3'-6"	3'-2"	2'-10"	2'-7"	2'-4"
200	12"	8'-0"	6'-3"	5'-0"	4'-2"	3'-6"	3'-1"	2'-9"	2'-6"	2'-3"	2'-1"

Conventional deck formwork. (*Courtesy of Dayton Superior Corporation.*)

Stay-in-place deck formwork. (*Courtesy of Topikal.*)

Precast concrete panels. The third type of deck forming is the use of stay-in-place precast, prestressed concrete deck panels.

This system, too, is designed to span the interior beam section and be self-supportive. The panels are supported on the beam structure by *grout pads, felt membrane,* or metal risers, which constructs the haunch section of the deck unit. When precast concrete panels are used, the thickness of the deck section is reduced by the amount of the concrete panels. As with the metal stay-in-place pans, these become a permanent part of the deck structure. A detailed account of these sections can be found under Forming Systems.

Overhang or fascia section. The *overhang* or *fascia* section of the bridge deck unit can be constructed only using a removable forming system, either wood or metal.

The most common is a wood system supported from the beam structure by metal *fascia jacks.* This is the section of the deck unit which cantilevers out from the exterior girders and of which the barrier sections rest. The jacks support the wood falsework and are hung from the exterior girders with embedded hangers and other assisted hardware. From these, the dimensional lumber and plywood are used to form the cantilever section of the deck unit.

The metal type of overhang forming system is a manufactured gang system, entirely of metal, that is supported from the bottom of the exterior girders. This system, like the interior gang system, is designed for speed and consistency and is normally used in large, repetitive situations. These, too, need some kind of mechanical picking system.

Precast, prestressed deck panels

Precast, prestressed deck panels are permanent forms which become an integral part of the deck slab in supporting both dead and live load. Panels may be made full depth resulting in a completely prefabricated deck. Panel widths may vary depending on manufacturing facilities. Deck panels must comply with AASHTO Standard Specifications for Highway Bridges, Articles 1.6.26 and 2.4.33 (p). (See also "Research, Application and Experience with Precast Prestressed Bridge Deck Panels" by James M. Barker, PCI JOURNAL, November-December 1975, pp. 66-85.) Details for anchoring panels during erection are not shown.

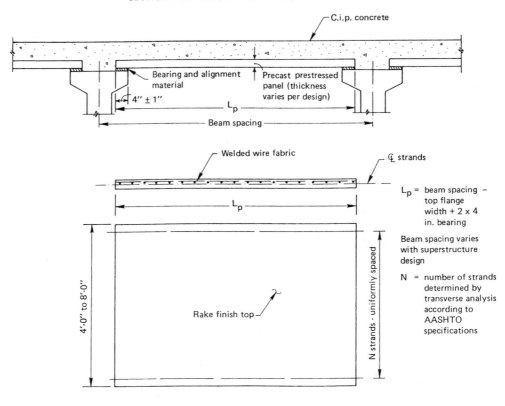

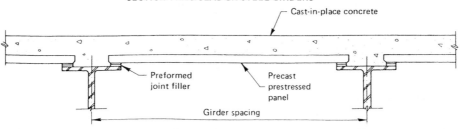

Concrete precast deck formwork. (*Courtesy of Prestressed Concrete Institute.*)

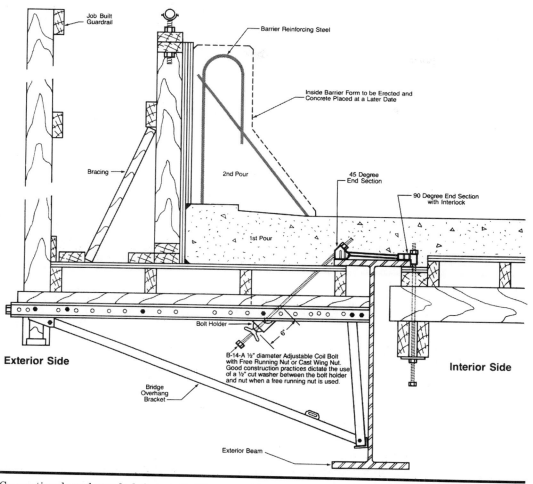

Conventional overhang deck formwork. (*Courtesy of Dayton Superior Corporation.*)

Edge fascia and transverse bulkheads. Once the deck is completely formed, the *edge fascia* and *transverse bulkheads* must be constructed. These are vertical forms which form the perimeter of the deck section and are used to designate the required thickness of the deck. This would be the point of *top slab elevations*. These are normally constructed of wood panels, but with the use of a metal overhang gang system, the edge perimeter forms can be of metal.

Once the deck and overhang components of the deck unit are constructed, there are other key elements of the superstructure which must be addressed prior to the placement of concrete.

Construction joints. Construction joints are designed for structural control of the deck unit and are designated for control of separate concrete pour sequences. The construction joints are located as shown on the plans. They are designed to be perpendicular to the principal lines of stress.

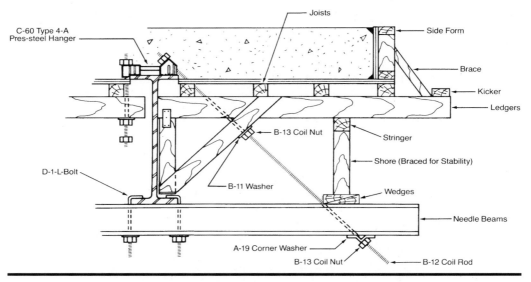

Edge and bulkhead fascia formwork. (*Courtesy of Dayton Superior Corporation.*)

All construction joints must be constructed true to line and grade and must maintain the proper elevation to control the finished screed elevation. In a (tied) construction joint or vertical joint, the reinforcing bars extend across the joint in a manner to make the structure monolithic.

Other necessary *dowels, load-transfer devices,* and *bonding* devices must be accurately placed as shown on the plans. *Keyed construction* joints or keyways must be constructed as shown.

Construction joints are designed to construct separate deck units monolithically to one complete and structural unit. *Bonding* agents such as epoxy or other approved preparatory methods may be necessary or designed to aid in the adhesion of the joint.

Expansion joints. These are designed to allow a deck unit a designated movement point during construction and expansion. This occurs during temperature changes which cause movement to the superstructure. In heat, the beam structure along with the superstructure may expand. In the cold, these will contract.

Care must be taken to construct these joints to the exact dimensions shown on the plans along with, in some cases, calculation of the designed temperature expansion opening to the current temperature during construction of this bulkhead. Expansion joints are constructed from a separate material, strip, or compound which is formed and/or inserted in the deck unit or portions of it prior to concrete placement.

Open joints. These are constructed in a deck unit from a removable form system. This forms a clear opening in the deck after the placement of concrete. This is a type of control or expansion joint which is designed without an

THE D.S. BROWN COMPANY

ANCHORAGE DESIGN

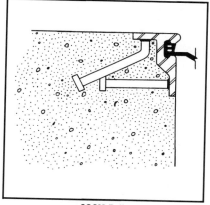

SSCM Rail
Texas Stud Anchorage

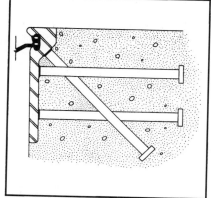

SSPA Rail
Pennsylvania Stud Anchorage

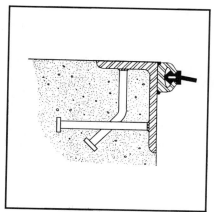

SSA2 Rail
Angle and Stud Anchorage

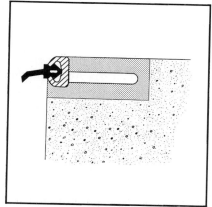

SSE2 Rail
Anchorage in Delcrete Elastomeric Concrete

Metal deck expansion joint. (*Courtesy of D.S. Brown Company.*)

additional mechanical device. This is formed like a construction or expansion joint bulkhead by an insertion with subsequent removal of wood, metal, or similarly approved material. This forms a template within the finished concrete to give an accurate dimensional opening in the deck. Normally, in open joint construction, no reinforcing or tie-bar system will extend through the joint.

422 Takeoff and Cost Analysis Techniques

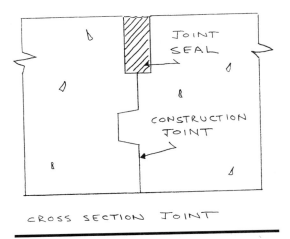

Filled deck expansion joint.

Filled joints. A filled joint or poured expansion joint is constructed like an open joint, the difference being a joint filler or *premolded* strip is inserted in the joint to form a seal.

As in other designed joints, the dimensions and location of the joint must be constructed accurately to assure the proper function of the joint.

Steel joints. These are a prefabricated joint system which is embedded in the concrete deck unit. They are constructed from plates, angles, and other structural shapes which conform to the design and shape of the concrete deck. Care must be taken to ensure that the surface and finished plane of this joint is true to line and grade and free from warping. Careful positive positioning methods are employed during installation to keep them in the proper and correct position during concrete placement. The opening of the joint, as designed on the plans, which as in other expansion joints must be corrected for correct temperatures, and must be accurate. Care must be taken to avoid impairment of the clearance in any manner.

Water stops. These are a mechanical device which is sometimes constructed in construction and expansion joints to impede the passage of water.

Compression joint seals. These are a continuous one-piece strip normally extending the full length of the transverse joint and longest practical length of a longitudinal joint. This is designed to serve as a watertight seal.

This joint is installed under compression to the concrete deck and is applied using a lubricant-adhesive compound. This seal must be installed as indicated on the plans including proper allowance for elongation and compliance with the manufacturer's recommendations. The dam of the seal must be free from dirt and debris and the concrete surface and contact area properly prepared.

Superstructure 423

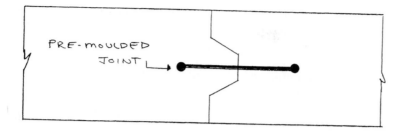

Waterstop device.

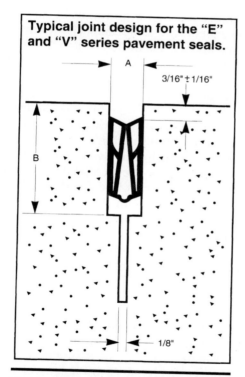

Neoprene deck expansion joint. (*Courtesy of D.S. Brown Company.*)

Elastomeric expansion seals. An elastomeric expansion seal is similar in composition but is designed for expansion control rather than compression control sealing.

Joint construction. Joint construction within a deck unit is crucial and is of great importance to the function and life of the superstructure unit. Proper procedures must be followed during construction and installation to assure its intended designed function. Because of the many types and functions of joints used, an estimator must be fully aware of the work tasks needed to construct and install the designed joint units. All joint devices must be firmly fixed to and against the concrete surface. The formed area or device must be secure so as not to be displaced during the concrete placement. All embedded items such as anchor bolts or other devices must be properly secured to assure the proper end result in designed functional use. All bulkhead formwork and templates used to secure these joints or anchor bolts must be true to line, grade, and dimensions or locations as designed. Joints must be recognized as a detailed and time-consuming item.

Deck joints can be measured and paid for in various ways. Mechanical devices and strip seals are normally measured and paid by the linear foot method. Construction joints and monolithic constructed joints are normally incidental. Individual project plans and specifications will outline this clearly. The estimator must be fully aware of the methods and properly account for the costs incurred.

Drainage devices or systems, *scuppers, drain castings,* and weep holes are units of construction within a deck unit and must be properly addressed as to location and installation procedures. The formwork and embedded items must be addressed and accounted for by the estimator.

Pipes, conduits, and ducts embedded within the concrete deck unit will be in place or allowed for prior to the placement of concrete. An estimator must be aware of the formwork and/or *block-outs* required for these. This type of formwork, along with the drainage devices, can, in some cases, be quite extensive and time-consuming. Any embedded materials such as collars and hangers must be taken into account.

Drainage systems, pipe, and conduits are measured and paid for by various methods depending on individual project requirements.

Barrier railing or parapet

Barrier railing or parapet, along with median or center barriers, are structural elements of the superstructure which contain the traffic within the bridge deck. They can be of reinforced concrete, metal, or a combination of both. They are permanently fastened to the bridge deck by means of reinforcing steel and construction joints or anchor bolts. These are all considered to be a form of barrier.

Bridge railings and barriers are constructed to the lines and grades as shown on the plans. Care must be taken not to show any unevenness caused by deflection of the structure. Railings and barriers are not constructed until

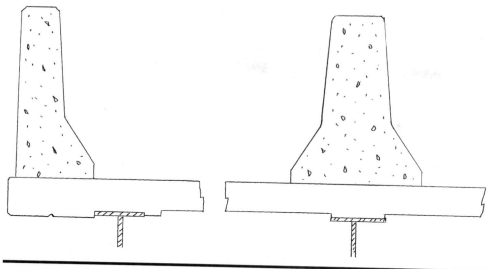

Barrier, parapet and median.

the deck forms and falsework have been removed, or until the concrete has cured, thus allowing the superstructure spans to be self-supporting.

Concrete railing will be constructed of the same class of concrete as the deck unit. The specifications regarding formwork for barrier construction will be followed as outlined in this section for deck construction. Forms will be secure, clean, and watertight. Concrete will not be placed until all embedded items are installed. Alignment and location of all joints within the deck unit must be matched precisely within the barrier unit. Concrete will be placed timely and continuously within a monolithic section.

Metal barriers will conform to the plans and specifications. Anchor bolts must be embedded properly in the concrete and conform to a proper template.

Barriers can be measured and paid for by either the linear foot or the cubic yard method. The payment of embedded materials and/or incidental materials must be researched and accounted for according to the project specifications. Depending on the design used, the predominant materials will include concrete, reinforcing steel, structural steel, beam-type steel railing, timber, or aluminum alloy railing.

Deck forming. The forms supporting the roadway slab or deck section of a structure will be supported on *wales* or similar supports fastened to the beam structure. The form system will not deflect more than $1/360$ of the span under full load.

Concrete forms will be mortartight and true to the dimensions, lines, and grades set forth in the plans. They will be designed with sufficient strength to support the load and prevent any appreciable deflection during the concrete placing operation. The falsework and formwork supporting the deck slab and overhang or girder bridge structures will be designed so that there will be no appreciable differential settlement between the girders and the deck forms during the placement of the deck concrete. This will include the stress and deflections for all load-supporting members and should consist of the sum of the dead and live vertical loads with an assumed horizontal load.

Dead load. The dead load will include the weight of the concrete, reinforcing steel, forms, and falsework. Normal concrete and reinforcing steel is assumed at a minimum of 150 pounds per cubic foot, but special mixes and designs may vary.

Live load. The live load will consist of the actual weight of any equipment that is to be supported by the formwork applied as concentrated loads at the points of contact along with a uniform load. This is usually assumed at 20 pounds per square foot applied over the supported area along with 75 pounds per linear foot applied to the outside edge of deck overhangs.

The assumed horizontal loads to be resisted by the formwork or falsework system will be the sum of the actual horizontal loads of equipment or construction sequencing with an allowance to wind.

This is discussed in further detail in the Formwork section.

The contact surface area of the forms will be cleansed of all dirt, loose debris, and foreign material prior to placing concrete. Removable forms will be thoroughly coated with an approved form oil that will aid in the release of the forms during stripping and not damage or discolor the concrete.

All construction and securing of the form system must be completed, including all incidental and permanently placed embedded items. The rate of concrete placement must be controlled as to an even displacement of weight on the form system.

The entire form system must be capable of supporting the loads imposed and be of necessary rigidity to produce the required finished product as indicated. All required *fillets* and *chamfers* must be properly installed. Owing to designed loading deflections in the beam structure, a *pouring sequence* is usually designed and must be adhered to, along with a minimum pour rate per hour.

The deck unit of the bridge structure is primarily composed of concrete and reinforcing steel. Concrete is classified by its compressive strength. Rebar is manufactured by grade. As in this design, the reinforced concrete deck section becomes an integral part of the bridge structure.

Common classes of superstructure concrete normally are 4000 and 4500 psi. Especially with modern technology, special mix designs are sometimes encountered. With prestressing, precast, and *posttensioned* structures, the normal minimum compressive mix design is 5000 psi.

Prior to the removal of the forming system, the concrete must obtain a minimum required compressive strength. This is verified by test *cylinders* or *beams* which are made of the same mix and at the same time of the pour.

Once strength is obtained, all removable forms and falsework must be removed from the finished structure and the exposed surface must be rubbed and patched.

Minimum Required Percentage of 28-Day Strength ($f'c$)

Structural element	Percentage of 28-day strength required for form removal
1. Box girders, T-beam girders, deck slab, cross beams	80
2. Deck slab supported on steel or precast, prestressed girders	70

Care should be taken not to damage the finished structure during the form removal. The forms should be well cleaned of all laitance and foreign material, oiled, and properly stored to preserve the integrity of the form.

An approved curing compound should be adequately applied to the exposed structural concrete to continue the appropriate curing period to achieve the designed strength and uniform appearance.

Superstructure deck sections require special attention to the curing of the top riding surface. This is detailed in the Concrete Placing and Curing section of the project specifications.

Superstructure concrete and rebar is normally measured and paid for by the unit of direct material. The concrete is paid by the cubic yard based on neat line measurement, and rebar is paid by the pound as obtained from a bar list. However, some payment schedules vary and should be determined by the estimator as to the method of payment for all permanent and indirect materials.

The construction of conventional superstructure units in regard to methods, techniques, equipment, and labor trades varies tremendously, by both constructive choice and demographic area. The takeoff procedures of the estimator must be consistent and precise to control the need for a competitive bid.

There are many additional but vitally important components to the superstructure unit other than the deck.

Introduction: Cast-in-Place Superstructure

Cast-in-place and *pile-supported* deck units are constructed without the use of the beam structure units. They are designed to be a self-supported reinforced concrete unit clear spanning the area from bearing point to bearing point without the aid of an independent structural unit such as a concrete or steel girder system.

This type of construction is normally produced with a temporary falsework or supportive system which supports the deck unit until the concrete has cured or, in some designs, a *posttensioning* system has been installed. A posttensioning system aids in the support of the span and is usually embedded within the pour.

Design

Many types of designs are used in cast-in-place construction of a superstructure system. In construction of the necessary falsework, bearing and structural support are key elements. In conventional deck construction, the deck

Cast-in-place superstructure. (*Courtesy of Dywidag Systems.*)

forms bear on the beam structure which is supported by the substructure. With cast-in-place construction, the falsework system must bear on either suitable ground, a temporary piling system, or the substructure itself, and in most cases all of these.

A cast-in-place superstructure has a monolithic composite which contains the reinforced element necessary to span the distance from substructure bearing point to substructure bearing point without the use of a conventional beam structure. It is a self-supported unit once completed.

Cast-in-place superstructures can have a variety of architectural shapes and span distances within the limitations of the intended design. They can be of an *arch* design, a *monolithic beam and slab* combination, a *voided* or *hollow* slab box design, or a solid deck section.

A cast-in-place superstructure can be of conventional reinforcement design or of a posttensioned design using both reinforcement and stressing cable strand.

Depending on the intended use, economics, and geographic concerns, a designer will consider and implement many concepts within this type of construction.

The estimator's bid must provide a safe and adequate falsework system that provides the necessary rigidity that supports the loads imposed and produces the finished structure to the lines and grades designed.

Falsework

Falsework must be designed to be supportive in many ways: the load imposed at ground level, whether piling, *matting,* or *cribbing*; the type of structural falsework needed for the imposed loads on itself, whether it be a form of scaffolding, timber, or structural steel; the method of spanning the falsework for the loads implied at the formwork or deck level; the wind resistance and conditions that may exist, and the type of safe anchoring system to the substructure, if any. Once the superstructure is constructed, a practical method of disassembling the system must be accounted for, both in time and monetarily.

A cast-in-place structure must be supported on falsework or other methods of a supporting apparatus. Thus falsework will be of sufficient load-bearing capacity to support safely its own weight plus the formwork, the load of reinforcement material, concrete, equipment, and labor force along with any other imposed loads to a safe value.

The estimator must derive a safe but economical method of constructing a falsework system that serves the purpose. Geographic, topographic, traffic, and existing or proposed conditions are major controlling items in the design and construction of a practical falsework system.

Once the falsework is accounted for, the normal formwork procedures, as outlined in this section, are followed.

The falsework or other temporary supports of the superstructure section will be supported on a suitable foundation or anchoring system. A detailed, feasible, and safe method outlining this temporary support system must be scheduled and approved. All falsework and temporary supports will be constructed according to these detailed drawings. All implied loads will be properly centered and balanced within this system unless otherwise specified.

The complete falsework system, if it is to be supported to the ground, will bear on a properly prepared foundation which is capable of supporting all loads implied. A satisfactory bearing foundation, whether compacted soil, piling, or concrete, must be substantial in capacity for this work. When the support system is to be anchored to a part of the existing structure, such as walls or piers, these parts of the structure must be capable of supporting themselves along with these additional loads of falsework and structure. Bearing points must be well detailed, along with means of attachment to the structure.

Elevation adjustment mechanisms must be installed within this falsework and temporary support system to allow for proper fine-tuned and precise dimensions to be maintained within the superstructure.

When a cast-in-place structure is constructed, the entire composition and section of the superstructure is poured. The shape and composite of the superstructure must be completely formed in advance of the concrete placement. Any designed deflection or camber must be built into the temporary support system along with any live load design deflection of the support system itself.

Falsework shore tower. (*Courtesy of Economy Forms Corporation.*)

Concrete placement

Within this type of construction usually lies a concrete placement schedule. This is detailed and designed for numerous reasons: (1) a balanced sequence for design loading within the structure, relating to deflection and camber or just sequential placement, (2) to minimize *creep* and *shrinkage* with the concrete itself throughout the entire pour and superstructure, and (3) for detailed tensioning and poststressing which may require, owing to design, the need for intermediate and partial construction. These types of pours, along with exact location, will be detailed by construction joints shown on the drawings. This construction sequencing of pours and location must be accurately followed. Of this scheduling, the actual placement of concrete within the placement must be carefully planned as for even, sequential lifts and sudden impacts of concrete imposed to the entire system.

Structural components

Cast-in-place superstructures have many internal components, with each having a significant value to the design of the entire structure. These major components are (but are not limited to): interior *diaphragms* and *baffles, anchorage blocks*; *reinforcing steel*; *pipes, conduits,* and *ducts,* stressing cable; *void forms*; and *drainage* and *weep hole piping*.

Interior diaphragms. These are constructed of structural concrete and are monolithic to the structure. They give the structure internal support and stability needed for design within integral parts of the bridge. As with a conventional superstructure, these diaphragms connect independent parts and unify them as one.

Drainage and weep holes. Drainage and weep holes will be located as shown on the plans. These will allow any moisture or water that may accumulate to be discharged from within the structure. Anchorage blocks and end details will be constructed as specified.

All reinforcing steel must be properly placed within the structure and be secured in place.

Pipes, conduits, and ducts. Pipes, conduits, and ducts that are to be encased within the concrete will be installed prior to the concrete placement. These will be of an approved, noncorrosive material. These pipes will be held, braced, or securely tied in place during concrete placing to prevent their displacement. The end or any openings must be protected to ensure the protection of any concrete or foreign material from entering the pipe.

Void forms. Void forms will be properly placed and supported to prevent movement during concrete placing operations or displacement of these tubes. Concrete will be placed in two layers around these tubes with the lower layer extending to the middiameter of the tube. After placing, each layer will be vibrated and allowed to settle before the next succeeding layer is placed. These subsequent layers will be placed while the lower lift is still plastic to permit the intermixing of the concrete by vibration.

All enclosures for posttensioning and reinforcement will be accurately placed at locations shown on the plans.

Duct enclosures for steel, including transition couplings, will be of a galvanized ferrous metal or other approved type, mortartight, and accurately placed within the structure.

When embedded steel is placed, all steel units will be accurately placed in the position shown on the plans, and firmly held during the placing and setting of the concrete.

Ducts will be fabricated with either welded or interlocking seams to prevent any separation. The ductwork will have sufficient strength to maintain correct alignment and shape during the placement of concrete. Joints between any sections of ducts will be positive connections which do not result

Anchorage device. (*Courtesy of Dywidag Systems.*)

in any angle or alignment changes and do not allow any intrusion of concrete, mortar, or other foreign material. A waterproof tape is allowable at connection points.

All ducts or anchorage assemblies will be provided with pipes or other suitable connections for the injection of nonshrink grout after the stressing operation.

The ducts embedded for stressing steel will be securely fastened in place to prevent any movement from their designed location.

After the installation within the formwork, the ends of the ducts will be kept covered at all times by necessary means to prevent the entry of water or debris. All ductwork will be vented at the high points of the duct profile and at any additional locations shown on the plans. The connection of the vent pipes will be performed in the same procedure as described in the above specifications of joining and sealing ducts. The vents will have a metallic connection, mortartight, taped as necessary, and will provide suitable means for the injection of nonshrink grout through the vents and for the sealing of the vents. After the grouting operation, the vent pipes will be removed to a specified dimension below the finished surface of the concrete.

The distance of the embedded ducts and pipes from formwork, or the exterior concrete surface, will be maintained by stays, blocks, ties, hangers, chairs, or other approved supports. Any type of support that comes in contact directly with the formwork of an exposed concrete surface will be either partially or fully coated with an epoxy resin to prevent rustification on the concrete surface.

Anchorage devices. These are used for posttensioning and will be capable of holding the prestressing steel at a load producing a stress of not less than 95 percent of the guaranteed minimum tensile strength of the prestressing steel. The load from the anchoring device will be distributed to the concrete by means of approved devices that will effectively distribute the load to the concrete. Approval of such devices will contain at minimum the following required data:

1. The final unit compressive stress on the concrete directly beneath the plate or assembly will not exceed the specified pounds per square inch, normally 3000 psi.
2. Bending stresses in the plates or assemblies induced by the pull of the stressing forces will not exceed the yield point of the material or cause visible distortion in the anchorage plate when 100 percent of the ultimate load is applied.

Anchoring devices which are significantly larger than required and work in conjunction with a steel *grillage* embedded in the concrete effectively distribute the compressive forces that may be addressed. All anchorage devices will be protected from corrosion.

All deviations in the concrete pad of which the device is seated will be corrected by the use of an approved epoxy resin or grout capable of withstanding the stressing forces applied during the load transfer. This material will be bonded to the concrete surface uniformly.

When steam curing is used, the steel for posttensioning will not be installed until the steam curing operation is completed. All prestressing and posttensioning steel will be protected against all forms of corrosion as described in the project specifications.

Suitable horizontal and vertical spacers will be provided to hold and support all stressing wires and cables in their true and designed locations within the enclosures.

The stressing elements will be accurately held in position and stressed by jacks using proper implemented and approved procedures. A record will be kept of the *jacking forces* and *elongations* produced thereby. Several precast units may be cast in one continuous line and stressed at one time, provided all spacing and *drape patterns* of the embedded steel are accurately placed. It is very crucial that no sharp bends or kinks be made in the strand.

Within the *casting beds,* sufficient space will be left between the ends of the precast units to permit access for cutting the strands. This procedure is completed after the concrete has obtained the sufficient and specified strength, and after the units are *detensioned.* No bond stress will be transferred to the concrete, nor end anchorages released, until the concrete has attained the specified compressive strength, as shown by test cylinders. The precast units will then be cut or released in such an order that *lateral eccentricity of prestress* will be at a minimum.

Pretensioning

The stress induced in the prestressing element as measured both by pressure gauges and by elongations of the prestressing steel will agree. The elongation computation will take into account strand anchorage, slippage, horizontal movement of abutments, and any change in temperature of the prestressing steel between the tensioning and the time when the concrete takes its initial set, only if this temperature change is expected to exceed 30 degrees Fahrenheit. The final pretensioning will not be done when temperatures are below 20 degrees Fahrenheit or when governed by other conditions within the specifications.

All strands of the pretensioned unit will be free of kinks or twists prior to the tensioning operations. Any unwinding of the strand in excess of one turn, after tensioning operations have begun, will not be permitted. All the strands will be tensioned to 20 percent of the final jacking force before the elongation readings are started, or by the governing specifications. This initial tension in any strand will not vary by more than 5 percent from that specified. The equipment used for producing this initial tension load will provide a means for accurately measuring the forces applied. When the initial tensioning load is applied by pressure jacks, they will be equipped with a proper and accurate gauging system to record the initial force.

The strands that are draped will be stressed no higher than the required design stress minus the stress increase in the strand from forcing it into a draped profile, or drag.

If the draped strands are tensioned in their draped position, they will be supported by rollers at the points of change in direction. These rollers will produce a minimum of friction. If the load for a draped strand, as determined by elongation measurements, is more than 5 percent less than that indicated by the jack gauges, the strand will be tensioned from both ends of the casting bed and the load as computed from the sum of elongation at both ends will agree within 5 percent of that indicated by the jack readings. Specific details showing the number, spacing, and method of draping pretensioned strands will be submitted for approval well in advance of the casting and stressing operations, as specifications imply.

The strand material used will have the same composite characteristics, metallic strength and properties, tensile requirements, and lay or twist.

Specifications may vary but the average working stress in the prestressing steel will not exceed 60 percent of the specified minimum ultimate tensile strength of the steel. The maximum temporary tensile stress, or jacking stress, in prestressing steel will not exceed 75 percent of the specified minimum ultimate tensile strength of the steel. The prestressing steel will be anchored at initial stresses that will result in the ultimate retention of working forces of not less than those detailed by specifications, but in no case will the initial stress after transfer exceed 70 percent of the specified minimum ultimate tensile strength of the prestressing steel.

Working force and working stress will be considered as the force and stress remaining in the prestressing steel after all losses, including *creep* and *shrinkage* of concrete, *elastic compression* of concrete, creep of steel, friction and take-up anchorages, and all other losses peculiar to the method or system of prestressing which has taken place or have been provided for.

It is anticipated that there may be a discrepancy in indicated stress between calibrated jack gauge pressure and elongation. With this condition, the load used will produce a slight overstress rather than an understress. If this discrepancy is greater than 5 percent, the entire operation will be reevaluated and the calculations will be checked for error prior to continuing. The prestressing elements will be tensioned to provide the required prestress specified. The prestressing steel strands within the pretensioned members, once tensioned individually, will be checked for loss of prestress within 3 hours of concrete placement. The method and equipment used for checking loss of prestress will be governed by the specifications. All strands that show a loss of prestress in excess of 3 percent will be pretensioned to the original computed jacking stress.

Upon completion, all pretensioned, prestressing steel will be cut flush with the end of the member. The end of the unit will then be properly cleaned of debris, prepared by specification, and painted with a protective material as specified.

Concrete will not be deposited within the formwork until the forms have been thoroughly cleaned of all debris and an accurate inspection has been made of all reinforcement, enclosures, anchorages, and prestressing steel. The concrete must be deposited and properly vibrated to avoid segregation and placed in a manner to avoid displacement of the reinforcement, conduits, or stressing wires. All ductwork must remain unobstructed.

Curing may consist of steam or water methods. The casting beds or formwork units with steam curing must be completely enclosed with a suitable type of housing, tightly constructed to contain the heat uniformly and to exclude the outside atmosphere. Normally, within 4 hours after the placement of the concrete and after the concrete has undergone its initial set, the first application of steam or radiant heat will be applied. This procedure may vary depending on the presence of concrete admixture retarders. During this time period, the temperature within the curing chamber will not be less than 50 degrees Fahrenheit. This minimum may be maintained by the use of live steam or radiant heat. The steam used will be at 100 percent relative humidity. The application of steam will not be directly to the concrete units. During the application of steam or radiant heat, the ambient temperature will increase at a rate not to exceed 40 degrees Fahrenheit per hour until the desired specified ultimate curing temperature is obtained. This specified maximum temperature will then be maintained until the concrete units have reached the desired design strength.

The application of live steam will not be directed onto the surface of the formwork so as to cause localized high temperatures which would be detrimental to the concrete units. Radiant heat may be applied by means of pipes, circulating steam, hot oil, or hot water, or by electric heating elements. Radiant heat curing will be done under a suitable enclosure to contain the heat transfer, and the moisture loss will be minimized by covering all exposed concrete surfaces. This may be accomplished with plastic sheeting or by the application of an approved liquid membrane curing compound. Care will be taken to avoid the deposit of membrane residue to any top concrete unit surfaces to be used in *composite* construction within the bonding design limits. This specification will also hold true to any other surfaces of the concrete members that will have other materials bonded to them.

The discontinuing of steam will be followed in much the same manner as increasing was accomplished. The heat must be decreased at a rate not to exceed 40 degrees Fahrenheit per hour until a temperature of 20 degrees above the ambient air temperature, which the concrete units will be exposed to, has been reached.

Detensioning will begin immediately after the steam or heat curing has been discontinued. The concrete units will not be exposed to any temperatures below freezing until at least 7 days after casting.

Posttensioning

When concrete units are to be *posttensioned,* the tensioning procedure will not commence until the concrete strength tests have been verified and the concrete members have been properly cured. The minimum compressive strength required will be specified by project specifications.

After the concrete has attained the required strength, the stressing reinforcement will be stressed by means of jacks to the desired and specified tension, with the stress being transferred to the end anchorage. Continuous posttensioned members will be tensioned by jacking forces applied at each end of the tendon. When single end stressing is detailed, tensioning may be performed by jacking at either or both ends of the tendon. In either condition, a detailed schematic of procedural methods along with exact jacking forces must be submitted and approved prior to performance of any work.

Cast-in-place concrete will not be posttensioned until a minimum of 10 days after the last concrete placement in a specific member, and until the concrete has reached the minimum compressive strength required for stressing. All formwork, except that of structural support, will be removed prior to posttensioning. The falsework and structural supportive formwork will not be released until the tensioning has been completed and a minimum of 48 hours has elapsed after the grouting of the posttensioned tendons has been completed, or as otherwise specified. The supportive falsework will be constructed in such a manner as to permit the concrete member to lift off the falsework during posttensioning. Interior formwork left in place will be detailed in a manner to offer minimum resistance to the member in shorting due to shrinkage and posttensioning.

The tensioning process will be conducted so that the tension being applied and the elongation of the prestressing elements may be measured at all times. The friction loss in the element, i.e., the difference between the tension at the jack and the minimum tension, will be determined in accordance with the specifications and AASHTO Standard Specifications for Highway Bridges. Suitable shims or other approved devices will be used to ensure that the specified *anchor set* loss is attained. Prestressing tendons in continuous posttensioned members will be tensioned by jacking. Documentation will be kept of all gauge pressures and elongations.

Grouting

Stressing steel within ducts and tubes will be bonded to the concrete member by grouting the voided space between the duct and the tendon with an approved nonshrink cementation material.

This material will be injected in the ducts by means of a pressure pump. The mix design of the grout will be of pumpability of approved consistency and compressive strength requirements. The grouting equipment will be capable of a pumping pressure as specified, normally a minimum of 100 pounds per square inch, and be furnished with a pressure gauge having full-scale readings in pound increments up to 300 psi. Flushing equipment capable of 250 pounds per square inch will be on standby and have sufficient capacity to flush out any partially grouted ducts.

Prior to grouting, all ducts will be clean and free of deleterious materials that would impair the bonding of grout or would interfere with the flowability and procedures of grouting as specified.

Ductwork. (*Courtesy of Dywidag Systems.*)

The grout injection pipes will be fitted with positive mechanical shutoff valves. Vents and ejection pipes will be fitted with valves, caps, or other devices capable of withstanding the pumping pressures. The valves and caps will not be removed until the grout has properly set. All posttensioned steel will be bonded to the concrete member. All prestressing steel will be free from dirt, loose rust, grease, or other deleterious substances.

Immediately after the concrete placement, the conduit ductwork will be blown out with compressed oil-free air to the extent necessary to break up and remove any intruded mortar before it sets. Prior to tensioning procedures, it will be demonstrated that the ductwork is unobstructed and the prestressing reinforcement is free and unbonded in the duct.

After the tendons have been stressed to the required tension, each conduit encasing the prestressing steel will be blown out with compressed oil-free air. The grouting procedure will then commence by filling the conduit from the low end with grout under pressure as specified. The grout will be pumped through the duct uninterrupted and continuously and wasted at the outlet end until no visible slugs of water or air are ejected. All vents and openings will then be closed and the grouting pressure at the injection end will be raised to a minimum pressure and time as specified.

Any steel that is specified to be unbonded to the concrete will be protected against corrosion with a tar coating, waterproofing, epoxy, or a specified galvanizing.

Extreme care will be exercised in the handling, moving, and transportation of precast, prestressed concrete members. Precast girders and slabs will be transported in an upright position and the points of support and directions of the reactions with respect to the member will be approximately the same during storage and transportation. These stress and moment points will remain in the same position as when the member is in its final position. Prestressed members will not be shipped until tests on the concrete cylinders, manufactured of the same concrete mix and cured under the same conditions as the member, indicate the member has attained the required minimum compressive strength by design.

Dimensional tolerances

The following dimensional tolerances will be specified under normal conditions and used as a guide for acceptance of members, but specific project designs and tolerances may vary.

1. Length (overall): $\frac{1}{8}$ inch per 10 ft or $\frac{1}{2}$ inch overall, whichever is greater
2. Width (flanges): $+\frac{3}{8}$ inch, $-\frac{1}{4}$ inch
3. Width (web): $+\frac{3}{8}$ inch, $-\frac{1}{4}$ inch
4. Depth (overall): $+\frac{1}{2}$ inch, $-\frac{1}{4}$ inch
5. Horizontal alignment: $\frac{1}{8}$ inch per 10 ft or $\frac{1}{2}$ inch overall, whichever is greater

6. Box girder webs and top slabs: $+\frac{1}{4}$ inch, $-\frac{1}{8}$ inch
7. Box girder bottom slab: $+\frac{1}{2}$ inch, zero
8. Tendon position: $\frac{1}{4}$ inch from center of gravity of strand group and/or individual tendons
9. Longitudinal position of deflection points (deflective strands): 12 inches
10. Bearing recess (center to end of beam): $\frac{1}{4}$ inch
11. Beam ends (deviation of square or designated skew): horizontal: $\frac{1}{4}$ inch from center of web to edge of flange; vertical: $\frac{1}{8}$ inch per foot of beam depth
12. Bearing area deviation (from plane in length or width of bearing): $\frac{1}{16}$ inch
13. Stirrup reinforcing spacing: $\frac{1}{2}$ inch
14. Stirrup protection (from top of beam): $\frac{3}{4}$ inch
15. Concrete cover (mild steel): $-\frac{1}{8}$ inch, $+\frac{3}{8}$ inch
16. Form joint deviation (straight plane of 5 ft either side of joint): $\frac{3}{16}$ inch
17. Differential camber of girders (individual spans): I girders: $\frac{1}{8}$ inch per 10 ft of beam length; T girders: $\frac{1}{8}$ to $\frac{1}{4}$ inch at midspan
18. Posttensioning plates and ducts: $\frac{1}{4}$ inch

All cast-in-place concrete constructed on falsework will be evenly loaded or loaded in such a manner as prescribed by detailed procedures to prevent unbalanced loading to the structure. This will ensure proper deflection and camber alignment. All falsework and shoring will be constructed in accordance with the construction drawings and specifications with suitable foundation bearing. Suitable wedges or jacks will provide for the raising and/or the lowering of the forms to the exact elevation shown, and for any adjustment for settlement occurrence during loading. Care should be taken to avoid any overstress of the entire system.

Sequential placements and adjacent loading to previously poured segments will be followed in accordance with the placement diagram and specifications with respect to curing and stressing procedures and loading.

The measurement and payment for structural members will be for the quantity of elements as specified by contract provisions and outlined in the bid schedule. This will include complete reinforcement and prestressing steel, anchorages, plates, nuts, elastomeric and structural bearings, concrete, and any other incidental element item, complete in place.

TAKEOFF QUANTIFICATION

Introduction

This section defines the various components and methods of construction associated with the superstructure unit. It establishes the required techniques and outlines the parameters for a detailed and accurate quantity takeoff. It highlights the procedures for the consistency of a formulated quantity takeoff system.

Definition

The superstructure unit is defined as the portion of the bridge structure that carries the roadway, or traveled section. The superstructure is the unit that is supported by the beam structure unit and is the last structural unit section of a bridge structure. Even though the superstructure bears on the beam structure, it contains its own structural integrity that enhances the beam unit, working in unison with design characteristics.

The superstructure unit comprises many individual components with various methods of construction operations, making the structural unit the most complicated.

A cast-in-place box girder superstructure unit represents a composite structural concrete unit which has a monolithic box girder beam and deck section. Each specific section consists of a reinforced unit which structurally supports itself for the given span.

Takeoff Procedure

The takeoff procedure of a superstructure unit begins with the identifying of the type of beam structure unit. This will be a controlling unit for the quantity takeoff and component operation performance. The next item to define is the requirement(s) of deck construction that exists, normally by specification. This, simply, is the method of decking, temporary or permanent, required as formwork between the interior beam or girder bays and the cantilevered overhang.

As with the other structure units, the quantity takeoff begins at the lowest point, or supporting subunit, and progresses upward and outward in the order of construction operation requirement.

The superstructure unit, unlike the substructure unit, has minimal component contact surface areas, thus requiring more internal support methods and formwork types for the required construction operations. Virtually the only component contact areas to stage the work operations from will be the top surfaces of the beam structure unit.

The quantity takeoff procedure begins with defining the method of formwork or form system attachment to the beam structure. This will be required for both the interior and exterior bay sections.

The section moves to the type of form systems required for the bay closures and the exterior overhang form system. The various methods and types of systems are the horizontal support for the poured-in-place concrete deck. Next are the vertical perimeter bulkhead forming systems, which define the thickness, width, and length of the deck units. The interior bulkheads define the joint requirements within the deck unit itself.

From this point, the permanent joint systems, which allow expansion and contraction movement, are outlined. These too consist of various types of systems, required for differing designs and intended uses.

Finally, the barrier and parapet components are covered, these being the vertical members that are constructed longitudinally along the deck unit. These are for protection, for both the structure sides and the median or center section.

Superstructure Unit

Interior deck

The interior deck section is the first defined parameter for the quantity takeoff of a superstructure unit.

Formwork systems. Within this parameter, there can exist three methods of formwork systems for the interior bay closures. They are (1) a conventional wood formwork system, (2) a permanent metal stay-in-place forming system, and (3) a permanent precast concrete deck panel system. These various systems must be structurally able to sufficiently support the required concrete load, reinforcing steel, and live workloads applied during the construction operations.

The quantity takeoff of the deck formwork system, both interior and exterior, will require computations to be made with respect to the imposed load per square foot of form contact area and the span dimensions for the formwork cross section. The following criteria, charts, and tables will be used in the formulation of the formwork requirements.

Loads. The first area to define, as mentioned in the beam hanger assembly section, is the *rated load,* or the maximum load that can be applied to a hanger assembly, overhang bracket. From this, a safe working load (SWL) must be established.

Next, the *ultimate load* must be identified. This is the average load, or force, at which an item fails or will no longer support or carry a load.

The next criteria to define are the *dead load* and *live load* weights. The dead load is defined as the weight of the concrete and reinforcing steel combined with the formwork itself. The concrete and steel load calculation is performed using an assumed weight of 150 pounds per cubic foot. The formwork must be calculated. The live load is defined as the additional loads imposed during construction such as material storage, workers, and equipment.

The *design load* is defined as the dead load plus the live load per square foot of net form contact area.

The *impact load* is defined as the resulted impact loading of discharged concrete, or the starting and stopping of construction equipment on the formwork. This load can be several times the dead load.

A *safety factor* is a term denoting the theoretical capability which has been determined by dividing the ultimate load of the product by its rated load. This is expressed as a ratio, an example being a 2 to 1 factor.

Interior spans. The bay span width is the actual dimension the formwork is required to carry a load between beams. The *overhang dimension* is the dimension which the proposed deck extends past the exterior beam.

The form systems, whether removable or permanent, are quantified by the required surface contact area of the interior bays for the horizontal forming system. The method of calculating is to establish an actual procedure for determining the surface contact area for the type of form system. The gross area of contact surface is calculated in the same manner for the various form systems listed, that method being the area from the centerline of the outside beam, across the deck section, to the centerline of the outside beam. The net area, or area of quantity determination, is calculated by a different method, depending on the chosen form system.

Haunch area. The haunch area of the deck section is calculated as part of the deck forming in regard to the deck grades and elevation of the horizontal formwork. The quantity takeoff of the actual rise, or the variable dimension of the haunch section, is performed by different methods, depending on the chosen system.

The required beam hanger device is dependent on the type of beam, the type of form system, the rise of the haunch, and attachment specifications.

Once the haunch rise section has been constructed, the deck forming remains a constant. From this point on, for the most, the balance of the quantity takeoff is a stand alone regardless of the type of system chosen or beam section unit.

The width of the opening of the interior bay section will have an impact on the support strength required for the cross section of the form system.

Overhang. The overhang form system is independent of the interior system with respect to type of system. The overhang system, in some conditions, shares the beam hanger devices with the adjacent bay. The overhang form system, however, is dependent on load and support requirements different from those of the interior bays. This is accomplished by the centerline spacing of the support brackets, which may differ from the centerline spacing of the interior bays or from the entire interior system. This requires different or additional beam hanger devices.

Bulkheads. The perimeter and interior bulkheads provide the vertical formwork requirement for the deck unit. The interior bulkheads can be either plain or undoweled for an open joint, or doweled for a construction joint. The bulkhead forms can be longitudinal or transverse in location.

Joints. Once the deck units are formed, the expansion joint assemblies, control joints, or blockouts must be accounted for and quantified. These perform the required movement within the superstructure deck unit.

Finishing. Most deck unit concrete placements are required to be finished and screeded by a mechanical finish machine. This machine is controlled in elevation by a metal rail that is placed adjacent and parallel to each pour. The screed rail must be quantified for each concrete placement.

The finish machine must be set up and verified as to profile and elevation, referred to as a dry run. This is an actual check of the X dimensions, a dimension to which the machine's finishing attachment is set in vertical distance, from various points on the deck surface. This is an assurance to the profile of the finish machine and final deck elevations, the required concrete cover over the reinforcing steel, and the required depth of concrete. This component is required by individual placement area.

Deck preparation. The deck preparation is the next component to define. This is defined as the entire horizontal surface concrete contact area of the deck unit. This component is the performance of the final cleaning and prepping of the deck surface, ready for the concrete placement.

Concrete placement. Concrete placement is the next component to define. This operation is the actual placement of the concrete by the placing method chosen. There are various placement procedures. The type chosen is quantified as the required operation to fully perform the placement, including setup, surface texturing if required, and cleanup, for the required cubic yards per pour. The concrete curing component is quantified in conjunction with the placement operation. This is the required procedure to properly perform the curing operation as outlined. This is defined as a quantity of required cubic yards and deck placement surface area.

Deck drain systems, miscellaneous blockouts, and other secondary components associated with the deck unit are identified at the time of the quantity takeoff procedure.

Parapet

The parapet and median barrier construction operations are defined and quantified next. These components are detailed as lineal feet required, first by type, then by form surface area and required cubic yards. The secondary operations required for the formwork and concrete placement are identified also. Another method of parapet and median wall construction is by slipform methods, which require no conventional formwork. The slipform placement procedure is quantified in the required lineal feet, by wall type, and required cubic yard volume. The secondary operations associated with this procedure are also defined.

Controlling components

The primary controlling components of the superstructure unit, along with the secondary components thereto, are categorized in two sections. The deck unit and the parapet and median rail unit are defined by beam section as:

Precast beam unit, deck unit

1. Beam type, total length, and section size
2. Haunch section (defining)

3. Interior form system
    a. Conventional wood
        (1) Haunch formwork
        (2) Beam hangers
        (3) Bay width
        (4) Formwork section requirement
    b. Metal stay-in-place form system
        (1) Haunch depth
        (2) Beam hangers (if required)
        (3) Support angle
        (4) Required conventional form system
    c. Precast deck panel form system
        (1) Haunch depth
        (2) Bearing area
        (3) Nonshrink grout support
        (4) Beam hangers (if required)
        (5) Support angle
        (6) Required conventional form system
4. Overhang form system
    a. Conventional wood system
        (1) Support bracket spacing
    b. Metal gang system
5. Perimeter bulkhead, formwork
6. Interior bulkhead, formwork
    a. Plain surface
    b. Doweled surface
7. Expansion joint devices, bulkhead formwork
8. Required concrete placements
9. Screed rail and yokes
10. Finish machine setup, dry run
11. Deck preparation
12. Deck drains
13. Miscellaneous blockouts
14. Deck concrete placements
15. Cure deck concrete
16. Concrete, rub and patch
17. Required protection
    a. False decking
    b. Safety nets

Precast beam unit, parapet and median rail unit

1. Parapet rail, formwork
2. Parapet rail, blockouts
3. Parapet rail, concrete placement
4. Parapet rail, cure concrete
5. Parapet rail, rub and patch

6. Parapet rail, slipform
7. Median rail, formwork
8. Median rail, blockouts
9. Median rail, concrete placement
10. Median rail, cure concrete
11. Median rail, rub and patch
12. Median rail, slipform

Structural steel beam unit, deck unit
1. Girder type, total length, and section size
2. Haunch section (defining)
3. Interior form system
    a. Conventional wood
        (1) Haunch formwork
        (2) Girder hangers
        (3) Bay width
        (4) Formwork section requirement
    b. Metal stay-in-place form system
        (1) Haunch depth
        (2) Girder hangers (if required)
        (3) Support angle
        (4) Required conventional form system
    c. Precast deck panel form system
        (1) Haunch depth
        (2) Bearing area
        (3) Nonshrink grout support
        (4) Girder hangers (if required)
        (5) Support angle
        (6) Required conventional form system
4. Overhang form system
    a. Conventional wood system
        (1) Support bracket spacing
    b. Metal gang system
5. Perimeter bulkhead, formwork
6. Interior bulkhead, formwork
    a. Plain surface
    b. Doweled surface
7. Expansion joint devices, bulkhead formwork
8. Required concrete placements
9. Screed rail and yokes
10. Finish machine setup, dry run
11. Deck preparation
12. Deck drains
13. Miscellaneous blockouts
14. Deck concrete placements
15. Cure deck concrete

16. Concrete, rub and patch
17. Required protection
    a. False decking
    b. Safety nets

Structural steel unit, parapet and median rail unit

1. Parapet rail, formwork
2. Parapet rail, blockouts
3. Parapet rail, concrete placement
4. Parapet rail, cure concrete
5. Parapet rail, rub and patch
6. Parapet rail, slipform
7. Median rail, formwork
8. Median rail, blockouts
9. Median rail, concrete placement
10. Median rail, cure concrete
11. Median rail, rub and patch
12. Median rail, slipform

Box Girder Unit

A cast-in-place reinforced concrete superstructure unit requires specific detailing of special operation components. As the superstructure is built, it must be temporarily supported until all the work operations have been performed. Unlike a conventional superstructure unit, there is no skeletal framework of a beam structure unit to support the deck unit. This type of structure incorporates the structural design and integrity of the beam unit within the deck unit, as a combined section.

The quantity takeoff of the cast-in-place superstructure begins with the temporary falsework requirements and progresses upward throughout the structural unit, defining and identifying each integral component.

The formwork system used, along with internal structural components, varies somewhat with this form of construction operation because of the design characteristics and complexity of the structure itself. The superstructure unit must be completed to a point of designed structural integrity in order to be self-supporting. This is defined not necessarily as the entire structure but at a minimum as a span unit from centerline to centerline of substructure bearing.

The quantity takeoff section describes the bottom deck requirements, continues to the wall units, both interior and exterior, and finalizes with the top deck section. Within the interior section are components of interior diaphragms, anchor blocks, and end wall units. The top deck section has the components of the interior deck section and exterior cantilevered wing sections. Finishing out the deck unit are the components of parapet and median rail.

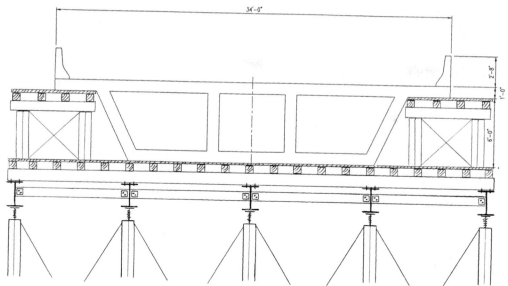

Box girder section and falsework.

Temporary falsework requirements

The temporary structural falsework and scaffolding is identified as a specific component for each condition. This varies based on the actual site conditions and the imposed design loads of the concrete superstructure unit, including formwork and equipment, and is designed accordingly.

The falsework and scaffolding component must be founded on a suitable and stable foundation, capable of supporting the imposed loads. This section defines and quantifies only the required falsework system, not the temporary supportive piling or grade work components associated below the temporary falsework. In other words, any required temporary piling, foundation preparation, or special grade requirements are quantified and calculated under the appropriate work operation as defined in the specific sections.

Situated on the structural falsework units is the temporary falsework deck. This is defined as the horizontal work platform to which the actual cast-in-place superstructure unit is constructed. The platform area is identified as the component contact area required for the bottom deck of the cast-in-place superstructure section itself, and the additional work area needed for the upper wing section and top deck support.

These operations conclude the required components necessary for a falsework system. The takeoff includes all operations required to construct and remove and dismantle the temporary falsework system, keeping in mind that the entire falsework system must be removable after construction of the permanent deck section without hindrance or damage to the structure.

Form system components

The form system components for the cast-in-place structure are identified and quantified next. The quantity takeoff for the cast-in-place box girder section is identified by component operation and defined by placement section. The box girder section can normally be constructed in stages, utilizing vertical and horizontal construction joints, and bulkheads. The formwork requirement is controlled by these defined joint areas.

Since the temporary falsework deck is constructed to the specific grade and profile requirements for the bottom of the superstructure, there is no requirement for horizontal bottom formwork for the bottom deck section. The sidewall formwork sets directly on the temporary falsework deck. A required formline may exist for the vertical edges of the bottom deck and wall sections. The quantity is defined as the contact surface area of the required formwork for the perimeter, or interior, bulkhead form for this component.

Prior to the concrete placement operation, the bottom deck area is prepared and cleaned. The bottom deck preparation component is defined in a calculated quantity of surface square feet, for the surface area of concrete contact.

The concrete volume requirement for the bottom deck is calculated and quantified in cubic yards of required concrete.

The sidewall formwork component is defined as the required concrete surface contact area of the sidewall section area. This can be identified as an exterior wall component or an interior wall component. The formwork is quantified by required surface area for both placement and removal.

The placement of concrete within the side and interior wall panels is quantified in a required volume measurement of cubic yards.

Interior components

The interior components of a cast-in-place superstructure deck unit are defined as diaphragms, anchor blocks, and end walls. The diaphragm panels are structural interior components that interlock or stiffen the box section, much like the diaphragms in the beam structure unit. The diaphragm components takeoff is performed by identifying the concrete surface contact area of each diaphragm component in a unit measurement of square feet.

The concrete placement for the diaphragm components is the required volume of specified concrete, in a unit measurement of cubic yards.

The end wall components are identified as the closure ends of each cast-in-place box girder section. The quantity is defined as the concrete surface contact area in a unit measurement of square feet, for the formwork placement and removal.

The volume of required concrete placed within the end wall components is defined and calculated in a unit measurement of cubic yards for each end wall panel.

The anchor block components are defined as the interior anchorage areas, required by design, for the posttensioned end anchorage assemblies. The formwork required is calculated by the concrete surface contact area for the placement and removal operations.

The concrete placement quantity is identified in a volume measurement of cubic yards for each required anchor block component.

Controlling components

The primary controlling components of the cast-in-place superstructure unit, along with the secondary components thereto, are categorized as this section.

Cast-in-place box girder unit

1. Temporary falsework and scaffolding
2. Temporary falsework deck section
3. Box section form system
   a. Wall panels
   b. Diaphragms
   c. End walls
   d. Anchor blocks

Superstructure Deck Section

The conventional wood formwork system requires a subunit, or skeletal system, to support the wood skin between the interior beam bays. This is normally constructed of structural wooden joist waler, or double ledger members and stringers. The framework system is supported from the beam members by special embeds and beam hangers. Keep in mind, the wood system is temporary and must be built to enable easy removal, from below, after the concrete placement and curing have been completed.

The quantity takeoff of the interior deck system for concrete beams is determined by the following procedures.

Conventional form system: haunch section

The haunch section of the conventional form system requires a calculation of the length of each beam, multiplied by the quantity of beams, multiplied by the number of beam sides requiring a haunch. This calculation method should be performed independently by span, owing to the variable haunch dimensions that may exist within different spans.

A category will be outlined for the calculation of the average rise dimension of the haunch, within each span, i.e., 0″–2″, 2″–4″, 4″–6″, etc. This will give a determination as to the craft-hour consumption requirement for each rise. The craft-hour production factor is based on the consumed time duration required to build and install the vertical haunch form, by category size. This is quantified in required lineal feet.

The outline requirement for the quantity takeoff of the conventional haunch component is listed as follows:

1. Beam type
2. Haunch type
3. Average haunch dimension per span
4. Lineal feet of required haunch, by rise

The quantity takeoff for the haunch is in required lineal feet, by average dimension per span, and summarized for the bridge structure, by rise, as:

1. Conventional haunch, 0"–2", lf
2. Conventional haunch, 2"–4", lf
3. Conventional haunch, 4"–6", lf

Conventional form system: beam hanger assemblies

The beam hangers are required devices that support, attach, or anchor the form system to the beam structure. They are classified as an embedded consumable item, incidental to the form system. There are two primary types of beam hanger assemblies, for both the full and half hanger devices. They are rated by the capacity of support to the imposed loads.

One type of full hanger assembly consists of a flat steel plate that transversely spans the top beam flange, a bracket or connection point that the waler support rods attach to, and two adjustable support coil rods with an upper and lower nut. The coil rod and lower adjusting nut are reusable and salvageable items. Another type, for precast beams, is a device that attaches to the reinforcing steel stirrups cast in the top flange. This is a singular device for each side of the beam and can be calculated as either a pair for a full hanger assembly or a singular device as a half hanger assembly.

A half hanger device is merely a half of a full hanger device, used for a one-sided forming requirement from the beam, which the overhang formwork system requires with a metal or precast interior form system. With this system, the metal or precast interior bay form systems support themselves from a different type of hardware rather than the beam hanger devices. The exterior overhang form system, however, still requires a form of hanger device.

The outline requirement for the quantity takeoff of the conventional beam hanger component is listed as follows:

1. Beam type
2. Hanger device type
3. Capacity load requirement
4. Required full hanger assemblies, each
5. Required half hanger assemblies, each
6. Length of support rods

The beam hanger device and support rods are required for every transverse waler location unless common longitudinal double ledger wales are used. The longitudinal waler system allows a single transverse joist option rather than the double needed for the conventional method. This option is analyzed as the most cost-efficient to both labor and material.

The spacing of the transverse waler is a variable controlled by a series of factors. The width of the interior deck bays is the controlling ingredient to the calculation. This span dimension, factored in with the imposed loading to the formwork from the deck, will determine (1) the dimensional size of waler required and (2) the spacing of the beam hanger devices that support the decking wales.

As part of the beam hanger device, the adjustable coil-rod assembly sets the required formwork elevation for the variable haunch dimension.

The quantity takeoff for beam hanger assemblies is calculated by the length of each beam requiring the full hangers, divided by the defined spacing of the transverse wales plus one, summarizing each required beam quantity.

The half beam hanger assemblies are calculated by the length of each beam (normally the exterior beams), requiring the half hangers, divided by the defined spacing of the overhang brackets plus one, summarizing each required beam quantity.

The calculating of the beam hanger assembly requirements is in conjunction with the interior and exterior form system calculations, based on load versus span requirements and criteria.

The required quantity is outlined as:

1. Conventional full beam hanger assembly, each
2. Conventional half beam hanger assembly, each

Conventional form system: interior deck formwork

The operation of the interior deck formwork includes, for quantity takeoff, the following components: the transverse walers and double ledgers, the longitudinal or transverse stringers, and the decking material. The interior bay forming is categorized based on the reuse ability of the formwork. Depending on the forming conditions of the deck, the amount of reusable lumber after cutting and fitting will constitute the scale. The factoring scale is set at 1 to 5, with 1 (reuse) being severe and 5 (reuses) being the ideal condition. For bidding purposes, I do not recommend the life expectancy of form lumber past five uses.

The outline requirements for the quantity takeoff of the interior deck forming system are listed as:

1. Bay width
2. Design load factor, dead plus live load, square foot contact

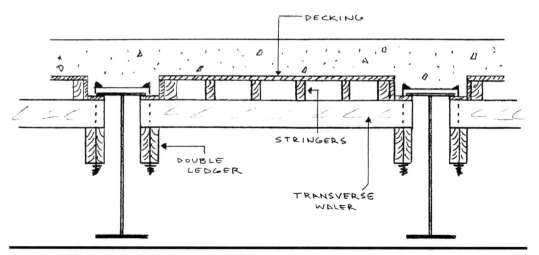

Conventional deck with wood formwork.

3. Joist waler or double ledger, dimension size and spacing
4. Stringer requirements, longitudinal and transverse
5. Square feet of required form surface (gross)

The quantity takeoff for the interior formwork system starts with defining the load criteria, to develop the required formwork cross section.

Walers, double ledger members, stringers. The order of construction for the skeletal framework form system is the beam hanger assemblies, the waler or double ledger members, the stringers, and the decking material. The determination for the transverse waler spacing and, directly, the beam hanger assemblies, is based on the imposed load criteria to the formwork surface. The first determination is the span dimension. To determine the waler spacing for the required design load, this span dimension is calculated based on the assumed dead load, of unit weight of the concrete and rebar, of 150 pounds per cubic foot. Added to this is an assumed live load factor of 50 pounds per square foot plus a formwork weight factor. From this cubic foot value, the imposed pounds for the thickness per square foot of deck is determined, including the impact and safety load factors.

The designed value for total pounds per square foot of form surface is now calculated into the formwork system. From the listed tables and charts in Chap. 3, the capacity requirement of beam hanger assemblies, the hanger and waler spacing dimensions, the dimensional size requirement for the waler members, and the dimensional size and spacing of the stringers can be determined. Once the skeletal framework for each bay has been analyzed, the $3/4''$ form grade plywood decking material can be summarized.

The quantity takeoff requirement, once the dimensional size and spacing criteria have been determined, is listed as:

1. Transverse waler, double ledger:
    Dimensional section _____
    Length per member
    Quantity per interior bay
    Summation of each interior bay

2. Longitudinal stringer members
    Dimensional section _____
    Length of each interior bay
    Quantity per interior bay
    Summation of each interior bay

3. Plywood decking material, net area
    Average width of each interior bay
    Length of each interior bay
    Summation of each interior bay (in square feet)

The calculations for the takeoff components are defined as:

The waler members are calculated for the required length (the span width dimension between beams), by the required number of walers per interior bay (the span length divided by the hanger spacing plus one), summarizing this calculation for each interior bay, listing the quantity requirement in a unit of each by dimensional section size.

The stringer members are calculated by the length of each bay (the length of the beam span, center to center of bearing), by the number of stringers per interior bay (the interior bay width divided by required spacing plus one), summarizing this calculation for each interior bay, listing the quantity requirement in a unit of lineal feet, by dimensional section size.

The plywood decking material is calculated by width of each interior bay (opening size between the beams), multiplied by the length of each interior bay (length of the beam span, center to center of bearing), summarizing this calculation for each interior bay, listing the quantity requirement in a unit of square feet.

This is the material quantity requirement calculation for embedded hardware and formwork lumber for the net square foot area, accounting for the reuse factor of formwork life.

NOTE: Other than custom or required site fabrication, the fabrication and building of waler or stringer members is justified for craft-hour consumption with the formwork itself, as a reusable asset, not the deck structure unit.

NOTE: If the lumber is calculated for the specific project, it should be realized that the entire deck unit may not be required to be formed at one time; this lumber may be recycled with respect to lumber allocation, cure requirements, and schedule of completion.

The quantity takeoff requirement for the craft-hour consumption is based on the gross formwork area in square feet. This calculation is performed by multiplying the transverse width dimension (the centerline of the exterior beam to the centerline of the exterior beam) by the length of the structure deck span (the centerline to centerline of beam bearing).

Interior deck surfaces. The summation of required interior deck surface is outlined in gross square feet, in a category by span dimension for placement and removal, as:

1. Conventional formwork system, interior deck formwork placement, span dimension to 4'
   a. Conventional formwork system interior deck, formwork removal, span dimension to 4'
2. Conventional formwork system, interior deck formwork placement, span dimension 4' to 6'
   a. Conventional formwork system interior deck, formwork removal, span dimension 4' to 6'
3. Conventional formwork system, interior deck formwork placement, span dimension 6' to 8'
   a. Conventional formwork system interior deck, formwork removal, span dimension 6' to 8'
4. Conventional formwork system, interior deck formwork placement, span dimension 8' to 10'
   a. Conventional formwork system interior deck, formwork removal, span dimension 8' to 10'

These takeoff quantities are then summarized, by dimensional section size, by span for each bridge structure. The removal of the haunch formwork is considered incidental to and is calculated as part of the interior deck removal factor.

For a more in-depth study and complete explanation of formwork design, planning, loads and pressures, design tables, and other related criteria, refer to ACI Publication SP-4, "Formwork for Concrete."

The metal stay-in-place deck forming system consists of structural corrugated metal pans, designed and custom-made for each specific interior bay, attached permanently to the beam members. The metal pans are attached to a steel angle, which is supported from the beam members by transverse beam hangers, or a weld to an embedded steel plate cast into the top of the beam. This system becomes a permanent part of the superstructure unit, spanning each interior bay, and requires no removal.

Metal stay-in-place formwork system: beam attachment device

There are two types of beam attachment methods for the support angle of the metal stay-in-place formwork system, both using a direct weld procedure for

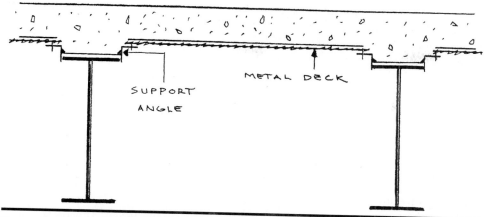
Permanent metal deck with stay-in-place formwork.

permanent attachment. With precast beams (the deck section being composite to the beam) embedded steel clips can be cast into the top surface of the beam for the direct weld attachment of the support angle to the beam, or a flat steel transverse plate spanning the top of the beam for the weld attachment of the support angle.

For structural steel girders, either a direct weld of the support angle to the girder (the deck section being composite to the girder) or a flat steel transverse plate spanning the top of the girder for the weld attachment of the support angle (the deck unit being noncomposite to the girder) is used. The required attachment points for each beam and girder are quantified as a unit of each, by type.

As defined with the conventional beam hanger assemblies, the spacing requirement for the metal stay-in-place deck form system is based on the implied load and span width calculation ratio and varies per deck unit.

Therefore, the attachment procedure for the support angle is identified and quantified as a independent component.

The welding procedure is performed equally, regardless of the beam hanger method.

The quantity outline requirement for the beam and girder attachment is listed as:

1. Type of beam structure
2. Width of top flange
3. Method of attachment
4. Quantity of direct welds

The quantity summation for the beam hanger attachment device and type is performed by:

1. Precast clip: the length of beam, divided by the required attachment spacing on the beam, plus one, summarizing each beam side requiring the clips, in a quantity of each.
2. Direct weld: the length of girder, divided by the required attachment spacing on the girder, plus one, summarizing each girder side requiring the attachment, in a quantity of each.
3. Support hanger: the length of beam and girder, divided by the required hanger spacing on the beam and girder, plus one, summarizing each girder requiring the hanger, in a quantity of each, and the width of the girder or beam must be identified for the length of the saddle hanger.

Metal stay-in-place formwork system: support angle

The metal support angle is the structural component which the metal deck pans are supported by and attached to. The support angle establishes the haunch area and formwork deck elevations. The angle is attached to the beam members by a weld to the beam clips, girder flanges, or hanger assemblies.

The metal support angle is attached to the beam and girder members fully around the perimeter of and within each interior bay, both longitudinally and transversely, including areas adjacent to any diaphragms composite to the deck and the surface edge contact area of precast beams, concrete diaphragms, structural steel girders, or steel diaphragms or cross frames that are in direct contact with the metal deck form system and require a structural permanent support.

The support angle is an L-shaped structural member. The dimensional size of the angle is governed by the required haunch dimension.

The takeoff quantity outline required for the structural support angle component is listed as:

1. Type of beam structure
2. Method of attachment
3. Lineal feet of required angle
4. Dimensional section size(s)

The takeoff procedure is accomplished by calculating the actual lineal feet of support angle required, by size, for each interior bay section. This is the complete interior perimeter of beam and diaphragm, or girder and cross frame contact surface for each bay.

The required production unit for the structural support angle is defined as the craft-hour consumption for the lineal feet of angle to be handled, by attachment method. The craft-hour usage of handling the angle components varies depending on the attachment method required. Added to this is the attachment procedure defined with the metal stay-in-place hanger component.

Within this component operation, any temporary scaffolding and staging required for access for the performance of the work is included as a usage factor.

The metal support angle quantity takeoff is summarized by dimensional section size relative to the haunch, by lineal feet, by attachment method, as:

1. Stay-in-place structural support angle, precast beams, embedded clip, angle size _____, lf
2. Stay-in-place structural support angle, precast beams, saddle hanger, angle size _____, lf
3. Stay-in-place structural support angle, structural girder, direct weld, angle size _____, lf
4. Stay-in-place structural support angle, structural girder, saddle hanger, angle size _____, lf

Metal stay-in-place formwork system: deck pans

The stay-in-place metal deck pan system is calculated by two methods of measurement, interior gross deck surface area and net interior deck surface area.

The gross interior deck surface area is defined and calculated as the width of each deck section from the centerline of one exterior girder to the centerline of the other exterior girder, by the length of the deck span from centerline of bearing to centerline of bearing. This unit measurement, in square feet, is required for the craft-hour production factoring procedure.

The net interior deck surface area is defined as the actual dimensions of each interior bay area. The quantity calculation is performed by multiplying the opening width of each interior bay by the length of each interior bay, from centerline of bearing to centerline of bearing, within each deck section. This unit of measurement, in square feet, is the actual material requirement of the deck pan system.

The required outline of the quantity takeoff for the metal stay-in-place deck pans is listed as:

1. Gross interior deck surface area
2. Net interior deck surface area
3. Type and section of metal pan
4. Attachment method
5. Conventional form requirements

The metal stay-in-place pan components are attached to the support angle by two methods, self-tapping screws and tack welding. The self-tapping screw method is the most used method. The required craft-hour consumption is factored by the actual time required to properly handle and attach the metal stay-in-place pans, by a unit of craft-hours per square feet.

For areas that cannot be formed using the metal stay-in-place system, it may be required to utilize the conventional formwork system. If this occurs, the procedures outlined in the appropriate section are implemented.

The quantity takeoff for the required metal stay-in-place pan components is summarized in a unit measurement of square feet by type of section as:

1. Metal stay-in-place deck pans: gross interior sf _____; net interior square feet _____

Precast panels

The precast concrete deck panel system also is a permanent component to the superstructure. Unlike the wood system and the metal stay-in-place system, the precast concrete deck panels are structural members, designed as part of the composite deck, replacing a portion of the cast-in-place concrete.

The concrete deck panels normally are supported directly from the top of the concrete beam members, spanning the interior bays between the beam members. Because of a somewhat irregular horizontal surface, or camber requirements, the precast members are set onto a structural support, a bituminous material bearing pad, or a structural grout pad. On occasion, however, depending on the haunch design, the precast deck panel forming systems may require a cantilevered type of support system. This is accomplished by a steel angle which is supported from the beam member by either transverse steel beam hanger devices or a weld attachment to an embedded steel plate cast within the beam.

Precast panel formwork system: bituminous bearing pads

The bituminous bearing pad components are strips of built-up bituminous material which act as a cushioned surface for the precast concrete panel members, and also an elevation or haunch buildup, to a certain dimensional limitation.

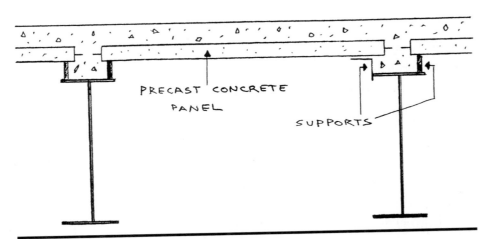

Permanent precast deck with concrete panel formwork.

Owing to camber variations and differing haunch dimensions, the bituminous material can be installed in a series of different sectional sizes to produce the variable elevation differences required throughout the length of the beam member.

The required quantity takeoff outline for the bituminous bearing pads is defined and listed as:

1. Type of beam unit
2. Dimensional section(s) of pad required
3. Required lineal feet of pad, bearing section

The requirement of lineal feet of pad, by bearing section, is defined as the total dimension in specified thickness of bearing pad, regardless of the number of actual layers. The rise categories are identified as: bituminous bearing pad, 0″–1″, and bituminous bearing pad, 1″–2″. When the elevation section is extended past the design criteria of the bituminous pad, a nonshrink grout support pad is built first.

The craft-hour consumption is factored by the unit measurement of lineal feet of bituminous bearing pad installed within a given span. The quantity takeoff is performed and calculated by the lineal feet of bearing pad section required, for each interior bay, by span. This calculation is formulated for the edge surface contact area required for precast panel support, on each beam member. This is normally one longitudinal edge of each exterior beam member and both longitudinal edges of each interior beam member.

The quantity takeoff for the bituminous bearing pad components is defined in a unit measure of lineal feet of required pad, categorized by section dimension, as:

1. Precast panel bituminous bearing pad, 0″–1″, lf
2. Precast panel bituminous bearing pad, 1″–2″, lf

Precast panel formwork system: nonshrink grout pads

The nonshrink grout pads are a structural component required for support to the precast panel members when the haunch section requires a thicker design than that which can be obtained with the bituminous bearing pads.

The quantity takeoff outline of the required nonshrink grout pads is defined as:

1. Type of beam structure
2. Dimensional section of grout pad
3. Required lineal feet of grout pad

The performance of the operation includes the longitudinal edge forming, the placement of the nonshrink grout, and the removal of the longitudinal formwork. These component operations, however, are not factored as separate items but are identified as the consumed craft-hours required to perform these operations, per lineal feet of required grout pad, categorized by height.

The quantity takeoff procedure is performed by calculating the lineal feet of grout pad section required, for each interior bay, by span. This calculation is formulated for the edge surface contact area required for precast panel support, on each beam member. This is normally one longitudinal edge of each exterior beam member and both longitudinal edges of each interior beam member.

The quantity takeoff for the grout bearing pad support components is defined in a unit measure of lineal feet of required pad, categorized by section dimension, as:

1. Precast panel grout pad support, 0″–3″, lf
2. Precast panel grout pad support, 3″–6″, lf

Precast panel formwork system: hanger assemblies and devices

There are two types of beam attachment methods for the support angle of the precast concrete deck panel system, both using a direct weld procedure for permanent attachment. With precast beams (the deck section being composite to the beam) embedded steel clips can be cast into the top surface of the beam for the direct weld attachment of the support angle to the beam, or a flat steel transverse plate spanning the top of the beam for the weld attachment of the support angle.

For structural steel girders, either a direct weld of the support angle to the girder (the deck section being composite to the girder) or a flat steel transverse plate spanning the top of the girder for the weld attachment of the support angle (the deck unit being noncomposite to the girder) is used. The required attachment points for each beam and girder are quantified as a unit of each, by type.

As defined with the conventional beam hanger assemblies, the spacing requirement for the precast deck panel system is based on the implied load and span width calculation ratio and varies per deck unit.

Therefore, the attachment procedure for the support angle is identified and quantified as an independent component.

The welding procedure is performed equally, regardless of the beam hanger method.

The quantity outline requirement for the beam and girder attachment is listed as:

1. Type of beam structure
2. Width of top flange
3. Method of attachment
4. Quantity of direct welds

The quantity summation for the beam hanger attachment device and type for the precast concrete deck panel system is performed by:

1. Precast clip: the length of beam, divided by the required attachment spacing on the beam, plus one, summarizing each beam side requiring the clips, in a quantity of each

2. Direct weld: the length of girder, divided by the required attachment spacing on the girder, plus one, summarizing each girder side requiring the attachment, in a quantity of each

3. Support hanger: the length of beam and girder, divided by the required hanger spacing on the beam and girder, plus one, summarizing each girder requiring the hanger, in a quantity of each

Precast panel formwork system: metal angle supports

The precast panel metal support angle, like the metal stay-in-place form system, is the structural component for which the precast concrete deck panels are supported. The support angle establishes the haunch area and formwork deck elevations. The angle is attached to the beam members by a weld to the beam clips, girder flanges, or hanger assemblies.

The metal support angle is attached to the beam and girder members fully around the perimeter of and within each interior bay, both longitudinally and transversely, including areas adjacent to any diaphragms composite to the deck. The surface edge contact area of precast beams, concrete diaphragms, structural steel girders, or steel diaphragms or cross frames that are in direct contact with the precast concrete deck panel form system and require a structural permanent support.

The support angle is an L-shaped structural member. The dimensional size of the angle is governed by the required haunch dimension.

The takeoff quantity outline required for the structural support angle component is listed as:

1. Type of beam structure
2. Method of attachment
3. Lineal feet of required angle
4. Dimensional section size(s)

The takeoff procedure is accomplished by calculating the actual lineal feet of support angle required, by size, for each interior bay section. This is the complete interior perimeter of beam and diaphragm, or girder and cross frame contact surface for each bay.

The required production unit for the structural support angle is defined as the craft-hour consumption for the lineal feet of angle to be handled, by attachment method. The craft-hour usage of handling the angle components varies depending on the attachment method required. Added to this is the attachment procedure defined with the precast concrete deck panel hanger component.

Within this component operation, any temporary scaffolding and staging required for access for the performance of the work is included as a usage factor.

The metal support angle quantity takeoff is summarized, by dimensional section size relative to the haunch, by lineal feet, by attachment method, as:

1. Precast concrete deck panel support angle, precast beams, embedded clip, angle size _____, lf
2. Precast concrete deck panel support angle, precast beams, saddle hanger, angle size _____, lf
3. Precast concrete deck panel support angle, structural girder, direct weld, angle size _____, lf
4. Precast concrete deck panel support angle, structural girder, saddle hanger, angle size _____, lf

Precast panel formwork system: deck panel units

The precast concrete deck panel system is calculated by two methods of measurement, interior gross deck surface area and net interior deck surface area.

The gross interior deck surface area is defined and calculated as the width of each deck section from the centerline of one exterior girder to the centerline of the other exterior girder, by the length of the deck span from centerline of bearing to centerline of bearing. This unit measurement, in square feet, is required for the craft-hour production factoring procedure.

The net interior deck surface area is defined as the actual dimensions of each interior bay area, plus the panel overhang dimension for each beam contact edge. The quantity calculation is performed by multiplying the full required width of each panel, for each interior bay, by the length of each interior bay, from centerline of bearing to centerline of bearing, within each deck section. This unit of measurement, in square feet, is the actual material requirement of the concrete deck panel components.

The required outline of the quantity takeoff for the precast concrete deck panel components is listed as:

1. Gross interior deck surface area
2. Required net interior deck surface area
3. Dimensional section of precast panel
*4. Reinforcing steel replacement quantity (if any)
*5. Concrete volume replacement quantity (if any)
6. Conventional formwork requirement

---

*Asterisks denote control items for operations.

NOTE: When precast concrete deck panels are designed as an alternate deck forming system, the deck section normally is designed and quantified as a conventional, full-depth concrete component. If the precast concrete deck panels are intended as an option, the designed section volume of concrete and reinforcing steel, replaced by the concrete panel components, must be deleted from the quantity calculations.

The precast deck panel components are erected by placement directly on the bituminous bearing pads, grout pads, or support angle. The required craft-hour consumption is factored by the actual time required to properly erect the precast concrete deck panel units, by a unit of craft-hours per square feet.

For areas that cannot be formed using the precast concrete deck panel system, it may be required to utilize the conventional formwork system. If this occurs, the procedures outlined in the appropriate section are implemented.

The quantity takeoff for the required concrete deck panel components is summarized in a unit measurement of square feet by type of section as:

1. Precast concrete deck panel unit: gross interior sf _____; net interior square feet _____.

Cantilevered deck

The exterior cantilevered forming system is not as optional as the interior system. The exterior concrete deck portion, which extends beyond the exterior beam members, is a cantilevered structural component of the deck, and aesthetically prohibits any formwork system to remain in place.

The overhang formwork system must be supported from the exterior of the beam unit members and must be structurally able to support not only the concrete and reinforcing load but the weight of the deck finishing equipment as well.

Two types of formwork systems are available for this operation. The first is the conventional overhang system, consisting of individual adjustable metal brackets that hang from, extend outward, and are supported from the exterior beam and girder. From these, a structural wooden form system, similar to the conventional interior bay system, is constructed. Again, this system must be removable once the concrete is placed and cured. This system is attached to the beam members by embedded item components of beam hanger assemblies or anchor devices.

The second type of exterior overhang form system is a metal gang system. This type of system also is hung from and supported by the exterior beam members but in modular units, usually in lengths of 10′. These are entire steel units, with a wide range of adjustability. They are attached to the bottom flange of the beam or girder member and must be removed in the same manner.

With the precast concrete beam unit, there must be a method of supporting the chosen form system from the beams. This can be accomplished by inserts that can be cast into the beam members during manufacturing, or by saddle hangers laid transversely across the top of the beam, or by rebar stirrup connectors that attach to the embedded rebar beam stirrups. Regardless of the method, these attachments and anchors are referred to as consumable embedded items, incidental to the form system.

Exterior overhang form system: haunch section

The haunch section of the exterior overhang form system requires a calculation of the length of each exterior beam member. This calculation method should be performed independently by span, owing to the variable haunch dimensions that may exist within different spans.

A category is outlined for the calculation of the average rise dimension of the haunch, within each span, i.e., 0"–2", 2"–4", 4"–6", etc. This gives a determination as to the craft-hour consumption requirement for each rise. The craft-hour production factor is based on the consumed time duration required to build and install the vertical haunch form, by category size. This is quantified in required lineal feet.

The outline requirement for the quantity takeoff of the exterior overhang haunch component is listed as follows:

1. Beam type
2. Haunch type
3. Average haunch dimension per span
4. Lineal feet of required haunch, by rise

The quantity takeoff for the haunch is in required lineal feet, by average dimension per span, and summarized for the bridge structure, by rise; as:

1. Exterior overhang haunch, 0"–2", lf
2. Exterior overhang haunch, 2"–4", lf
3. Exterior overhang haunch, 4"–6", lf

Exterior overhang form system: beam hanger assemblies

The beam hanger assemblies are required devices that support, attach, or anchor the exterior overhang form system to the beam structure. They are classified as an embedded consumable item, incidental to the form system. There are two primary types of beam hanger assemblies, for both the full and half hanger devices. They are rated by the capacity of support to the imposed loads.

One type of full hanger assembly can consist of a flat steel plate that transversely spans the top beam flange, a bracket or connection point that the bracket support rods attach to, and two adjustable support coil rods with an upper and lower nut. The coil rods and lower adjusting nut are a reusable and salvageable item. Another type, for precast beams, is a device that attaches to the reinforcing steel stirrups cast in the top flange. This is a singular device for each side of the beam and can be calculated as either a pair for a full hanger assembly or a singular device for a half hanger assembly.

A half hanger device is merely a half of a full hanger device, used for a one-sided forming requirement from the beam, which the overhang formwork system requires with a metal or precast interior form system. With this system, the metal or precast interior bay form systems support themselves from a different type of hardware rather than the beam hanger devices. The exterior overhang form system, however, still requires a form of hanger device.

The outline requirement for the quantity takeoff of the exterior beam hanger component is listed as follows:

1. Beam type
2. Hanger device type
3. Capacity load requirement
4. Required full hanger assemblies, each
5. Required half hanger assemblies, each
6. Length of support rods

NOTE: If a required full beam hanger assembly has been calculated with the interior conventional formwork system for the exterior beam members, do not calculate this item component again as an exterior overhang device.

The exterior beam hanger device and support rods are required for every overhang support bracket location. The spacing of the overhang support brackets is a variable and is controlled by a series of factors. The interior deck bays are the controlling ingredient to the calculation when full beam hangers are used in conjunction with a conventional interior system. Provided that the interior hanger spacing can support the imposed loads of the exterior spacing, the overhang brackets can utilize this spacing of hanger assemblies.

As part of the beam hanger device, the adjustable coil-rod assembly sets the required exterior formwork elevation for the variable haunch dimension.

The quantity takeoff for the exterior beam hanger assemblies is calculated by the length of each exterior beam requiring the hangers, divided by the defined spacing of the overhang brackets plus one, summarizing each required exterior beam quantity.

The half beam hanger assemblies are calculated by the same method.

The calculating of the exterior beam hanger assembly requirements is in conjunction with the interior and exterior form system calculations, based on load versus span requirements and criteria.

The required quantity is outlined as:

1. Exterior overhang full beam hanger assembly, each
2. Exterior overhang half beam hanger assembly, each

Exterior overhang form system: support brackets

The quantity takeoff for the exterior formwork system starts with defining the load criteria, to develop the required formwork and support of the exterior cross section.

The order of construction for the skeletal framework of the exterior overhang system is the exterior hanger assemblies, the overhang support brackets, the stringers, and the decking material. The determination for the exterior support brackets and, directly, the beam hanger assemblies, is based on the imposed load criteria to the formwork surface. The first determination is the overhang dimension. To determine the support bracket spacing for the required design load, the overhang dimension is calculated based on the assumed dead load, of unit weight of the concrete and rebar, of 150 pounds per cubic foot. Added to this is an assumed live load factor of 50 pounds per square foot plus a formwork weight factor. Added to the factor is the imposed load criterion of the deck finish machine to each exterior overhang section (normally the full operating weight of the finish machine is transferred to each overhang section equally). From this cubic foot value, the imposed pounds for the thickness per square foot of deck is determined, including the impact and safety load factors. This weight factor calculation directly determines the spacing requirement for both the exterior hanger assemblies and the exterior support brackets.

There does exist a possibility of compounding conflicts to the hanger spacing requirements with regard to the interior formwork load requirement versus the exterior requirements. This is due predominantly to the added weight of the finish machine to the exterior overhang system. When this situation occurs, the exterior beam hanger assembly requirement should be increased to accommodate this, rather than adding to the interior waler and form section requirements.

The rollover movement of the exterior beam must be analyzed when adding the imposed live and dead load weight to the overhang section. This can be overcome by temporary bracing to the beam structure, quantified under the beam structure section.

The outline requirements for the quantity takeoff of the exterior overhang brackets are listed as:

1. Overhang width
2. Design load factor, dead plus live load, square feet contact
3. Overhang support bracket spacing
4. Bracket type and rating

The takeoff procedure for calculating the quantity is identical to that of the beam hanger assembly requirement. The quantity takeoff requirements for the exterior support brackets, once the spacing criteria have been determined, are listed in the unit measurement of each, by bracket capacity, as:

1. Exterior support bracket, type _____, each

Exterior overhang form system: deck formwork

The operation of the exterior deck formwork includes, for quantity takeoff, the following components: the wooden bracket member, the longitudinal or transverse stringers, and the decking material. The exterior overhang forming is categorized based on the reuse ability of the formwork. Depending on the forming conditions of the deck, the amount of reusable lumber after cutting and fitting constitutes the scale. The factoring scale is set at 1 to 5, with 1 (reuse) being severe, and 5 (reuses) being the ideal condition. For bidding purposes, I do not recommend the life expectancy of form lumber past five uses.

The outline requirements for the quantity takeoff of the exterior deck forming system are listed as:

1. Overhang width, including work platform and handrail
2. Design load factor, dead plus live load, square foot contact
3. Stringer requirements, longitudinal and transverse
4. Square feet of required form surface (gross)

The designed value for total pounds per square foot of form surface is now calculated into the formwork system. From the listed tables and charts in Chap. 3, the capacity requirement of beam hanger assemblies, the overhang

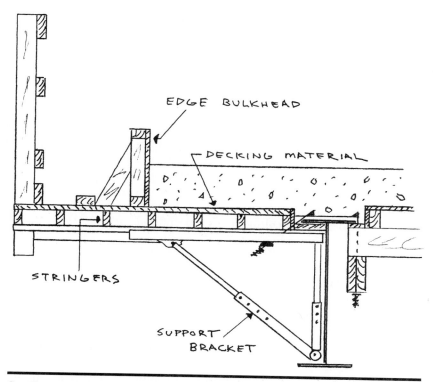

Cantilevered overhang section with wood formwork.

bracket spacing dimensions, the dimensional size requirement for the bracket member, and the dimensional size and spacing of the stringers can be determined. Once the skeletal framework for each overhang has been analyzed, the $\frac{3}{4}''$ form grade plywood decking material can be summarized.

The overhang system requires an additional operation performance requirement once the actual decking system has been completed. This is the walkway and handrail components. The walkway is calculated as additional deck formwork, and the handrail is calculated based on the length of the overhang system.

The quantity takeoff requirement, once the dimensional size and spacing criteria have been determined, is listed as:

1. Bracket member
   Dimensional section _____
   Length per member plus walkway
   Quantity per span side (equal to bracket)
   Summation of each span
2. Longitudinal stringer members
   Dimensional section _____
   Length of each exterior bay
   Quantity per exterior bay
   Summation of each exterior bay
3. Plywood decking material, net area
   Average width of each overhang plus walkway
   Length of each exterior overhang
   Summation of each exterior overhang

The required additional calculations for the walkway and handrail are considered as a singular component. The normal recommendation for the additional dimension requirements is 2.0′, horizontally past the vertical edge of the concrete deck section. This dimension must include the required walkway area for the work area and the area for the vertical formwork bulkheads, plus the vertical handrail section.

The calculations for the takeoff components are defined as:

The bracket members are calculated for the required length (the exterior width dimension plus the walkway dimension), by the required number of support brackets per side (the span length divided by the hanger spacing plus one); summarizing this calculation for each exterior overhang bracket, listing the quantity requirement in a unit of each by dimensional section size.

The stringer members are calculated by the length of each bay (the length of the beam span, center to center of bearing), by the number of stringers per overhang section (the exterior overhang width, plus the walkway, divided by required spacing plus one); summarizing this calculation for each exterior overhang, listing the quantity requirement in a unit of lineal feet, by dimensional section size.

The plywood decking material is calculated by width of each overhang section plus the walkway (concrete surface contact area plus the walkway), multiplied by the length of each exterior overhang section (length of the beam span, center to center of bearing); summarizing this calculation for each span, listing the quantity requirement in a unit of square feet.

This is the material quantity requirement calculation for embedded hardware and formwork lumber for the net square foot area, accounting for the reuse factor of formwork life.

NOTE: Other than custom or required site fabrication, the fabrication and building of bracket or stringer members is justified for craft-hour consumption with the form work itself, as a reusable asset, not the deck structure unit.

NOTE: If the lumber is calculated for the specific project, it should be realized that the entire deck unit may not be required to be formed at one time; this lumber may be recycled with respect to lumber allocation, cure requirements, and schedule of completion.

The quantity takeoff requirement for the craft-hour consumption is based on the gross formwork area in square feet, for each overhang. This calculation is performed by multiplying the required overhang width dimension (the centerline of the exterior beam to the exterior edge of the required formwork, including the walkway area) by the length of the structure deck span (the centerline to centerline of beam bearing).

The width of the gross overhang will affect the production unit of craft-hour consumption and, ultimately, the uniform loading requirements. For this, the overhang section is defined in two categories, overhang sections from 0' to 5', and overhang sections over 5' in width.

The summation of required overhang deck surface is outlined in gross square feet, in a category of overhang width:

1. Exterior overhang formwork system, deck formwork placement, width dimension to 5'
    a. Exterior overhang formwork system, deck formwork removal, width dimension to 5'
2. Exterior overhang formwork system, deck formwork placement, width dimension over 5'
    a. Exterior overhang formwork system, deck formwork removal, width dimension over 5'

These takeoff quantities are then summarized, by category, by span for each bridge structure. The removal of the haunch formwork is considered incidental to and is calculated as part of the exterior overhang system removal factor.

The overhang metal gang formwork system is quantified by the required unit components needed to perform the operation. The sectional units are defined in units of measurement in lineal feet of complete section, as required. The operation includes the placement, elevation adjustment, and removal of the required system.

Vertical bulkhead formwork system

The vertical bulkhead formwork system provides for the dimensional thickness of the structural concrete deck, defining the perimeter outline requirements of each continuous deck span and the interior joint areas and pouring sequences.

Perimeter bulkhead: formwork

The perimeter bulkhead is defined as the outside perimeter line of the total deck unit of the bridge structure. This is the vertical formwork located within the overhang section, being the edge of the deck surface and the end of the deck unit at each abutment.

The requirement of the quantity takeoff outline, for the perimeter vertical bulkhead formwork, is listed as:

1. Concrete deck thickness, at the form line
2. Perimeter dimensions

The required calculation procedure to determine the perimeter vertical formwork is performed by multiplying the thickness of the concrete deck, at the required form line, by the length of each perimeter line of the deck. This calculation is performed and summarized in a unit measurement of required square feet. The craft-hour consumption unit is factored for the full operation performance required to complete the task.

The quantity takeoff of the vertical perimeter formwork component is defined, in required square feet, as:

1. Perimeter bulkhead, formwork placement, sf
2. Perimeter bulkhead, formwork removal, sf

Interior bulkhead: plain, formwork

The interior vertical bulkheads, plain, are defined as the required interior deck form lines, or open deck joints, that require a nontied joint. These are considered as transverse or longitudinal, interior end of deck joints, or plain bulkhead joints.

The requirement of the quantity takeoff outline, for the plain, interior vertical bulkhead formwork, is listed as:

1. Concrete deck thickness, at the form line
2. Length of each required interior joint

The required calculation procedure to determine the vertical plain interior formwork is performed by multiplying the thickness of the concrete deck, at the required form line, by the length of each form line within the deck. This calculation is performed and summarized in a unit measurement of required square feet. The craft-hour consumption unit is factored for the full operation performance required to complete the task.

The quantity takeoff of the plain interior formwork component is defined, in required square feet, as:

1. Interior bulkhead, plain, formwork placement, sf
2. Interior bulkhead, plain, formwork removal, sf

Interior bulkhead, doweled: formwork

The vertical interior bulkheads, doweled, are defined as the required interior deck doweled construction joint form lines, or doweled deck joints, that require a tied joint. These are considered as transverse or longitudinal interior deck joints, or doweled construction bulkhead joints.

The requirements of the quantity takeoff outline, for the doweled, interior vertical bulkhead formwork, are listed as:

1. Concrete deck thickness, at the form line
2. Length of each required interior joint
3. Dowel, tie bar spacing (vertical and horizontal)

The required calculation procedure to determine the doweled, vertical interior formwork is performed by multiplying the thickness of the concrete deck, at the required form line, by the length of each form line within the deck. This calculation is performed and summarized in a unit measurement of required square feet. The craft-hour consumption unit is factored for the full operation performance required to complete the task.

The quantity takeoff of the vertical doweled interior formwork component is defined, in required square feet, as:

1. Interior bulkhead, doweled, formwork placement, sf
2. Interior bulkhead, doweled, formwork removal, sf

Deck unit: embedded hardware

The embedded item quantifying defines any incidental embeds required within the superstructure deck unit. These are summarized in a unit of each as:

1. Superstructure deck, embeds, ea

Expansion joint device: bulkhead formwork

The vertical expansion joint bulkheads are defined as the required expansion joint deck form lines. These are considered as transverse expansion deck bulkhead joints.

The requirement of the quantity takeoff outline, for the expansion joint vertical bulkhead formwork, is listed as:

1. Concrete deck thickness, at the form line
2. Length of each required expansion joint
3. Special formwork requirement
4. Type of joint required

The required calculation procedure to determine the vertical expansion joint formwork is performed by multiplying the thickness of the concrete deck, at the required form line, by the length of each expansion joint within the deck. This calculation is performed and summarized in a unit measurement of required square feet.

With this are the quantity takeoff of preliminary joint attachment, blockout, or embedded bolt locations. This operation is the performance of presetting the expansion joint device within the formwork, or the joint blockout with the bolt locations. This calculation is performed and summarized in a unit measurement of required lineal feet. The craft-hour consumption unit is factored for the full operation performance required to complete the task.

The quantity takeoff of the vertical expansion joint formwork component is defined, in required square feet, as:

1. Expansion joint bulkhead, formwork placement, sf
2. Expansion joint bulkhead, formwork removal, sf
3. Expansion joint, preset blockout, lf

Required placement locations

This quantity takeoff is merely the operation of defining each placement location, or pour sequence area, within the deck unit. These differ from the actual span locations owing to multiple placements within a specific span, or the continuous placements encompassing or continuing through to multiple spans.

These defined pour locations can be identified as required, being those designated owing to design criteria with respect to controlled beam and deck unit deflection, or as optional. The optional placement locations give the contractor the choice to make the actual defined, independent placement or combine with another. This is normally defined as an option merely for volume control of the placement rate of cubic yards per hour. When an optional placement area is defined, there is also the optional component of the construction joint, or interior bulkhead associated with the placement. This must be added to or deleted from that quantity component operation.

The defined area for each placement is each independent surface deck segment, located between and within any vertical bulkhead form lines.

The outline requirement for the quantity takeoff procedure is identified as:

1. Number of required or optional placement areas

The quantity takeoff of the deck concrete placement location component is defined, in a requirement of each, as:

1. Placement locations, each

Screed rail

The operation of the quantity takeoff for the required finish machine screed rail component is performed by first identifying the placement location areas within the deck unit. From this, the length of each placement location is defined.

Normally, it will be required to set a finish machine screed rail parallel to each placement location, parallel to the finish machine direction of placement. The screed rail is seated on yokes, or saddles, which are set to a preliminary grade elevation during this procedure.

The screed rail with yokes sets directly on the defined bulkhead formwork (perimeter or interior) or the beams and girders for the pilot placements, or directly on the adjacent cured concrete for the filler placement areas of the deck unit.

The operation of the finish machine screed rail is categorized in two components: (1) screed rail, bulkhead placement, or (2) screed rail, deck surface placement. This separate defining accounts for the additional work required to place the screed rail on the bulkhead formwork and the yoke placement within the bulkhead formwork.

For the approach area and runoff area of each placement section, it is recommended to add 20' of screed rail, per side, to each end of the placement length. This additional length must be determined and quantified for either each or specific placements, or once for the structure.

The main consumption requirement for this operation includes the outlined work necessary to set the yokes and saddles within the bulkhead formwork or deck, place and rough grade the screed rail, and remove the system after the placement.

The quantity outline procedure for the finish machine screed rail component is defined as:

1. Required placement locations
2. Length of placement area
3. Additional rail required
4. Type of screed rail component, bulkhead or surface

The calculation for the required quantity of screed rail is performed by category component type, by summarizing the length of each side of the specific placement area and adding the additional required rail for the approach and runoff area, normally 80 lineal feet. Then each placement area is summarized together for the complete rail quantity requirement of the deck unit.

The quantity takeoff for the requirement of the finish machine screed rail component is defined, in lineal feet by category type, as:

1. Screed rail, bulkhead mount, lf
2. Screed rail, surface mount, lf

Finish machine: setup, dry run

The operation of the deck finish machine setup and dry run is performed for quantity takeoff by first defining the required placement locations within the deck unit. The actual surface square feet within each placement location are calculated next, by multiplying the width of the placement surface area by the length.

This operation includes the required and necessary work operations to set the finish machine on the screed rail for the proper width, set the profile of the machine to that required within the placement area, and dry-run the finish machine over the placement area. The dry run verifies the profile of the finish machine to that of the specified deck elevation grades. This dry-run procedure also checks and verifies the reinforcing steel depth below the surface of the concrete, referred to a cover, and the X dimension, or the variable profile dimensions relative to the beam camber.

The outline requirement for the quantity takeoff of the finish machine component is defined as:

1. Required placement areas
2. Square feet surface area and placement area

The craft-hour consumption for the finish machine operation is calculated by two methods. The first method is the setup requirement. This is defined as the work necessary to set the finish machine on the rail, to its proper width, and set the required profile to the machine. The performance of this operation is calculated and recorded by a unit of each, for each required placement location.

The dry-run operation of this component is quantified in a unit of placement surface area in square feet. This is the necessary performance of setting the machine to the proper elevations for deck screed points.

The quantity takeoff for the finish machine requirements is listed by a dual component operation, in locations and square feet, as:

1. Finish machine, set up, each
2. Finish machine, dry run, sf

Deck preparation

The requirement for the quantity takeoff of the deck preparation operation is performed by a calculation in the unit measure of square feet of actual deck surface area. The takeoff outline requirement is defined as:

1. Deck surface area, square feet

This operation is the cleaning and final preparation of the gross deck surface area immediately prior to the concrete placement operation. The quantity summation is performed for each placement area. The calculation required for the deck preparation component is performed by multiplying actual deck unit width (fascia bulkhead to fascia bulkhead) by the length of the deck unit (transverse bulkhead to transverse bulkhead) for each placement area.

The quantity takeoff for the deck preparation component requirements is listed, by location in square feet, as:

1. Deck preparation, sf

Deck drains

This quantity takeoff procedure is merely the operation of defining each deck drain component, within the deck unit. These differ within the actual span locations owing to the design criteria and designated drainage areas within a specific span.

These defined deck drain locations are identified as each required, being those designated per span.

The outline requirement for the deck drain quantity takeoff procedure is identified as:

1. Number of required deck drain units

The operation performance is the necessary consumed craft-hours to properly set, and attach, the deck drain unit within the deck formwork.

The quantity takeoff of the deck drain component is defined, in a requirement of each, as:

1. Deck drain units, each

Deck blockouts

The quantity takeoff procedure for the deck blockout operation defines each deck blockout component, within the deck unit. These differ within the actual span locations owing to the design criteria and designated blockout areas within a specific span.

These defined deck blockout locations are identified as each required, being those designated per span.

The outline requirement for the deck blockout component quantity takeoff procedure is identified as:

1. Number of required deck blockouts

The operation performance is the necessary consumed craft-hours to properly set, form and strip, or attach, the deck blockout unit within the deck formwork.

476   Takeoff and Cost Analysis Techniques

The quantity takeoff of the deck blockout component is defined, in a requirement of each, as:

1. Deck blockout units, each

Concrete placement: deck unit

The deck concrete placement procedure of quantity takeoff is considered one operation with multiple component functions. Within the deck unit exist the main components of the deck section, the haunch section, and the cantilevered overhang section. The placement operation incorporates these as a monolithic deck unit, but for quantity calculations dealing with the variable dimensions, they are identified and quantified separately. The individual component quantities are then summarized into the primary deck placement component.

1. Haunch section. The haunch section is defined as the cross section of area directly associated with the beam or girder, normally the thickened bearing area above or adjacent to the beam unit.

The calculation for the concrete volume quantity takeoff defines two fixed dimensions and one variable dimension. The fixed dimensions are the width of the haunch section at the beam unit, and the length of the beam unit. The variable dimension encompasses the rise or thickness section, which is relative to the camber design. This variable dimension is normally detailed on the plan specifications as the minimum and maximum haunch dimension requirements, associated with the beam and girder camber and the X dimension. Averaging the haunch dimensions aids with the volume calculations.

The quantity outline for the haunch calculation requirement is defined as:

1. Haunch width each beam flange section
2. Haunch length associated with each beam

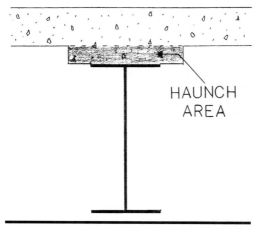

Haunch section.

3. Average haunch thickness per beam
4. Required number of beams and span
5. Yield and overrun loss factor

The volume of concrete for the beam haunch component is calculated for each beam and girder section separately owing to the variables of differing haunch thicknesses per beam unit and the differing lengths of each beam unit.

The individual calculation is performed by multiplying the width of the haunch section by the length of the beam unit requiring the haunch, multiplied by the average thickness of the haunch section, converting to a cubic yard unit quantity calculation. Added to this net volume quantity is a yield loss and overrun factor, giving the gross required quantity volume. The cubic yard quantity, both net and gross, is then summarized for each beam member within each deck unit, for the bridge structure.

The takeoff quantity for the concrete volume of the beam haunch component is summarized, by a gross and net cubic yard unit measurement, as:

1. Concrete placement, haunch, net cy
2. Concrete placement, haunch, gross cy

2. Cantilevered overhang section. The cantilevered overhang haunch section is defined as the cross section of additional deck thickened area over that of the planned deck thickness directly associated with each cantilevered section of the deck unit, normally the thickened area beyond the exterior beam member. This calculation is not performed for the total deck thickness of the cantilever overhang component, only for the additional thickened area, that in addition to the uniform section.

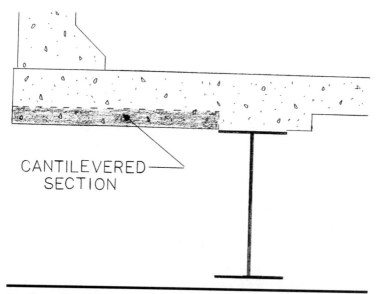

Cantilever haunch section.

The calculation for the concrete volume quantity takeoff defines two fixed dimensions and one variable dimension. The fixed dimensions are the width of each cantilevered section (beyond the exterior beam member) and the length of the cantilevered section. The variable dimension encompasses the tapered or thickness section change of the cantilever. This variable dimension is normally detailed on the plan specifications as the tapered or variable section dimension associated with the cantilevered component.

The quantity outline for the thickened cantilever calculation requirement is defined as:

1. Cantilever overhang width (average)
2. Cantilever overhang thickened dimension (average)
3. Cantilever overhang length
4. Yield and overrun loss factor

The volume of concrete for the thickened cantilevered overhang component is calculated for each cantilever section separately owing to the variables of differing thicknesses per overhang unit and the differing lengths of each exterior beam unit.

The individual calculation is performed by multiplying the width, or average width, of each cantilever section (beyond the exterior beam member) by the length of each exterior beam unit, multiplied by the average thickness of the variable section of the cantilever unit, converting to a cubic yard unit quantity calculation. Added to this net volume quantity is a yield loss and overrun factor, giving the gross required quantity volume. The cubic yard quantity, both net and gross, is then summarized for each cantilever section within each deck unit, for the bridge structure.

The takeoff quantity for the concrete volume of the cantilever overhang component is summarized, by a gross and net cubic yard unit measurement, as:

1. Concrete placement, thickened cantilever, net cy
2. Concrete placement, thickened cantilever, gross cy

3. **Uniform deck section.** The uniform deck section is defined as the cross section of the main deck component, normally the uniform thickness section of the deck, from edge of fascia to edge of fascia. This calculation is not performed for the additional deck thickness of the cantilever overhang component or for the beam haunch sections.

The calculation for the concrete volume quantity takeoff defines the variable dimensions and normally the fixed dimension of this component. The fixed dimensions are the uniform thickness of the deck section. The variable dimensions encompass the deck width or deck length. These variable dimensions are detailed on the plan specifications.

The width of each specified deck unit may vary in cross section dimension owing to design criteria of skews and variable pattern requirements and should be calculated to an average width by specific deck unit. The length of

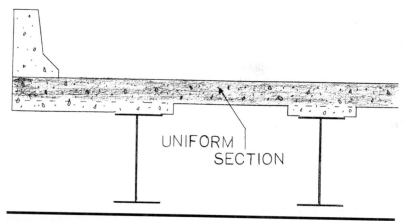

Uniform deck section.

each specified deck unit may vary in dimension owing to design criteria of skews and variable span dimensions and should be calculated to an average length measurement by deck unit.

The quantity outline for each uniform deck unit component calculation requirement is defined as:

1. Uniform deck component thickness
2. Deck unit cross section width (average)
3. Deck unit length (average)
4. Yield and overrun loss factor
5. Placement method

The volume of concrete for the uniform deck component is calculated for each deck unit separately owing to the variables of differing dimensional characteristics of each deck unit.

The individual calculation, for each deck unit, is performed by multiplying the width (average) of each deck unit by the length (average) of each deck unit, multiplied by the uniform thickness of the deck unit, converting to a cubic yard unit quantity calculation. Added to this net volume quantity will be a yield loss and overrun factor, giving the gross required quantity volume. The cubic yard quantity, both net and gross, is then summarized for each deck unit within the bridge structure.

The takeoff quantity for the concrete volume of the uniform deck component is summarized, by a gross and net cubic yard unit measurement, as:

1. Concrete placement, uniform deck, net cy
2. Concrete placement, uniform deck, gross cy

Each concrete volume quantity for the individual components of the deck placement operation is summarized for total craft-hour and volume requirement, by category. This provides the net and gross required concrete volume, and the total consumed craft-hour requirement for the placement operation. All required blockouts are deducted from the appropriate quantity summaries.

Concrete deck unit placement summary

Haunch section:	Net volume, cy	Gross volume, cy	Craft-hours
Cantilever section:	Net volume, cy	Gross volume, cy	Craft-hours
Deck section:	Net volume, cy	Gross volume, cy	Craft-hours
Total requirement:	Net volume, cy	Gross volume, cy	Craft-hours

Curing: concrete placement

The operation of curing of the concrete deck unit incorporates the gross volume of placed concrete and the gross square foot surface area of the bridge deck unit.

The curing procedure for the deck component may be specified as water curing, covering, curing compound application, or a combination.

The gross cubic yard volume is derived from the concrete placement component, listed as cubic yards, and the gross deck square footage is derived from the deck preparation component, listed as square feet.

The curing component is the required operations necessary to perform the curing procedures as specified. If the requirement of heating or cooling the concrete mixture is relevant, the gross cubic yard volume provides this quantity.

The outline for takeoff quantity of the curing operation is listed as:

1. Required gross cubic yard volume
2. Required gross deck square feet
3. Curing method required
4. Heating and cooling mix requirement

The takeoff quantity for the concrete deck curing component is summarized by a gross cubic yard unit measurement and gross deck surface square feet as:

1. Concrete curing, deck, gross cy
2. Concrete curing, deck, gross sf

Concrete deck unit: rub and patch

The rubbing and patching operation entails the quantity takeoff of the required component areas of the deck superstructure for which the project specifications require the application, and for the type of application required. Normally, the application is applied to the vertical cantilevered overhang fascia edge, and the horizontal underside surface (soffit), back to the exterior beam member.

If the specifications require the underside horizontal surface of the interior bays to be treated, this required quantity needs to be identified along with the type of application.

The outline requirement for the quantity takeoff of the rub and patch component is listed as:

1. Required cantilever surface square feet
2. Required interior deck surface square feet
3. Required procedure and application

The quantity takeoff procedure of calculation, for the cantilevered section, is performed by identifying the dimensional surfaces, in square feet, required for the procedure. The vertical edge surface of the cantilevered overhang is calculated by multiplying the deck thickness (at the edge form line) by the length of the deck unit, for each cantilevered overhang section of the structure. The horizontal underside of the cantilevered overhang section (soffit) is calculated by multiplying the bottom surface dimension (from the exterior beam member to the fascia edge) by the length of the deck unit, for each cantilevered section of the structure. The fascia edge quantity and the soffit quantity are then summarized for the required cantilevered section quantity.

The rubbing and patching operation for the interior bay components is calculated by specific bay, by multiplying the span width (beam edge to beam edge) by the length of the bay, summarizing each interior bay together.

The quantity takeoff for the concrete deck rub and patch component is summarized, by a square feet unit measurement, as:

1. Cantilevered deck section, rub and patch, sf
2. Interior deck section, rub and patch, sf

Protection: deck unit

The protection requirement for a deck unit is quantified by the requirement and type of protective component. Normally, a deck unit protection consists of wood decking staged between the beam members, false decking, or fiber netting draped from the superstructure unit. The false decking and/or netting component also aids as a safety feature for the workforce.

Protection: false decking

The false decking component consists of the required area, in square feet, for the decking material to be placed. The area of required false decking is defined as the part of the deck structure that infringes over or into a sensitive area that is required to be protected or sheltered, i.e., traffic lanes, pedestrian areas, etc. This may not be the entire span of the structure. The calculation outline is listed as:

1. Gross area of required decking
    a. Interior bay
    b. Cantilever overhang
2. Type of required decking

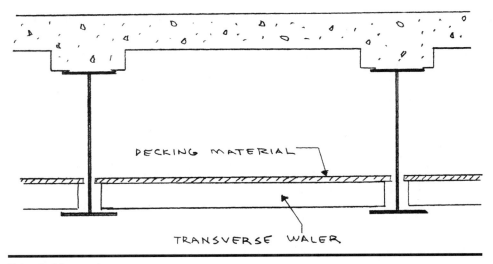

False decking section.

The quantity takeoff calculation is performed by a gross area measurement. The calculation of the interior section is the gross width of the beam unit (centerline to centerline of exterior beam members) multiplied by the length of the required protected area. The length requirement includes any point of ingress and egress of the workforce for insurance of maximum protective area.

The cantilevered section is defined and calculated as the width from the centerline of the exterior beam member to the required overhang point, multiplied by the length of the required protected area, summarizing both cantilever sections. This section is identified separately from the interior owing to the differing method of construction and craft-hour consumption associated with the cantilever section.

The type requirement of false decking is defined as the method of construction and material type, whether plyform is used with transverse walers and stringers or a more detailed built-up section is required.

The takeoff quantity summary for the component of false decking, interior or cantilever, is measured in a unit of gross square feet, by type, as:

1. False decking, deck unit, interior placement, type, sf
   a. False decking, deck unit, interior removal, type, sf
2. False decking, deck unit, cantilever placement, type, sf
   b. False decking, deck unit, cantilever removal, type, sf

Protection: netting

The netting component consists of the required area, in square feet, for the netting material to be placed. The area of required netting is defined as the part of the deck structure that infringes over or into a sensitive area that is required to be protected or sheltered, i.e., traffic lanes, pedestrian areas, etc. The netting component also aids as a safety feature for the workforce. This may not be the entire span of the structure. The calculation outline is listed as:

1. Gross area of required netting
   a. Interior bay
   b. Cantilever overhang

The quantity takeoff calculation for netting is performed by a gross area measurement. The calculation of the interior section is the gross width of the beam unit (centerline to centerline of exterior beam members) multiplied by the length of required protected area. The length requirement includes any point of ingress and egress which is infringed by the workforce for insurance of maximum protective area.

The cantilevered section is defined and calculated as the width from the centerline of the exterior beam member to the required overhang point, multiplied by the length of the required protected area, summarizing both cantilever sections. This operation includes any and all supports required for the netting. This section is identified separately from the interior owing to the differing method of construction and craft-hour consumption associated with the cantilever section.

The takeoff quantity summary for the component of netting, interior or cantilever, is measured in a unit of gross square feet, by type, as:

1. Protective netting, deck unit, interior, placement, sf
   a. Protective netting, deck unit, interior, removal, sf
2. Protective netting, deck unit, cantilever placement, type, sf
   a. Protective netting, deck unit, cantilever removal, type, sf

Box Girder Section

Temporary falsework and scaffolding

The takeoff and design of the temporary falsework or scaffolding is based on the spacing of the vertical members, the supportive foundation on which it sets, and the deflection and spacing of the horizontal stringer members, all controlled by the cast-in-place structure and working loads.

Between the construction operations of the supportive foundation system and the temporary falsework system, the final top elevation follows the horizontal grade elevation and contours of the cast-in-place superstructure section as closely as possible, easing and minimizing further component operations. Each unit can be stair-stepped in order to maintain the level horizontal consistency at the top of the falsework component.

The operation of temporary falsework and scaffolding is defined as the required and necessary vertical system needed to support the cast-in-place superstructure component during the construction stage. The quantity take-

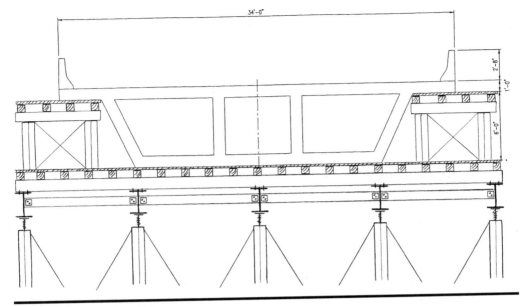

Box girder section (falsework).

off of this component is the number of vertical structural members, in a quantity of each unit, required for each span.

The takeoff procedure includes all the required components of the falsework system from the supportive foundation component, upward to the area of the temporary decking required for the construction of the superstructure section.

Within this operation are the internal components of the vertical structural members (scaffold legs), the required X bracing for stability, the transverse horizontal structural strut or cap members required to support the longitudinal stringers, and the longitudinal stringer beams. The structural longitudinal stringer beams are the final internal component of the temporary falsework system, on which the falsework deck is placed. Any cross bracing required for the stringer beams is included with this component. Also within this component section is the quantifying of support jacks and screw jacks required for the towers, for the final adjustment of elevation to the stringer members for permanent profile.

The outline requirement for the quantity takeoff of the temporary falsework and scaffold component is listed as follows:

1. Foundation condition
2. Scaffold unit spacing and height
3. Cap and strut requirement
4. Support and screw type
5. Longitudinal stringer beam type and spacing
6. Final grade

The takeoff is controlled by the height of each scaffold unit and the number of supportive units required, by span, summarizing the entire structure.

The quantity takeoff for the temporary falsework and scaffolding towers will be (1) in required units, by type and span; (2) by horizontal longitudinal stringer beam, by section type and lineal feet; (3) by support or screw jack, by type and quantity; and (4) by fine grading the elevation and profile of the system, by each jack, summarized for the bridge structure as:

1. Temporary falsework, scaffold unit, type, each
2. Temporary falsework, longitudinal beam member, type, lf
3. Temporary falsework, support and screw jack, type, each
4. Temporary falsework, fine grade, type, each

Temporary falsework deck section

The takeoff operation the falsework deck system consists of quantifying the internal components of transverse steel stringer beam members and the wood decking system. The wood deck system consists of longitudinal wood stringer members, placed on the transverse steel members, and a structural plyform deck.

The quantity takeoff is composed of identifying the required steel stringer beams by cross section size and calculating in a lineal feet requirement for the transverse spacing needed for horizontal support of the box girder. The wood deck system is identified by two components: calculating the wood stringer members by dimensional size in a quantity of required lineal feet, and the wood plyform deck in a quantity of required square feet for the gross area of the temporary falsework deck.

The outline requirement for the quantity takeoff of the temporary falsework deck component is listed as follows:

1. Transverse steel stringer beam type and spacing
2. Wood stringer member dimensional size and spacing
3. Gross square feet requirement deck

The spacing of the transverse stringer members is a variable and is controlled by a series of factors. The width of the spacing of the longitudinal stringer members, versus the imposed loads, and the section modulus of the transverse stringer, are the controlling ingredient to the calculation. This span dimension, factored in with the imposed loading to the falsework deck, determines (1) the section size of stringer required and (2) the spacing of the transverse stringer member that will support the decking and the box girder section.

The quantity takeoff for transverse stringer members is calculated by the length of each falsework deck section, requiring the steel stringer members, divided by the defined spacing of the transverse stringers plus one, summarizing each required span quantity. The unit of measurement is lineal feet, by section size, and length of each steel stringer.

The calculating of the stringer member requirements is in conjunction with the wood deck system calculations, based on load versus span requirements and criteria.

The quantity takeoff for longitudinal wood stringer members is calculated by the width of each falsework deck section, requiring the wood stringer members, divided by the defined spacing of the longitudinal wood stringers plus one, summarizing each required span quantity. The unit of measurement is lineal feet, by dimensional size, of each wood stringer.

The quantity takeoff for the gross plyform deck is performed by the width of the deck area (including the added work area) by the length of the deck area for each span, in a unit measurement of square feet, summarizing each required span quantity.

The required takeoff quantity is outlined as:

1. Falsework deck section, steel stringer, type, lf
2. Falsework deck section, wood stringer, type, lf
3. Falsework deck section, gross deck area, sf

Cast-in-place box girder: bottom

The cast-in-place box girder bottom component is the first order of operation once the falsework is completed. The bottom section unit rests and is founded directly on the falsework deck area; therefore, there is no requirement of formwork for the bottom surface area. This is referred to as the component contact area. The grade and elevation of the falsework deck contact area are verified as to defined tolerances for control of the box girder construction. These checks should include elevation and dimension verification.

The box girder bottom component consists of numerous work operations.

1. Deck placement preparation
2. Formwork placement
3. Embed item installation
4. Concrete placement
5. Formwork removal
6. Rub and patch concrete

These work operations are independent functions required to complete the component of box girder bottom sections.

The perimeter of the box girder bottom component is then calculated for formwork contact surface area. This dimension calculation requires the continued perimeter of the box girder bottom measurement multiplied by the thickness measurement of the bottom section. This computation is then quantified in a unit of square feet of required formwork, summarized as:

1. Box girder, bottom perimeter formwork, sf

The doweled bulkhead takeoff requirement is summarized at this time. This calculation consists of the length of the bulkhead multiplied by the thickness, multiplied by the number of bulkheads required by plans, and summarized as:

1. Box girder, doweled bulkhead, sf

The embedded item quantifying defines any incidental embeds required within the box girder bottom section. These are summarized in a unit of each as:

1. Box girder, bottom embeds, each

The quantity of formwork stripping, or removal, is identical for the required formwork placement quantity, summarized in a unit measurement of square feet as:

1. Box girder bottom, formwork removal, sf

The requirement for the quantity takeoff of the bottom deck preparation operation for the concrete placement is performed by a calculation in the unit measure of square feet of actual bottom deck surface area. The takeoff outline requirement is defined as:

1. Bottom deck surface area, square feet

This operation is the cleaning and final preparation of the net deck surface area immediately prior to the concrete placement operation. The quantity summation is performed for each placement area. The calculation required for the deck preparation component is performed by multiplying actual deck unit width (fascia bulkhead to fascia bulkhead) by the length of the deck unit (centerline of span bearing to centerline of span bearing).

The quantity takeoff for the deck preparation component requirement is listed, by location in square feet, as:

1. Box girder bottom section, deck preparation, sf

The concrete requirements are calculated by a simple volume determination, the width, multiplied by the depth, multiplied by the length, of the total area of the box girder bottom section to be filled by the concrete. When there are quadrants of variable volumes and sizes, each rectangular area is calculated independently, then summarized together. This quantity, of cubic feet, is then translated to a volume measurement of required cubic yards, adding the required waste and yield factor, as:

1. Box girder bottom, concrete placement, cy

Rubbing and patching is calculated by the surface area, in square feet, of the component that is required to be completed. This is the perimeter sides and the underside of the bottom section. The quantity is then summarized in a unit measurement of square feet as:

1. Box girder bottom, rub and patch, sf

Cast-in-place box girder: wall

The side and interior wall components of the box girder superstructure rest on the bottom or falsework deck component. With this known, no formwork requirement exists for the bottom horizontal surface of the wall section or component contact area.

The box girder wall components consist of numerous work operations.

1. Formwork placement
2. Embed item installation
3. Concrete placement
4. Formwork removal
5. Rub and patch concrete

These work operations are independent functions required to complete the component of box girder wall sections.

The walls themselves, whether exterior or interior, are comprised of a front side, back side, and end areas. The formwork requirement is calculated by multiplying the length of the wall by the "average" height, to determine the square foot surface area. This calculation is performed for the front face as well as the back face, for each wall. The average height in elevation is determined for the wall section, which sometimes may have variable vertical dimensions. The component unit is summarized as required square feet of formwork as:

1. Box girder wall, interior or exterior formwork placement, sf

If an intermediate construction joint exists within the wall, it requires an internal bulkhead forming system. This is summarized as a required square foot:

1. Box girder wall, internal bulkhead formwork placement, sf

The vertical ends of the box girder wall section are classified as end bulkheads. The calculation of this operation is the width of the end multiplied by the height of the end by the number of ends required. This is completed and summarized in square feet as:

1. Box girder wall, end bulkhead formwork placement, sf

The embedded item quantifying defines any incidental embeds required within the box girder wall. These are summarized in a unit of each as:

1. Box girder wall, embeds, each

The quantity of formwork stripping, or removal, is identical for the required formwork placement quantity, summarized as:

1. Formwork stripping, box girder wall or bulkhead, sf

The concrete requirements are calculated by simple volume methods, multiplying the wall length by the wall height by the depth or thickness, averaging dimensions when required. This cubic foot volume is then converted to a unit measurement of cubic yards, adding the required waste and yield factor, summarized for both net and gross as:

1. Concrete placement, box girder wall, cy

Rubbing and patching is calculated by the surface area, in square feet, of the component that is required to be completed. The rubbing and patching operation for the box girder wall is a calculation of required surface area to be treated. The summation is in square feet of:

1. Box girder wall, rub and patch, sf

Cast-in-place box girder: end wall

The box girder end wall is the structural component of the cast-in-place superstructure located at the end of each box girder section. The wall is comprised of a front face and back face and is situated between the exterior and interior sidewalls.

The box girder end wall component is constructed dependent on the box girder unit, thus being composite to the sidewalls.

The box girder end wall components consist of numerous work activities:

1. Formwork placement
2. Embed item installation
3. Concrete placement
4. Formwork removal
5. Rub and patch concrete

These work operations are independent functions required to complete the component of box girder end wall sections.

The takeoff of the end wall component is similar to that of the sidewall component. The bottom surface of the end wall bears on the bottom or falsework deck section, thus not requiring formwork for this horizontal component contact area.

The end wall may also have a variance in elevation which would require an averaged height calculation. The formwork requirement is the surface contact area of the front and back faces of the end wall component.

The formed surface area for a back or outside face of a box girder end wall calculates as the length or average length from the left side to the right of the wall, multiplied by the average height, giving square feet of wall area. The inside of the end wall is calculated in a similar manner, except that the surface areas between the side of the exterior or interior walls are deducted from the surface calculation for the formed area.

The requirement for the removal of the formwork system is the identical square feet as required with the placement of the formwork.

The quantity summation is listed in a unit measurement of square feet as:

1. Box girder end wall, formwork, placement, sf
2. Box girder end wall, formwork, removal, sf

The end sections of the end wall component, if required, are classified as end or closure bulkheads. The surface contact area calculation required for the formwork is the average thickness of the end wall section multiplied by the height of each end section. The summation is categorized as: end bulkhead square feet, box girder end wall.

The embedded item quantifying defines any incidental embeds required within the box girder end wall sections. These are summarized in a unit of each as:

1. Box girder end wall, embeds, each

Concrete volume requirements for placement calculations are compiled using volumetric methods.

The box girder end wall is calculated for concrete volume by simply multiplying the averaged width by the height by the thickness, then converting the resulting cubic feet to cubic yards adding the required waste and yield factor.

The quantity summation volume requirement is classified in a volume of cubic yards for both net and gross as:

1. Box girder end wall, concrete placement volume, cy

A calculation for surface area of required rubbing and patching is the surface area required for this procedure. This is reported in a square foot dimension of:

1. Box girder end wall, rub and patch, sf

Cast-in-place box girder: diaphragm component

Once the wall panels are cast or the exterior and interior formwork is set, the diaphragm components are the next operation of construction. The diaphragms are structural components that bridge or link each wall unit together and give a complete structural uniformity to the box girder unit. The purpose is to take the independent structural capacity of each wall section and unite them into a dependent system, distributing the implied loads evenly over the area section rather than over just one wall, creating a composite structure.

The diaphragms are constructed in a row, transversely across the box girder section, between the exterior and interior wall sections.

The concrete diaphragms are cast in place between the wall sections, at their designated location.

Owing to the variability of design, the length of the diaphragm components may vary per bay.

Types of diaphragms are defined as interior or midspan diaphragms and pier composite diaphragms.

The interior diaphragms are located at the midpoint or throughout longitudinally to the span. Normally the interior diaphragms rest directly on the bottom deck, requiring no horizontal formed surface. The interior diaphragms are connected to each wall section, which means that a formed surface will be required on two sides.

The pier or composite diaphragms connect two adjacent spans, creating a continuous box girder unit of multiple spans. The composite diaphragms are located between wall sections and normally rest directly on the bottom deck component surface area beneath each diaphragm.

All diaphragms contain a vertical component contact area, which abuts the surface of the wall. The diaphragm components consist of numerous work operations:

1. Formwork placement
2. Embed item installation
3. Concrete placement
4. Formwork removal
5. Rub and patch concrete

These work operations are independent functions required to complete the component of box girder diaphragms.

The formwork requirements are the quantity of surface contact area. The required unit calculation of square feet is each side area, as required. This formwork quantity is summarized for each type of diaphragm.

The calculation requirement to arrive at the surface contact area for the formwork is done by determining the horizontal dimension between the box girder walls and the vertical height or elevation difference from the bottom horizontal elevation of the diaphragm to the top of the proposed diaphragm (averaging the width and height if required). These are the dimensions used for each bay calculation of the diaphragms, for each type of diaphragm. The width of each interior bay is calculated independently owing to the variances in the bay widths.

The quantity summary for the formwork placement requirements of box girder diaphragms is categorized in square feet as:

1. Box girder diaphragm, interior, formwork placement, sf
2. Box girder diaphragm, composite, formwork placement, sf

The removal of the formwork system is identical to the quantity of square feet placed and is summarized as:

1. Box girder diaphragm, interior, formwork removal, sf
2. Box girder diaphragm, composite, formwork removal, sf

The embed or form hardware components are identified for the installation of each required, for each type of diaphragm. The quantity summation for the hardware embeds for concrete diaphragms is listed as:

1. Box girder diaphragm, embed installation, interior, ea
2. Box girder diaphragm, embed installation, composite, ea

The concrete volume requirement is the calculation of cubic yards of concrete needed for each type of diaphragm, summarized by diaphragm type.

Concrete volume requirements for placement calculations are compiled using volumetric methods.

A concrete diaphragm component is calculated for concrete volume by simply multiplying the averaged width by the height by the thickness for each diaphragm per bay, summarizing the calculations, then converting the resulting cubic feet to cubic yards and adding the required waste and yield factor, following the format discussed.

The quantity summation volume requirement is classified in a unit measurement of cubic yards for both net and gross as:

1. Concrete placement volume, cubic yards, interior diaphragm, cy
2. Concrete placement volume, cubic yards, composite diaphragm, cy

The operation of rubbing and patching if required is the quantity of required surface area, in square feet, to be treated. This area, usually identical to the required formwork surface area, is summarized as:

1. Concrete diaphragm, interior, rub and patch, sf
2. Concrete diaphragm, composite, rub and patch, sf

Cast-in-place box girder: anchor blocks

The quantity takeoff procedure for the box girder anchor block operation defines each box girder block component within the box girder unit. These differ within the actual span locations owing to the design criteria and designated anchor block areas within a specific girder span.

Superstructure

These defined box girder anchor block locations are identified as each required, being those designated per span.

The outline requirement for the box girder anchor block component quantity takeoff procedure is identified as:

1. Number of required anchor blocks
2. Formwork placement
3. Embed item installation
4. Concrete placement
5. Formwork removal
6. Rub and patch concrete

The operation performance is the necessary consumed craft-hours to properly set, form and strip, or attach, the anchor block components within the box girder formwork.

The quantity takeoff of the box girder anchor block component is defined, in a requirement of each, as:

1. Box girder anchor block units, each

Deck work follows normal deck operations outlined earlier in the chapter.

Parapet and Median Rail

The parapet and median rail sections of the superstructure deck unit are defined as the vertical concrete protective barriers of the deck unit that contain the traffic to its designated location within the superstructure.

The parapet rail is identified as the vertical barrier located at the cantilevered fascia edge of the deck unit. The median rail is identified as the vertical barrier

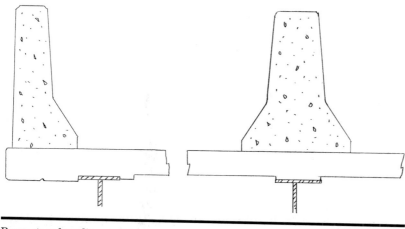

Parapet and median section.

494 Takeoff and Cost Analysis Techniques

located within the interior or center portion of the deck unit. The parapet and median rail units are comprised of various components and work operations.

The outline requirement for the quantity takeoff of the parapet and median rail is listed by component as:

1. Type of required rail
2. Lineal feet of required rail
   a. Main deck
   b. End transition
3. Vertical dimensions, front and back face
4. Cross section area

Parapet and median: formwork

The formwork requirement of the parapet and median rail is defined as the vertical concrete surface contact area of required formwork. The component contact surface area is the horizontal surface between the deck unit and the parapet and median rail unit; this is not a formed surface area.

The quantity takeoff for the formwork requirement is performed by multiplying the vertical contact surface, for both sides of the rail, by the length of required rail. This calculation is categorized by rail type as to cross section, and by area as the main deck rail and the end transition rail.

The formwork removal is summarized as the identical quantity as the formwork requirement for placement.

The quantity takeoff for the formwork requirements of the parapet and median rail are summarized in square feet of contact area, by type, by area, as:

1. Parapet and median deck rail: formwork placement, sf, type _____
   a. Parapet and median deck rail: formwork removal, sf, type _____
2. Parapet and median transition rail: formwork placement, sf, type _____
   b. Parapet and median transition rail: formwork removal, sf, type _____

Parapet and median: bulkhead

The bulkhead takeoff requirement is summarized at this time. This calculation consists of the square feet of the bulkhead end area cross section, multiplied by the number of bulkheads required by plans, and is summarized as:

1. Parapet and median deck rail, bulkhead, sf, type _____
2. Parapet and median transition rail, bulkhead, sf, type _____

Parapet and median: embedded hardware

The embedded item quantifying defines any incidental embeds required within the parapet and median rail. These are summarized in a unit of each as:

1. Parapet and median deck rail, embeds, ea, type _____
2. Parapet and median transition rail, embeds, ea, type _____

Parapet and median: concrete placement

The concrete requirements are calculated by a simple volume determination, the width in section, multiplied by the depth, multiplied by the length of the total area of the parapet and median rail to be filled by the concrete. When there are quadrants of variable volumes and sizes, each area is calculated independently, then summarized together. This quantity, of cubic feet, is then multiplied by the yield and loss overrun factor, and is translated to a volume measurement of required cubic yards, concrete placement, by required type and area of wall.

1. Parapet and median deck rail, concrete placement, cy, type _____
2. Parapet and median transition rail, concrete placement, cy, type _____

Parapet and median: rub and patch

Rubbing and patching is calculated by the surface area, in square feet, of the component that is required to be completed. This quantity should be equal to the formwork placement component. The quantity is then summarized as:

1. Parapet and median deck rail, rub and patch, sf, type _____
2. Parapet and median transition rail, rub and patch, sf, type _____

Parapet and median: slipform

The quantity takeoff for the operation of slipforming the parapet and median rail component is defined as a unit measure of lineal feet for the portion of rail to be performed by the slipforming procedure.

Slipforming consists of molding the parapet and median rail by extruding the fresh concrete mix through a mold mounted on a self-propelled machine, requiring no formwork.

The operation of slipforming constitutes the following outlined operational components under one progressing operation. The slipform operation includes the placement of concrete, required joint work, and finishing or texturing of the wall during the procedure.

Slipforming may not encompass the entire parapet and median rail quantity as detailed and may require conventional construction methods combined with this. With this condition, the slipform operation is summarized with the conventional and/or transition rail summary costs.

The quantity takeoff requires the lineal feet of required wall to be slipformed, the volume of concrete required for the cross section along with the yield loss and overrun factor, and the conditions that exist for the operation.

This component is summarized in a unit measurement of required lineal feet of rail, by type, for this operation as:

1. Parapet and median rail, slipform, lf, type ―――――

Procedure

The procedure of quantity takeoff for a bridge superstructure is a detailed and systematic outline detailing the required components within the deck unit. The superstructure component operations vary depending on many factors, design conditions, and construction procedures.

A well-detailed quantity takeoff not only provides for the required component information but also gives a complete understanding of the required construction procedures needed for the performance and execution of the work tasks.

The deck units' main variables, as to the formwork components, are the beam unit type, the span dimensions and implied loads, both live and dead, along with additional point loading of special equipment. The detailed takeoff defines these requirements and conditions.

The first step of the procedure is to identify the type of beam structure unit and its conditions and restrictions. Once this component has been defined, the formwork method, types of support, haunch details, and construction techniques are determined. From this point the upper deck unit components can be categorized and quantified.

Questions of Estimate

The performance of an accurate quantity takeoff procedure demands an outline of requirements to be defined. This outline defines the parameters of components and construction operations, along with any special requirements of the project itself.

1. What is the beam structure type?
2. What are the deck formwork requirements and options?
3. What is the haunch and bearing area detail?
4. What is the cantilever detail?
5. Are there any special formwork and embed requirements?
6. What is the design and loading requirement of formwork?
7. Is the deck unit composite or noncomposite?
8. Is the deck unit continuous?
9. What are the unit components?

10. What is the concrete placement schedule?
11. Are there optional construction joints?
12. What is the method of concrete placement?
13. What is the placement rate per hour?
14. What is the concrete yield requirement?
15. Is there a need of falsework, shoring, or netting?
16. What specialty equipment is required?
17. Are traffic conditions or limitations present?

Application

The following example details a given superstructure quantity takeoff. This defines the production requirements, the craft-hour production factoring method, and crew determination and identification, and produces a detailed summary listing of required formwork, concrete volumes, material needs, both permanent and incidental, and equipment usage.

A given dual span structure has the following superstructure units:

Span 1 composite

Span 2 composite

Within these span units are the following deck components of:

Haunch form system

Beam member hanger assemblies

Interior form system

Cantilever form system

Vertical bulkhead system

Expansion devices

Deck component

Incidental blockout and drain assemblies

Parapet and median rail

For these components, the required work operations are:

Formwork placement and removal

Installation of form system hardware, embeds

Determination of beam type

Screed rail installation

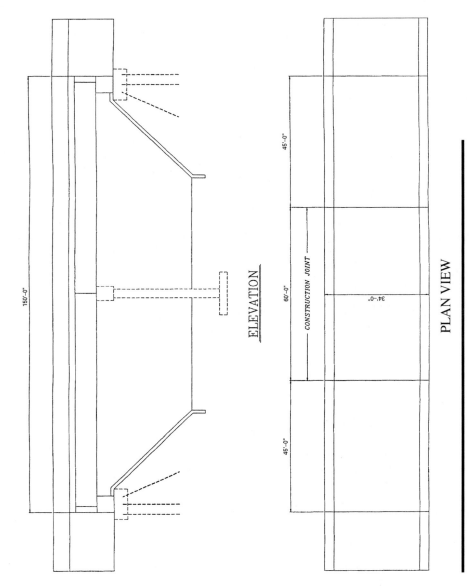

Two-span deck unit.

Finish machine setup, dry run

Concrete placement

Rub and patch concrete surface

The following is a detailed listing of required component dimensions per the project plans:

Beam type	Precast concrete type IV
Beam spacing	4 each @ 8′ centers
Required beam members	8 each
Deck unit type	Continuous composite
Total deck dimension:	8″ thick, 34′ wide, 150′ long
Deck unit	3 placements
Placement 1, span 1	8″ thick, 34′ wide, 45′ long
Placement 2, span 2	8″ thick, 34′ wide, 45′ long
Placement 3, pier	8″ thick, 34′ wide, 60′ long
Design load	175 lb/cubic foot
Haunch dimension	2″–4″ thick, 1′8″ wide
Cantilever section	12″ thick
Overhang dimension	5′ (from centerline of beam)
Form system	Conventional
Hanger assemblies spacing	2′ centers
Interior form system	Design load, 175 lb/sf
	Wood, 2″ × 8″ waler @ 2′ centers
	2″ × 6″ stringer @ 1′ centers
	¾″ plyform
Cantilever form system	Design load, 225 lb/sf
	Support brackets @ 2′ centers
	Wood, 4″ × 4″ × 4′ bracket member
	2″ × 4″ stringer @ 1′ centers
	¾″ plyform
Perimeter bulkhead	Fascia 12″ thick
	Ends 8″ thick
Interior bulkhead, plain	None required
Interior bulkhead, doweled	8″ high
Expansion devices, 2″ open joint	Span 2, abutment 2, @ 34′
Deck component	Concrete class 4500 psi
Screed rail	Bulkhead condition
Yield loss and overrun factor	12%
Blockout and drain assemblies	None required
Parapet and median rail	Conventional 32″ high
	Vertical joints @ 20′ centers
	Concrete class 4500 psi

Example: Operation Analysis

The structure used in the example is comprised of a continuous composite deck unit (34′w by 150′l), which is defined as one monolithic concrete deck continuous through both spans, being composite or connected to the beam structure. The detail identifies three separate placements using defined construction transverse joints. The expansion area is at the abutment located in span 2.

Takeoff and Cost Analysis Techniques

This example uses a conventional form system for both the interior and cantilevered sections. The haunch section and beam hanger assemblies are uniformly symmetrical for both the interior and exterior cantilevered form system requirements.

Superstructure deck unit

Operation. Beam haunch, 2″–4″. Unit: Lineal feet.
The operation of forming and stripping the beam haunch component is calculated in the following manner for the required unit area:

Required: Haunch depth, 2″–4″ (avg 3″)

Beam length, 75′ each, symmetrical

The length of each beam member, multiplied by two sides, multiplied by the number of beam members.

$$(75' * 2) * 8 = 1200.0 \text{ lf form and strip}$$

Operation. Beam hanger assembly. Unit: Each.
The operation of installing the beam hanger assembly component is calculated in the following manner for the required unit area:

Required: Beam width, 1′8″

Beam length, 75′ each, symmetrical

Full hanger assembly

Spacing, 2′ centers

Load, 175 lb/cubic foot

Coil rod length, 24″

The length of each beam member divided by the hanger spacing plus 1, multiplied by the number of beam members.

$$((75'/2) + 1)\text{rounded} * 8 = 312.0 \text{ each hanger}$$

Operation. Interior deck system, form and strip. Unit: Square feet.
The operation of forming and stripping the interior deck component is calculated in the following manner for the required unit area:

Required: Reuse factor, 4

Waler spacing, 2′ centers

Load, 175 lb/cubic foot

Waler and double ledger, 2"×8". The length of each double waler member (between beams). The beam member length, divided by the waler spacing plus one, summarizing each interior bay.

$$8' - ((1'8'' + 1'8'')/2) = 6.33' \text{ long}$$

$$((75'/2) + 1)\text{rounded} * 6 = 234.0 \text{ each}$$

Stringer, 2"×6". The width of each interior bay, divided by the stringer spacing plus one, multiplied by the beam member length for each interior bay, summarizing each interior bay.

$$((6.33'/1') + 1)\text{rounded} = 8 \text{ each}$$

$$(8 * 75') * 6 = 3600.0 \text{ lf}$$

Decking, ¾". The width of each interior bay, multiplied by the length of each interior bay, summarizing each interior bay.

$$(6.33' * 75') * 6 = 2848.50 \text{ net sf}$$

Gross formwork area. The transverse width of beam structure (center of exterior beam to center of exterior beam), multiplied by the length of each span unit, summarizing each span.

$$(24' * 75') * 2 = 3600.00 \text{ gross sf form}$$

$$= 3600.00 \text{ gross sf strip}$$

Operation. Cantilever deck system, form and strip. Unit: Square feet.

The operation of forming and stripping the cantilever deck component is calculated in the following manner for the required unit area:

Required: Reuse factor, 4

Support bracket spacing, 2' centers

Load, 225 lb/cubic foot

Walkway dimension, 2'

Support bracket. The exterior beam member length, divided by the support bracket spacing plus one, summarizing each exterior beam member.

$$((75'/2) + 1)\text{rounded} * 4 \text{ each} = 156.0 \text{ each}$$

Bracket member, 4"×4". The number of required support brackets and the cantilever width dimension (from edge of exterior beam member) plus walkway dimension, multiplied by the required brackets.

$$\text{Bracket member} = 156.0 \text{ ea}$$

$$(5' + 2') * 156 = 1092.0 \text{ lf}$$

Stringer, 2″×4″. The width of each exterior bay plus the walkway, divided by the stringer spacing plus one, multiplied by the beam member length for each exterior bay, summarizing each exterior bay.

$$(((4.5' + 2')/1') + 1)\text{rounded} = 8 \text{ each}$$

$$(8 * 75') * 4 \text{ each} = 2400.0 \text{ lf}$$

Decking, ¾″. The width of each exterior bay, from beam edge, plus the walkway, multiplied by the length of each exterior bay, summarizing each exterior bay.

$$((4.17″ + 2') * 75') * 4 = 1851.0 \text{ net sf}$$

Gross formwork area. The width of the exterior bay, from the center of the beam member, plus the walkway, multiplied by the length of each exterior bay, summarizing each exterior bay.

$$((5' + 2') * 75') * 4 = 2100.00 \text{ gross sf form}$$

$$= 2100.00 \text{ gross sf strip}$$

$$\text{Overhang brackets} = 156.0 \text{ each/bracket}$$

Operation. Perimeter bulkhead, form and strip. Unit: Square feet.

The operation of forming and stripping the perimeter bulkhead component is calculated in the following manner for the required unit area: The thickness of the exterior fascia bulkhead multiplied by each span fascia component, plus the thickness of the abutment end bulkheads multiplied by the width of the deck unit ends (including skew).

$$((1' * 75') * 4) + (0.67' * 34') * 2 = 345.56 \text{ sf form}$$

$$= 345.56 \text{ sf strip}$$

Operation. Interior bulkhead, doweled, form and strip. Unit: Square feet.

The operation of forming and stripping the interior bulkhead component is calculated in the following manner for the required unit area:

Required: Dowel spacing, bottom mat @ 8″
top mat @ 10″

The thickness of the interior bulkhead multiplied by the length of each interior bulkhead, summarizing the quantity.

$$((0.67″ * 34') * 2) = 45.56 \text{ sf form}$$

$$= 45.56 \text{ sf strip}$$

Operation. Expansion device bulkhead, form and strip. Unit: Square feet.

The operation of forming and stripping the expansion device bulkhead component is calculated in the following manner for the required unit area:

Required: Bolt spacing, 12"

The thickness of the expansion device bulkhead multiplied by the length of each expansion device bulkhead, summarizing the quantity.

$$((0.67" * 34') * 1) = 22.78 \text{ sf form}$$

$$= 22.78 \text{ sf strip}$$

$$= 34.0 \text{ lf}$$

Operation. Required placement locations. Unit: Each.

$$\text{Placements} = 3.0 \text{ each}$$

Operation. Screed rail. Unit: Lineal feet.

The operation of installing the finish machine screed rail component is calculated in the following manner for the required unit area:

Required: Bulkhead condition

Approach and runoff   20' * 4 ea, structure

Placement length   45', 45', 60'

The length of each placement, multiplied by the two sides, summarizing the quantity for each placement, plus the runoff dimension (80').

$$((45' * 2) + (45' * 2) + (60' * 2)) + 80' = 380.0 \text{ lf rail}$$

Operation. Finish machine, dry run, set up. Unit: Square feet, each.

The operation of the finish machine dry-run setup component is calculated in the following manner for the required unit area:

Required: Placement areas 3 ea

The length of each placement area multiplied by the width of each placement area, summarizing the quantity for each placement area, giving the required surface area, plus the quantity of placement areas, giving the required setup locations.

$$(45' * 34') + (45' * 34') + (60' * 34') = 5100.0 \text{ sf dry run}$$

$$\text{Location area setup} = 3.0 \text{ each setup}$$

Operation. Deck preparation. Unit: Square feet.

The operation of the deck preparation component is calculated in the following manner for the required unit area:

Required: Placement areas 3 ea

The length of each placement area multiplied by the width of each placement area, summarizing the quantity for each placement area, giving the required surface area.

$$(45' * 34') + (45' * 34') + (60' * 34') = 5100.0 \text{ sf prep}$$

Operation. Concrete placement, deck unit. Unit: Cubic yard.

The operation of the deck concrete placement component is calculated in the following manner for the required unit area:

Required: Haunch volume 0.25'×1.67'

Thickened cantilevered volume 0.34'×4.17'

Uniform deck volume 0.67'×34'

Yield loss 12%

Haunch volume. The width of the haunch section, multiplied by the length of each beam member, multiplied by the average thickness of the haunch section, plus the yield loss factor, converting to a cubic yard measurement and summarizing the quantity for each placement area (beam member), giving the required volume.

$$(((0.25' * 1.67' * 75') * 1.12)/27) * 8 = 10.39 \text{ cy gross}$$

$$((0.25' * 1.67' * 75')/27) * 8 = 9.28 \text{ cy net}$$

Cantilever thickened section. The width of the cantilevered section, multiplied by the length of each exterior beam member, multiplied by the average thickness of the thickened section, plus the yield loss factor, converting to a cubic yard measurement and summarizing the quantity for each placement area (cantilevered unit), giving the required volume.

$$(((4.17' * 75' * 0.34') * 1.12)/27) * 4 = 17.64 \text{ cy gross}$$

$$(((4.17' * 75' * 0.34'))/27) * 4 = 15.75 \text{ cy net}$$

Uniform deck section. The width of each deck unit placement, multiplied by the length of each deck unit placement, multiplied by the average thickness of the uniform deck section, plus the yield loss factor, converting to a cubic yard measurement and summarizing the quantity for each placement area, giving the required volume.

$$(((((34' * 45') + (34' * 60') + (34' * 45')) * 0.67) * 1.12)/27) = 141.74 \text{ cy gross}$$

$$(((34' * 45') + (34' * 60') + (34' * 45')) * 0.67)'/27) = 126.56 \text{ cy net}$$

Deck placement summary

Component	Net volume	Gross volume
Haunch	9.28 cy	10.39 cy
Cantilever	15.75 cy	17.64 cy
Uniform deck	126.56 cy	141.74 cy
Total	151.59 cy	169.77 cy placement

Operation. Curing, concrete placement. Unit: Cubic yard.

The operation of curing, of the concrete placement component, is calculated in the following manner for the required unit area:

$$\text{Gross concrete volume requirement} = 169.77 \text{ cy cure}$$

$$\text{Deck surface area} = 5100.0 \text{ sf cure}$$

Operation. Rub and patch, concrete surface. Unit: Square feet.

The operation of the rubbing and patching component is calculated in the following manner for the required unit area:

Required: Cantilevered section

The vertical fascia edge calculation; the deck thickness (at perimeter bulkhead line) multiplied by the length of each longitudinal bulkhead, summarizing the quantity for each surface area: plus the horizontal soffit calculation; the soffit width (beam member to fascia edge), multiplied by the length of each cantilevered section, summarizing the quantity for each surface area. These quantities, the vertical and soffit are summarized, giving total required surface area of application.

$$((1' * 75') * 4) + ((4.17' * 75') * 4) = 1551.0 \text{ sf rub and patch}$$

Superstructure parapet and median rail unit

Operation. Parapet rail, form and strip. Unit: Square feet.

The operation of forming and stripping the parapet rail component is calculated in the following manner for the required unit area:

Required: Height 2.67' back face

3.02' front surface (contour area)

Required: Cross section area 2.646 sq ft (381.0 sq in)

The height of the back face plus the height of the front face, multiplied by the length of the required main rail (both sides of the structure), giving the required square feet of formwork.

$$(2.67' + 3.02') * (150' * 2) = 1707.0 \text{ sf form}$$

$$= 1707.0 \text{ sf strip}$$

Operation. Parapet rail, bulkhead. Unit: Square feet.

The operation of forming and stripping the parapet rail bulkhead component is calculated in the following manner for the required unit area:

$$\text{Cross section area} = 2.646 \text{ sq ft } (381.0 \text{ sq in})$$

The square feet of each bulkhead, multiplied by the number of required bulkheads within the main rail (length of rail divided by the joint spacing plus one, rounded, for each side), giving the required square feet of bulkhead.

$$2.646' * (((150'/20) + 1)\text{rounded} * 2) = 47.63 \text{ sf form and strip}$$

Operation. Parapet rail, embedded hardware. Unit: Each.
None required

Operation. Parapet rail, concrete placement. Unit: Cubic yard.

The operation of concrete placement for the parapet rail component is calculated in the following manner for the required unit area:

$$\text{Required: Yield/loss factor } 11\%$$

$$\text{Cross section area } 2.646 \text{ sq ft } (381.0 \text{ sq in})$$

The square feet of end area cross section, multiplied by the lineal feet of main rail, multiplied by the yield and loss factor, converting to a cubic yard measurement, giving the required concrete volume.

$$((2.646' * 300') * 1.11)/27 = 32.63 \text{ cy gross}$$

$$(2.646' * 300')/27 = 29.40 \text{ cy net}$$

Operation. Parapet rail, rub and patch. Unit: Square feet.

The operation of the parapet rail rub and patch component is calculated in the following manner for the required unit area: The square feet of surface area equal to the formwork requirement, giving the required rub and patch surface area.

$$\text{Contact surface area} = 1707.0 \text{ sf rub and patch}$$

Unit quantity summary

Superstructure quantity summary: deck unit

Deck unit	Span 1 and 2	
	Unit	Quantity
Beam haunch, form and strip	lf	1200.00
Hanger assembly	ea	312.00
Interior deck, form	sf	3600.00
Interior deck, strip	sf	3600.00
Overhang brackets	ea	156.00
Cantilever deck, form	sf	2100.00
Cantilever deck, strip	sf	2100.00
Perimeter bulkheads, form	sf	345.56
Perimeter bulkheads, strip	sf	345.56
Interior bulkhead, doweled form	sf	45.56
Interior bulkhead, doweled strip	sf	45.56
Expansion bulkhead, dowel form	sf	22.78
Expansion bulkhead, dowel strip	sf	22.78
Hardware and embeds	ea	0.00
Concrete placement locations	ea	3.00
Screed rail	lf	380.00
Finish machine, dry run	sf	5100.00
Finish machine, set up	ea	3.00
Deck preparation	sf	5100.00
Concrete placement (gross)	cy	169.77
Concrete cure	cy	169.77
Concrete cure	sf	5100.00
Rub and patch	sf	1551.00

Superstructure quantity summary: parapet and median unit

Parapet and median unit	Span 1 and 2	
	Unit	Quantity
Parapet rail, form	sf	1707.00
Parapet rail, strip	sf	1707.00
Joint bulkhead, form and strip	sf	47.63
Embedded hardware	ea	0.00
Concrete placement (gross)	cy	32.63
Rub and patch	sf	1707.00

From the quantity summary list, the time duration and production factoring are compiled. The quantity summary list is formed by grouping the operations of similar components.

Example:

Production Factoring, Time Duration

Superstructure deck unit formwork placement

    Operation. Interior deck, formwork placement, conventional.

Crew members	= 10.0 each
Shift duration	= 8.0 hours
Production unit	= square feet
Project quantity requirement	= 3600.0
Formwork type	= wood
Forming condition	= good, 4
Unit cycle time	= 0.70 min/sf
Productivity factor	= 50 minutes
Cycles per hour, 50/0.70	= 71.43 cycles
Production per hour	= 71.43 sf/hour
Production per shift: 71.43 * 8	= 571.44 sf/shift
Required shifts, 3600.0 sf/571.44 sf	= 6.30 shifts
Total craft-hours, 6.3 * 8 * 10	= 504.0 craft-hours
Operation production factor, 504.0/3600.0	= 0.140
Operation factor	= 0.140 craft-hour/sf

Superstructure deck unit formwork removal

    Operation. Interior deck, formwork removal, conventional.

Crew members	= 6.0 each
Shift duration	= 8.0 hours
Production unit	= square feet
Project quantity requirement	= 3600.0
Formwork type	= wood
Forming condition	= good, 4
Unit cycle time	= 0.2 min/sf
Productivity factor	= 55 minutes
Cycles per hour: 55/0.20	= 275.0 cycles
Production per hour	= 275 sf/hour
Production per shift: 275.0 * 8	= 2200.0 sf/shift
Required shifts: 3600.0 sf/2200.0 sf	= 1.64 shifts
Total craft-hours: 1.64 * 8 * 6	= 78.72 craft-hours
Operation production factor: 78.72/3600.0	= 0.022
Operation factor	= 0.022 craft-hour/sf

Superstructure parapet and median unit: concrete placement

Operation. Concrete placement, parapet rail.

Crew members	= 9.0 each
Shift duration	= 8.0 hours
Production unit	= cubic yards
Project quantity requirement	= 32.63
Formwork type	= wood
Forming condition	= n/a
Unit cycle time	= 15.0 min/cy
Productivity factor	= 50 minutes
Cycles per hour: 50/15.0	= 3.33 cycles
Production per hour	= 3.33 cy/hour
Production per shift: 3.33 * 8	= 26.64 cy/shift
Required shifts: 32.63 cy/26.64 cy	= 1.23 shifts
Total craft-hours: 1.23 * 8 * 9	= 88.56 craft-hours
Operation production factor: 88.56/32.63	= 2.714
Operation factor	= 2.714 craft-hours/cy

Superstructure parapet and median unit: rub and patch

Operation. Rub and patch, parapet rail.

Crew members	= 2.0 each
Shift duration	= 8.0 hours
Production unit	= square feet
Project quantity requirement	= 1707.0
Formwork type	= wood
Forming condition	= n/a
Unit cycle time	= 1.5 min/sf
Productivity factor	= 50 minutes
Cycles per hour: 50/1.5	= 33.33 cycles
Production per hour	= 33.33 sf/hour
Production per shift: 33.33 * 8	= 266.64 sf/shift
Required shifts: 1707.0 sf/266.64 sf	= 6.40 shifts
Total craft-hours: 6.40 * 8 * 2	= 102.40 craft-hours
Operation production factor: 102.40/1707.0	= 0.060
Operation factor	= 0.060 craft-hour/sf

Example: Material Allocation

The material costs associated with each operation are the direct permanent and indirect or incidental materials and supplies required per specification.

The cost of the formwork itself is allocated as formwork, which is treated as an asset to the company. The formwork is treated as equipment and has an ownership rate applied to it for its use. This allocation is detailed in the formwork section of the book and deals with the cost associated with the formwork purchase or construction and reuse disbursement.

The consumable formwork required, that will not be reused, is considered direct to the job as an incidental supply.

If there is a requirement for formwork that is not in inventory or is classified as a specialty system, the allocation of price or rental should be factored in the operation over any direct allocation.

The material listed is summarized by operation and allocated to the direct pay item, which is the main component operation required*, for the direct and incidental related items, and the cubic yards of neat line concrete required for the concrete items.

The material allocation is performed for each operation for each type of material required, permanent direct and indirect incidental supplies. The material cost is disbursed against the control operation item quantity*, not the work item quantity.

Superstructure

Operation. Construct uniform deck unit.

Direct material:		
*Concrete, 4500 psi (gross)		= 169.77 cy
Indirect material:		
Beam hanger assembly		= 312.0 ea
24″ coil rod with nuts (2 ea)		= 624.0 ea
Incidental material:		
Form oil (2 oz per sf)		
((3600.0 + 2100.0 + 345.56 + 45.56 + 22.78) * 2 oz)/128		= 95.53 gal
Consumed lumber 2″ dimensional		= 400.0 lf
Consumed ¾ plyform		= 320.0 sf
Nails: 4.5 lb per 100 sf = (6113.9/100) * 4.5		= 275.13 lb
Mortar, rub and patch (94 lb sack)		
0.5 lb per sf = (1551.0 * 0.5)/94		= 8.25 bag
Formwork:		
Beam haunch	1″ dimensional lumber	= 1200.00 lf
Interior deck	¾″ plyform	= 2848.50 sf
Double waler	2″ × 8″ @ 6.33′	= 234.00 ea
Stringer	2″ × 6″	= 3600.00 lf
Cantilever deck	¾″ plyform	= 1851.00 sf
Support bracket	54″	= 156.00 ea
Bracket member	4″ × 4″ @ 7′6″	= 1092.00 lf
Stringer	2″ × 4″	= 2400.00 lf
Bulkhead formwork:		
Perimeter	2″ × 4″ wood panel	= 345.56 sf
Interior	1″ × 8″ dimensional	= 45.56 sf
Expansion	2″ × 4″ wood panel	= 22.78 sf

*Asterisks denote control items for operation.

Operation. Construct parapet and median unit.

Direct material:	
*Concrete, 4500 psi (gross)	= 32.63 cy
Indirect material:	= n/a
Incidental material:	
Form oil (2 oz per sf)	
$((1707.0) * 2\ oz)/128$	= 26.67 gal
Consumed lumber 2″ dimensional	= 50.0 lf
Nails: 2.5 lb per 100 sf = $(1707.0/100) * 2.5$	= 42.68 lb
Mortar, rub and patch (94 lb sack)	
0.5 lb per sf = $(1707.0 * 0.5)/94$	= 9.08 bag
Formwork:	
Wall 2″×4″ wood panel forms	= 1707.00 sf
Bulkhead $\tfrac{3}{4}$″ plyform	= 47.63 sf

Material summary

Operation. Construct superstructure deck.

Direct material:	
*Concrete, 4500 psi (gross)	= 202.40 cy
Indirect material:	
Beam hanger assembly	= 312.0 ea
24″ coil rod with nuts	= 624.0 ea
Incidental material:	
Form oil	= 122.20 gal
Consumed lumber 2″ dimensional	= 450.0 lf
$\tfrac{3}{4}$″ plyform	= 320.0 sf
Nails	= 317.81 lf
Mortar, rub and patch (94 lb sack)	= 17.33 bag
Formwork:	
1″ dimensional	= 1200.00 lf
$\tfrac{3}{4}$″ plyform	= 4699.50 sf
Double waler 2″ × 8″ @ 6.33′	= 234.00 ea
Stringer 2″ × 6″	= 3600.00 lf
Stringer 2″ × 4″	= 2400.00 lf
Support bracket 54″	= 156.00 ea
Bracket member 4″ × 4″ @ 7′6″	= 1092.00 lf
2″ × 4″ wood panel	= 2075.34 sf
1″ × 8″ dimensional	= 45.56 sf
$\tfrac{3}{4}$″ plyform bulkhead	= 47.63 sf

*Asterisks denote control items for operation.

This material list is comprised of the direct permanent and indirect and incidental material required to perform the operations within the controlling item. The direct and indirect material is outlined, and required by specification, and the incidental supplies are required by the operation to be performed. All materials are priced with all applicable sales and use taxes included.

Any required formwork is detailed separately owing to the different required forming systems and the intended reuse, or life of the panels.

The ratio of surface contact area to the control cubic yards gives a prospective for a relationship of cost, both for the current job and for future reference. This is taken from the total formwork surface area of the operation (gross deck area) against the neat line cubic yards for the operation. The same can be done with the cost ratio.

Surface ratio, sf per cy

$$5100.0 \text{ sf}/180.99 \text{ cy} = 28.18 \text{ square feet per cubic yard}$$

Takeoff Quantification Section Summary

The superstructure unit is comprised of various components and operations that will greatly affect the takeoff and costing of the entire unit. Dependent on the structure, deck design, and beam structure unit, these listed components vary in complexity and intended use. These operations must be identified and thoroughly detailed to produce an accurate quantity takeoff and ultimately an accurate and competitive bid analysis.

COMPONENT TAKEOFF

Superstructure Unit

The following outline contains the normal work operation accounts for the performance and cost coding of superstructure construction components.

1. **Deck conventional, form and strip**

   All work related to the conventional wood forming of the interior section of the bridge superstructure deck to its proper line, grade, and dimension, and the stripping of the same. This item also covers all work involved in the forming and stripping of haunches. This includes all embedded hardware items and necessary supports, excluding the beam hangers, needed to construct the form system. Quantity should be reported in gross square feet of the concrete surface contact area (deck area) from the centerline of the exterior girder transversely to the centerline of the exterior girder along with the net area (concrete contact area) required. This measurement should be calculated times the length of each individual span for both forming and stripping, plus the number of bays per span formed, plus the type and size of girders. The nominal size and spacing of walers used should be reported.

2. **Beam hangers**

   All work related to the installation of the metal hanger system needed to support the interior wood forming system and the exterior fascia jacks. Quantity should be reported for the number of each hanger installed, the spacing of the hanger, the lineal feet of beams requiring hangers, and the type of beam or girder.

3. **Fascia jacks**

   All work related to the installation of the conventional fascia or overhang jacks for a wood forming system to their proper line, grade, and spacing, and the removal of same. Quantity should be reported for the number of each jack installed, the spacing of the jacks, the lineal feet of girders requiring the jacks, and the type of beams or girders.

4. **Overhang conventional, form and strip**

    All work related to forming a conventional wood cantilevered system to the proper line, grade, and dimension, including safety handrail and the stripping of same. This includes all embedded hardware items, excluding beam hangers, needed to support the form system. Quantity should be reported in square feet of concrete surface contact area, plus one-half of the width of exterior girder, plus a minimum of 2 feet for walkways for both forming and stripping, along with the net area (concrete contact area) required.

5. **Bulkhead, plain, form and strip**

    All work related to forming a plain longitudinal or transverse vertical bulkhead to its proper line, grade, and dimension, and the stripping of same. Quantity should be reported in square feet of concrete surface contact area.

6. **Bulkhead, doweled, form and strip**

    All work related to forming a doweled rebar longitudinal or transverse vertical bulkhead to its proper line, grade, and dimension, and the stripping of same. Quantity should be reported in square feet of concrete surface contact area.

7. **Deck, metal, stay-in-place**

    All work related to the installation of permanent metal stay-in-place decking on prepared metal support angle, excluding the angle. Quantity should be reported in gross square feet of deck concrete surface contact deck area from the centerline of exterior girder to the centerline of exterior girder times the length of each individual span, along with the net area of metal required.

8. **Deck, metal, support angle**

    All work related to the installation of the metal support angle for the metal decking, to the proper line, grade, and dimension. Quantity should be reported in lineal feet of angle installed on each side of girders, diaphragms, and back walls, along with the type of beam, number of beam straps for structural steel girders, and spacing of either straps or clips (for precast beams).

9. **Precast concrete deck panels**

    All work related to the installation of prestressed, precast concrete deck panels on a prepared haunch buildup, excluding the buildup. Quantity should be reported in gross square feet of deck concrete surface contact area from the centerline of exterior girder to the centerline of exterior girder times the length of each individual span, along with the net area of the panels. This item should include on-site trucking and handling.

10. **Grout pads, form and strip**

    All work related to the forming of the grout pad haunch section of the beam structure in preparation of prestressed, precast concrete deck panels to the proper line, grade, and dimension, and the stripping of same. Quantity should be reported in actual lineal feet of grout pad formed on individual beams and appurtenances for both forming and stripping.

11. **Grout pads, pour**

    All work related to the placement of concrete grout of the specified type. Quantity should be reported in lineal feet of grout placed along with cubic feet including overrun and waste, for each individual placement.

12. **Precast panel support angle**

    All work related to the installation of metal support angle for the haunch section of the beam structure in preparation of prestressed, precast concrete deck panels to the proper line, grade, and dimension. Quantity should be reported in actual lineal feet of support angle installed on individual beams and appurtenances including the method of attachment.

13. **Felt pads**

    All work related to the installation of bituminous felt strip pads for the haunch section of the beam structure in preparation of the prestressed, precast concrete deck panels to the proper line, grade, and dimension. Quantity should be reported in actual lineal feet of bituminous material placed on individual beams and appurtenances.

14. **Overhang gang system**

    All work related to the installation of prebuilt cantilevered overhang system to its proper line, grade, and dimension, and the stripping of same. Quantity should be reported in square feet of horizontal surface placed (actual work times length system), lineal feet of system placed, type of beams attached, for each individual span and the moving to adjacent site. Separate quantity should be reported for assembly of sections.

15. **Half beam hanger**

    All work related to the installation of the metal hanger system needed to support the exterior fascia jacks. These are used with conventional exterior fascia jacks in conjunction with a metal stay-in-place or prestressed, precast concrete deck interior system. Quantity should be reported for the number of each hanger installed, the spacing of the hanger, the lineal feet of beams requiring hangers, and the type of beam.

16. **False decking, forming and stripping**

    All work related to the installation of a protective false deck system on the beam structure and the stripping of same. Quantity should be reported in gross square feet of decking placed, including support members, from the centerline of exterior girder to the centerline of exterior girder times the length of each individual span for both installation and removal, plus the cantilevered portion if required.

Superstructure

17. Safety netting

All work related to the installation of safety netting. Quantity should be reported in gross square feet of netting installed including support members.

18. Finish machine, set screed rail

All work related to the installation of the screed rail to its proper line, grade, including saddles, yokes, and embedded hardware and the removal of same. Quantity should be reported in actual lineal feet of rail placed for each individual concrete placement, including runoff area.

19. Finish machine, setup

All work related to the setup, grade adjustments, dry run, and advancement to the next pour. Quantity should be reported for each individual concrete placement.

20. Prep deck

All work related to deck preparation and cleaning prior to the placement of concrete. Quantity should be reported in gross square feet of concrete surface deck area for each individual span.

21. Deck concrete placement

All work related to the placement of concrete of the specified type. Quantity should be reported in cubic yards of concrete placed and gross square feet of deck area, including overrun and waste, and the net amount of each individual placement, categorized by placement area, i.e., haunch, overhang, deck.

22. Concrete curing

All work related to the curing of concrete of the specified type. Quantity should be reported in cubic yards of concrete placed and gross square feet of deck area, including overrun and waste, and the net amount of each individual placement.

23. Deck, rub and patch

All work related to the rubbing and patching of all concrete areas of the superstructure unit. Quantity should be reported in square feet of surface area prepared.

24. Texture surface finish

All work related to the application of a preapproved material compound to the concrete surface. Quantity should be reported in actual square feet of surface application.

25. Barrier and median railing, form and strip

All work related to the forming of a concrete barrier railing to its proper line, grade, and dimension, and the stripping of same. Quantity should be reported in square feet of concrete surface contact area and lineal feet of barrier section along with hardware embedded items and bulkheads.

26. Barrier and median railing, pour

All work related to the placement of concrete of the specified type for a conventionally formed rail. Quantity should be reported in cubic yards of concrete placed, including overrun and waste, and the net amount of each individual placement, for each individual placement.

27. Barrier and median railing, rub and patch

All work related to the rubbing and patching of all concrete areas of the barrier unit. Quantity should be reported in square feet of surface area prepared.

28. Other

This account is used for other operation accounts not specified here. Quantity should be reported in a representative format relating to the work performed and the appropriate unit measure.

29. Flat slab deck, form and strip

All work related to the erection of a special form deck system required for a cast-in-place deck (without conventional beams) and the removal of same. This includes all embedded items necessary to support the form system. Quantity should be reported in gross square feet of system required, with specific details, for both installation and removal. This includes metal and wood systems.

30. Flat slab deck, falsework

All work related to the system required to support a flat slab deck form system and the removal of same. This includes all necessary structural steel, scaffolding, and shore towers, along with the required hardware and accessories. Quantity should be reported and detailed as per the specific item of work required. This item of work is limited to the deck forming system and requires all foundation work necessary.

Cast-in-Place Superstructure Unit

The following outline contains the normal work operation accounts for the performance and cost coding of cast-in-place superstructure construction components.

1. Falsework (structural)

All work related to the erecting and dismantling of the necessary falsework, scaffolding, and shoring needed to support the concrete structure to its proper line and grade. This also includes soil stabilization and foundation work required. Foundation support piling required should be accounted for under the appropriate piling accounts. Quantity for this item of work should be reported in specific structural components required for the work, against the craft-hours expended.

2. **Falsework (decking)**

    All work related to the decking and profiling of the deck or bottom formwork required to form or support the bottom of the concrete member or structure. Quantity should be reported in square feet of deck formwork completed for both installation and removal along with a detailed material list.

3. **Formwork***

    All work related to forming of the structure above the deck line to its proper line, grade, and dimension. This will include all necessary embedded items and specialty hardware needed to support and anchor the form system. Quantity should be reported in square feet of concrete surface contact area for both forming and stripping of all components. Due to this type of work being of a specialty and normally a custom nature, a detailed summary of the various quantities of components should be separately recorded.

4. **Prestressing steel reinforcement**

    All work required to install the required stressing reinforcement to its designed location, alignment, and profile within the concrete member. This includes all necessary embedded items required to properly support and contain the stressing cables. Quantity should be reported in lineal feet of strand cable installed, detailed by category and size, along with any required waste or overrun needed for the stressing operation.

5. **Ductwork**

    All work related to the installation of the tubing or ductwork, required for posttensioning procedures, to their proper line, grade, dimension, and profile. This would include all necessary embedded hardware needed to support and fasten the ducts in their permanent position. Quantity should be reported in lineal feet of ducts or tube of symmetrical size installed.

6. **Anchorages**

    All work related to the installation of steel anchorages and end plates required for stressing and tensioning procedures. This includes any temporary devices required. Quantity should be reported in number of each of the specified units installed with the structure or member.

7. **Temporary tension and stressing devices**

    All work related to the installation and removal of temporary devices required to support a structure or member until the permanent tensioning reinforcement is in place. Quantity should be reported in the unit of measure required by the specific item.

8. **Concrete placement***

    This work item should be detailed under the provisions set forth in the concrete placement section of the Superstructure Unit.

---

*In addition to this, all component operations listed under Superstructure Unit will apply.

9. **Posttension steel reinforcement**

   All work related to the installation of stressing steel reinforcement wire cable within the ductwork previously placed. Quantity should be reported in total lineal feet installed by designated size of cable, and number of individual cables within each duct, along with any waste or overrun required for stressing procedures.

10. **Stressing**

    All work related to the stressing and tensioning of steel wire reinforcement within the concrete structure or member. Quantity should be reported in lineal feet stressed, number of strands and/or groups, size of cable, and a detailed account of the required force with the number of pulls required to fully stress the cable.

11. **Detensioning**

    All work related to the detension operation required for either the stripping of members or the removal of any stress or tension cables. Quantity should be reported in lineal feet of cable destressed, number of strands or groups, size of cable, and a detailed account of any series of procedures required.

12. **Grouting**

    All work related to the permanent filling of the ducts and tubes with non-shrink grout material. Quantity should be reported in cubic feet of grout placed, lineal feet and size of tube filled, and any waste or overrun of grout needed to complete the operation.

13. **Other**

    This account is used for other operation accounts not specified here. Quantity should be reported in a representative format relating to the work performed and the appropriate unit of measure.

Chapter 16

Approach Structure Unit, Concrete

TECHNICAL SECTION

Introduction

The concrete approach slab is a structural component of the roadway section which lies between the bridge structure and the conventional portion of the roadway pavement.

Approach slabs normally are reinforced and rest on the pavement seat of the abutment backwall for additional support. The approach usually extends off of the bridge structure to a point, normally, at the top of the slope transition (20 to 40 feet).

Technical: Concrete Approach Slab

This work consists of constructing a concrete approach slab composed of air-entrained portland cement concrete, with or without reinforcements, and of structural strength specified. The slab is constructed in accordance with project specifications and in reasonably close conformity with the lines, grades, thicknesses, and typical cross sections established by the plans.

The roadbed subsoil and base are constructed as shown and compacted to a suitable ratio, as specified, to support the roadway approach slab without settlement.

On the prepared subgrade, formwork is placed to form the perimeter of the slab and to establish the proper elevation at these points. Also any required construction or expansion joint lines are formed at this time.

Within these cells, the specified reinforcement and embedded items are installed at their designated location and to project specifications.

520 Takeoff and Cost Analysis Techniques

Prior to the placement of concrete, the area must be prepared and debris or foreign material must be removed.

The concrete must be placed by a mechanical finisher, to its proper line and grade, be of the specified class, and be finished and cured according to project specifications, and as outlined in the Concrete Section: Mixing, Placing, Finishing, and Curing.

TAKEOFF QUANTIFICATION

Introduction

The structural approach unit is the reinforced concrete slab section that "bridges" the area from the abutment section to the conventional roadway section to the bridge structure. It is designed and constructed so the structural integrity of the cross section of the approach unit is not solely reliant on the earthen fill section beneath it but is capable of support within itself.

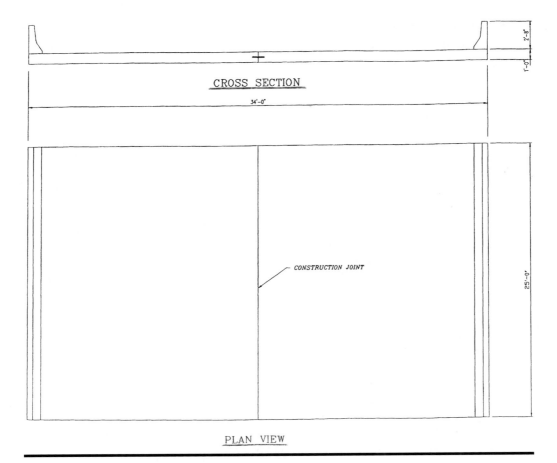

Typical section, approach slab.

Definition

The approach unit to the bridge structure is the structural concrete section located between the bridge structure and the typical roadway portion. It is comprised of various components and work operations.

Takeoff

The takeoff of the approach unit is performed by identifying the specific components and work operations within the defined area of work, by project specifications and plans, and by understood and known construction techniques.

The structural approach unit, normally, is constructed after the bridge structures' adjacent deck units but can be placed prior to the roadway section.

Primary and secondary components

The quantity takeoff procedure of the structural approach unit is compiled by the identification of the primary components first, then the secondary operations and components. The primary component of the approach unit is the concrete slab. The operations associated with the concrete slab are the formwork operation, embed installation, the placing and curing of the concrete, and cleanup and restoration relative to the slab.

The secondary components are defined as the grade preparation, safety or parapet walls, and curbs or gutters directly relative to the approach unit itself.

The quantity takeoff for the approach unit provides and associates the components with the required craft-hours necessary to produce the function, establish the productivity factors, and identify the equipment requirements along with the material needs, both direct and indirect incidental.

The techniques and procedures are performed like the superstructure deck unit, the difference being that the approach unit is constructed on grade, whereas the superstructure deck is constructed on the beam structure. The bulkhead formwork, placement, finishing, and curing procedures are treated like the deck unit components. The use of the conventional placing method of the deck unit and the use of the superstructure finish machine is strongly recommended owing to the adjacent deck profile and defined crowns and logistics of access.

The normal components associated directly with the approach unit are identified as:

1. Grade and base preparation
2. Formwork requirements
3. Approach slab preparation
4. Screed rail
5. Finish machine setup, dry run
6. Concrete placement

7. Parapet and median rail
8. Curb and gutter, miscellaneous section

Within these components exist the secondary operations required for the performance and execution of the direct component activity. The quantity takeoff is outlined for the proper procedure and allocation of the appropriate and relative work operation.

The grade preparation quantity is the required component contact area for the concrete slab itself. The operation item includes the preparatory work required to ensure that the grade and elevation of the subgrade or base is within specification and to prevent yield loss overrun of the concrete placement component. The grade preparation component is defined only as the fine-grade operation; the backfill or earthwork operations are defined under the foundation preparation category.

Once the component contact area has been graded and prepared, the required formwork can be identified. The two classifications of formwork are defined as the perimeter formwork and the interior bulkhead formwork. As with the superstructure unit, the interior bulkhead joint component operation is identified by the design characteristics of the approach unit with respect to designated interior construction joints and optional interior construction joints, plain or doweled. The perimeter bulkhead form system is the defined perimeter of the approach unit requiring formwork.

The embed component is calculated as the actual quantity of embeds or formwork hardware item required.

Most approach unit concrete placements are required to be finished and screeded by a mechanical finish machine, similar to the superstructure deck unit. This machine is controlled in elevation by a metal rail that is placed adjacent to and parallel to each pour. The screed rail must be quantified for each concrete placement.

The finish machine must be set up and verified as to profile and elevation, referred to a dry run. This is an actual check of the elevation dimensions, a dimension to which the machine's finishing attachment is set in vertical distance, from various points on the grade surface. This is an assurance to the profile of the finish machine and final approach unit elevations, the required concrete cover to the reinforcing steel, and the required depth of concrete. This component is required by individual placement area.

The approach slab preparation is the next component to define. It is defined as the entire horizontal surface concrete contact area of the approach unit, identical to the grade preparation. This component is the performance of the final cleaning and prepping the approach slab, ready for the concrete placement.

Concrete placement is the next component to define. This operation is the actual placement of the concrete by the placing method chosen. There are various placement procedures. The type chosen is quantified as the required operation to fully perform the placement, including setup, surface texturing if required, and cleanup, for the required cubic yards per pour.

Approach Structure Unit, Concrete

The concrete placement operation is defined by the designated square yard measurement of actual approach slab unit, categorized as the net and gross required cubic yard concrete volume. The controlling unit of measure is in square yards of required approach unit.

The concrete curing component is quantified in conjunction with the placement operation. This is the required procedure to properly perform the curing operation as outlined. It is defined as a quantity of required gross cubic yards and approach slab placement surface area.

Adjacent to the approach unit are components which may or may not be present for all designs and/or applications. These components are defined as secondary items and are identified as parapet and median rail, safety rail, or curb and gutter sections. These components are quantified by the specific work operation associated with the unit. The work operations are the grade preparation, formwork requirements, concrete placement and curing, embedded items, and the rubbing and patching requirements. The controlling unit of measure is in lineal feet of required rail or curb.

The parapet and median barrier construction operations are first detailed as lineal feet required, by type, then by form surface area and required cubic yards. The secondary operations required for the formwork and concrete placement are identified also.

Another method of parapet and median wall construction is by slipform methods, which require no conventional formwork. The slipform placement procedure is quantified in the required lineal feet, by wall type and required cubic yard volume. The secondary operations associated with this procedure are also defined.

Approach unit: grade preparation

The quantity takeoff of the required grade preparation and final grading for the approach unit are performed by calculating the component contact surface area of the concrete slab.

The requirement of the quantity takeoff outline, for the grade and base preparation component, is listed as:

1. Component contact surface area

The quantity takeoff calculation is performed by multiplying the width of the concrete slab by the length of the concrete slab, maintaining a quantity of square feet. In addition to the slab component, the surface quantity for the component contact area of any subsequent components, such as parapet and median rail or curb and gutter sections, is calculated.

When the width or length dimension is variable, owing to a skew or transition, the dimensions are averaged for ease of calculation.

The outlined procedure for the grade preparation formulated, by a cumulative unit of square feet, is listed as:

1. Approach structure, grade preparation, slab unit, sf
2. Approach structure, grade preparation, parapet and median rail, sf
3. Approach structure, grade preparation, curb and gutter section, sf

Approach unit: formwork, perimeter bulkhead

The perimeter bulkhead is defined as the outside perimeter line of the approach structure unit requiring formwork. This is the vertical formwork located at the required form line of the perimeter edge.

The requirement of the quantity takeoff outline, for the perimeter vertical bulkhead formwork, is listed as:

1. Concrete slab thickness, at the form line
2. Perimeter dimensions

The required calculation procedure to determine the perimeter vertical formwork is performed by multiplying the thickness of the concrete slab, at the required form line, by the length of each perimeter line of the approach unit. This calculation is performed and summarized in a unit measurement of required square feet. The craft-hour consumption unit is factored for the full operation performance required to complete the task.

The quantity takeoff of the vertical perimeter formwork component is defined, in required square feet, as:

1. Approach slab, perimeter bulkhead, formwork placement, sf
2. Approach slab, perimeter bulkhead, formwork removal, sf

NOTE: Any adjacent abutting pour areas are not calculated.

Approach unit: formwork, interior bulkhead, plain

The interior vertical bulkheads, plain, are defined as the required interior slab form lines that require a nontied joint. These are considered as transverse or longitudinal, interior plain bulkhead joints.

The requirement of the quantity takeoff outline, for the plain, interior vertical bulkhead formwork, is listed as:

1. Concrete slab thickness, at the form line
2. Length of each required interior joint

The required calculation procedure to determine the vertical plain, interior formwork is performed by multiplying the thickness of the concrete slab at the required form line by the length of each form line within the approach unit. This calculation is performed and summarized in a unit measurement of required square feet. The craft-hour consumption unit is factored for the full operation performance required to complete the task.

The quantity takeoff of the plain, interior formwork component is defined, in required square feet, as:

1. Approach slab, interior bulkhead, plain, formwork placement, sf
2. Approach slab, interior bulkhead, plain, formwork removal, sf

Approach unit: formwork, interior bulkhead, doweled

The vertical interior bulkheads, doweled, are defined as the required interior slab, doweled construction joint form lines, that require a tied joint. These are considered transverse or longitudinal interior slab joints, doweled construction bulkhead joints.

The requirement of the quantity takeoff outline, for the doweled, interior vertical bulkhead formwork, are listed as:

1. Concrete slab thickness, at the form line
2. Length of each required interior joint
3. Dowel, tie bar spacing

The required calculation procedure to determine the doweled, vertical interior formwork is performed by multiplying the thickness of the concrete slab, at the required form line, by the length of each form line within the approach unit. This calculation is performed and summarized in a unit measurement of required square feet. The craft-hour consumption unit is factored for the full operation performance required to complete the task.

The quantity takeoff of the vertical doweled, interior formwork component is defined, in required square feet, as:

1. Approach slab, interior bulkhead, doweled, formwork placement, sf
2. Approach slab, interior bulkhead, doweled, formwork removal, sf

Approach unit: embedded hardware

The embedded item quantifying defines any incidental embeds required within the approach structure unit. These are summarized in a unit of each as:

1. Approach unit, embeds, ea

Approach unit: required placement locations

This quantity takeoff is merely the operation of defining each placement location, or pour sequence area, within the approach unit.

These defined pour locations can be identified as required, being those designated owing to design criteria with respect to the correlation of the deck unit, the roadway section, or as optional construction choices. The optional placement locations give the contractor the choice to make the actual defined, independent placement or combine with another. This is normally defined as an option merely for volume control of the placement rate of cubic yards per hour, or project logistics. When an optional placement area is defined, there will also be the optional component of the construction joint, or interior bulkhead associated with the placement. This must be added to or deleted from that quantity component operation.

The defined area for each placement is an independent approach slab segment, located between and within any vertical bulkhead form lines.

The outline requirement for the quantity takeoff procedure is identified as:

1. Number of required or optional placement areas

The quantity takeoff of the approach unit concrete placement location component is defined, in a requirement of each, as:

1. Approach unit, placement locations, each

Approach unit: screed rail

The operation of the quantity takeoff for the required finish machine screed rail component, as with the superstructure deck unit, is performed by first identifying the placement location areas within the approach unit. From this, the length of each placement location is defined.

Normally, it is required to set a finish machine screed rail, parallel to each placement location, and to the finish machine direction of placement. The screed rail is seated on yokes, or saddles, which are set to a preliminary grade elevation during this procedure.

The screed rail with yokes will set directly on the defined bulkhead formwork (perimeter or interior) or grade, for the pilot placements, or directly on the adjacent cured concrete for the filler placement areas of the approach unit.

The operation of the finish machine screed rail is categorized in two components: (1) screed rail, bulkhead placement; or (2) screed rail, surface placement. This separate defining accounts for the additional work required to place the screed rail on the bulkhead formwork and the yoke placement within the bulkhead formwork.

For the approach slab area and runoff area for each placement segment, it is recommended to add 20′ of screed rail, per side, to each end of the placement length. This additional length must be determined and quantified for either each specific placement or once for the approach structure.

The main consumption requirement for this operation includes the outlined work necessary to set the yokes and saddles within the bulkhead formwork or approach surface, place and rough grade the screed rail, and remove the system after the placement.

The quantity outline procedure for the finish machine screed rail component is defined as:

1. Required placement locations
2. Length of placement area

3. Additional rail required
4. Type of screed rail component; bulkhead or surface

The calculation for the required quantity of screed rail is performed by category component type, by summarizing the length of each side of the specific placement area and adding the additional required rail for the approach slab and runoff area, normally 80 lineal feet. Then each placement area is summarized for the complete rail quantity requirement of the approach unit.

The quantity takeoff for the requirement of finish machine screed rail component is defined, in lineal feet by category type, as:

1. Approach unit, screed rail, bulkhead mount, lf
2. Approach unit, screed rail, surface mount, lf

Approach unit: finish machine, setup, dry run

The operation of the approach finish machine setup and dry run is performed for quantity takeoff by first defining the required placement locations within the approach unit. The actual surface square feet within each placement location are calculated next by multiplying the width of the placement surface area by the length.

This operation includes the required and necessary work operations to set the finish machine on the screed rail for the proper width, set the profile of the machine to that required within the placement area, and dry-run the finish machine over the placement area. The dry run verifies the profile of the finish machine to that of the specified approach slab elevation grades. This dry-run procedure also checks and verifies the reinforcing steel depth below the surface of the concrete, referred to a cover, and the proper depth of the concrete placement.

The outline requirement for the quantity takeoff of the finish machine component is defined as:

1. Required placement areas
2. Square feet surface area and placement area

The craft-hour consumption for the finish machine operation is calculated under two functions. The first function is the setup requirement. This is defined as the work necessary to set the finish machine on the rail to its proper width and set the required profile to the machine. The performance of this operation is calculated and recorded by a unit of each, for each required placement location.

The second function, the dry-run operation of this component, is quantified in a unit of placement surface area in square feet. This is the necessary performance of setting the machine to the proper elevations for approach slab screed points.

The quantity takeoff for the finish machine requirements is listed by a dual component operation, in locations and square feet, as:

1. Approach slab, finish machine, setup, each
2. Approach slab, finish machine, dry run, sf

Approach unit: placement preparation

The requirement for the quantity takeoff of the approach slab placement preparation operation is performed by a calculation in the unit measure of square feet of actual approach slab surface area. The takeoff outline requirement is defined as:

1. Approach slab surface area, square feet

This operation is the cleaning and final preparation of the gross approach slab surface area immediately prior to the concrete placement operation. The quantity summation is performed for each placement area. The calculation required for the approach slab preparation component is performed by multiplying actual slab unit width (fascia bulkhead to fascia bulkhead) by the length of the slab unit (centerline of abutment bearing to the roadway side of the approach slab).

When the width or length dimension is variable, owing to a skew or transition, the dimensions are averaged for ease of calculation.

The quantity takeoff for the approach slab preparation component requirement is listed by location, in a unit of square feet, as:

1. Approach slab placement preparation, sf

Approach unit: concrete placement

The concrete placement procedure of quantity takeoff, for the approach structure, is considered one operation.

The uniform approach slab section is defined as the cross section of approach component, normally the uniform thickness section of the slab, from edge of fascia to edge of fascia.

The calculation for the concrete volume quantity takeoff defines the variable dimensions and normally the fixed dimension of this component. The fixed dimensions are the uniform thickness of the slab section. The variable dimensions encompass the approach slab width or length. These variable dimensions are detailed on the plan specifications.

The width of each specified approach unit may vary in cross section dimension because of design criteria of skews and variable pattern requirements and should be calculated to an average width by specific approach unit. The length of each specified approach unit may vary in dimension owing to design criteria of skews and variable station dimensions and should be calculated to an average length measurement by approach unit.

The quantity outline for each uniform approach slab unit component calculation requirement is defined as:

Approach Structure Unit, Concrete

1. Uniform slab component thickness
2. Slab unit cross section width (average)
3. Slab unit length (average)
4. Yield and overrun loss factor
5. Placement method

The volume of concrete for the uniform slab component is calculated for each approach unit separately owing to the variables of differing dimensional characteristics of each approach unit.

The individual calculation, for each approach unit, is performed by multiplying the width (average) of each slab unit by the length (average) of each slab unit, multiplied by the uniform thickness of slab unit, converting to a cubic yard unit quantity calculation. Added to this net volume quantity is a yield loss and overrun factor, giving the gross required quantity volume. The cubic yard quantity, both net and gross, is then summarized for each approach slab unit within the approach structure.

The takeoff quantity for the concrete volume of the uniform approach component is summarized, by a gross and net cubic yard unit measurement, as:

1. Approach slab, concrete placement, net cy
2. Approach slab, concrete placement, gross cy

Approach unit: curing, concrete placement

The operation of curing of the concrete approach unit incorporates the gross volume of placed concrete and the gross square foot surface area of the approach structure unit.

The curing procedure for the approach component may be specified as water curing, covering, curing compound application, or a combination.

The gross cubic yard volume is derived from the concrete placement component, listed as cubic yards, and the gross approach slab square footage is derived from the slab placement preparation component, listed as square feet.

The curing component is the required operations necessary to perform the curing procedures as specified. If the requirement of heating or cooling the concrete mixture is relevant, the gross cubic yard volume provides for this quantity.

The outline for takeoff quantity of the curing operation is listed as:

1. Required gross cubic yard volume
2. Required gross approach slab, square feet
3. Curing method required
4. Heating and cooling mix requirement

The takeoff quantity for the concrete approach curing component is summarized by a gross cubic yard unit measurement and gross slab surface square feet as:

1. Approach slab, concrete curing, gross cy
2. Approach slab, concrete curing, gross sf

Parapet and Median Rail

The parapet rail is identified as the vertical barrier located at the approach slab unit, connecting with the deck parapet rail. The median rail is identified as the vertical barrier located within the interior or center portion of the approach unit, connecting with the deck median rail. The parapet and median rail units are comprised of various components and work operations.

The outline requirement for the quantity takeoff of the parapet and median rail is listed by component as:

1. Type of required rail
2. Lineal feet of required rail
   a. Main slab
   b. End transition
3. Vertical dimensions, front and back face
4. Cross section area

Approach unit: parapet and median, formwork

The formwork requirement of the parapet and median rail is defined as the vertical concrete surface contact area of required formwork. The component contact surface area is the horizontal surface between the slab unit and the parapet and median rail unit; this is not a formed surface area.

The quantity takeoff for the formwork requirement is performed by multiplying the vertical contact surface, for both sides of the rail, by the length of required rail. This calculation is categorized by rail type as to cross section and by area as the main slab rail and the end transition rail.

The formwork removal is summarized as the identical quantity of the formwork requirement for placement.

The quantity takeoff for the formwork requirements of the parapet and median rail is summarized in square feet of contact area by type, by area, as:

1. Approach unit, parapet and median main rail: formwork placement, sf, type _____
   a. Approach unit, parapet and median main rail: formwork removal, sf, type _____
2. Approach unit, parapet and median transition rail: formwork placement, sf, type _____
   a. Approach unit, parapet and median transition rail: formwork removal, sf, type _____

Approach unit: parapet and median, bulkhead

The bulkhead takeoff requirement is summarized at this time. This calculation consists of the square feet of the bulkhead end cross section area multiplied by the number of bulkheads required by plans, and summarized as:

1. Approach unit, parapet and median main rail, bulkhead, sf, type _____
2. Approach unit, parapet and median transition rail, bulkhead, sf, type _____

Approach unit: parapet and median, embedded hardware

The embedded item quantifying defines any incidental embeds required within the parapet and median rail. These are summarized in a unit of each as:

1. Approach unit, parapet and median main rail, embeds, ea, type _____
2. Approach unit, parapet and median transition rail, embeds, ea, type _____

Approach unit: parapet and median, concrete placement

The concrete requirements are calculated by a simple volume determination, the width in section, multiplied by the depth, multiplied by the length, of the total area of the parapet and median rail to be filled by the concrete. When there are quadrants of variable volumes and sizes, each area is calculated independently, then summarized together. This quantity of cubic feet is then multiplied by the yield and loss overrun factor, and is then translated to a volume measurement of required cubic yards, concrete placement, by required type and area of wall.

1. Approach unit, parapet and median main rail, concrete placement, cy, type _____
2. Approach unit, parapet and median transition rail, concrete placement, cy, type _____

Approach unit: parapet and median, rub and patch

Rubbing and patching is calculated by the surface area, in square feet, of the component that is required to be completed. This quantity should be equal to the formwork placement component. The quantity is then summarized as:

1. Approach unit, parapet and median, main rail, rub and patch, sf, type _____
2. Approach unit, parapet and median, transition rail, rub and patch, sf, type _____

Approach unit: parapet and median, slipform

The quantity takeoff for the operation of slipforming the parapet and median rail component is defined in a unit measure of lineal feet for the portion of rail to be performed by the slipforming procedure.

Slipforming consists of molding the parapet and median rail by extruding the fresh concrete mix through a mold mounted on a self-propelled machine, requiring no formwork.

The operation of slipforming constitutes the following outlined operational components under one progressing operation. The slipform operation includes the placement of concrete, required joint work, and finishing or texturing of the wall during the procedure.

Slipforming may not encompass the entire parapet and median rail quantity as detailed and may require conventional construction methods combined with this. With this condition, the slipform operation is summarized with the conventional and/or transition rail summary costs.

The quantity takeoff encompasses the lineal feet of required wall to be slipformed, the volume of concrete required for the cross section along with the yield loss and overrun factor, and the conditions that exist for the operation. This component is summarized in a unit measurement of required lineal feet of rail, by type, for this activity as:

1. Approach unit, parapet and median rail, slipform, lf, type _____

Curb and Gutter Section

The curb and gutter section is identified as the closure end section located at the side or edge of the approach slab unit, connecting to the approach slab. The curb and gutter units are comprised of various components and work operations.

The outline requirement for the quantity takeoff of the curb and gutter section is listed by component as:

1. Type of required curb and gutter
2. Lineal feet of required curb and gutter
3. Vertical dimensions, front and back face
4. Cross section area

The curb and gutter, miscellaneous concrete components, are similar to those of barrier walls, and detailed and categorized accordingly.

Procedure

The quantity takeoff procedure for the structural approach unit is performed under the basis of order of construction operation. The grade preparation is the first item of quantity identification, followed by the formwork, placement procedures, and secondary work item operations.

Approach Structure Unit, Concrete

The takeoff is performed in a systematic manner of accuracy and detail, for both major and minor work operations and components.

Questions of Estimate

In order to perform and properly identify the outline of priority order and operation sequencing for the structural approach unit, the following questions are posed and verified to accurately ensure a proper and competitive quantity takeoff.

1. What type of grade is present?
2. Is the approach slab a uniform thickness?
3. How many placement areas are required?
4. What are the optional or defined joint patterns?
5. What type of construction access is available?
6. What type of concrete is required?
7. What are the placement and construction conditions?

Application

The application of the quantity takeoff techniques, for a structural approach unit, is performed using the following example and conditions of a proposed approach unit.

This example of actual conditions and dimensions shows the production requirements produce the required craft-hour production factors, along with a detailed summary listing of required formwork, concrete volumes, material needs, both permanent and incidental, and equipment usage.

A given structural approach unit, being symmetrical for both abutments, contains the required control components of:

Approach slab

Parapet rail

Within these components, the required work operations are:

Grade preparation

Formwork and formwork removal

Installation of form hardware, embeds

Screed rail installation

Finish machine setup, dry run

Slab placement preparation

Concrete placement

Parapet and median, miscellaneous components

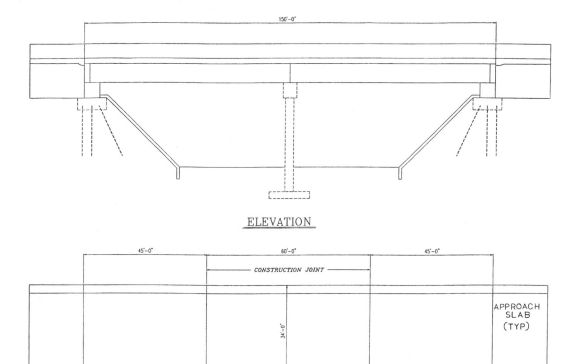

Plan view. North approach. South approach.

The following is a listing of required component dimensions as shown by the project plans:

Structural approach unit (north and south symmetrical)

Approach slab, each	12″ thick, 42′ wide, 25′ long
Construction joint	Required, centerline
Parapet	32″ high, 25′ long

Example: Operation Analysis
Approach structure, north unit

Operation. Grade preparation. Unit: Square feet.
The operation of grade preparation for the approach slab unit is calculated in the following manner for the required surface area, by placement zone:

Required: Placement panels @ 21′w × 25′l, 2 ea

The width (average) of the placement area, multiplied by the length (average) of the placement area, summarizing the concrete placement areas of all approach slab units, in a measurement of square feet.

$$(21' * 25') * 2 = 1050.0 \text{ sf preparation}$$

Operation. Perimeter bulkhead, formwork, form and strip. Unit: Square feet.

The operation of formwork placement of the perimeter bulkhead for the approach slab unit is calculated in the following manner for the required contact surface area. The removal of the formwork is calculated as an equal quantity.

The length of each form line at the required perimeter area, multiplied by the thickness of the slab at the form line, summarizing the perimeter bulkheads of the approach slab units, in a measurement of square feet.

$$(25' + 42' + 25') * 1.0' = 92.0 \text{ sf place formwork}$$

$$= 92.0 \text{ sf remove formwork}$$

Operation. Interior bulkhead, doweled, form and strip. Unit: Square feet.

The operation of forming and stripping the interior bulkhead component is calculated in the following manner for the required unit area:

Required: Dowel spacing, bottom mat @ 8″

top mat @ 10″

The thickness of the interior bulkhead multiplied by the length of each interior bulkhead, summarizing the quantity.

$$(1.0' * 25') = 25.0 \text{ sf form}$$

$$= 25.0 \text{ sf strip}$$

Operation. Required placement locations. Unit: Each.

Placements = 2.0 each

Operation. Screed rail. Unit: Lineal feet.

The operation of installing the finish machine screed rail component is calculated in the following manner for the required unit area:

Required: Bulkhead condition

Approach and runoff 20' * 4 ea, approach

Placement length: 25', 25'

Required: Surface condition

Approach and runoff 20' * 2 ea, approach

Placement length: 25'

The length of each placement, for the bulkhead condition, summarizing the quantity for each placement, plus the runoff dimension (80′), plus the length of the placement for the surface condition, plus the runoff dimension (40′).

$$(25' + 25') + 80' = 130.0 \text{ lf rail, bulkhead}$$

$$(25') + 40' = 65.0 \text{ lf rail, surface}$$

Operation. Finish machine, dry run, setup. Unit: Square feet, each.

The operation of the finish machine dry-run setup component is calculated in the following manner for the required unit area:

Required: Placement areas, 2 ea

The length of each placement area multiplied by the width of each placement area, summarizing the quantity for each placement area, giving the required surface area; plus the quantity of placement areas, giving the required setup locations.

$$(25' * 21') + (25' * 21') = 1050.0 \text{ sf dry run}$$

$$\text{Location area setup} = 2.0 \text{ each setup}$$

Operation. Placement preparation. Unit: Square feet.

The operation of the approach slab placement preparation component is calculated in the following manner for the required unit area:

Required: Placement areas, 2 ea

The length of each placement area multiplied by the width of each placement area, summarizing the quantity for each placement area, giving the required surface area.

$$(25' * 21') + (25' * 21') = 1050.0 \text{ sf preparation}$$

Operation. Concrete placement, approach unit. Unit: Cubic yard.

The operation of the approach slab concrete placement component is calculated in the following manner for the required unit area:

Required: Uniform slab volume

1.0′ thick

Yield loss 10%

The width of each slab unit placement, multiplied by the length of each slab unit placement, multiplied by the average thickness of the uniform slab section, plus the yield loss factor, converting to a cubic yard measurement and

Approach Structure Unit, Concrete

summarizing the quantity for each placement area, giving the required volume in both gross and neat line cubic yards.

$$(((((21' * 25') + (21' * 25')) * 1.0') * 1.10)/27) = 42.78 \text{ cy gross}$$

$$((((21' * 25') + (21' * 25')) * 1.0')/27) = 38.89 \text{ cy neat}$$

Operation. Curing, concrete placement. Unit: Cubic yard.

The operation of curing, of the concrete placement component, is calculated in the following manner for the required unit area:

$$\text{Gross concrete volume requirement} = 42.78 \text{ cy cure}$$

$$\text{Approach slab surface area} = 1050.0 \text{ sf cure}$$

Approach slab parapet and median rail unit

Operation. Parapet rail, form and strip. Unit: Square feet.

The operation of forming and stripping the parapet rail component is calculated in the following manner for the required unit area:

Required: Height 2.67' back face

3.02' front surface (contour)

Cross section area 2.646 sq ft (381.0 sq in)

The height of the back face plus the height of the front face, multiplied by the length of required main rail (both sides of the approach slab unit), giving the required square feet of formwork.

$$(2.67' + 3.02') * (25' * 2) = 284.50 \text{ sf form}$$

$$= 284.50 \text{ sf strip}$$

Operation. Parapet rail, bulkhead. Unit: Square feet.

The operation of forming and stripping the parapet rail bulkhead component is calculated in the following manner for the required unit area:

Cross section area 2.646 sq ft (381.0 sq in)

The square feet of each bulkhead, multiplied by the number of required bulkheads within the main rail (length of rail divided by the joint spacing plus one, rounded, for each side), giving the required square feet of bulkhead.

$$2.646' * (((25'/15) + 1)\text{rounded} * 2) = 15.88 \text{ sf form and strip}$$

Operation. Parapet rail, embedded hardware. Unit: Each.
None required.

Operation. Parapet rail, concrete placement. Unit: Cubic yard.

The operation of concrete placement for the parapet rail component is calculated in the following manner for the required unit area:

Required: Yield and loss factor 11%

Cross section area 2.646 sq ft (381.0 sq in)

The square feet of end area cross section, multiplied by the lineal feet of main rail, multiplied by the yield and loss factor, converting to a cubic yard measurement, giving the required concrete volume in both gross and neat line cubic yards.

$$((2.646' * 50') * 1.11)/27 = 5.44 \text{ cy gross}$$

$$(2.646' * 50')/27 = 4.90 \text{ cy neat}$$

Operation. Parapet rail, rub and patch. Unit: Square feet.

The operation of the parapet rail rub and patch component is calculated in the following manner for the required unit area:

The square feet of surface area equal to the formwork requirement, giving the required rub and patch surface area.

$$\text{Contact surface area} = 284.50 \text{ sf rub and patch}$$

Approach structure, south unit

Operation. Grade preparation. Unit: Square feet.

The operation of grade preparation for the approach slab unit is calculated in the following manner for the required surface area, by placement area:

Required: Placement panels @ 21'w × 25'l, 2 ea

The width (average) of the placement area, multiplied by the length (average) of the placement area, summarizing the concrete placement area of all approach slab units, in a measurement of square feet.

$$(21' * 25') * 2 = 1050.0 \text{ sf preparation}$$

Operation. Perimeter bulkhead, formwork, form and strip. Unit: Square feet.

The operation of formwork placement of the perimeter bulkhead for the approach slab unit is calculated in the following manner for the required contact surface area. The removal of the formwork is calculated as an equal quantity.

The length of each form line at the required perimeter area, multiplied by the thickness of the slab at the form line, summarizing the perimeter bulkheads of the approach slab units, in a measurement of square feet.

$$(25' + 42' + 25') * 1.0' = 92.0 \text{ sf place formwork}$$

$$= 92.0 \text{ sf remove formwork}$$

Operation. Interior bulkhead, doweled, form and strip. Unit: Square feet.

The operation of forming and stripping the interior bulkhead component is calculated in the following manner for the required unit area:

Required: Dowel spacing, bottom mat @ 8″

top mat @ 10″

The thickness of the interior bulkhead multiplied by the length of each interior bulkhead, summarizing the quantity.

$$(1.0' * 25') = 25.0 \text{ sf form}$$

$$= 25.0 \text{ sf strip}$$

Operation. Required placement locations. Unit: Each.

$$\text{Placements} = 2.0 \text{ each}$$

Operation. Screed rail. Unit: Lineal feet.

The operation of installing the finish machine screed rail component is calculated in the following manner for the required unit area:

Required: Bulkhead condition

Approach and runoff 20′ * 4 ea, approach

Placement length, 25′, 25′

Required: Surface condition

Approach and runoff 20′ * 2 ea, approach

Placement length, 25′

The length of each placement, for the bulkhead condition, summarizing the quantity for each placement, plus the runoff dimension (80′) plus the length of the placement for the surface condition, plus the runoff dimension (40′).

$$(25' + 25') + 80' = 130.0 \text{ lf rail, bulkhead}$$

$$(25') + 40' = 65.0 \text{ lf rail, surface}$$

Operation. Finish machine, dry run, setup. Unit: Square feet, each.

The operation of the finish machine dry-run setup component is calculated in the following manner for the required unit area:

Required: Placement areas, 2 ea

The length of each placement area multiplied by the width of each placement area, summarizing the quantity for each placement area, giving the required surface area, plus the quantity of placement areas, giving the required setup locations.

$$(25' * 21') + (25' * 21') = 1050.0 \text{ sf dry run}$$

$$\text{Location area setup} = 2.0 \text{ each setup}$$

Operation. Placement preparation. Unit: Square feet.
The operation of the approach slab placement preparation component is calculated in the following manner for the required unit area:

$$\text{Required: Placement areas, 2 ea}$$

The length of each placement area multiplied by the width of each placement area, summarizing the quantity for each placement area, giving the required surface area.

$$(25' * 21') + (25' * 21') = 1050.0 \text{ sf preparation}$$

Operation. Concrete placement, approach unit. Unit: Cubic yard.
The operation of the approach slab concrete placement component is calculated in the following manner for the required unit area:

$$\text{Required: Uniform slab volume}$$

$$1.0' \text{ thick}$$

$$\text{Yield loss } 10\%$$

The width of each slab unit placement, multiplied by the length of each slab unit placement, multiplied by the average thickness of the uniform slab section, plus the yield loss factor, converting to a cubic yard measurement and summarizing the quantity for each placement area, giving the required volume in both gross and neat line cubic yards.

$$(((((21' * 25') + (21' * 25')) * 1.0') * 1.10)/27) = 42.78 \text{ cy gross}$$

$$((((21' * 25') + (21' * 25')) * 1.0')/27) = 38.89 \text{ cy neat}$$

Operation. Curing, concrete placement. Unit: Cubic yard.
The operation of curing, of the concrete placement component, is calculated in the following manner for the required unit area:

$$\text{Gross concrete volume requirement} = 42.78 \text{ cy cure}$$

$$\text{Approach slab surface area} = 1050.0 \text{ sf cure}$$

Approach slab parapet and median rail unit

Operation. Parapet rail, form and strip. Unit: Square feet.

The operation of forming and stripping the parapet rail component is calculated in the following manner for the required unit area:

Required: Height 2.67' back face

3.02' front surface (contour)

Cross section area 2.646 sq ft (381.0 sq in)

The height of the back face plus the height of the front face, multiplied by the length of required main rail (both sides of the approach slab unit), giving the required square feet of formwork.

$$(2.67' + 3.02') * (25' * 2) = 284.50 \text{ sf form}$$

$$= 284.50 \text{ sf strip}$$

Operation. Parapet rail, bulkhead. Unit: Square feet.

The operation of forming and stripping the parapet rail bulkhead component is calculated in the following manner for the required unit area:

Cross section area 2.646 sq ft (381.0 sq in)

The square feet of each bulkhead, multiplied by the number of required bulkheads within the main rail (length of rail divided by the joint spacing plus one, rounded, for each side), giving the required square feet of bulkhead.

$$2.646' * (((25'/15) + 1) \text{rounded} * 2) = 15.88 \text{ sf form and strip}$$

Operation. Parapet rail, embedded hardware. Unit: Each.
None required.

Operation. Parapet rail, concrete placement. Unit: Cubic yard.

The operation of concrete placement for the parapet rail component is calculated in the following manner for the required unit area:

Required: Yield and loss factor 11%

Cross section area 2.646 sq ft (381.0 sq in)

The square feet of end area cross section, multiplied by the lineal feet of main rail, multiplied by the yield and loss factor, converting to a cubic yard measurement, giving the required concrete volume in both gross and neat line cubic yards.

$$((2.646' * 50') * 1.11)/27 = 5.44 \text{ cy gross}$$

$$(2.646' * 50')/27 = 4.90 \text{ cy neat}$$

Operation. Parapet rail, rub and patch. Unit: Square feet.

The operation of the parapet rail rub and patch component is calculated in the following manner for the required unit area:

The square feet of surface area equal to the formwork requirement, giving the required rub and patch surface area.

$$\text{Contact surface area} = 284.50 \text{ sf rub and patch}$$

Unit quantity summary

Approach slab quantity summary

Approach unit	Unit	North	South
Grade preparation	sf	1050.00	1050.00
Perimeter bulkheads, form	sf	92.00	92.00
Perimeter bulkheads, strip	sf	92.00	92.00
Interior bulkhead, doweled form	sf	25.00	25.00
Interior bulkhead, doweled strip	sf	25.00	25.00
Hardware and embeds	ea	0.00	0.00
Concrete placement locations	ea	2.00	2.00
Screed rail, bulkhead	lf	130.00	130.00
Screed rail, surface	lf	65.00	65.00
Finish machine, dry run	sf	1050.00	1050.00
Finish machine, setup	ea	2.00	2.00
Placement preparation	sf	1050.00	1050.00
Concrete placement (gross)	cy	42.78	42.78
Concrete cure	cy	42.78	42.78
Concrete cure	sf	1050.00	1050.00

Parapet and median unit	Unit	North	South
Parapet rail, form	sf	284.50	284.50
Parapet rail, strip	sf	284.50	284.50
Joint bulkhead form and strip	sf	15.88	15.88
Embedded hardware	ea	0.00	0.00
Concrete placement (gross)	cy	5.44	5.44
Rub and patch	sf	284.50	284.50

From the quantity summary list, the time duration and production factoring are compiled. The quantity summary list is formed by grouping the operations of similar components.

Example: Production Factoring, Time Duration

Operation. Grade preparation.

Crew members	= 2.0 ea
Shift duration	= 8.0 hours
Production unit	= square feet
Project quantity requirement	= 2100.00
Unit cycle time	= 0.30 min/sf
Productivity factor	= 50-minute hour
Cycles per hour: 50/0.30	= 166.67 cycles
Production per hour	= 166.67 sf/hour
Production per shift: 166.67 * 8	= 1333.36 sf/shift
Required shifts: 2100.0/1333.36	= 1.575 shifts
Total craft-hours: 1.575 * 8.0 * 2	= 25.20 craft-hours
Operation production factor: 25.20/2100.0	= 0.012
Operation factor	= 0.012 craft-hour/sf

Operation. Perimeter bulkhead, formwork place.

Crew members	= 3.0 each
Shift duration	= 8.0 hours
Production unit	= square feet
Project quantity requirement	= 184.00
Formwork type	= wood
Forming condition	= good, 4
Unit cycle time	= 2.0 min/sf
Productivity factor	= 50 minutes
Cycles per hour: 50/2.0	= 25.0 cycles
Production per hour	= 25.0 sf/hour
Production per shift: 25.0 * 8	= 200.0 sf/shift
Required shifts: 184.00 sf/200.0 sf	= 0.920 shift
Total craft-hours: 0.92 * 8 * 3	= 22.08 craft-hours
Operation production factor: 22.08/184.0	= 0.120
Operation factor	= 0.120 craft-hour/sf

Example: Material Allocation

The material costs associated with each operation are the direct permanent, and indirect or incidental materials and supplies required per specification. The cost of the formwork itself is allocated as formwork, which is treated as an asset to the company. The formwork is treated as equipment and has an ownership rate applied to it for its use. This allocation is detailed in the formwork section of the book and deals with the cost associated with the formwork purchase or construction and reuse disbursement.

The consumable formwork required, that will not be reused, is considered direct to the job as an incidental supply.

If there is a requirement for formwork that is not in inventory or is classified as specialty, the allocation of price or rental should be factored in the operation over any direct allocation.

The material listed is summarized by operation and allocated to the direct pay item, which is square yards of concrete approach slab required, and the cubic yards of neat line concrete required for the concrete items.

The material allocation is performed for each operation, for each type of material required, permanent direct and indirect incidental supplies. The material cost is disbursed against the control operation item quantity,(*) not the work item quantity.

Operation. Construct, uniform approach unit.

Direct material:	
*Concrete, 3500 psi (gross)	= 85.56 cy
Indirect material:	
n/a	
Incidental material:	
Form oil (2 oz per sf)	
((92.0 + 25.0 + 92.0 + 25.0) * 2 oz)/128	= 3.66 gal
Consumed lumber 2″ dimensional	= 50.0 lf
Nails: 2.5 lb per 100 sf = (234.0/100) * 2.5	= 5.85 lb
Formwork:	
Bulkhead formwork	
Perimeter: 2″ × 4″ wood panel	= 184.0 sf
Interior: 1″ × 12″ dimensional	= 50.0 sf

Operation. Construct, parapet and median unit

Direct material:	
*Concrete, 3500 psi (gross)	= 10.88 cy:
Indirect material:	
n/a	
Incidental material:	
Form oil (2 oz per sf)	
((284.50 + 15.88 + 284.50 + 15.88) * 2 oz)/128	= 9.39 gal
Consumed lumber 2″ dimensional = 50.0 lf	
Nails: 2.5 lb per 100 sf = (600.76/100) * 2.5	= 15.02 lb
Mortar, rub and patch: 94-lb sack	
0.5 lb per sf = (569.0 * 0.5)/94	= 3.03 bags
Formwork:	
Wall 2″×4″ wood panel forms	= 569.0 sf
Bulkhead ¾″ plyform	= 31.76 sf

*Asterisks denote control items for operation.

Material summary

Operation. Construct, approach unit.

Direct material:	
*Concrete, 3500 psi (gross)	= 96.44 cy
Indirect material:	
n/a	
Incidental material:	
Form oil	= 13.05 gal
Consumed lumber 2″ dimensional	= 100.0 lf
Nails	= 20.87 lb
Mortar, rub and patch (94-lb sack)	= 3.03 bags
Formwork:	
2″ × 4″ wood panel forms	= 753.0 sf
Bulkhead ¾″ plyform	= 31.76 sf
1″ × 12″ dimensional	= 50.0 sf

This material list is comprised of the direct permanent and indirect and incidental material required to perform the operations within the controlling item. The direct and indirect material are outlined and required by specification, and the incidental supplies are required by the operation to be performed. All materials are priced with all applicable sales and use taxes included.

Any required formwork is detailed separately owing to the different required forming systems and the intended reuse, or life of the panels.

The ratio of surface contact area to the cubic yards gives a prospective for a relationship of cost, both for the current job and for future reference. This is taken from the total component surface area of the operation against the neat line cubic yards for the operation. The same can be done with the cost ratio.

Surface ratio, sf per cy. Approach unit.

2934.76 sf/87.58 cy = 33.51 square feet per cubic yard

Takeoff Quantification Section Summary

The structural approach unit has many operations similar to those of the superstructure unit and is quantified and production factored by utilizing the same component identities and crews. The approach units do, however, have a better accessibility than the deck units, which alters the production somewhat. The key to an accurate quantity takeoff is accurately identifying the required components and work operations and applying a realistic production unit and duration.

---

*Asterisks denote control items for operation.

COMPONENT TAKEOFF

Approach Structure Unit

The following outline contains the normal work operation accounts for the performance and cost coding of approach structure construction components.

1. **Foundation grading**

    All work related to the grading, compacting, and preparation of the subbase to the proper line, grade, and cross section as shown. This item of work is considered after the installation of the subbase material has been completed. Quantity should be reported in square yards of area graded and compacted within the parameter of the approach slab.

2. **Form setting**

    All work related to the installation of the perimeter joint forms required for the approach slab to their proper line, grade, and location as shown and detailed on the plans. This account also includes the stripping of said forms. Quantity should be reported in lineal feet of forms set and stripped along with square feet of formed contact area.

3. **Finish machine setup**

    All work related to the setup and dry run of the finish machine. Quantity should be reported in craft-hours expended per bay of placement along with the surface square feet per bay.

4. **Finish machine rail**

    All work related to the installation and removal of the screed rail needed for the finish machine, along with the installation of yokes and elevation grading and any falsework needed to support the rail. Quantity should be reported in lineal feet of rail installed.

5. **Concrete placement**

    All work related to the gross placement of concrete within the approach slab area. The work includes the finishing and texturing of the concrete. Quantity should be reported in cubic yards placed and the waste or overrun, and square feet of flat surface area with the thickness specified.

6. **Concrete curing**

    All work related to the curing of concrete within the approach slab area. The work includes the covering, watering, and curing of the concrete. Quantity should be reported in gross cubic yards placed including any waste or overrun, and square feet of flat surface area with the thickness specified.

7. Other

This account is used for other operation accounts not specified here. Quantity should be reported in a representative format relating to the work performed and the appropriate unit of measure.

Chapter 17

Reinforcing Steel

TECHNICAL SECTION

Introduction

Steel bar reinforcing materials for use in concrete structures are used to structurally enhance plain concrete in many ways. Structural capabilities are achieved with reinforcing steel embedded within the concrete mass, being of a detailed design of various sizes and grades. It is used to reinforce or add strength to concrete.

The reinforcing steel enables concrete to be used in greater clear span distances, to carry greater loads, and to achieve many different architectural and supportive shapes. It serves as an inner tie to gather the concrete as one structural unit. Without this, the concrete would be more susceptible to collapse or fracture. The steel bar reinforcement is a solid material form.

Steel wire rope or strand material is another reinforcing material that is used to enhance the strength of concrete. This material is most commonly used in conjunction with reinforcing steel bar in *posttension* and *prestressing* capacities, which develop great compressive and tension forces.

These types of reinforcing materials give the concrete mass the inner grid of strength that allows the concrete materials to be used in structural and supporting roles. Without these integral components, the concrete, even though it has its own structural capacities, could not be used in most of today's technology.

Steel reinforcing material is manufactured in various sizes, grades, and properties, depending on its designed use. This section discusses these various types and the work consisting of furnishing and installing the reinforcing steel.

All reinforcing steel, depending on its designed use and purpose, must be furnished as per project specifications and installed to its proper location as detailed in the drawings or established by specific documents.

The most common type of steel reinforcement is *deformed round steel bar* of a specified grade. The two most common *grades* are 40 and 60. The grading, or yield strength, of deformed steel is determined by ASTM and AASHTO standards, which govern industry practices. These guidelines are used to define the strength of the steel based upon the mixture of ingredients of raw materials at the steel mill manufacturing facility.

This type of reinforcing steel is manufactured in rigid form of various lengths and diameters. The *deformation,* or ridges on the steel bar, gives it its gripping capabilities to the concrete mass and disallows any slippage under stress and force.

Deformed bar reinforcing steel is not only controlled by grade but sized and referred to by diameter. The increment of sizing used is by $\frac{1}{8}$-inch intervals. The sizing of deformed bar reinforcing steel begins with $\frac{3}{8}$-inch size, referred to as number three bar. As the size increases, so does the reference number. The weight per lineal foot of deformed bar is also of a constant factor depending on its referenced size.

Reinforcing Steel Properties

Bar	Diameter	Weight/lf	Cross section	Circumference
#3	$\frac{3}{8}''$	0.376 lb	0.11"	1.178"
#4	$\frac{1}{2}''$	0.668 lb	0.20"	1.571"
#5	$\frac{5}{8}''$	1.043 lb	0.31"	1.936"
#6	$\frac{3}{4}''$	1.502 lb	0.44"	2.356"
#7	$\frac{7}{8}''$	2.044 lb	0.60"	2.749"
#8	1"	2.670 lb	0.79"	3.142"
#9	$1\frac{1}{8}''$	3.400 lb	1.00"	3.544"
#10	$1\frac{1}{4}''$	4.303 lb	1.27"	3.990"
#11	$1\frac{3}{8}''$	5.313 lb	1.56"	4.430"
#14	$1\frac{3}{4}''$	7.650 lb	2.25"	5.320"
#18	$2\frac{1}{4}''$	13.60 lb	4.00"	7.090"

Reinforcing Steel and Wire Rope

Reinforcing steel shall conform to the following applicable requirements:

Type	Standard
Deformed Billet-Steel Bars	AASHTO M32 (ASTM A615, grade 60)
Rail-Steel Deformed and Plain Bars	AASHTO M42 (ASTM A616)
Axle-Steel Deformed and Plain Bars	AASHTO M53 (ASTM A617)
Epoxy-Coated Reinforcing Bars	AASHTO M284 (ASTM D3963)
Deformed Steel Wire	AASHTO M225 (ASTM A496)
Cold-Drawn Steel Wire	AASHTO M32 (ASTM A82)
Welded Deformed Steel Wire Fabric	AASHTO M221 (ASTM A497)
Fabricated Steel Bar or Rod Mats	AASHTO M54 (ASTM A184)
Low Alloy Steel Deformed	ASTM A706

Deformed bar reinforcing steel

This is designed within a concrete structure to impose certain loads and forces within a specific location. Thus it must be fabricated to specified shapes. This process is known as bending.

A fabricating facility produces these custom-shaped and -sized bars from the standard stock lengths and sizes produced from the mill. The bar is detailed as shown on the drawings and listed on an order form showing the bends, lengths, size of bar, and overall measurements needed to produce this custom-shaped bar needed for a specific project. This is known and referred to as a *bar list* or *detailing*. Each designated bar is (or similar bundles are) then tagged with a specific identifying number which corresponds to the bar list defining system. This is usually an industrywide, common denoting system.

Depending on the order of each number and letter, this system defines the type, size, and length of each bar. A common reference of reinforcing bar steel is either *bent bar* or *straight bar*. The straight bar could be only of specific cut length.

These bar lists and bending diagrams must be thoroughly detailed for approval as to the responsibility and conformation of proper fabrication.

The bending of deformed bar steel reinforcing must adhere to standard practices. They shall be bent by cold methods unless otherwise permitted.

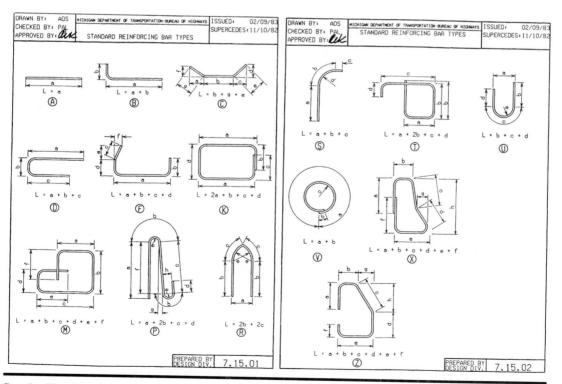

Standard bending details. (*Courtesy of Michigan Department of Transportation.*)

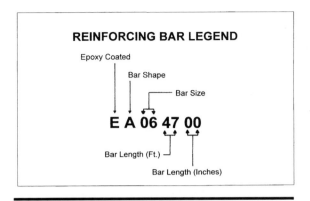

Bar coding and classification system. (*Courtesy of Barnes & Sweeny Enterprises.*)

Any heat that is applied to the steel may cause damage to its intended strength and design. Normally, bar fabricating facilities have machines that can be set up to bend the steel bar to its proper dimensions and shape. In some instances, however, field or job bending and cutting may be required. Bending specifications can be referred to in ACI Publication 318.

An enhancement to deformed bar steel reinforcing that has been developed to add to the life of the steel and the integrity of the concrete structure is *epoxy coating* applied to the circumference of the steel. This coating is used for corrosive protection to the steel and to stop rust. This is most commonly used in areas of salt and high corrosive atmospheres. This will also be detailed on the bar list and by specifications. The epoxy coating is applied to the steel bar under controlled manufacturing conditions, but touch-up procedures are detailed for job site corrective measures of damage repairs.

Once fabricated and detailed, the steel reinforcement shall be stored in a protective manner. It shall be kept above ground level upon platforms, skids, or other supports and be protected, as practical methods provide, from mechanical damage and surface deterioration caused by condition exposures that cause rust.

When placed within the concrete structure, the reinforcement bar must be free from dirt, detrimental and flaking rust, loose scales, paint, grease, oils, and other foreign materials which would cause improper adhesion to the concrete.

The deformed steel reinforcement shall be free from injurious defects such as cracks, lamination, or breaks. In the case of rust and surface irregularities, the minimum dimensions, cross-section area, and *tensile* properties must remain intact.

Epoxy coated deformed reinforcing steel must be stored on padded supports which shall not cause the separation of the epoxy coating from the steel bar. Handling and lifting procedures must follow the same format and intent. Areas of small damage may be permitted (according to project specifications) but specified to be field repaired, within certain limitations.

The placing and fastening of bar reinforcing steel must be precisely followed according to the drawings and specifications. The structure has a design which accounts for specific-sized bars in specific locations with a detailed *lap splice* and a certain amount of intersection *tying*.

When bar reinforcing is placed within a structure, it is usually placed with a *mat* or *grid* series of some defined orientation. The intersections of these mats or splices must be tied to prevent displacement and disorientation by the concrete, thus keeping the reinforcement in its proper and designed location. The percentage of the intersections that are required to be tied will be detailed in the project specifications. This will have a great impact on the estimation of time and tie materials needed to place the reinforcing steel.

The deformed bar reinforcing steel must maintain certain distances from the top surface of the concrete, the bottom surface, and between the individual mats of reinforcing. This is accomplished by *bar supports, chairs,* and *bolsters*. These supports can be of either precast mortar blocks or metal conforming to specific requirements and specifications. These supports must be attached to the reinforcing bar by methods or tying to prevent any movement by either the concrete or placement procedures. These supports shall only be used and intended for the support of the reinforcing steel, not for construction and placing operations, runways, or equipment loads.

These supports must also be properly placed as dimensional as possible to prevent any undue movement or reinforcing mat dislocation between the supports. Depending on bar size and location, the intended support structure must be located properly in vertical, transverse, and longitudinal location to assure the specified and proper final designed position of reinforcing steel.

Epoxy-coated reinforcing steel supports, chairs, tie wire, and other devices used to support shall be coated with a dielectric material. These are required to prevent damage to the coating of the bar steel until the final set of the concrete, which is the only intended use of any type of bar support.

Splicing of reinforcing will not usually be permitted except as shown on the plans. The length of the lap splices shall be as specifically detailed.

Number of Bar Diameters (Minimum Lap)	REINFORCING BAR - LAP CRITERIA								
	Inches of Lap Corresponding to Number of Bar Diameters * REINFORCING BAR SIZE								
	# 3	# 4	# 5	# 6	# 7	# 8	# 9	# 10	# 11
20	-	-	13"	15"	18"	20"	23"	26"	29"
22	-	-	14"	17"	20"	22"	25"	28"	32"
24	-	12"	15"	18"	21"	24"	28"	31"	34"
30	12"	15"	19"	23"	27"	30"	34"	39"	43"
32	12"	16"	20"	24"	28"	32"	37"	41"	46"
36	14"	18"	23"	27"	32"	36"	41"	46"	51"
40	15"	20"	25"	30"	35"	40"	46"	51"	57"
48	18"	24"	30"	36"	42"	48"	55"	61"	68"
Minimum Lap = 12"	(* Rounded to next larger whole inch)								

Standard lap lengths per bar size. (*Courtesy of Barnes & Sweeny Enterprises.*)

554 Takeoff and Cost Analysis Techniques

Reinforcing Bar Supports **DAYTON SUPERIOR®**

BC—Bar Chair

Bar Chairs are generally used to support miscellaneous reinforcing, and occasionally as a substitute for Slab Bolster. Bar Chairs are well designed, strong and available in heights of ¾" to 2", in increments of ¼".

JC—Joist Chair

The Joist Chair is a specialized support designed to support two reinforcing bars in 4", 5" and 6" wide joists or beams. Available in ¾", 1" and 1½" heights.

CHC—Continuous High Chair

The Continuous High Chair is similar to the Slab Bolster except it is fabricated out of heavier gauge wire. Legs are spaced at 7½" centers. Available in heights of 2" and over in ¼" increments.

CHCU—Continuous High Chair Upper

Two longitudinal wire runners are resistance welded to the bottom of Continuous High Chairs to form the upper support. Available in heights of 2" and over in ¼" increments.

Chair with Plates

20 Gauge Sheet Metal Plates are welded to the legs of individual or continuous supports to provide bearing on fill materials. Not recommended for use on decks.

Reinforcing Bar Supports **DAYTON SUPERIOR**

Shown below are several reinforcing bar supports which are typically used in bridge construction.

Dayton Superior manufactures and stocks a complete line of rebar supports from slab bolster to high chairs in all heights from ½" to over 30" to custom specials, if needed. These rebar supports are manufactured to the requirements and specifications of the Concrete Reinforcing Steel Institute (CRSI), Class 1 Plastic Protected, Class 2 Type A Stainless Steel Protected, Class 2 Type B Stainless Steel Protected, Class 3 Bright Basic or Galvanized Wire and Entirely Plastic Coated. They are shipped in convenient cartons, bundles, or on skids and are clearly identified.

Baked On Plastic Feet   Entirely Plastic Coated   Plastic Tipped Feet   Stainless Steel Tips

SB—Slab Bolster

Slab Bolsters support the lower mat of reinforcing steel in bridge deck slabs. Available in heights of ¾" to 3" in increments of ¼" and stocked in 5 foot lengths. Legs are spaced at 5" on center.

SBU—Slab Bolster Upper

Slab Bolster Upper is fabricated with runner wires "resistance" welded to the bottom side of the legs. SBU is generally used to sit on top of the lower mat of reinforcing steel so that the upper mat of reinforcing steel is properly positioned. Available in heights of ¾" to 3" and stocked in 5 foot lengths.

BB—Beam Bolster

Beam Bolster Lower is used to support the lower reinforcing bars in beams and girders. Available in heights of 1½" to 5" in ¼" increments. Shipped in stock lengths of 5 foot and cut to beam width by the steel setter. Legs are spaced at 2½" centers.

BBU—Beam Bolster Upper

Beam Bolster Upper acts as a separator between layers of reinforcing bars in beam and girders. Available in heights of 1½" to 5" in increments of ¼". Shipped in 5 foot stock lengths and cut to required length in the field.

HC—Individual High Chair

Available in heights of 2" and over in ¼" increments. High Chairs are used, depending on their height, to support the bottom or top mat of reinforcing bars in bridge decks

Special variations of this chair are available for use on Stay-In-Place Metal Decking.

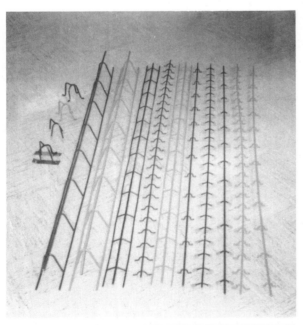

Bar support devices. (*Courtesy of Dayton Superior and Barnes & Sweeny Enterprises.*)

Welding of the bar reinforcing steel will not be permitted unless authorized. All welding, if permitted, shall conform to AWS D14 specifications and standards, and the chemical composition of the steel that is to be welded shall contain the proper carbon (c), manganese (ma), and carbon equivalent (c.e.) properties.

CAD welding is a recognized method of properly welding reinforcement bar. The welds shall be tested using *magnetic particles, radiography,* or other *nondestructive* inspection techniques.

Another form of splicing or connecting of bars is with *mechanical splices* or *couplers.* They must provide 125 percent of the specified yield strength of the deformed reinforcing bar and must be specified and approved.

All reinforcing bar must be in its final and secured position prior to the placement of concrete. Certain tolerances of maximum and minimum variance to spacing and coverage requirements will be specified by each project. In areas of bridge decks or other large open-topped areas, a template or screed of the final top concrete surface must be provided to assure the position of the reinforcing as to the final concrete cover.

Deformed bar reinforcement steel will be measured by the unit of pounds in place as per the detailed bar list. Normally, tie wire and supports will not be measured and paid for separately. Epoxy coated steel reinforcement will normally be of separate measure from that of plain bar reinforcement.

Other types

Other types of steel reinforcement materials are *welded wire fabric, wire rope* or *cable,* and *stressing strand.*

Welded wire fabric. The welded wire fabric is a solid-gauge wire of specified diameter and yield strength which is fabricated in a grid system and welded at the intersection points. The grid spacing of the wires, along with gauge size, determines its structural capacities. This type of reinforcement is manufactured in a rigid sheet form or in a rolled, somewhat flexible form.

Wire rope or cable. Wire rope or cable consists of small strands of steel wire of specific gauge and yield strength wrapped together into a rope or cable form. The number, size, and physical properties of strands used determine its ultimate strength class.

Stressing strand. Stressing steel is high-tensile wire or steel alloy bars conforming to specific standards of AASHTO and ASTM. Depending on properties and applied forces, these reinforcing steel materials are designed and used for either *prestressing* or *posttensioning.* Probably more so than normal reinforcing steel, stressing steel must be thoroughly protected from rust, corrosion, and physical damage due to the implied forces from the time of manufacture to concrete cover or grouting. A corrosion inhibitor may be required

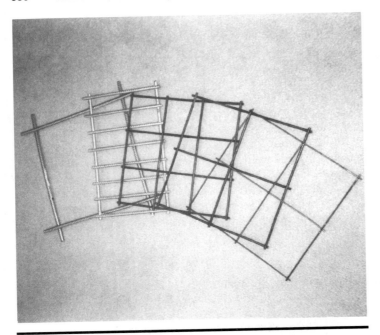

Wire fabric. (*Courtesy of Barnes & Sweeny Enterprises.*)

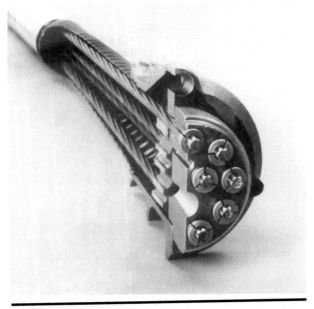

Wire cable. (*Courtesy of Dywidag Systems International.*)

and used during shipping and storage. As in all *embedded* products, reinforcement materials will be sampled and tested by lot.

Rebar legend (sample)

Epoxy bar		EA 06 47 00
	E =	epoxy coated
	A =	bar shape
	06 =	bar size
	47 =	bar length, feet
	00 =	bar length, inches
Plain bar		(A 06 47 00)

TAKEOFF QUANTIFICATION

Introduction

The deformed steel bar is the most commonly used for cast-in-place concrete components within the substructure and superstructure units. The steel bar is sized by its diameter, in increments of $\frac{1}{8}$ inch, starting with #3 bar ($\frac{3}{8}''$). The deformed reinforcing steel is graded by composition, determining the tensile strength. For concerns of corrosion, the deformed bar is coated with an epoxy resin material.

The deformed reinforcing steel is either fabricated, bent, and shaped for each specific area and component for which the reinforcement is required; or straight rebar for continuous. The deformed bar is then calculated, by bar size, for its weight by required length, per piece. The quantity takeoff is performed by calculating the weight per lineal foot of bar for the designated configuration, by bar size, summarizing the concrete component for the required total pounds.

Deformed reinforcing steel is placed and secured by tying each piece together with steel tie wire, at the intersections or splice laps of the deformed bar. This assures the deformed reinforcing steel will remain in its proper and designed location during the placement of concrete. The specification will identify the percentage of intersections to be tied within the component.

The welded wire fabric is most commonly used for miscellaneous flatwork areas, where a reinforcement is necessary for a grid-type support rather than a rigid structural support and provides a wider range of smaller-diameter wire size functional uses. The wire fabric is designated by wire size and spacing and quantified in a unit measure of square feet or square yards.

The cable-type reinforcement is normally used in precast or cast-in-place concrete operations for pretensioning and posttensioning procedures. The steel cable is designated by diameter size, number of segments, and number of strands per segment and is quantified in a unit measure of lineal feet.

Definition

The reinforcing steel components for a concrete structure are comprised of deformed steel bars, welded wire steel fabric mesh, and steel cable. The steel reinforcing components give the concrete greater structural capabilities.

Takeoff

The deformed reinforcing bar is categorized by bar size and quantified by a unit measurement of pounds. There are identifiers associated with the category of either plain or uncoated, or epoxy coated. Therefore, the quantity takeoff is summarized by identity, bar size, and total pounds, for each component area of the concrete structure.

The quantity takeoff of deformed reinforcing steel is first identified by plain or coated, then the component area of the concrete structure for which it is to be placed, followed by the bar sizing and weight summarization. This designates the order of control for the craft-hour usage and time duration.

The operation of reinforcing steel placement normally works in conjunction with the formwork placement operation. Therefore, the quantity takeoff and production factoring must realize and define areas of component congestion in conjunction with the formwork operations. In other words, will the reinforcing steel be placed prior to the formwork, during the formwork, or after the formwork placement? When reinforcing steel is placed afterward, the formwork may be somewhat of an obstruction and slow the production of the rebar placement.

Deformed steel reinforcing bar

Deformed reinforcing steel is calculated by weight per lineal foot of deformed bar required. This calculation is performed by determining the total length of each bar, straight or shaped, by size, within the specific concrete component, referred to as a bar list. The fabrication diagram shows the bending details for every bar type within the structure.

Each deformed bar is detailed by size and length, then calculated by the determined weight per lineal foot, giving the total weight for each bar, then summarizing for the component. Within this calculation will be the allowances for the specified and required overlapping and splicing, to assure the proper gross quantity requirements.

The quantity takeoff for deformed reinforcing bar will also identify the ratio of large bars to smaller bars, the percent of ties required for each intersection, the amount of tied-in-place versus pretying, and the ratio of fabricated deformed bar to that of straight.

Bar ratio. The ratio of large bar to small is a handling identifier referred to as the *bar ratio*. This means the placing labor can normally maneuver a larger bar as effectively as a smaller bar, thus placing more weight per unit craft-hour, performing more cost-effectively. The classification of this will be numbers 3 bar to 5 being considered small, and number 6 bar and over being classified as large. This is normally the condition of most bridge projects.

	TYPE	BAR	DIMENSIONS							NO REQD.	TOTAL WGHT.
			LENGTH	A	B	C	D	E	F	G		
PIER FOOTING	S	FT10296	29'-6"								18	2285
	S	FT08146	14'-6"								30	1161
	S	FT06296	29'-6"								18	798
	S	FT05146	14'-6"								45	681
	1	FT07060	6'-0"	2'-0"	4'-0"						36	442
										SUBTOTAL		5367
PIER COLUMNS	S	CL07200	20'-0"								36	1472
	C	CL06084	8'-4"	7'-10"	0'-6"	1'-3"					66	826
										SUBTOTAL		2298
PIER CAP	S	CP09356	35'-6"								15	1810
	2	CP06136	13'-6"	1'-7"	3'-6"	3'-0"	3'-6"				43	872
	3	ECP04060	6'-0"	1'-6"	3'-0"	1'-6"					16	64
										SUBTOTAL		2682
										EPOXY SUBTOTAL		64

S — STRAIGHT

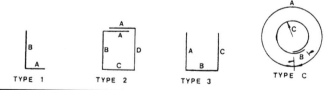

Typical bar list. (*Courtesy of Barnes & Sweeny Enterprises.*)

Tie ratio. The tying requirement, or tie ratio, determines and controls the labor consumption for the placement of the reinforcing bar. A mat of steel that is required to be tied 100 percent, or every intersection, will require more time than one requiring 50 percent, or every other intersection. The unit production factor will account for the tie requirements.

Tie method. Pretying reinforcement steel at a staging area, versus tying in-place, allows a more efficient and controlled method of labor usage but does require hoisting equipment and double handling for its final positioning. This determines the tie method. When a mat or cage is pretied, it can be supported by a rack or jigs and allows better mobility of the workforce rather than hoisting and tying individual pieces in place within the component. This method has limited conditions and should be researched, and also should be estimated to an absolute average that will be pretied (not abusing the operation taking into account realistic field conditions). The production factor will require the summation of time for both the pretie operation and final placement operation.

Production base. The last requirement for the labor factor will be the ratio of fabricated bar to straight bar. This identifier will define the quantity of straight, nonfabricated rebar which is more efficient to place than the fabricated or shaped reinforcing steel. This ratio, referred as the *production base,* will establish the basic factor controlling the production of the component. The production base, which is controlled by the ratio of "fabricated rebar" to that of "straight rebar," determines the actual in place handling of each bar.

Normally straight rebar requires less layout and offers a more efficient placement operation whereas fabricated rebar normally deals with smaller pieces requiring a more detailed placement procedure. Often, the greater the quantity of fabricated rebar the slower the placement operation is. Therefore, the production base ratio will be the main controlling function in factoring and establishing the overall production for the component. The recommended range for the production base is between 0.20 and 0.80.

Within the category of reinforcing steel, there will exist these prime components listed by structural unit. The deformed bar list will be produced following this outlined format.

Foundation structure unit:

1. Augered caissons
2. Cast-in-place pile
3. Miscellaneous pile

Substructure unit:

1. Footer
2. Abutment
3. Backwall
4. Wing
5. Columns
6. Struts
7. Caps
8. Closure walls

Beam structure unit:

1. Concrete diaphragms

Superstructure unit:

1. Deck
2. Parapet and median

Approach structure unit:

1. Slab
2. Parapet and median

Cast-in-place box girder segment:

1. Bottom deck unit
2. Wall and interior unit
3. Diaphragms and closures
4. Deck unit
5. Parapet and median

Within this outlined format, the deformed steel reinforcing bar can be designated as plain or coated.

For each component there may be the performance operation of:

1. Rebar layout
2. Rebar unloading and rehandling
3. Rebar installation, tie and placement
4. Rebar support installation
5. Rebar pretie

The craft-hour unit production factor is the required craft-hours to properly install the reinforcement steel, by method, size, and condition, for the appropriate component.

The quantity takeoff for deformed steel reinforcing bar is summarized, in a unit measure of pounds, by component area, tied in place or pretied, identifying a factor of tie percentage, size ratio, and determining a production base, as:

1. Plain deformed reinforcing steel, component _____, lb:
    Bar ratio _____ lb
    Tie ratio _____ %
    Tie method _____
    Production base _____

2. Epoxy deformed reinforcing steel, component _____, lb:
    Bar ratio _____ lb
    Tie ratio _____ %
    Tie method _____
    Production base _____

Reinforcement unloading, storage, rehandling

Reinforcing steel is normally delivered in truckload quantity. It is recommended that an operation of unloading, storage, and rehandling be utilized to account for the craft-hour consumption for this item.

The quantity takeoff for the unloading and handling operation is summarized by category of plain and epoxy, in a unit measure of pounds.

1. Unload, store, rehandle plain bar, lb
2. Unload, store, rehandle epoxy bar, lb

Bar supports

Bar supports are incidental components required to maintain the spacing between two mats of steel, proper distance and elevation from the bottom of the structure, vertical clearance at the formwork surface, or "cover," which is the detailed dimension the rebar is from the concrete surface.

Bar supports can be identified in two types, individual chairs and continuous runners. An individual chair is normally used for a condition of higher dimensional support requirement or wider spacing, and the continuous runner is normally used in a condition of lower dimensional support requirement or closer spacing.

The quantity takeoff for the bar supports is in a unit measurement of each, for the individual chairs, and in lineal feet for the continuous runners. The bar supports are also categorized by required height, or distance of support or separation.

The bar supports are calculated by the actual unit of each required to support a given unit of reinforcing steel. In other words, an individual chair is required to support a mat of steel at one chair per 4 square feet of mat area, or a continuous runner is required to support a mat of steel at 1 lineal foot of runner per 4 square feet of mat area. In the condition of multiple mats being stacked within a given component, the requirement of support would be at each level starting at the bottom. The craft-hour consumption is calculated for the amount of time required to install the necessary support.

The bar supports required for the deformed reinforcing steel are summarized in a unit measurement of lineal feet for the continuous runner type or in each for the individual chair type, as detailed, by specified size and identifier of plain or coated.

1. Bar supports, plain runners, size _____, lf
2. Bar supports, plain chairs, size _____, ea
3. Bar supports, coated runners, size _____, lf
4. Bar supports, coated chairs, size _____, ea

Reinforcing layout

The layout and positioning of the deformed steel reinforcement bar is considered incidental to the appropriate component. This entails the required craft-hours to perform the required layout for the proper positioning and required spacing of the reinforcing steel component.

The layout for the reinforcing bar components is summarized in a unit measurement of craft-hours required to perform the operation and is correlated back to a cost per pound of rebar.

1. Layout, plain reinforcement, _____ component, hours
2. Layout, coated reinforcement, _____ component, hours

Procedure

The detailed procedure for an accurate quantity takeoff for the deformed reinforcement identifies the specific component areas, specification requirements, and conditions for which the placement operations control. An accurate bar list must be established. The craft-hour usage and time duration are defined for the performance of all required operations for the detailed installation of the reinforcing steel.

Questions of Estimate

The questions that should be answered for a detailed and accurate quantity takeoff of reinforcing steel operations are listed as:

1. What are the site conditions?
2. What are the total pounds of plain bar?
3. What are the total pounds of epoxy bar?
4. What are the bar sizes?
5. What is the bar list, by category, per component?
6. What percentage can be pretied? (Tie method.)
7. What is the ratio, in pounds, of fabricated bar versus straight bar? (Production base)
8. What is the ratio, in pounds, of small bar versus large bar? (Bar ratio.)
9. What is the tie requirement for each component? (Tie ratio.)
10. What are the support requirements?
11. What influence will the formwork have to production?

Application: Structural Per Unit

The following example details a given deformed reinforcing steel component quantity takeoff. This defines the production requirements, the craft-hour production factoring method, crew determination and identification, and produces a detailed summary listing of the required reinforcing component, detailing material needs both permanent and incidental, and the equipment usage.

A given structural pier unit has the following detail of components with the deformed steel reinforcement bar requirements:

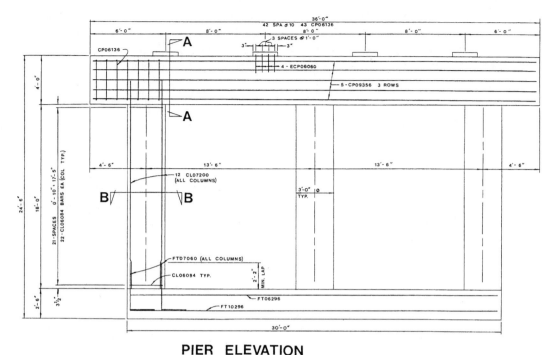

PIER ELEVATION

Pier unit details.

Pier unit: structural component dimensions

Pier footer	2.5' thick, 15' wide, 30' long
Pier columns	3' diameter, 18' high, 3 ea
Pier cap	3.5' wide, 4' high, 36' long
Beam seats	3.5' × 2' ea

Within the pier unit, there will exist the required deformed reinforcing bar for each structural component, along with bar sizes and fabrication configurations detailing each specific bar.

Footer component

The footer component consists of two horizontal mats of plain reinforcing steel. The bottom mat is specified to be supported 8″ off of the bottom surface, and the top mat is specified to have a spacing of 12″ between the two mats. The bottom mat consists of #10 bars, longitudinally, spaced at 10″ centers (FT10296), and #8 bars, transversely, spaced at 12″ centers (FT08146). The top mat consists of #6 bars, longitudinally, spaced at 10″ centers (FT06296), and #5 bars, transversely, spaced at 8″ centers (FT05146).

Vertical L bars or dowels extending upward into the pier columns are detailed as footer component rebar. The vertical rebar requirement is specified as #7 bars, spaced at 10″ centers (FT07060).

Unless otherwise specified, all clearance dimensions are 4″ from the formwork, or concrete, surfaces. The steel reinforcing bar is tied at a minimum of 50% at the intersecting joints.

Pier footer component specifications, bottom mat:

1. Plain reinforcement
2. #10 bars @ 10″ center spacing
3. #8 bars @ 12″ center spacing
4. 8″ chair supports @ 1 per 3 square feet

Footer component specifications, top mat:

1. Plain reinforcement
2. #6 bars @ 10″ center spacing
3. #5 bars @ 8″ center spacing
4. 12″ chair supports @ 1 per 5 square feet

Footer component specifications, vertical column bar:

1. Plain reinforcement
2. #7 bars @ 10″ center spacing, vertically

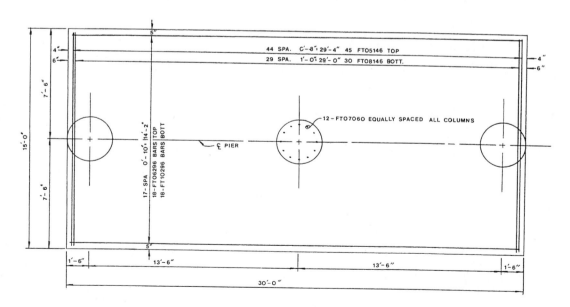

PIER FOOTING PLAN

Pier unit bar list, footer.

Column component

The reinforcing steel column component, three individual column units, consists of two component bars, the horizontal circular reinforcing bars and vertical straight reinforcing bars, both specified as plain reinforcing steel. The vertical bars are specified to be supported from the bottom vertical L footer bar components, and the horizontal circular bar components are tied to the vertical column bars. The vertical bar components, for each column, consist of #7 bars, spaced at 8″ centers (CL07200), around the perimeter of each column. The cover or setback distance from the form face is specified at 4″. The circular horizontal bar components are detailed as #6 bars spaced at 10″ centers for each column (CL06084).

The vertical column bars that are detailed to extend upward into the pier cap component are detailed as column component rebar.

Unless otherwise specified, all clearance dimensions are 4″ from the formwork, or concrete, surfaces. The steel reinforcing bars are tied at a minimum of 100% at the intersecting joints.

Pier column component specifications, vertical bars:

1. Plain reinforcement
2. #7 bars @ 8″ center spacing

Pier column component specifications, horizontal bars:

1. Plain reinforcement
2. #6 bars @ 10″ center spacing (circular)

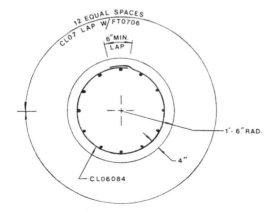

SECTION B-B

Pier unit bar list, column.

Cap component

The reinforcing steel cap component consists of multiple reinforcing steel components, the horizontal reinforcing bars, the vertical loop bars, and vertical straight reinforcing bars extending into the beam seat units, all specified as plain reinforcing steel. The horizontal bars, which are tied to the vertical column rebar dowels, are specified to be supported off of the bottom pier cap formwork 4", utilizing continuous runner bar supports. The vertical loop bars are tied to the horizontal bars, forming a cage. The vertical beam seat bar components are tied to the cage at the specified dimensions.

The horizontal bar components, for each pier cap, consist of three vertical rows of #9 bars, spaced at 10" centers (CP09356). The cover or setback distance from the form face is specified at 4". The vertical loop bar components are detailed as #6 bars spaced at 10" centers (CP06136), for the cap unit.

The vertical stirrup bars that are detailed as #4 epoxy coated bars extend upward into the beam seat component, and they are spaced at 6" centers (ECP04060). They are detailed as cap component rebar.

Unless otherwise specified, all clearance dimensions are 4" from the formwork, or concrete, surfaces. The steel reinforcing bars are tied at a minimum of 100% at the intersecting joints.

Pier cap component specifications, horizontal bars:

1. Plain reinforcement
2. #9 bars @ 10" center spacing

Pier cap component specifications, vertical loop bars:

1. Plain reinforcement
2. #6 bars @ 10" center spacing
3. 4" continuous supports, 3 runs @ 35'6" ea

Pier cap component specifications, vertical stirrup bars:

1. Epoxy coated reinforcement
2. #4 bars @ 6" center spacing

For these components, the required work operations are:

1. Rebar unloading and rehandling
2. Rebar layout
3. Rebar installation and tying
4. Rebar support installation
5. Rebar pretie, pier columns and caps

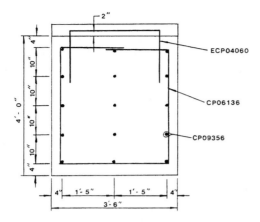

SECTION A-A

Pier unit bar list, cap.

The following is a detailed listing of required reinforcing steel component bar listed as per the project plans for the structural pier unit:

Bar list, pier footer

Mark	Size	Quantity	Length	Space	Type	A	B	C	D
FT10296	10	18.0	29'6"	10"	STR				
FT08146	8	30.0	14'6"	12"	STR				
FT06296	6	18.0	29'6"	10"	STR				
FT05146	5	44.0	14'6"	8"	STR				
FT07060	7	30.0	6'0"	10"	1	2.0'	4.0'		

Bar weight calculation, pier footer component

Mark	Size	Quantity		Length		Lb/lf		Total weight, lb
FT10296	10	18.0	*	29'6"	*	4.303	=	2284.893
FT08146	8	30.0	*	14'6"	*	2.670	=	1161.45
FT06296	6	18.0	*	29'6"	*	1.502	=	797.562
FT05146	5	45.0	*	14'6"	*	1.043	=	680.558
FT07060	7	36.0	*	6'0"	*	2.044	=	441.504
Total weight (plain)							=	5365.967 lb
Bar ratio: Small							=	12.7%
Large							=	87.3%
Fabricated percentage							=	7%

Bar list, pier columns

Mark	Size	Quantity	Length	Space	Type	A	B	C	D
CL07200	7	36.0	20'0"	8"	STR				
CL06084	6	66.0	8'4"	10"	CIRC				

Bar weight calculation, pier column component

Mark	Size	Quantity	Length	Lb/lf	Total weight, lb
CL07200	7	36.0 *	20'0" *	2.044 =	1471.680
CL06084	6	66.0 *	8'4" *	1.502 =	826.10
Total weight (plain)				=	2297.78 lb
Bar ratio: Small				=	0.0%
Large				=	100.0%
Fabricated percentage				=	35.0%

Bar list, pier cap

Mark	Size	Quantity	Length	Space	Type	A	B	C	D
CP09356	9	15.0	35'6"	10"	STR				
CP06136	6	43.0	13'6"	10"	2	1'7"	3'6"	3'0"	3'6"
ECP04060	4	16.0	6'0"	6"	3	1'6"	3'0"	1'6"	

Bar weight calculation, pier cap component

Plain

Mark	Size	Quantity	Length	Lb/lf	Total weight, lb
CP09356	9	15.0 *	35'6" *	3.400 =	1810.500
CP06136	6	43.0 *	13'6" *	1.502 =	871.911
Total weight (plain)				=	2682.411 lb
Bar ratio: Small				=	0.0%
Large				=	100.0%
Fabricated percentage				=	33.0%

Epoxy

Mark	Size	Quantity	Length	Lb/lf	Total weight, lb
ECP04060	4	16.0 *	6'0" *	0.668 =	64.128
Total weight (epoxy)				=	64.128 lb
Bar ratio: Small				=	100.0%
Large				=	0.0%
Fabricated percentage				=	100%

Reinforcing steel summary, pier component

Total weight (plain)	=	10,346.158 lb
Total weight (epoxy)	=	64.128 lb

NOTES: FT = footer bar (plain)
CL = column bar (plain)
CP = cap bar (plain)
ECP = cap bar (epoxy)

Sample: FT10296.
FT = area, 10 = bar size, 29 = length in feet, 6 = length in inches.

Example: Operation Analysis

The pier structure used in the example is comprised of three structural component units, which are defined as the footer, columns, and cap. The detail identifies the separate reinforcing steel components within each unit.

This example uses a conventional placement procedure, or tied-in-place operation, for the footer rebar components. The column and cap rebar components will, in general, be pretied prior to placement in the structure, and are placed as a singular unit within each related form system. The column rebar can be placed prior to the form system since it will be tied to the footer steel. The round column form system is set over the in-place reinforcing steel. The cap reinforcing steel must be placed after the formwork due to support requirements and formwork restrictions. The cap rebar must be supported from the formwork bottom soffit panels.

Unload and store reinforcing bar, plain reinforcing steel

Operation. Unload rebar, plain. Unit: Pounds.

The operation of unloading and storing the reinforcing steel bar components is calculated in the following manner for the total required quantity of plain reinforcing steel bar:

Required: plan quantity, plan.

The plan quantity of plan reinforcing steel summarized from the bar list and calculated for the craft-hour production factor.

Bar list summary = 10,346.158 lb

Epoxy coated reinforcing steel

Operation. Unload rebar, epoxy. Unit: Pounds.

The operation of unloading and storing the reinforcing steel bar components is calculated in the following manner for the total required quantity of coated reinforcing steel bar:

Required: Plan quantity, coated.

The plan quantity of coated reinforcing steel summarized from the bar list and calculated for the craft-hour production factor.

Bar list summary = 64.128 lb

Pier footer component

Operation. Layout and mark reinforcing steel. Unit: Pounds.

The operation of layout and marking the reinforcing steel bar for the pier footer components is calculated in the following manner for the total required quantity of plain and coated reinforcing steel bar:

Required: Plain bar, component quantity.

The required quantity of reinforcing steel summarized from the bar list calculated for the required time to perform the layout.

Plain rebar

$$\text{Plain rebar layout, footer} = 5365.967 \text{ lb}$$

Operation. Install bar supports, chairs. Unit: Each.

The operation of installation of the reinforcing steel bar supports for the pier footer components is calculated in the following manner for the total quantity of plain and coated reinforcing steel bar supports required by size, continuous runners or chairs:

Component quantity of reinforcing steel	= 5365.967 lb
Required: Bottom mat	
Type of support	= 8″ chairs
Spacing ratio	= 1 @ 3 sf
Gross surface area of pier footer	= 450.0 sf
Required: Top mat	
Type of support	= 12″ chairs
Spacing ratio	= 1 @ 5 sf
Gross surface area of pier footer	= 450.0 sf

The required quantity of reinforcing steel, summarized from the bar list, is calculated for the quantity of supports needed. The calculation required is the spacing ratio of one support chair for the defined square feet, divided into the surface area of the bar mat, performed for each mat. This unit is then divided by the required component quantity for a unit measure of support for each pound of reinforcement steel. This summary is then calculated for the required craft-hour production to perform the operation.

Bottom rebar mat:

$$(450.0/3.0) = 150.0 \text{ ea } 8″ \text{ chairs, plain}$$

$$(150.0/5365.967) = 0.0280 \; 8″ \text{ chairs per pound}$$

Top rebar mat:

$$(450.0/5.0) = 90.0 \text{ ea } 12″ \text{ chairs, plain}$$

$$(90.0/5365.967) = 0.0168 \; 12″ \text{ chairs per pound}$$

Operation. Place and tie reinforcing steel. Unit: Pounds.

The operation of placing and tying the reinforcing steel bar for the pier footer component is calculated in the following manner for the total required quantity of plain and coated reinforcing steel bar:

Required: Plain bar
  Component quantity
  Pretie quantity (Tie method)
  Tie factor (Tie ratio)
  Bar size ratio (small, large)
  Production base (fabricated ratio)

The required quantity of reinforcing steel is summarized from the bar list. The pretie quantity is derived from the defined amount to be pretied, with the calculation for both pretie, and conventional tying for installation. The footer component requires no pretie operation. The tie factor is applied as required. The bar ratio is factored in with the production factor.

Plain rebar

Pretie quantity	none
In-place tie	100%
Tie factor	50%
Bar ratio	12.7–87.3
Production base	0.07 (7%)
Plain rebar	= 5365.967 lb

Pier column component

Operation. Layout and mark reinforcing steel. Unit: Pounds.

The operation of layout and marking the reinforcing steel bar for the pier column components is calculated in the following manner for the total required quantity of plain and coated reinforcing steel bar:

Required: Plain bar, component quantity.

The required quantity of reinforcing steel summarized from the bar list calculated for the required time to perform the layout.

Plain rebar

$$\text{Plain rebar layout, column} = 2297.78 \text{ lb}$$

Operation. Place and tie reinforcing steel. Unit: Pounds.

The operation of placing and tying the reinforcing steel bar for the pier column components is calculated in the following manner for the total required quantity of plain and coated reinforcing steel bar:

Required: Plain bar
  Component quantity
  Pretie quantity (Tie method)
  Tie factor (Tie ratio)
  Bar size ratio (small, large)
  Production base (fabrication ratio)

The required quantity of reinforcing steel is summarized from the bar list. The pretie quantity is derived from the defined amount to be pretied, with the calculation for both pretie and conventional tying for installation. The column component permits a pretie operation of the complete quantity of steel, except for the area of attachment to the footer L bars, which are tied in place. The pretie quantity is rounded for estimating purposes. The tie factor is applied as required. The bar ratio is factored in with the production factor.

Plain rebar

Pretie quantity	95%
In-place tie	5%
Tie factor	100%
Bar ratio	0.0–100.0
Production base	0.35 (35%)
Plain rebar	= 2297.78 lb

Pier cap component

Operation. Layout and mark reinforcing steel. Unit: Pounds.

The operation of layout and marking the reinforcing steel bar for the pier cap components is calculated in the following manner for the total required quantity of plain and coated reinforcing steel bar:

Required: Plain bar
  Component quantity
Required: Epoxy bar
  Component quantity

The required quantity of reinforcing steel is summarized from the bar list calculated for the required time to perform the layout.

Plain rebar

Plain rebar layout, cap = 2682.411 lb

Epoxy rebar

Epoxy rebar layout, cap = 64.128 lb

Operation. Install bar supports, continuous. Unit: Lineal feet.

The operation of installation of the reinforcing steel bar supports for the pier cap components is calculated in the following manner for the total quantity of plain and coated reinforcing steel bar supports required, continuous runners or chairs:

Component quantity of reinforcing steel = 2682.411 lb

Required: Pier cap soffit

Type of support = 4″ runners

Spacing ratio = 1′ centers, @ 35′6″

Gross surface area of pier cap soffit = 126.0 sf

The required quantity of reinforcing steel, summarized from the bar list, is calculated for the quantity of supports needed. The calculation required is the lineal feet of continuous support runner for the given area of support, performed for each rebar area. This unit is divided by the required component quantity for a unit measure of support for each pound of reinforcement steel. This summary is then calculated for the required craft-hour production to perform the operation.

$$3 \text{ runs} * 35.5' = 106.5 \text{ lf } 4'' \text{ runners, plain}$$

$$106.5/2682.411 = 0.0397 \text{ lf } 4'' \text{ runner per pound}$$

Operation. Place and tie reinforcing steel. Unit: Pounds.

The operation of placing and tying the reinforcing steel bar for the pier cap components is calculated in the following manner for the total required quantity of plain and coated reinforcing steel bar:

Required: Plain bar
        Component quantity
        Pretie quantity (Tie method)
        Tie factor (Tie ratio)
        Bar size ratio (small, large)
        Production base (fabricated ratio)
Required: Epoxy bar
        Component quantity
        Pretie quantity (Tie method)
        Tie factor (Tie ratio)
        Bar size ratio (small, large)
        Production base (fabricated ratio)

The required quantity of reinforcing steel is summarized from the bar list. The pretie quantity is derived from the defined amount to be pretied, with the calculation for both pretie and conventional tying for installation. The cap component permits a pretie operation of the complete quantity of steel, except for the area of attachment to the column vertical bars and the stirrup bars, which are tied in place. The pretie quantity is rounded for estimating purposes. The tie factor is applied as required. The bar ratio is factored in with the production factor.

Plain rebar

Pretie quantity	90%
In-place tie	10%
Tie factor	100%
Bar ratio	0.0–100.0
Production base	0.33 (33%)
Plain rebar	= 2682.411 lb

Epoxy rebar

Pretie quantity	0%
In-place tie	100%
Tie factor	100%
Bar ratio	100.0–0.0
Production base	1.0 (100%)
Epoxy rebar	= 64.128 lb

Pier unit reinforcement steel quantity summary

Unload reinforcing steel:
  Plain bar   = 10,346.158 lb
  Epoxy bar   = 64.128 lb
Reinforcing steel layout:
  Plain rebar
    Footer component   = 5365.967 lb
    Column component   = 2297.78 lb
    Cap component   = 2682.411 lb
  Epoxy rebar
    Cap component   = 64.128 lb
Bar supports:
  Plain rebar
    Footer component 12" chairs   = 150.0 each
                          8" chairs   = 90.0 each
    Cap component 4" runners   = 106.5 lf

Plain reinforcing placing:

Component	Quantity	Pretie	In-place tie	Tie factor	Bar ratio	Production base
Footer	5,365.967	0.0%	100.0%	50.0%	12.7–87.3	0.07
Column	2,297.78	95.0%	5.0%	100.0%	0.0–100.0	0.35
Cap	2,682.411	90.0%	10.0%	100.0%	0.0–100.0	0.33
Total	10,346.158 lb					

Epoxy reinforcing placing:

Component	Quantity	Pretie	In-place tie	Tie factor	Bar ratio	Production base
Cap	64.128	0.0%	100.0%	100.0%	100.0–0.0	1.0
Total	64.128 lb					

From the quantity summary list, the time duration and production factoring are compiled. The quantity summary list is formed by grouping the operations of similar components.

Basic factoring. The production factoring for deformed steel reinforcement bar is performed by calculating the unit production cycle time, based on standard conditions of an equal bar ratio of 50–50 = 1.0, a 100% tie factor, and an in-place tie method equating to a fixed base of 1.0. A *production base* of 0.50 is also assumed (the ratio of 50% fabricated rebar to 50% of straight rebar).

From this, the variable multipliers for factoring are applied to the tie method (pretie or in-place), the tie ratio, and the bar size ratio.

To arrive at a bar ratio multiplier factor, first determine the time gained or lost by the operation (less than 1 equals a gain, greater than 1 equals a loss), then invert the gain/loss factor to its reciprocal to determine the multiplier (the factor divided into 1.0). This is represented by decimal form, using the chart below.

Example: A 0.70 factor = 1/0.70 = 1.4286 multiplier factor.

Bar ratio factor		Gain or loss factor	Multiplier
Small	large		
0–25	100–75	= 1.30	0.7692
26–49	74–51	= 1.15	0.8696
50	50	= 1.0	1.00
51–75	49–25	= 0.85	1.1765
76–100	24–0	= 0.70	1.4286

To arrive at the *tie ratio multiplier*, determine the intersection tie requirement for the rebar component, by percentage. Following the chart, choose the appropriate production gain/loss factor produced by the change in tying from the established base of 1.0 or 100%. Convert the decimal equivalent of the gain/loss factor to its reciprocal to determine the tie ratio multiplier.

Tie ratio, %:	Gain	Multiplier
25	= 0.55	1.8182
50	= 0.70	1.4286
75	= 0.85	1.1765
100	= 1.00	1.00

Tie method. To arrive at the *tie method multiplier*, begin with the fixed in-place base multiplier of 1.0, meaning the assumption is made that 100% of the rebar will be tied in place. Next, determine the quantity of rebar within the component (by percentage) to be pretied. If there is no quantity of rebar to be pretied within the component, use the pretie base multiplier of 1.0, 0% pretie, from the Pretie Base Factor chart. This pretie base multiplier of 1.0 is then added to the fixed in-place base of 1.0, creating a tie method multiplier of 2.0 for the in-place tied rebar component.

If there is a portion of the component that can be pretied, first determine the percentage quantity that can be pretied. The *remaining* in-place portion, in decimal form, becomes the pretie multiplier for the component. This pretie base multiplier is added to the fixed in-place base tie method multiplier of 1.0, creating the tie method multiplier for the pretied rebar component. An example being a pretie multiplier of 0.7, meaning that 30% of the component can be pretied and the remaining 70% is tied in place, is added to the fixed in-place base of 1.0 equaling a tie method multiplier of 1.7.

In-place factor

	In-place multiplier
Fixed in-place base	1.00

Pretie base factor

% of pretie	% of remaining in-place	Pretie multiplier
0	100	1.00
10	90	0.90
20	80	0.80
30	70	0.70
40	60	0.60
50	50	0.50
60	40	0.40
70	30	0.30
80	20	0.20
90	10	0.10

Note: It is not recommended to use a factor of more than 90% pretie. It should be understood that an additional 10% be allotted for in-place tying and component contingency.

The multiplier from each section, the bar ratio, tie ratio, and tie method, is chosen and multiplied together to arrive at the *combined multiplier factor*. A *production base* must also be calculated. This is performed by first determining the ratio of fabricated rebar to that of straight or nonfabricated rebar, in decimal form. This number becomes the "usable" production base factor for the component. From this, the *crew cycle time* (in minutes) will be established by dividing the production base factor by the chosen crew size (number of personnel). Example: 0.33/3 workers = 0.11 unit cycle.

Production base (rebar fabrication ratio)

Fabricated rebar, %	Straight rebar, %	Production base
20	80	0.20
30	40	0.30
40	60	0.40
50	50	0.50
60	40	0.60
70	30	0.70
80	20	0.80
90	10	0.90
100	0	1.00

Note: It is not recommended to use a percentage factor of less than 20 for fabricated rebar. Any percentage less than 20 should be increased to a 0.20 base.

The final production factor for the "place-and-tie" rebar component is calculated in the following manner. The *productivity factor per hour* (the productive work

Takeoff and Cost Analysis Techniques

minutes per crew hour, i.e., 50), divided by the *crew cycle time,* multiplied by the *combined multiplier,* will produce the anticipated estimated units per crew hour for the given component. Example: (50/0.11) * 0.8461 = 384.59 cycles/hour.

The required information needed to perform these calculations is outlined in the following guide.

Bar ratio	= gain/loss production factor
Tie ratio	= percent of tied intersections
Tie method	= percent of in-place or pretied quantity
Combined multiplier	= multiple of bar ratio, tie ratio, tie method
Production base	= ratio of rebar, fabricated to straight

The factoring methods and multipliers are for informational use; actual methods and factors should be developed for specific conditions and actual use.

Example: Production Factoring, Time Duration

Operation. Unload reinforcing steel, plain.

Crew members	= 3.0 each
Shift duration	= 8.0 hours
Production unit	= pounds
Project quantity requirement	= 10,346.158
Unit cycle time	= 0.005 min/lb
Productivity factor	= 50.0 minutes
Cycles per hour: 50/0.005	= 10,000.0 cycles
Production per hour	= 10,000.0 lb/hour
Production per shift: 10,000.0 * 8	= 80,000.0 lb/shift
Required shifts: 10,346.158 lb/80,000.0 lb	= 0.129 shift
Total craft-hours: 0.129 * 8 * 3	= 3.096 craft-hours
Operation production factor: 3.096/10,346.158	= 0.0003
Operation factor	= 0.0003 craft-hour/lb

Operation. Layout plain reinforcing steel, footer component.

Crew members	= 1.0 each
Shift duration	= 8.0 hours
Production unit	= pounds
Project quantity requirement	= 5365.967
In-place tie quantity	= 5365.967
Unit cycle time	= 0.019 min/lb
Productivity factor	= 50.0 minutes
Cycles per hour: 50/0.019	= 2631.58 cycles
Production per hour	= 2631.58 lb/hour
Production per shift: 2631.58 * 8	= 21,052.64 lb/shift
Required shifts: 5365.967 lb/21,052.64 lb	= 0.255 shift
Total craft-hours: 0.255 * 8 * 1	= 2.040 craft-hours
Operation production factor: 2.040/5365.967	= 0.0004
Operation factor	= 0.0004 craft-hour/lb

Operation. Place and tie plain reinforcing steel, cap component.

Crew members		= 3.0 each
Shift duration		= 8.0 hours
Production unit		= pounds
Project quantity requirement		= 2682.411
Pretie quantity: 90%		= 2414.17
In-place tie quantity 10%		= 268.24
*Tie factor = 100%	Multiplier	= 1.00 A
*Bar ratio factor = 0.0–100.0 (1.30)	Multiplier	= 0.7692 B
*Tie method factor: 1.00 + 0.10	Multiplier	= 1.10 C
Combined multiplier A * B * C		= 0.8461
Production base		= 0.33
Crew cycle time: 0.33/3 workers		= 0.11 min/lb
Productivity factor		= 50.0 minutes
Cycles per hour: (50/0.11) * 0.8461		= 384.59 cycles
Production per hour		= 384.59 lb/hour
Production per shift: 384.59 * 8		= 3076.72 lb/shift
Required shifts: 2682.411 lb/3076.72 lb		= 0.872 shift
Total craft-hours: 0.872 * 8 * 3		= 20.928 craft-hours
Operation production factor: 20.928/2682.411		= 0.0078
Operation factor		= 0.0078 craft-hour/lb

* = multipliers

Example: Material Allocation

The material costs associated with each operation are the direct permanent and indirect or incidental materials and supplies required per specification.

The material listed is summarized by operation and allocated to the direct pay item, which is reinforcement steel, plain, and reinforcement steel, epoxy coated.

The material allocation is performed for each operation, for each type of material required, permanent direct and indirect incidental supplies. The material cost is disbursed against the control operation item quantity(*), not the work item quantity.

Operation. Plain reinforcement steel, grade 60.

Direct material:	
*Deformed steel reinforcing bar	= 10,346.158 lb
Indirect material:	
12" rebar chair supports	= 150.00 ea
8" rebar chair supports	= 90.00 ea
4" continuous runner support	= 106.50 lf
Incidental material:	
Tie wire (1 roll per 1000.0 lb)	
(10,346.158/1000.0)	= 10.35 rolls

*Asterisk denotes control items for operation.

Operation. Epoxy coated reinforcement steel, grade 60.

Direct material:	
*Deformed steel reinforcing bar	= 64.128 lb
Indirect material:	
None required	= 0.00
Incidental material:	
Epoxy tie wire (1 roll per 1000.0 lb)	
(64.128/1000.0)	= 0.07 roll

This material list is comprised of the direct permanent and indirect and incidental material required to perform the operations within the controlling item. The direct and indirect material is outlined and required by specification, and the incidental supplies are required by the operation to be performed. All materials are priced with all applicable sales and use taxes included.

The ratio of reinforcing steel pounds to the control cubic yards gives a prospective for a relationship of cost, both for the current job and for future reference. This is taken from the total reinforcing steel within the component, or unit, against the neat line cubic yards for the component. The same can be done with the cost ratio.

The controlling concrete neat line cubic yard measurement for the pier unit is 116.90. The ratio includes the total reinforcing steel quantity for the unit, both plain and epoxy coated.

Reinforcing steel ratio, lb per cy

10,410.286 lb/116.90 cy = 89.05 rebar pounds per cubic yard

Takeoff Quantification Section Summary

The reinforcing steel section details the various component operations and costs associated with each structure component. Defining the different types of reinforcing categories and associated components, both plain and epoxy, that are required within each singular structure component will affect the end result cost. These operations and components, some minor, are required to be defined and accounted for to accurately produce a detailed quantity takeoff and cost summary.

COMPONENT TAKEOFF

Reinforcing Steel Unit

The following outline contains the normal work operation accounts for the performance and cost coding of reinforcing steel construction components.

1. **Reinforcing bar steel, substructure**

    All work related to the installation of reinforcing bar steel for all substructure components, including accessories. Quantity should be reported in pounds of steel placed according to the bar size and area of placement and categorized by coated and uncoated, defining the appropriate multipliers.

2. **Reinforcing bar steel, superstructure**

    All work related to the installation of reinforcing bar steel for all superstructure deck components, including accessories. Quantity should be reported in pounds of steel placed according to the bar size and area of placement and categorized by coated and uncoated, defining the appropriate multipliers.

3. **Reinforcing bar steel, diaphragms**

    All work related to the installation of reinforcing bar steel for diaphragm components, including accessories. Quantity should be reported in pounds of steel placed according to the bar size and area of placement and categorized by coated and uncoated, defining the appropriate multipliers.

4. **Reinforcing bar steel, parapet and median railings**

    All work related to the installation of reinforcing bar steel for parapet and median railing components, including accessories. Quantity should be reported in pounds of steel placed according to the bar size and area of placement and categorized by coated and uncoated, defining the appropriate multipliers.

5. **Reinforcing bar steel, retaining wall**

    All work related to the installation of reinforcing bar steel for all retaining wall components, including accessories. Quantity should be reported in pounds of steel placed according to the bar size and area of placement and categorized by coated and uncoated, defining the appropriate multipliers.

6. **Reinforcing bar steel, box culverts**

    All work related to the installation of reinforcing bar steel for all box culvert and drainage components, including accessories. Quantity should be reported in pounds of steel placed according to the bar size and area of placement and categorized by coated and uncoated, defining the appropriate multipliers.

7. **Reinforcing bar steel, other**

    This account is used for other operation accounts not specified here. Quantity should be reported in a representative format relating to the work performed and the appropriate unit of measure, defining the appropriate multipliers.

8. **Reinforcing steel wire fabric**

    All work related to the installation of welded wire steel fabric, including accessories. Quantity should be reported in square yards of material placed along with the size of the appropriate material.

9. **Stressing strand**

    All work related to the installation and stressing of prestressing or posttension strand, including accessories. Quantity should be reported in lineal feet of strand installed, including accessories, along with the size of the strand.

10. **Stressing bar steel**

    All work related to the installation and stressing of prestressing or posttension bar steel, including accessories. Quantity should be reported in pounds of bar installed, including accessories, along with the size and location component of the bar.

Chapter 18

Structure Drainage

TECHNICAL SECTION

Introduction

This section pertains to structure drainage systems, which include *subsurface, ground,* and *superstructure* or *deck* drain systems.

The purpose of *subsurface* and *ground* drain systems is to capture, relieve, and drain the subsurface and surface areas of a bridge structure. Trapped water within the foundation and substructure portions of the structure can be very detrimental and damaging. Structure design and construction must fully realize the lay of the existing land, the current drainage contours, any subaquifer conditions, and the sequence of construction that will alter these contours.

Surface water may be controlled as to the direction of flow after containment. This can be accomplished by slope control methods, drainage ditches, and inlet structures and piping. The surface water must be captured, from the natural course of flow, at the point where it meets the new construction. From this point it can be diverted away from the structure. The surface water or runoff cannot be allowed to cause any soil erosion or saturate the foundation.

Subsurface water must be contained and diverted more than captured, as in most cases it develops from a large area rather than from a directional flow. This containment is achieved by a series of piping and drainable aggregate filter blankets, and filter fabrics that collect the subsurface water and dispose of it away from the structure foundation.

Technical

Subsurface drain systems

The *subsurface drain systems* can be constructed from a variety of approved construction materials, which will be specified by the project specifications. The piping can be manufactured of polyvinyl chloride (PVC) or corrugated metal pipe (CMP). With either case the pipe shall be perforated to allow the subsurface water to enter the pipe at numerous areas. The perforations shall be spaced based on a formulated calculation of volume of water to pipe surface area.

The perforated piping shall be placed in the immediate area of the foundation structure, as detailed by the project plans, which is to be drained. Subsequent piping may also be needed to divert the subsurface water in advance of the structure. From a designated point outside the drainage area of the structure, to the outlet point, the piping shall be constructed of a solid or nonperforated pipe. The purpose of this type of system is to, at this point, contain and direct all of the water collected to a proper drainage area.

The aggregate filter blanket serves as a protective covering for the pipe from infiltration of silt, which would restrict the water flow in the pipe. It also serves as a reservoir to add in the available surface area of collected water for the pipe to gather. The aggregate used shall be of the size and type specified and shall be clean from dirt and silt. The gradation will determine the minimum sizing of fine aggregates permitted as these would also aid in the restriction of the water flow.

The membrane filter fabric serves as a bedding material to contain the water in the intended drainage area and to prevent the settlement and dislodging of the drain system. This fabric is normally placed directly beneath the drainage structure that is to be constructed.

After the subsurface drainage system construction, normal backfilling procedures would take place as detailed by specifications and construction drawings.

Superstructure drain systems

The *superstructure* or *deck drain systems* are designed and intended to capture, control, and properly dispose of the surface water collected on the superstructure deck and roadway. One type of system consists of a series of deck inlets or *scuppers*, piping and downspouting, and outlet pipes. A second type of system is a series of openings or pipe outlets at the base of the parapet wall which allows the surface water to run off the edge of the superstructure.

The surface water is collected within the limits of the roadway deck surface and directed to "low points" in elevation by design. At these points, which are not normally on the actual traveled area but usually at gutter or shoulder lines, the inlets or scuppers, or outlet pipes are placed.

An enclosed drain system, that of scuppers and downspouts, collects the water through a controlled system of piping and disposes of the water at a predetermined location.

The water is collected from the deck surface into drainage basin inlets, or scuppers. The scuppers, or deck drains, must be installed flush or slightly below the finished deck surface, $\pm \frac{1}{4}''$. The deck drains must be anchored so that the placement of concrete does not displace them from their proper alignment and grade. From this point, the surface water can either drain freely to the ground below or be collected through a piping system to a controlled outlet point.

An enclosed pipe system can be constructed of either steel, cast iron, or PVC pipe, as detailed by the project plans and specifications.

The connection points must be an approved and watertight fixture. The pipe system is to be securely anchored to the structure at points shown on the plans. The anchoring devices shall be attached to the concrete or steel structure with an approved embed capable of supporting the drain system.

All piping must be constructed in a manner as to drain and not prevent a flow restriction or backup to the system. Cleanout structures shall be installed as detailed. The outlet end of the collection system shall be constructed to the point of discharge, either surface ground or to a subsurface collection system.

An open drain system is that which utilizes either deck drain scuppers or an open drain hole through the deck. This system then discharges the collected surface water freely to the area below the structure. This type of system is constructed by a formed cast-in-place opening in the deck, a prefabricated inlet housing, or piping cast into the deck structure.

This system shall be constructed at points shown and detailed by the project plans.

The drain system design and construction shall be fully implemented to prevent erosion and foundation saturation to the structure.

TAKEOFF QUANTIFICATION

Introduction

This section categorizes the structure drainage systems into three units, the subsurface drainage system, the downspout and collection system, and the deck drain system.

Definition

The quantity takeoff of the structure drainage system entails defining and identifying the exposed and unexposed drainage components. This includes the drain pipe (both ground and subsurface), downspout and collection systems, and the surface deck drains.

Takeoff

The quantity takeoff of the control units of the structure drainage system, as defined, is performed by first identifying the specific components of each category. These components are defined and outlined within the project plans and specifications.

The subsurface or ground drainage system consists of all drain lines and appurtenances specified and detailed to carry the water from the inlet point of a structure collection system to a defined outlet point. This drainage system is located either on or under the ground surface, normally within the limits of the bridge structure.

The structure collection system consists of the downspouts, carrier lines, and appurtenances specified and detailed to carry the water from the superstructure deck drains to the subsurface or ground collection system. This system is physically located within the substructure and beam structure units of the bridge.

The deck drain assemblies are the actual fixtures, castings, or inlet devices that collect the deck surface water and deposit it in the structure collection system. These components are located within the superstructure deck unit.

The drainage units consist of primary and secondary components and operations.

Within the subsurface or ground drainage system, the primary components are identified as:

1. Pipe excavation
2. Pipe bedding
3. Pipe installation
4. Pipe backfill

The secondary components of the subsurface or ground drainage system are identified as the incidental operations of pipe fitting and connection components. These are required to be identified separately from the actual pipe component.

The primary component of the structure collection system is identified as:

1. Structure drainage pipe

Within the primary component of structure drainage pipe there exist the secondary components of pipe fitting and connection components, and structure hanger and attachment devices.

The deck drain unit consists of the primary component identified as:

1. Superstructure deck drain fixture

The material composition and type that will be encountered with a structure drainage system may range from plastic ABS or PVC, steel pipe or cast iron pipe and appurtenances, or a combination of these. The designated material must be identified during the quantity takeoff procedure to properly determine the production factors and techniques required.

Subsurface, ground unit

The first category unit to identify is the subsurface or ground drainage system within the parameters defined. These operations are listed by material type and specific component requirement.

Drainage trench excavation (pipe)

The quantity takeoff for the excavation component is performed by calculating the volume area, in cubic yards, of the trench to be dug. The dimension will be, at a minimum, the dimensions shown on the plans. By definition of the minimum dimension, the excavating unit or method may exceed this dimension, and the calculation will have to account for this.

The method of calculation will be multiplying the length of the trench by the depth of the trench by the width of the trench, converting the result to a cubic yard unit.

The craft-hour consumption is based on the cycle time and capacity size of the excavation unit per cubic yard of excavated material, including the disposal or stockpiling of the material.

The quantity summation of required volume for the trench excavation is outlined in a unit measurement of cubic yards, in a category defined by pipe type and diameter dimension, as:

1. Drainage trench excavation, pipe size _____, cubic yards

Pipe bedding

The quantity takeoff for the pipe bedding component is identified as the volume of material, in a unit measure of cubic yards. This component is defined as the quantity of specified material required to bed and seat the drainage pipe within the pipe trench.

The quantity calculation is performed by multiplying the length of the trench by the width of the trench or required bedding by the volume depth of bedding material, including the yield loss for compaction and waste. This quantity is converted into a unit measure of cubic yards and required tons.

The craft-hour consumption for this operation is based on the required time needed to perform the placement, shaping, and compaction of the bedding material in place, in a craft-hour per cubic yard ratio. The control of the operation is the shaping and placement of each cubic yard of material within the trench.

The quantity summation of required volume for the pipe bedding material is outlined in a unit measurement of cubic yards and tons, in a category defined by material type, and pipe type and diameter dimension, as:

1. Pipe bedding material, type _____, pipe size _____, cubic yards and tons

Drainage pipe installation

The takeoff quantity operation for the component of drainage pipe installation is performed by calculating the lineal feet of pipe required by specific type and diameter size. Along with this primary operation are the identifying and quantifying of the required pipe fittings, connections, and appurtenances.

The craft-hour consumption is defined as the required time needed to properly install the required pipe by the installation method and procedure specified and outlined. In addition to this, the time duration for the necessary fittings, connections, and appurtenances is allocated.

The quantity summation of required footage of pipe material is outlined in a unit measurement of lineal feet, in a category defined by material type and diameter dimension. The quantity summation for the pipe fittings, connections, and appurtenances is identified and categorized as independent components in a unit measurement of each, by pipe category, as:

1. Drainage pipe installation, type _____, pipe size _____, lf
   a. Pipe fittings, type _____, size _____, ea
   b. Pipe connections, type _____, size _____, ea
   c. Pipe appurtenances, type _____, size _____, ea

Drainage trench backfill

The quantity takeoff for the backfill component is performed by calculating the volume area, in cubic yards, of the trench to be backfilled. The dimension is, at a minimum, the dimensions shown on the plans of the proposed trench. By definition of the minimum dimension, the actual trench dimension may exceed the planned dimension based on excavation methods, and the calculation will have to account for this.

This component is defined as the quantity of specified material required to backfill the remaining volume area of the pipe trench, excluding the pipe and pipe bedding.

The quantity calculation is performed by multiplying the area of the trench to be backfilled by the length of the trench by the width of the trench by the volume depth of required backfill material, including the yield loss for compaction and waste. This quantity is then converted into a unit measure of cubic yards, and required tons for off site purchase.

The craft-hour consumption for this operation is based on the required time needed to perform the placement and compaction of the specified backfill material in place, in a craft-hour per cubic yard ratio. The control of the operation is the delivery and placement of each cubic yard of material within the trench.

The quantity summation of required volume for the pipe trench backfill material is outlined in a unit measurement of cubic yards and tons, in a category defined by material type, and pipe type and diameter dimension, as:

1. Pipe backfill, material type _____, pipe size _____, cubic yards and tons

Structure collection pipe

The takeoff quantity operation for the component of structure collection pipe installation is performed by calculating the lineal feet of pipe required by specific type and diameter size within the bridge structure unit. Along with this primary operation are the identifying and quantifying of the required pipe fittings, connections, appurtenances, and hanger assemblies and attachment devices.

The craft-hour consumption is defined as the required time needed to properly install the required collection pipe by the installation method and procedure specified and outlined. In addition to this, the time duration for the necessary fittings, connections, and hanger devices is allocated, within the structure unit.

The quantity summation of required footage of collection pipe material is outlined in a unit measurement of lineal feet, in a category defined by material type and diameter dimension. The quantity summation for the pipe fittings, connections, and appurtenances is identified and categorized as independent components in a unit measurement of each, by pipe category, as:

1. Collection pipe installation, type _____, pipe size _____, lf
   a. Collection pipe fittings, type _____, size _____, ea
   b. Collection pipe connections, type _____, size _____, ea
   c. Collection pipe appurtenances, type _____, size _____, ea
   d. Collection pipe hanger assemblies, type _____, size _____, ea
   e. Collection pipe attachment devices, type _____, size _____, ea

Deck drain fixture

The quantity takeoff of the superstructure deck drain fixtures is performed by identifying the actual number of specific drains required within the superstructure deck unit.

The craft-hour consumption is defined as the required time needed to properly install and set the required drain fixtures by the installation method and procedure specified and outlined, within the deck unit.

The quantity summation of required deck drain fixtures is outlined in a unit measurement of each, in a category defined by material type and size, as:

1. Deck drain fixture installation, type _____, size _____, ea

Procedure

The detailed procedure for the performance of an accurate quantity takeoff for the structure drainage system identifies each specific component area and operation, the specification requirements, and conditions for which the individual operations control. An accurate material component list must be established.

The craft-hour usage and time duration are defined for the performance of all required operations for the detailed installation of the structure drainage system within the bridge unit.

Questions of Estimate

The questions that should be answered for a detailed and accurate quantity takeoff of the structure drainage system operations are listed as:

590 Takeoff and Cost Analysis Techniques

1. What are the parameters of the structure drainage system?
2. Is there a subsurface system?
3. Is there a collection system?
4. What are the permanent material requirements?
5. What is the pipe type and size specified?
6. What are the fitting and appurtenance details?
7. What is the method of installation and connection?

Application

The application of this takeoff method is performed in the following example using the given specified components. This example defines the production requirements and installation methods, the labor-hour factoring method, and the crew determination and identification, and produces a detailed summary listing of the required structure drainage components. It also details the material needs, both permanent and incidental, and the equipment usage requirements.

A given structure drainage system has the following detail of operations along with the primary and secondary components. This system is detailed to

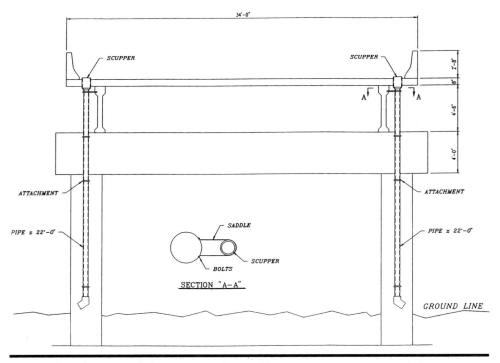

Center pier drainage system.

be outleted onto the ground surface, adjacent to the pier structure, thus not requiring a subsurface drainage system.

Collection pipe, appurtenances, deck drain fixtures

The collection pipe detailed will be 6" ID steel pipe, schedule 40. There will be a downspout system on each side of the pier structure. The outlet end of each downspout drain will have a 45-degree elbow fitting discharging the collected water away from the pier structure. The inlet portion of each downspout drain will have a bolted connection fitting into each deck drain fixture.

Each downspout pipe will be attached to the structure pier cap in one location, and to the pier column in two locations, by a pipe saddle and an incidental embedded bolted concrete insert anchor device.

Between the fittings and connection points, there will be a continuous length of 6" schedule 40 steel pipe.

The deck drain fixtures will be specified as cast iron scupper units, 24" wide by 30" long, each.

The operation of connecting the collection pipe to the specific fitting, connection points, and deck drain fixtures will be identified with each appurtenance, not the pipe itself.

Structure drain component system summary

6" schedule 40, steel pipe	
2 each @ 22.0'	= 44.0 lf
45-degree elbow fittings	= 2.0 ea
6" connection fittings	= 2.0 ea
Deck drain scupper fixtures	= 2.0 ea
Pipe saddles, hanger devices, 6"	= 6.0 ea

Example: Operation analysis

Operation. Pipe hanger devices. Unit: Each.

The operation of installing the pipe saddle devices and concrete anchors, for the structure drainage system, is calculated in the following manner for the total required quantity of pipe saddles.

Required: 2 each bolts and anchors per saddle.

The quantity calculation for the pipe saddles is performed by multiplying the number of required saddles by each location.

$$3.0 \text{ each} * 2.0 \text{ columns} = 6.0 \text{ ea saddles}$$

The quantity calculation for the saddle anchors is performed by calculating the required pipe saddles and the number of bolts and concrete anchors per saddle.

$$6.0 \text{ saddles} * 2.0 \text{ bolts and anchors} = 12.0 \text{ ea anchors and bolts}$$

Operation. Pipe fittings, 45-degree elbow. Unit: Each.

The operation of installing the pipe fitting devices for the structure drainage system is calculated in the following manner for the total required quantity of pipe fittings, by type.

Required: 6 each, bolts per fitting.

The quantity calculation for the pipe elbow fittings is performed by multiplying the number of required fittings by type by each location.

$$1.0 \text{ each} * 2.0 \text{ columns} = 2.0 \text{ ea elbow fittings}$$

Operation. Pipe connection. Unit: Each.

The operation of installing the pipe connection devices for the structure drainage system is calculated in the following manner for the total required quantity of pipe connections by type.

Required: 6 each, bolts per connection.

The quantity calculation for the pipe connection fittings is performed by multiplying the number of required connections by type by each location.

$$1.0 \text{ each} * 2.0 \text{ drain fixtures} = 2.0 \text{ ea connections}$$

Operation. Collection pipe installation, downspout. Unit: Lineal feet.

The operation of installing the collection pipe for the structure drainage system is calculated in the following manner for the total required quantity of collection pipe, by type.

Required: 6″ schedule 40, steel pipe.

The quantity calculation for the structure collection pipe is performed by identifying the number of required lineal feet of collection pipe by type by each location.

$$22.0 \text{ lf} * 2.0 \text{ each location} = 44.0 \text{ lf 6″ pipe}$$

Operation. Deck drain fixture, installation. Unit: Each.

The operation of installing the deck drain fixtures for the structure drainage system is calculated in the following manner for the total required quantity of deck fixtures by type.

Required: Cast iron scupper assemblies.

The quantity calculation for the deck drain fixtures is performed by multiplying the number of required fixtures by type by each location.

$$2.0 \text{ each location} * 1.0 \text{ drain fixtures} = 2.0 \text{ ea fixtures}$$

From the quantity summary list, the time duration and production factoring are compiled. The quantity summary list is formed by grouping the operations of similar components.

Example: Production factoring, time duration

Operation. Collection pipe, 6″ steel.

Crew members	= 3.0 each
Shift duration	= 8.0 hours
Production unit	= lineal feet
Project quantity requirement	= 44.0
Unit cycle time	= 10 min/lf
Productivity factor	= 50 minutes
Cycles per hour: 50/10	= 5.00 cycles
Production per hour	= 5.00 lf/hour
Production per shift: 5.00 * 8	= 40.0 lf/shift
Required shifts: 44.00 lf/40.00 lf	= 1.10 shifts
Total craft-hours: 1.10 * 8 * 3	= 26.40 craft-hours
Operation production factor: 26.40/44.00	= 0.600
Operation factor	= 0.600 craft-hour/lf

Operation. Deck drain fixtures.

Crew members	= 2.0 each
Shift duration	= 8.0 hours
Production unit	= each
Project quantity requirement	= 2.0
Unit cycle time	= 120 min/ea
Productivity factor	= 50 minutes
Cycles per hour: 50/120	= 0.417 cycle
Production per hour	= 0.417 ea/hour
Production per shift: 0.417 * 8	= 3.336 ea/shift
Required shifts: 2.0 ea/3.336 ea	= 0.60 shift
Total craft-hours: 0.60 * 8 * 2	= 9.600 craft-hours
Operation production factor: 9.600/2.0	= 4.800
Operation factor	= 4.800 craft-hours/ea

Example: Material allocation

The material costs associated with each operation are the direct permanent and indirect or incidental materials and supplies required per specification.

The material listed is summarized by the operation and allocated to the direct pay item, which is structure drainage system, each required.

The material allocation is performed for each operation, for each type of material required, permanent direct and indirect incidental supplies. The material cost is disbursed against the control activity item quantity(*), not the work item quantity.

594   Takeoff and Cost Analysis Techniques

Operation.   Install, structure drainage system.

Unit measure	= 2.0 each
Direct material:	
*6" schedule 40, steel pipe	= 44.0 lf
Deck drain scupper fixtures	= 2.0 ea
Indirect material:	
45-degree elbow fittings	= 2.0 ea
6" connection fittings	= 2.0 ea
Pipe saddles, hanger devices: 6"	= 6.0 ea
Incidental material:	
1" concrete anchors and bolts	= 12.0 each

This material list is comprised of the direct permanent and indirect and incidental material required to perform the operations within the controlling item. The direct and indirect material is outlined, and required by specification, and the incidental supplies are required by the operation to be performed. All materials are priced with all applicable sales and use taxes included.

Takeoff Quantification Section Summary

The structure drainage system consists of various operations and components. The quantity takeoff and cost method procedures incorporate all the necessary functions required to perform each task, including any temporary scaffolding requirements. Each operation can be summarized or grouped according to the requirements of the project specifications.

COMPONENT TAKEOFF

The following outline contains the normal work operation accounts for the performance and cost coding of structure drainage construction components.

1. **Deck drain scuppers**

   All work related to the installation of the deck drain fixtures to their proper line, grade, and dimensions of the type specified. Quantity should be reported in each installed.

2. **Deck drain outlets**

   All work related to the installation of the deck outlet opening to the proper line, grade, and dimension of the type specified. Quantity should be reported in square feet of concrete surface contact area (if formed), for both forming and stripping, or each if an individual fixture is installed.

3. **Downspout piping**

   All work related to the installation of all piping, located between the deck inlets and any subsurface outlet point or structure, to the proper line, grade, and dimension of the type specified. Quantity should be reported in lineal feet and type of piping installed. The quantity and type of anchors and joint connection fixtures should be reported by each installed.

4. **Subsurface drainage**

   All work related to the installation of any subsurface drain system associated with structure drainage to the proper line, grade, and dimension of the type specified. Quantity should be reported in lineal feet and type of pipe installed along with the depth of excavation, cubic yards of excavation, and volume and type of backfill required.

5. **Outlet fixture**

   All work related to the installation of the outlet device associated with the appurtenant drain system. Quantity should be reported in each of the types specified.

6. **Structure drainage system, other**

   This account is used for other operation accounts not specified here. Quantity should be reported in a representative format relating to the work performed and the appropriate unit of measure.

Chapter 19

Slope Protection

TECHNICAL SECTION

Introduction

Slope paving and protection is intended for the protection of the underside of abutment, slope, and embankment areas of structures against erosion and scouring of the soils. It also aids in the ease of maintenance of these areas. Depending on certain soil types, the constructed slope materials also add a structural value to the foundation areas within the slope.

Slope paving is normally constructed within the limits of the face of the abutment of a bridge and generally wraps around the slope to the backside of the wings, perpendicular to the roadway.

Technical

This work shall consist of the construction of slope paving and slope control in accordance with the project plans and specifications, and constructed to within reasonably close conformity with the lines, grades, and dimensions as outlined and established.

Components

The types of slope pavement and control shall consist of cast-in-place plain or reinforced concrete; masonry, precast concrete, or stone; loose rubble stone or machined riprap; or grouted rock.

Site preparation shall be performed within the limits of the slope pavement construction area. All debris and undesirable material shall be removed from the area. The entire slope area shall be properly filled, compacted, and graded to obtain reasonable conformity with the grades and cross sections established. The final grade for the prepared foundation shall obtain a smooth and uniform surface.

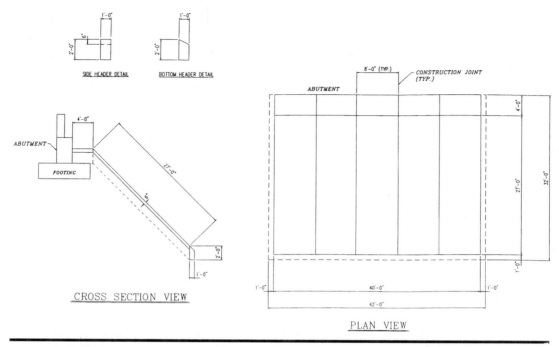

Slope protection.

Cast-in-place concrete slope paving. This paving (plain or reinforced) shall be constructed in accordance with the project plans and specifications. The dimensional size, class of concrete, and surface finish shall conform to project requirements. With monolithic type of construction, the joint pattern detailed must be closely adhered to. With alternate panel type of construction, the pilot panel slabs shall be formed to their designed dimensions and poured. The alternate panels shall then be poured after the appropriate cure time has been achieved on the pilot panel pours.

Reinforcing steel, either deformed bar steel or welded wire fabric, shall be placed according to project specifications and location. When shown, it shall be placed continuously through the *dummy joints* or *construction joints*.

Curing of cast-in-place concrete slope paving shall conform to project specifications and guidelines.

Toe headers. These shall be constructed as shown for perimeter containment and structural support to the slope pavement.

They shall be constructed to the proper line, grade, and dimensions shown and of the class of concrete required. The formwork used shall be such as to properly contain the concrete to its designed configuration.

All concrete shall be of such consistency that it will not flow on the slope and can be finished to the grade and thickness shown. The joints shall be scored, sawed, or formed as designed; and the joint filler, if specified, shall be of an approved type as set forth by project provisions. All areas of slope pavement which is contained within the perimeter of a concrete structure or abuts a structure shall be separated by a required expansion device.

Masonry, precast concrete block, or stone slope protection shall be constructed to the dimensions shown on the plans. This material shall be placed on a prepared foundation. Each unit shall be bedded with the depth perpendicular to the surface upon which it is set or placed and oriented as the conditions require. Each unit shall be placed against the adjoining unit with the sides and ends in contact. The joints may or may not be staggered based on design or individual unit conformity.

The completed surface shall not vary from the designed or anticipated theoretical plane by more than the allowable tolerances. Joint fillers or material grout may be required. The base or bottom or perimeter of the slope protection may be required to be founded on a header system for stability.

Rubble stone or random riprap. This slope protection shall be founded on a prepared foundation with the stones placed as close together as practicable to ensure interlocking for stability and to reduce voids with the variable-sized material.

When the rubble stone riprap is constructed in multiple layers it shall be placed so it is thoroughly tied together, with the larger stones protruding from the lower layer into the upper. The depth of this protection varies based on design and physical characteristics of the rock. The thickness, however, should maintain an average depth with a workable variation as indicated.

Each stone shall be placed so that the depth will be perpendicular to the surface upon which it is set. The length of the stone shall be placed as directed but will be always against the adjoining stones. The joints shall be staggered as far as possible and practicable.

The stones shall be placed with the larger ones first and thoroughly *chinked* and filled with the smaller ones to ensure interlocking and stability. This shall continue with the progress of the slope construction.

Machined riprap. This slope protection shall be constructed as defined by rubble riprap except that the material is specified as a dimensional size with respect to minimum and maximum gradation.

With regard to the construction of all riprap slope protected surfaces, the material shall be placed and dumped by the use of appropriate power equipment or hand methods that will produce a uniform and practicable surface.

Grouted riprap. This shall be constructed as described for rubble stone or machined riprap except that the voided areas of the rock shall be grouted with a cementitious grout. After chinking and filling, the voids between the stones shall be completely filled with grout to the limits shown by specification.

Care shall be taken to prevent earth or sand from filling the voided spaces between the stones, during construction, before the grout is placed. The grout used shall be of the specific mix design determined by the project requirements. This grout mix shall be of such a consistency that it will flow into and completely fill the voids but still contain the proper and specified water-cement ratio. The grout shall be mixed in advance only of the amount that can be placed within the time limits set forth for grout to remain plastic.

Immediately before the grout is placed, the stones shall be wetted by sprinkling. The grout shall be carefully poured into the voids between the stones. This operation shall begin at the lower portions of the riprap and progress up grade to prevent any tearing or separating of the grout. The entire bottom line of voids shall be filled (transversely) before progressing upward. The placing of the grout shall be accomplished by the use of vessels of adequate size, shape, and volume. Broadcasting, sloping, or spilling of grout from the vessels on the surface of the riprap will not be permitted. The grout mix shall be contained within the voided spaces only.

The progress of the placement shall be sufficiently slow to prevent the grout from oozing from the voids and flowing over the rock surface. During the pouring operation and continuing until the grout has assumed its final set, a uniform distribution and consistency must be maintained within the voided areas, and until such a time all of the voids have been completely filled and the grout has set even with the riprap surface.

As soon as any section of grouted riprap has hardened sufficiently, it shall be cured by sprinkling with water until it has been cured or covered by the required procedure and adequate protection. As with any water used for curing or wetting, it shall be free from any salt or alkali.

Normal methods of measurement of riprap will be of the type specified with a unit measure of either square yards of surface area covered or cubic yards placed by volume.

TAKEOFF QUANTIFICATION

Introduction

The quantity takeoff section for structure slope protection deals with identifying and defining the components for the related tasks and functions required for the performance of the specified operations.

Definition

The structure slope protection is defined as the components and operations directly related with the stabilization and protection of the underlying and adjoining slopes within the limits of the bridge structure unit.

Takeoff

The quantity takeoff procedures detailed within this section define and identify the four basic types of structure slope protection.

Cast-in-place concrete slope protection consists of the placement of concrete in a defined area of the structure. The area is formed and poured in either individual panels or a continuous placement with controlled joints. Within this primary component is a secondary component of slope headers.

The cast-in-place concrete headers consist of a perimeter foundation around the slope pavement for erosion control and bearing within the soils.

These headers usually are designed for the bottom of the slope for keying into the soil, and the vertical sides upward to the abutment.

The cast-in-place quantity takeoff procedure consists of defining the individual components required along with the formwork and concrete placement and curing parameters. The control unit of measurement for the cast-in-place concrete slope protection is the square feet of defined surface area for the specified thickness.

Premanufactured or natural units, a second type of structure slope protection, are placed individually within the limits defined. These units can consist of masonry units, precast concrete units, and individual-sized stone members. This type of structure slope protection is summarized in a unit measure of required surface area in square feet for the individually placed members.

Rubble stone and machine riprap structure slope protection are random-sized natural rock loosely placed within the defined area of the structure. This quantity is defined in a unit measure of required surface area in square yards for the material size specified.

Grouted stone riprap uses a quantity takeoff like that for the rubble stone and machine riprap component, but with the added operation of concrete grouting. The grouting is a secondary operation consisting of placing and curing a specified grout material within the voids and crevices of the loosely placed rock to produce a solid mass of slope protection.

The control unit of measure is defined as the surface area of slope protection required summarized in square yards for the defined thickness, with a secondary unit of measure of cubic feet of required grout.

In addition to all the methods of slope protection, the operation of foundation surface preparation is identified and quantified for each. This is considered an incidental component, for the fine grading and compaction, required for each method, summarized in unit measure of square feet for the required surface area.

Slope protection, foundation preparation

The foundation preparation for the structure slope protection is calculated in a square foot unit measurement, for the surface area required to be prepared, component contact area. The quantity is calculated by multiplying the width of the area by the length of the area.

The slope protection preparation is summarized in a unit measure of square feet, for the specified method of slope protection, identified for the operation of:

1. Concrete slope protection, foundation preparation, sf

Cast-in-place concrete slope protection

The cast-in-place structure slope protection is summarized by calculating the different components within the primary unit.

The quantity takeoff for the concrete header is calculated by defining the lineal feet of header required along with its controlling dimensions, width and thickness.

602   Takeoff and Cost Analysis Techniques

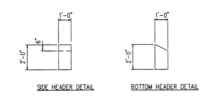

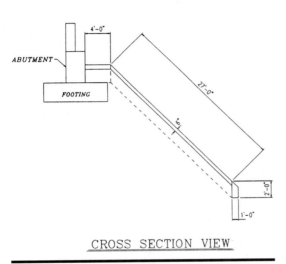

Slope protection section.

The required calculation for the formwork within the concrete headers is performed by multiplying the vertical thickness of the header by both sides by the required length of each specified header. The slope protection headers are calculated and summarized in a unit measure of square feet, identified for the activities of:

1. Concrete slope protection, header formwork placement, sf
2. Concrete slope protection, header formwork removal, sf

The concrete placement requirement for the slope protection headers is calculated in a volume of cubic yards, for the defined area of the header components. The quantity calculation is performed by multiplying the width of each individual header by the thickness of the header by the lineal feet of the header, and converting to a unit measurement of cubic yards. Included within this calculation is any required yield loss or waste overrun factor for the material.

The operation includes all necessary work for the placement and curing of the required concrete.

The slope protection headers are summarized in a unit measure of cubic yards, for the specified type of concrete, identified for the operation of:

1. Concrete slope protection, header concrete placement, cy; gross and net

The operation of any required rubbing and patching for the slope headers is summarized as the identical square footage of formwork requirements listed as:

1. Concrete slope protection, rub and patch, sf

The quantity takeoff for the formwork requirements of the cast-in-place concrete slope protection is calculated for the perimeter or interior formwork needed to contain and define the slope panel, or panels. This is defined as the vertical edge surface required to be formed, which would not include any adjoining surfaces to the structure or headers. The form edge area is referred to as the concrete surface contact area; the adjoining component surfaces are referred to as component contact area (no formwork requirement).

The calculation is performed by multiplying the specified thickness of the concrete panels by the lineal feet of required formwork, both perimeter and interior. The slope panel formwork is calculated and summarized in a unit measure of required square feet, identified for the operations of:

1. Concrete slope protection, panel formwork placement, sf
2. Concrete slope protection, panel formwork removal, sf

The concrete placement component for the structure slope protection is calculated in cubic yard volume measurement, for the surface area required. The calculation for the defined area is performed by multiplying the width of each placement by the length by the depth of the placement, converting the area to a cubic yard unit. Included within this calculation is any required yield loss or waste overrun factor for the material.

The concrete placement for the slope protection concrete panels is summarized in a unit measure of cubic yards, for the specified type of concrete, identified for the operation of:

1. Concrete slope protection, panel concrete placement, cy; gross and net

Masonry, precast concrete units, stone slope protection

The quantity takeoff for the individual members of this type of slope protection is calculated for the surface area required to be protected. The labor duration is defined as the time required for the placement of the individual members, by a ratio of number of units per square foot of surface area.

The surface calculation is performed by multiplying the width of the surface by the length of the surface area, in a unit measurement of square feet, and calculating the required number of member units required by multiplying the number of each required per square foot by the required square footage. Included within this calculation is any required yield loss or waste overrun factor for the material.

The placement for the slope protection is summarized in a unit measure of square feet for the surface area, and in the appropriate unit measure for the specified type of members required, identified for the operation of:

604   Takeoff and Cost Analysis Techniques

1. Concrete slope protection, masonry unit, placement, sf
   a. Concrete slope protection, masonry unit, material, sf
2. Concrete slope protection, precast concrete, placement, sf
   a. Concrete slope protection, precast concrete, material, sf
3. Concrete slope protection, stone unit, placement, sf
   a. Concrete slope protection, stone unit, material, sf

Rubble stone and riprap slope protection

The quantity takeoff for the loose rubble stone or riprap designated for this type of slope protection is calculated for the surface area required to be protected. The labor duration is defined as the time required for the placement of the loose stone within the area defined for the specified thickness.

The surface calculation for placement is performed by multiplying the width of the surface by the length of the surface area, converting to a unit measurement of square yards. The calculation for the required volume of stone is performed by multiplying the surface area square footage by the required thickness of stone. This volume is then converted to cubic yards, accounting for the designated yield loss and overrun or waste quantity required. If required, a conversion of cubic yards to tons is calculated by determining the specific gravity, unit weight per cubic foot, of the specified material, multiplied by the required cubic feet, dividing by 2000.0 to convert to required tons.

The placement for the rubble stone or riprap slope protection is summarized in a unit measurement of square yards for the surface area and in the appropriate unit measure for the specified type of material required, identified for the operations of:

1. Concrete slope protection, loose rubble stone, placement, sy
   a. Concrete slope protection, loose rubble stone, material, cy or ton
2. Concrete slope protection, riprap, placement, sy
   a. Concrete slope protection, riprap, material, cy or ton

Grouted riprap slope protection

The quantity takeoff for the grouted riprap designated for this type of slope protection is calculated for the surface area required to be protected. The labor duration is defined as the time required for the placement of the loose stone within the area defined for the specified thickness, and for the placement of the concrete grout.

The surface calculation for stone riprap placement is performed by multiplying the width of the surface by the length of the surface area, converting to a unit measurement of square yards. The calculation for the required volume of stone is performed by multiplying the surface area square footage by the required thickness of stone. This volume is then converted to cubic yards, accounting for the designated yield loss and overrun or waste quantity required. If required, a conversion of cubic yards to tons is calculated by determining the specific gravity, unit weight per cubic foot, of the specified material, multiplied by the required cubic feet, dividing by 2000.0 to convert to required tons.

The calculation for the grout placement is performed by defining the volume of grout required, in a unit measurement of cubic feet, for filling the anticipated voids in one average square yard of riprap area. This cubic foot volume, including yield loss and waste overrun, is then multiplied by the required square yards of slope surface, summarizing in a unit measurement of cubic feet of concrete grout.

The placement of the stone for the grouted riprap stone slope protection is summarized in a unit measure of square yards for the surface area and in the appropriate unit measure for the specified type of material required. The placement of the grout for the grouted riprap stone slope protection is summarized in a unit measure of cubic feet for the volume required for the defined surface area, identified for the operations of:

1. Concrete slope protection, riprap, placement, sy
    a. Concrete slope protection, riprap, material, cy or ton
2. Concrete slope protection, grout, placement, cf
    a. Concrete slope protection, grout, material, cf

Procedure

The performance of an accurate and detailed quantity takeoff of structure slope protection entails the identification and defining of the components and operations. The type of slope protection, along with the method of construction, controls the component of the takeoff.

The placement of the protection, whether concrete or rock, is a key and important factor in the crew type and costing, with both the labor crew and the equipment crew allocating.

The condition of whether the structure has been built over the area requiring slope protection, or if it has an unobstructed access, is a consideration for constructability. The most efficient time to construct the slope protection is prior to the beam and superstructure unit placements.

Another key factor to be realized during the quantity takeoff and plan review is the conditions of the site—ease of access, waterway concerns, or traffic staging requirements.

Questions of Estimate

When analyzing the slope protection takeoff, these questions formatted in this method best outline the work operations required for the given components:

1. What site conditions exist?
2. What access is available for equipment?
3. Is staging and part width construction required?
4. What is the type of slope protection?
5. What type of construction methods are required?
6. What are the required unit components?
7. What is the method of concrete placement?
8. What will the concrete or stone yield be?
9. What specialty equipment is required?
10. Are there traffic conditions within the work area?

Application

The following example details a given nonreinforced, cast-in-place concrete slope protection quantity takeoff. This shows the production requirements, producing a craft-hour production factor, along with a detailed summary listing of required formwork, concrete volumes, material needs, both permanent and incidental, and equipment usage.

A given multispan structure has the following cast-in-place concrete slope protection unit areas:

South abutment unit

North abutment unit

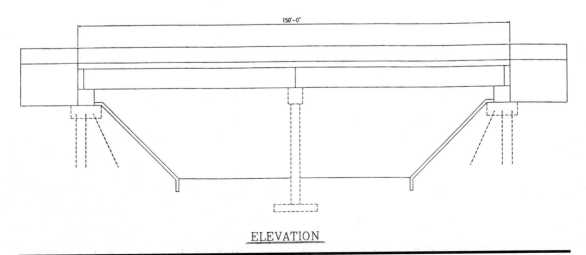

Two-span structure with slope protection section.

Within each of the symmetrical abutment units there are the slope protection components of:

Slope paving headers
    Bottom header, 1 each
    Side header, 2 each

Slope paving panels, 5 each

For these components, the required work operations are:

Formwork and formwork removal

Installation of form hardware, embeds

Installation of reinforcing material

Concrete placement

Rub and patch concrete surface

NOTE: The slope panels abut the slope headers and bridge abutment, creating a component contact surface and requiring no perimeter formwork for the slope panels.

The following lists required component dimensions per project plans:

Cast-in-place concrete component dimensions (north and south symmetrical)

Slope headers			
Bottom:	2' thick	1' wide	42' long
Sides:	2' thick	1' wide	31' long
Slope pavement:	6" thick	40' wide	31' long
Panel joints:	8' wide	interior formwork	
Reinforcement:	n/a		

The beam and superstructure units have not been constructed as of this time; therefore, there are no obstructions or restrictions for the concrete placement within the slope protection area.

Example: Operation analysis, north abutment slope protection

Operation. Grade preparation. Unit: Square feet.

The operation of grade preparation for the slope protection panel units is calculated in the following manner for the required surface area, by placement area:

Required: Placement panels @ 8'w × 31' l, 5 ea

The width (average) of the placement area, multiplied by the length (average) of the placement area, summarizing the concrete placement area of all slope panel units, in a measurement of square feet.

$$(8' * 31') * 5 = 1240.0 \text{ sf preparation}$$

Operation. Slope headers, formwork, form and strip. Unit: Square feet.

The operation of formwork placement of the toe and side headers for the slope protection unit is calculated in the following manner for the required contact surface area. The removal of the formwork is calculated as an equal quantity:

Required: Bottom @ 42'
Sides @ 31'
Height @ 2'

The length of each formline at the required perimeter area, multiplied by the thickness of the headers at the formline, multiplied by both sides, summarizing the perimeter headers of the slope protection units, in a measurement of square feet.

$$(31' + 42' + 31') * 2 * 2' = 416.0 \text{ sf place}$$

$$= 416.0 \text{ sf remove}$$

Operation. Concrete placement, slope headers. Unit: Cubic yard.

The operation of the slope header concrete placement component is calculated in the following manner for the required unit area:

Required: 2.0' thick
1.0' wide
Yield loss 10%

The width of each slope header, multiplied by the thickness of each slope header, multiplied by the length of the header section, plus the yield loss factor; converting to a cubic yard measurement and summarizing the quantity for each placement area, giving the required volume.

$$(((1' * 2') * (31' + 42' + 31')) * 1.10)/27 = 8.47 \text{ cy gross}$$

$$((1' * 2') * (31' + 42' + 31'))/27 = 7.70 \text{ cy neat}$$

Operation. Slope headers, rub and patch. Unit: Square feet.

The operation of the slope header rub and patch component is calculated in the following manner for the required unit area:
The square feet of surface area equal to the formwork requirement, giving the required rub and patch surface area.

$$\text{Contact surface area} = 416.0 \text{ sf rub and patch}$$

Operation. Slope panels, interior formwork, form and strip. Unit: Square feet.

The operation of interior formwork placement of the pavement panels of the slope protection unit is calculated in the following manner for the required contact surface area. The removal of the formwork is calculated as an equal quantity:

Required: Thickness @ 6"

The length of each interior formline at the required joint area, multiplied by the thickness of the slope pavement at the formline, summarizing the interior formlines of the slope protection units, in a measurement of square feet.

$$(31' * 0.5') * 4 \text{ each} = 62.0 \text{ sf place}$$

$$= 62.0 \text{ sf remove}$$

Operation. Required placement locations. Unit: Each.

$$\text{Placements} = 5.0 \text{ each}$$

Operation. Concrete placement, slope pavement panels. Unit: Cubic yard.

The operation of the slope pavement panel concrete placement component is calculated in the following manner for the required unit area:

Required: Uniform slab volume
    0.5' (6") thick
    Yield loss 10%

The width of each slope panel unit placement, multiplied by the length of each panel unit placement, multiplied by the average thickness of the uniform slab section, plus the yield loss factor; converting to a cubic yard measurement and summarizing the quantity for each placement area, giving the required volume.

$$((((8' * 31') * 0.5') * 1.10) * 5 \text{ ea})/27 = 25.26 \text{ cy gross}$$

$$(((8' * 31') * 0.5') * 5 \text{ ea})/27 = 22.96 \text{ cy neat}$$

South abutment slope protection

Operation. Grade preparation. Unit: Square feet.

The operation of grade preparation for the slope protection panel units is calculated in the following manner for the required surface area, by placement area:

Required: Placement panels @ 8' w × 31' l, 5 ea

The width (average) of the placement area, multiplied by the length (average) of the placement area, summarizing the concrete placement area of all slope panel units, in a measurement of square feet.

$$(8' * 31') * 5 = 1240.0 \text{ sf preparation}$$

Operation. Slope headers, formwork, form and strip. Unit: Square feet.

The operation of formwork placement of the toe and side headers for the slope protection unit is calculated in the following manner for the required contact surface area. The removal of the formwork is calculated as an equal quantity:

Required: Bottom @ 42'
    Sides @ 31'
    Height @ 2'

The length of each formline at the required perimeter area, multiplied by the thickness of the headers at the formline, multiplied by both sides, summarizing the perimeter headers of the slope protection units, in a measurement of square feet.

$$(31' + 42' + 31') * 2 * 2' = 416.0 \text{ sf place}$$
$$= 416.0 \text{ sf remove}$$

Operation. Concrete placement, slope headers. Unit: Cubic yard.

The operation of the slope header concrete placement component is calculated in the following manner for the required unit area:

Required: 2.0' thick
    1.0' wide
    Yield loss 10%

The width of each slope header, multiplied by the thickness of each slope header, multiplied by the length of the header section, plus the yield loss factor; converting to a cubic yard measurement and summarizing the quantity for each placement area, giving the required volume.

$$(((1' * 2') * (31' + 42' + 31')) * 1.10)/27 = 8.47 \text{ cy gross}$$
$$((1' * 2') * (31' + 42' + 31'))/27 = 7.70 \text{ cy neat}$$

Operation. Slope headers, rub and patch. Unit: Square feet.

The operation of the slope header rub and patch component is calculated in the following manner for the required unit area:
The square feet of surface area equal to the formwork requirement, giving the required rub and patch surface area.

$$\text{Contact surface area} = 416.0 \text{ sf rub and patch}$$

Operation. Slope panels, interior formwork, form and strip. Unit: Square feet.

The operation of interior formwork placement of the pavement panels of the slope protection unit is calculated in the following manner for the required contact surface area. The removal of the formwork is calculated as an equal quantity:

Required: Thickness @ 6"

The length of each interior formline at the required joint area, multiplied by the thickness of the slope pavement at the formline, summarizing the interior formlines of the slope protection units, in a measurement of square feet.

$$(31' * 0.5') * 4 \text{ each} = 62.0 \text{ sf place}$$
$$= 62.0 \text{ sf remove}$$

Operation. Required placement locations. Unit: Each.

$$\text{Placements} = 5.0 \text{ each}$$

Operation. Concrete placement, slope pavement panels. Unit: Cubic yard.

The operation of the slope pavement panel concrete placement component is calculated in the following manner for the required unit area:

Required: Uniform slab volume
 0.5′ (6″) thick
 Yield loss 10%

The width of each slope panel unit placement, multiplied by the length of each panel unit placement, multiplied by the average thickness of the uniform slab section, plus the yield loss factor, converting to a cubic yard measurement and summarizing the quantity for each placement area, giving the required volume.

$$((((8' * 31') * 0.5') * 1.10) * 5 \text{ ea})/27 = 25.26 \text{ cy gross}$$

$$(((8' * 31') * 0.5') * 5 \text{ ea})/27 = 22.96 \text{ cy neat}$$

Unit quantity summary, cast-in-place concrete slope protection quantity summary

Slope protection unit		North	South
Grade preparation	sf	1240.0	1240.0
Header formwork place	sf	416.0	416.0
Header formwork removal	sf	416.0	416.0
Hardware and embeds	ea	0.0	0.0
Header concrete placement	cy	8.47	8.47
Header rub and patch	sf	416.0	416.0
Panel formwork place	sf	62.0	62.0
Panel formwork removal	sf	62.0	62.0
Concrete placements	ea	5.0	5.0
Panel concrete placement (gross)	cy	25.26	25.

From the quantity summary list, the time duration and production factoring are compiled. The quantity summary list is formed by grouping the operations of similar components.

Example: Production factoring, time duration

Operation. Grade preparation, slope protection.

Crew members	= 2.0 each
Shift duration	= 8.0 hours
Production unit	= square feet
Project quantity requirement	= 2480.0
Formwork type	= n/a
Forming condition	= n/a
Unit cycle time	= 0.5 min/sf
Productivity factor	= 50 minutes
Cycles per hour: 50/0.5	= 100.0 cycles
Production per hour	= 100.00 sf/hour
Production per shift: 100.0 * 8	= 800.00 sf/shift
Required shifts: 2480.0 sf/800.0 sf	= 3.10 shifts
Total craft-hours: 3.10 * 8 * 2	= 49.60 craft-hours
Operation production factor: 49.60/2480.0	= 0.020
Operation factor	= 0.020 craft-hour/sf

Operation. Slope header, concrete placement and cure.

Crew members	= 5.0 each
Shift duration	= 8.0 hours
Production unit	= cubic yards
Project quantity requirement	= 16.94
Formwork type	= n/a
Forming condition	= n/a
Unit cycle time	= 15 min/cy
Productivity factor	= 50 minutes
Cycles per hour: 50/15.0	= 3.33 cycles
Production per hour	= 3.33 cy/hour
Production per shift: 3.33 * 8	= 26.64 cy/shift
Required shifts: 16.94/26.64	= 0.636 shift
Total craft-hours: 0.636 * 8 * 5	= 25.44 craft-hours
Operation production factor: 25.44/16.94	= 1.502
Operation factor	= 1.502 craft-hour/cy

Example: Material allocation

The material costs associated with each operation are the direct permanent and indirect or incidental materials and supplies required per specification. The cost of the formwork itself is allocated as formwork, which is treated as an asset to the company. The formwork is treated as equipment and has an ownership rate applied to it for its use. This allocation is detailed in the formwork section of the book and deals with the cost associated with the formwork purchase, or construction, and reuse disbursement.

The consumable formwork required that will not be reused is considered direct to the job as an incidental supply.

If there is a requirement for formwork that is not in inventory or is classified as specialty, the allocation of price or rental should be factored in the operation over any direct allocation.

The material listed is summarized by operation and allocated to the direct pay item, which is slope protection, cast-in-place concrete, for the square feet related item.

The material allocation is performed for each operation, for each type of material required, permanent direct and indirect incidental supplies. The material cost is disbursed against the control operation item quantity(*), not the work item quantity.

Operation. Slope protection, cast-in-place concrete 6″.

Direct material: 3500 psi concrete	
* Concrete, class 3500 psi (gross)	= 67.46 cy
Indirect material: n/a	
Incidental material:	
Form oil (2 oz per square foot formwork)	
(956.0 sf * 2 oz)/128 oz/gal	= 14.94 gal
Consumed lumber 2″ dimensional	= 100.00 lf
Nails: 4.5 lb per 100 sf = (956.0/100) * 4.5	= 43.02 lb
Mortar, rub and patch: Headers only (94-lb sack)	
0.5 lb per sf = (832.0 * 0.5)/94	= 4.43 bags
Cure compound: white, 100.0 surface sf per gallon	
Header: @ (832.0/100.0) +	
Panel surface: @ (2480.0/100.0)	= 33.12 gal
Formwork:	
Header formwork 2″ × 4″ wood panels	= 832.0 sf
Panel formwork 2″ × 6″ dimensional	= 124.0 sf

This material list is comprised of the direct permanent and indirect and incidental material required to perform the operations within the controlling item. The direct and indirect material is outlined and required by specification, and the incidental supplies are required by the operation to be performed. All materials are priced with all applicable sales and use taxes included.

Any required formwork is detailed separately owing to the different required forming systems and the intended reuse, or life of the panels.

The ratio of surface contact area to the control cubic yards gives a prospective for a relationship of cost, both for the current job and for future reference. This is taken from the total component surface area of the operation against the neat line cubic yards for the operation. The same can be done with the cost ratio.

Surface ratio, sf per cy

2480.0 sf/61.32 cy = 40.44 square feet per cubic yard

Takeoff Quantification Section Summary

The slope protection for structures varies in component units, design, and function. The quantity takeoff must realize and define all operations and conditions to accurately detail and perform the task.

A key item of control for the performance of an accurate estimate is access to the work area with respect to the restriction of the beam and superstructure above. As stated previously, the most efficient time to perform the work operations of the slope protection is prior to the erection of the beam structure and construction of the superstructure. This gives unobstructed overhead access to the work area.

COMPONENT TAKEOFF

The following outline contains the normal work operation accounts for the performance and cost coding of slope protection construction components.

1. **Cast-in-place concrete slope paving, form and strip**

    All work related to the forming of concrete slope paving to its proper line, grade, and dimension and the stripping of the same. Quantity should be reported in square feet of concrete surface contact area for both forming and stripping of formwork.

2. **Cast-in-place concrete slope paving, placing**

    All work related to the placement and curing of concrete, of the type specified, within the formlines. Quantity should be reported in cubic yards placed, including waste and overrun, and the net amount required. Also the surface area in square feet and designed thickness.

3. **Cast-in-place header, form and strip**

    All work related to the forming of concrete slope headers to their proper line, grade, and dimensions and the stripping of the same. Quantity should be reported in square feet of concrete surface contact area for both forming and stripping of formwork.

4. **Cast-in-place header, placing**

    All work related to the placement and curing of concrete, of the type specified, within the formlines. Quantity should be reported in cubic yards placed, including overrun and waste, and the net amount required. Also the dimensional area and lineal feet of header.

Slope Protection

5. **Masonry, precast concrete, stone slope protection**

    All work related to the placement of slope protection, of the type specified, to the proper line, grade, and dimensions specified. Quantity should be reported in square feet of units placed and volume or quantity required.

6. **Rubble stone and machined riprap**

    All work related to the placement of stone riprap slope protection to the proper line, grade, and dimensions. Quantity should be reported in square yards of surface area covered along with unit tons or cubic yards of material used and size gradation, being either random variable or required dimension.

7. **Grouted riprap**

    All work related to the placement of stone or machined riprap as described here plus the function of grouting as specified. Quantity should be reported in square yards of surface area covered, unit tons or cubic yards of material used, and cubic feet of concrete grout placed.

8. **Foundation surface preparation**

    All work related to the preparation, compaction, and minor elevation grading of the slope area to the proper line, grade, and dimensions. This account is not for any major excavation or backfill required. Quantity should be reported in square feet of surface area prepared.

9. **Slope protection, other**

    This account is used for other operation accounts not specified here. Quantity should be reported in a representative format relating to the work performed and the appropriate unit of measure.

Chapter 20

Waterproofing and Joint Fillers

TECHNICAL SECTION

Introduction

The waterproofing of structures is required for the protection of the surfaces and/or joints within the structure from water, ice, or other elements which may be detrimental to the concrete surface.

This protection may be required to an entire surface or a specified jointed area of the structure. Under certain conditions, concrete is very susceptible to water and chemicals which would deteriorate the component area and decrease the life of the structure.

The intent of waterproofing is not only for protection, but in some designs it aids with the control of water and moisture to a directed area.

Technical

This work shall consist of the waterproofing, dampproofing, and joint protection of concrete surfaces in accordance with the project specifications and in reasonably close conformity with the details of the plans and drawings.

The *waterproofing* materials shall conform to those specified of the following types: *membrane* waterproofing, *bridge deck* waterproofing, waterproofing with *mortar* protection, and waterproofing with *asphalt plank* protection.

Dampproofing shall conform to that specified of the following types: tar and asphaltic-type materials.

The *joint fillers* and *waterstops* shall conform to those specified of the following types: copper, rubber, and plastic waterstops; and, mortar, flexible watertight gaskets, and oakum joint fillers.

Waterproofing

Membrane waterproofing. This shall be a firmly bonded membrane composed of, normally, two layers of asphalt-treated fabric and three moppings of asphalt or three layers of tar-treated fabric and four moppings of tar, together with a coating of primer.

All concrete surfaces which are to be waterproofed shall be smooth and free of projections or depressions that would cause puncture of the membrane material. The surface shall be dry and free of dust and loose materials and should be cleaned immediately before the application of the waterproofing. Waterproofing shall not be done in wet or freezing weather conditions and should be protected against the same for its specified curing period.

Normally, asphalt shall be applied at a product temperature between 300 and 350 degrees Fahrenheit, and tar between 200 and 250 degrees Fahrenheit. A coating of primer shall be applied and allowed to dry before any mop coat is applied. Fabric shall be laid that any drainage will be over, and not against or along, the laps. It shall be installed in a "shingled" manner so there will be three thicknesses at all points. Edge laps shall be at least 4 inches and end laps at least 12 inches. The fabric must be thoroughly bonded between each of the coated moppings. This method shall begin at the lowest point of the surface to be waterproofed. Starting with a horizontal section, $\frac{1}{2}$ the width of a roll of fabric, a thorough coat of hot asphaltic tar is to be applied. A half strip of fabric shall be immediately placed over the asphaltic material and worked in to expel any entrapped air, obtaining close contact with the surface.

This strip and an adjoining horizontal area of concrete surface, again slightly more than $\frac{1}{2}$ of a roll in width, shall be mopped with asphalt material and a full-width strip of fabric applied, completely overlapping the first half layer. This second strip and an adjoining area of the concrete surface shall be mopped with asphalt material and the third strip of fabric applied, overlapping the top of the first strip by 4 inches.

This process shall continue until the entire concrete surface is completed. Once this is completed, the surface shall receive a complete coating of hot asphaltic material. Care shall be taken to ensure all laps and edges are thoroughly sealed down. A normal application rate for the asphaltic material is 12 gallons for each 100 square feet of horizontal surface, and 15 gallons for each 100 square feet of vertical surface.

Along all edges and at any *appurtenances*, such as drains or pipes, suitable provisions shall be made for the membrane to be sealed against the intrusion of water to the concrete surface. All flashing at curbs, girders, spandrel walls, etc., shall be done with separate sheets of fabric lapping the membrane not less than 12 inches in all directions. Any counterflashing shall be completed in the same manner.

Joints that are essentially open but which are not designed to provide for expansion shall be first caulked with oakum and wool and then filled with an approved hot joint filler. Expansion joints, both horizontal and vertical, shall be constructed with approved waterstops in accordance with the project plans and specifications. The waterproofing membrane shall then be installed across all expansion joints as detailed. At the ends of any structures, the membrane fabric shall be well tucked or layered to prevent any separation.

Bridge deck waterproofing. This normally is composed of a prime-coat coal tar emulsion and two layers of asphalt-coated glass fabric, and three moppings in the following sequence: prime coat, coal tar emulsion, glass fabric, coal tar emulsion, glass fabric, coal tar emulsion, then the normal asphalt concrete wearing surface.

The prime coat shall be applied with a brush or hand roller so as to fully penetrate the concrete surface and provide a bond between the concrete and the waterproofing material. The primer shall be applied as specified, but a minimum of 0.05 to 0.10 gallons per square yard. The primer shall not be applied when the *ambient* air temperature is below 50 degrees Fahrenheit and shall be allowed to cure a minimum of 4 hours under specified conditions prior to the application of the waterproofing material.

The application surface must be dry to prevent the formation of condensation and steam when the coal tar emulsion is applied. The bituminized glass fabric shall be applied in a shingled manner to provide at least two thicknesses at all points. Edge laps shall be a minimum of 2 inches. A thorough mopping of coal tar emulsion must be maintained between each layer of fabric and the concrete surface.

The first strip of fabric shall be one-half the width of a roll; the second shall be full width lapped 100 percent over the first horizontal roll; the third shall be full width lapped one-half over the second roll horizontally, with each succeeding strip lapping one-half over the preceding layer. Between each layer shall be a thorough mopping of coal tar emulsion. At any point throughout this procedure, a minimum fabric strip width of 12 inches shall be maintained.

On horizontal surfaces an average application of 3 to 4 gallons of asphalt mop coat per 100 square feet shall be used of finished net area, and on vertical surfaces an average application of 4 to 5 gallons of mop coat per 100 square feet of finished net area shall be used. Each strip of fabric shall be thoroughly pressed into each mopping to eliminate trapped air and to provide a flat surface.

When waterproofed surfaces change abruptly in direction, they shall be reinforced with an extra layer of mopping and fabric of suitable dimensions at these points.

Care shall always be taken to prevent damage to the waterproofed surfaces during other construction operations and backfilling procedures. Any damage that may occur must be immediately repaired following the guidelines established in this section.

Special waterproofing junctures shall be done according to the specific project plans and details.

Membrane waterproofing with mortar protection. This shall be completed as specified here except that the entire membrane surface shall have a protective course of reinforced mortar applied.

This course shall have a nominal thickness of 2 inches and shall be reinforced with a welded wire fabric with the grid and wire gauge specified. The mesh reinforcing shall lie midway between the thickness of the mortar compound.

The mortar shall be troweled to a smooth finish and cured by burlap curing methods. On undercut surfaces (which make an angle of less than 90 degrees), the membrane shall be protected with a layer of rolled asphalt or coal tar instead of the mortar. This procedure would follow the membrane mopping method.

Membrane waterproofing with asphalt plank protection. The plans indicate which membrane waterproofing is to be protected with asphalt plank. The plank shall be laid in straight, regular courses. Whole planks shall be used in all cases except otherwise required for closures and fitting around obstacles and openings. Closing and trimming pieces shall be carefully cut to size.

Prior to plank installation, all talc or other powder must be removed by stiff brushing or brooming. Each plank shall be laid in a mopping of hot asphalt or tar with the edges and ends coated with the same material prior to installation of the adjoining plank. Each individual piece must be firmly pressed and tightly fitted to the previously placed plank. The completed surface shall have a smooth and uniform texture without any open joints. Any damaged areas must be repaired to a permanent watertight condition.

Dampproofing

Dampproofing shall consist of coating concrete or masonry surfaces with a specified tar or asphalt in accordance with the project plans and details.

After the proper, specified concrete curing time has elapsed, the surfaces to be dampproofed shall be allowed to air dry at least 10 days. The surface shall be thoroughly cleaned of laitance and debris. A specified primer shall be applied in four equal and uniform coats by either brush or spray methods. Each coat shall be allowed ample time to dry and absorb into the surface before the succeeding coat is applied. After the absorption of the fourth coat, a heated seal coat shall be applied. This specified material must be done according to the ambient temperature requirements and can be either brushed or sprayed onto the prepared surface. The seal coat shall be allowed proper curing time prior to any backfilling or severe weather exposure. Each of these five coatings must be applied to a dry surface, and necessary precautions must be made for unsuitable and inclement weather.

Waterstops

Waterstops shall be constructed in accordance with the project plans and specifications. They consist of devices installed at and in *expansion joints* and *construction joints* for the purpose of stopping water seepage.

Copper sheet waterstops, or flashing. These shall be of the thickness, weight, width, and shape shown by the project plans. The sheet at each joint shall be of a continuous form, either by a single sheet or by soldering multiple sheets together into one segmented sheet over the joint in a continuous watertight unit.

Rubber waterstops. These are made of a rubberized material composition designed to be embedded into the structure at a specified joint location. They shall be of a continuous strip of the shape and size specified. Splicing shall be done either by vulcanizing or by an approved mechanical method. The waterstop device must be securely fastened and held in its proper alignment to the formwork or the bulkhead prior to the concrete placement and monitored during the placement to ensure its final location remains intact.

Plastic waterstops. These are also a manufactured device and are normally installed much the same as the rubber devices. They shall be continuous either by a singular unit or by adhering multiple segments together into a single unit.

Rubber and plastic waterstops shall be carefully placed and supported at the locations specified. Precautions shall be taken to avoid displacement or damage to the waterstop devices during installation procedures and related construction operations. All surfaces of the device shall be kept free from oil, grease, mortar, or any other foreign matter while the waterstop device is being embedded into the concrete structure. The concrete must be fully consolidated around the perimeter of the device to ensure proper embedment.

Structural bonding. This shall consist of crack repair within concrete structures by the pressure injection of a specified epoxy material according to the procedural specifications.

Backer rod material. This shall be used in the crack when a temporary or void filler is required. It must have sufficient strength and adhesion properties to confine the injected epoxy material until cured.

The area, size, type, and location of the cracks to be repaired will be identified in detail. The designated cracks shall be pressure injected using a two-component epoxy system with an in-line metering and mixing device. The joints shall be prepared and cleaned by compressed air, vacuuming, or other methods to ensure that all dirt, debris, laitance, and other loose matter are removed and that the joint is dry and free from any accumulated moisture.

Entry ports. Prior to injection of epoxy, a surface seal material shall be applied to the face of the crack to prevent escaping of the epoxy filler. Openings, or entry ports, shall be established along the crack. The distance of the entry ports shall not be less than the thickness of the concrete member being repaired.

The injection of the adhesive into each crack shall begin at the entry port at the lowest elevation. Injection shall continue at the first port until the adhesive begins to flow out of the port at the next highest elevation. The first port shall then be plugged and injection should immediately start at the second port until the adhesive begins to flow from the next port. This same sequence is to continue until the entire crack is filled. After the injection adhesive has cured, the surface seal shall be removed. The face of the repaired crack shall then be finished flush with the adjacent concrete. There shall be no indentations or protrusions caused by the placement of entry ports.

Injection equipment. The equipment used to meter and mix the two injection adhesive components and pressure inject the mixed adhesive into the cracks shall be of a positive-displacement type. The injection equipment shall have a capacity and capability of discharging the mixed adhesive at a pressure of up to 200 psi and maintaining a constant and equal flow at even pressures. The equipment shall also have the capability of maintaining the mix ratio for the injection adhesive of differing consistencies as prescribed by specifications. The tolerances at discharge shall be ±5 percent by volume at any pressure.

Tests. A material ratio test shall be performed for each injection unit at the beginning and end of each day that the unit is used for crack injection. The mixing head of the two adhesive components shall be disconnected and the two components shall be pumped simultaneously through the ratio check device. The ratio check device shall consist of two independent valved nozzles capable of controlling the flow rate and back pressure by opening or closing the valve to restrict material flow. A pressure gauge located behind each valve will sense and display the amount of back pressure. The discharge pressure should be adjusted to the maximum for the equipment for both adhesive components. Both adhesive components shall be simultaneously discharged into separate calibrated containers. The amounts discharged shall compare to the predetermined mix ratio. After this test has been completed, the discharge shall be set to 0 psi and the procedure repeated.

A pressure check shall be completed for each injection unit used at the beginning and end of each day for which the unit is to be used for crack repair.

The mixing head of the injection equipment shall be disconnected and the two adhesive component delivery lines shall be attached to the pressure check device. This device shall consist of two independent valved nozzles capable of controlling flow rate and pressure by opening or closing the valve. There shall be a pressure gauge capable of sensing the pressure buildup behind each valve.

The valves on the pressure check device shall be closed and the equipment operated until the gauge pressure on each line reads the maximum for the equipment. The pumps shall be stopped and the gauge pressure shall not drop by more than 5 percent within 3 minutes.

Accurate recorded information shall be maintained of these daily ratio and pressure check tests for each piece of injection equipment used.

After the injection process has been completed, a test for epoxy penetration shall be performed. These tests shall be extracted at intervals of no more than 50-foot increments. The test cores shall be a minimum of 2 inches in diameter and shall be done in accordance with the project specifications and ASTM C42.

When a core demonstrates that the epoxy has a penetration value of less than 90 percent of the crack volume within the specific core segment, the complete segment shall be retreated to achieve a minimum of 90 percent penetration value.

Upon completion of the testing, all core holes shall be filled with an epoxy grout and the surface finished to blend with the adjacent concrete. The grout mixture shall consist of an approved two-component epoxy bonding agent, fine aggregate meeting specifications, and any other admixtures required or specified. These components shall be combined in accordance with the mix ratio prescribed by the project specifications.

The embedded materials used within the prescribed methods of this section shall conform to these minimum specifications.

Joint fillers. These must be of the type specified and designated and shall conform to the following AASHTO and ASTM standards.

Poured filler shall conform to AASHTO M173 or AASHTO M282 as specified.

Preformed fillers shall conform to AASHTO M33 for bituminous type; AASHTO M153 for sponge rubber (type I), cork (type II), and self-expanding cork (type III); AASHTO M213 for nonextruding and resilient bituminous types; and AASHTO M220 for preformed elastomeric types as specified.

When dowels are specified, the filler material shall be punched out to receive the dowels as designed by the project plans. The joint filler shall be supplied in a single piece for the depth and width required for the joint. When more than one piece is required, the abutting ends shall be securely fastened and held accurately to shape by a positive fastening system.

Foam filler shall be expanded polystyrene filler having a minimum compressive strength of 10 psi.

Hot poured sealants for concrete and asphaltic surfaces shall conform to ASTM D3405.

Hot poured elastomeric-type sealant for concrete surfaces shall conform to ASTM D3406.

Cold poured silicone type sealant for concrete surfaces shall conform to Federal Specification TTS-1543, Class A. This sealant shall be 1 part low-modulus silicone rubber with an ultimate elongation of 1200 percent.

Joint mortar. This shall consist of 1 part portland cement and 2 parts approved sand with the necessary water added to obtain the required and specified consistency. This mix shall conform to the specified project mix designs outlined for this work.

Flexible watertight gaskets or ring gaskets. Used for rigid pipe, these shall conform to AASHTO M198 type A or type B. Ring gaskets for flexible metal pipe shall conform to ASTM C443, and continuous flat gaskets for flexible metal pipe shall conform to ASTM D1056. For flat bands and bands with projections grade SCE 41 is to be used, and for corrugated bands grade SCE 43 shall be used.

Oakum. Used for rigid pipe joints, this shall be made from hemp line (*Cannabis sativa*) or Benares Sunn fiber, or a combination of these fibers. The oakum shall be thoroughly corded and reasonably free from lumps, dirt, and extraneous matter.

Mortar for masonry beds and joints. These shall be composed of 1 part portland cement or air-entrained portland cement and 2 parts fine aggregate by volume. Hydrated lime or fly ash may be added in a percentaged amount of proportionate cement rate by that specified. An air-entraining admixture may be used in lieu of the air-entrained cement. The mix design and the raw materials must be approved by specification.

Copper waterstops and flashing. These shall be of sheet type and shall conform to AASHTO M138 (ASTM B152), light cold rolled, soft anneal.

Rubber waterstops. These may be molded or extruded and shall have a uniform cross section, free from porosity or other defects, and conforming to the nominal dimensions shown on the plans. The waterstop may be compounded from natural rubber, synthetic rubber, or a blend of the two, together with other compatible materials. The waterstops shall meet the specified properties of hardness, compression set, tensile strength, elongation of rupture, tensile strength at maximum elongation, water absorption by weight, and tensile strength after aging.

Plastic waterstops. These shall be fabricated with a uniform cross section, free from porosity or other defects, to the nominal dimensions detailed by the project plans. The material for the waterstop shall be a homogeneous, elastomeric, and plastic compound of basic polyvinyl chloride. The waterstops shall meet the specified properties of tensile strength, elongation at breaking, hardness, specific gravity, resistance to alkali, water absorption, cold bending, and volatile loss.

The measurement of these specified types of waterproofing and stops shall be of the unit of measure determined by specification.

TAKEOFF QUANTIFICATION

Introduction

This section details the required methods and applications required for a detailed and accurate takeoff. It defines the procedures along with the scope and limits of the operation. The waterproofing and joint filler components are constructed for various structure units and have numerous functions.

Definition

The quantity takeoff of the waterproofing and joint filler components is defined as identifying the required operations, both primary and secondary, necessary for the performance of the work.

Takeoff

The quantity takeoff of the waterproofing and joint filler components is performed by identifying the type of application required for the specific structural unit.

The primary components related to these operations are the performance and application of a waterproof surface coating of various methods and material uses, and waterstop joint sealers of various methods and material uses.

The takeoff procedures for these components outline the type of application required and the designated unit of measure. This quantity summary is defined and categorized to the corresponding structural unit.

The structure waterproofing components and operations are categorized into the primary units of:

Surface waterproofing

Joint waterproofing

Within the primary components will be the secondary operations for each unit which may consist of:

Sandblasting

Membrane application

Asphaltic application

Mortar application

Joint filler and waterstop installation

The type, method, and application of waterproofing and waterstop consist of the operations dictated by the scope of work. The quantity takeoff of these various methods is outlined by specific units of measure dependent on the structural component and category. The takeoff is performed by first defining the primary category, then each component within the category, then the method operation either chosen or specified. The takeoff procedure must identify the component area and accessibility due to the degree of difficulty and time duration restraints associated with different portions of the bridge structure.

The quantity takeoff of the waterproofing component of membrane sealing consists of the necessary operations required for the complete installation of the component. This includes all material surface preparation, product applications, and incidental items required. The membrane component is summarized in a unit measurement of square feet according to the structural component that it is subject to.

The quantity takeoff of the waterstop component defines all joint lines that require the procedure. The component summary is summarized by structural component, in a unit measurement of lineal feet.

The joint filler quantity takeoff operation is similar to the waterstop component and defines all joint lines that require the procedure. The component summary is summarized by structural component, in a unit measurement of either lineal feet or square feet, specified by width.

The quantity takeoff of the joint mortar component is performed by identifying the required structural joints requiring the procedure and is summarized in a unit measurement of either square feet or lineal feet, specified by width.

The gasket takeoff quantity is identified by the specified type of gasket, per structure component, and summarized in a unit measurement of each unit required.

Membrane waterproofing

The membrane waterproofing component is performed in a procedural system according to material type and application.

The quantity takeoff operation for the component of membrane waterproofing is performed by calculating the square feet of surface area required for each application of the waterproofing procedure, within the bridge structure unit. Along with this primary operation are the identifying and quantifying of the incidental and interim coating components and operations.

The labor consumption is defined as the required time needed to properly apply each specified layer of waterproofing material, by the installation method and procedure specified and outlined. In addition to this, the time duration for the necessary remobilizations and move-ins back to the component area for the additional coatings are allocated within the structure unit.

The quantity calculation for the membrane waterproofing is calculated for the surface area required, multiplying the length by the width, arriving at a unit measurement of square feet, for each required application including laps and splices. Each individual application is allocated for the labor, equipment, and material cost procedures, then summarized for the complete component item of work. All overrun and waste calculations are factored in with the material quantity summary list.

The quantity summation of required surface area of membrane waterproofing material is outlined in a unit measurement of square feet, in a category defined by material type and application method, as:

1. Membrane waterproofing, type _____, square feet

Asphalt plank waterproofing

The asphalt plank waterproofing component is performed in a procedural system similar to the asphalt plank waterproof component, according to material type and application.

The takeoff quantity operation for the component of asphalt plank waterproofing is performed by calculating the square feet of surface area required for each application of the waterproofing procedure along with laps, within the bridge structure unit. Along with this primary operation are the identifying and quantifying of the incidental and interim coating components and operations.

The labor consumption is defined as the required time needed to properly apply each specified layer of waterproofing material, by the installation method and procedure specified and outlined. In addition to this, the time duration for the necessary remobilizations and move-ins back to the component area for the additional coatings is allocated within the structure unit.

The quantity calculation for the asphalt plank waterproofing is calculated for the surface area required, multiplying the length by the width, arriving at a unit measurement of square feet for each required application including laps and splices. Each individual application is allocated for the labor, equipment, and material cost procedures, then summarized for the complete component item of work. All overrun and waste calculations are factored in with the material quantity summary list.

The quantity summation of required surface area of asphalt plank waterproofing material is outlined in a unit measurement of square feet, in a category defined by material type and application method, as:

1. Asphalt plank waterproofing, type _____, square feet

Joint waterstop

The joint waterstop component is performed in a procedural system according to material type and application.

The takeoff quantity operation for the component of joint waterstop is performed by calculating the lineal feet of joint, for each required installation type and material size, within the bridge structure unit.

The labor consumption is defined as the required time needed to properly install, within the formwork bulkhead, each specified type of joint waterstop material, by the installation method and procedure specified and outlined.

The quantity calculation for the joint waterstop is calculated by summarizing each joint required, for the length of each joint waterstop requirement. All overrun and waste calculations are factored in with the material quantity summary list.

The quantity summation of required joint waterstop material is outlined in a unit measurement of lineal feet, in a category defined by material type and size, as:

1. Joint waterstop, type _____, size _____, lineal feet.

Joint filler

The joint filler component is performed in a procedural system according to material type and application.

The takeoff quantity operation for the component of joint filler is performed by calculating the lineal feet of joint, for each required installation type and material size or square feet of joint area for each required installation type and material size, within the bridge structure unit.

The labor consumption is defined as the required time needed to properly install each specified type of joint filler material, by the installation method and procedure specified and outlined.

The quantity calculation for the joint filler is calculated by summarizing each joint required, for the length of each joint filler requirement, to determine the lineal feet of filler required or by calculating the surface area of joint filler by multiplying the required width by the determined length, in a unit measurement of square feet. All overrun and waste calculations are factored in with the material quantity summary list.

The quantity summation of required joint filler material is outlined in a unit measurement of lineal feet in a category defined by material type and size or in a unit measurement of square feet in a category defined by material type and size as:

1. Joint filler, type _____, size _____, lineal feet
2. Joint filler, type _____, size _____, square feet

Joint mortar

The joint mortar component is performed in a procedural system similar to the joint filler, according to material type and application.

The takeoff quantity operation for the component of joint mortar is performed by calculating the square feet of joint area for each required installation type and material size or the lineal feet of joint for each required installation type and material size, within the bridge structure unit.

The labor consumption is defined as the required time needed to properly install each specified type of joint mortar material, by the installation method and procedure specified and outlined.

The quantity calculation of the joint mortar is calculated for the surface area of joint mortar required by multiplying the specified width by the determined length in a unit measurement of square feet or is calculated by summarizing each joint area for the length of each joint mortar application requirement to determine the lineal feet of mortar required. All overrun and waste calculations are factored in with the material quantity summary list for each application required.

The quantity summation of required joint mortar material is outlined in a unit measurement of square feet in a category defined by material type or in a unit measurement of lineal feet in a category defined by material type and application width as:

1. Joint mortar, type _____, square feet
2. Joint mortar, type _____, width _____, lineal feet

Joint gaskets

The joint gasket component is performed in a procedural system according to material type and application.

The takeoff quantity operation for the component of joint gaskets is performed by calculating the quantity of each required, for each installation type and material size, within the bridge structure unit.

The labor consumption is defined as the required time needed to properly install each specified type of joint gasket material, by the installation method and procedure specified and outlined.

The quantity calculation of the joint gasket component is calculated for the quantity of each joint gasket required by the specifications. All overrun and waste calculations are factored in with the material quantity summary list.

The quantity summation of required joint gasket material is outlined in a unit measurement of each, in a category defined by material type, as:

1. Joint gasket, type _____, each

Procedure

The procedure for the summarization and calculation of the waterproofing components is performed by properly and accurately defining the required procedures, specifications, and application methods of the component required.

The takeoff is performed by identifying the appropriate unit of measure and the quantity, per material type, of required applications for the specified procedure.

The required secondary operations are calculated and summarized according to their appropriate unit of measure with respect to the performance operation; an example is sandblasting, to be quantified in a unit measurement of square feet of surface area prepared corresponding to a control operation of lineal feet of joint waterproofing.

Questions of Estimate

The questions outlined for a proper and accurate quantity takeoff are:

1. What type of waterproofing component is required?
2. What is the application and installation method?
3. What are the secondary operations?
4. What is the buildup or layer requirement?
5. What are the structure components detailed for this?
6. How many move-ins are required?

Application

The quality takeoff of waterproofing is performed in the following example using the given specified components. This example defines the production requirements and installation methods, the craft-hour factoring method, and the crew determination and identification, and produces a detailed summary listing for the required waterproofing components. It also details the material needs both permanent and incidental, and the equipment usage requirements.

A given retaining wall structure has two vertical construction joints, with the following detail of joint waterstop and membrane waterproofing operations along with the primary and secondary components. This waterstop is detailed to be installed within each vertical construction joint of the wall, and the membrane waterproofing is detailed to be applied over each construction joint, on the back side of the wall.

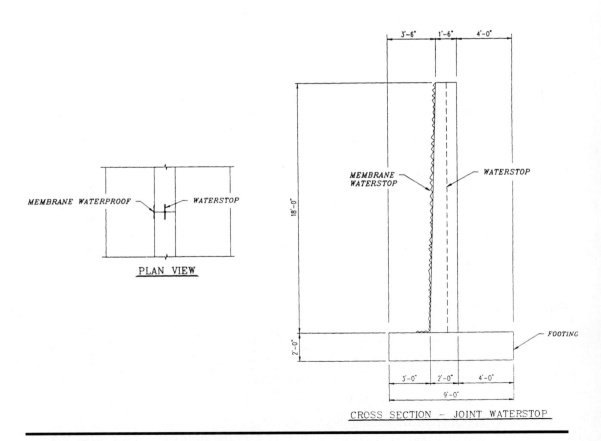

Joint waterstop, plan view and cross section.

Waterproofing and Joint Fillers

The preformed neoprene joint waterstop is a 6″ component, installed at the center of each vertical construction joint.

The membrane waterproofing covers a surface area of 1.0 foot on each side of the vertical joint, for the full length of the vertical construction joint. The application consists of a coating of asphaltic tar compound at an application rate of 0.1 gallon per square foot, a layer of membrane fabric, and a second coating of asphaltic tar compound at an application rate of 0.15 gallon per square foot. The membrane waterproofing overlaps the footer component horizontally 1.0 lineal foot.

All material overrun and waste is factored in with the material quantity summary.

Structure waterproofing system summary

6″ preformed neoprene waterstop 2.0 ea @ 18.0′	= 36.0 lf
Membrane waterproofing (2.0′ * 19.0′) * 2 each	
Asphaltic tar (0.10 gal per sf)	= 76.0 sf and 7.6 gal
Membrane fabric (6 mil)	= 76.0 sf
Asphaltic tar (0.15 gal per sf)	= 76.0 sf and 11.4 gal
Sandblast	= 0.0 sf

Example: Operation analysis

Operation. Joint waterstop. Unit: Lineal feet.

The operation of installing the joint waterstop device for the wall structure construction joints is calculated in the following manner for the total required quantity of waterstop.

Required: 6″ preformed neoprene waterstop

The quantity calculation for the joint waterstop is performed by summarizing the lineal feet of each required waterstop device per joint, by each location.

$$2 \text{ each} * (18.0') = 36.0 \text{ lf } 6'' \text{ waterstop}$$

Operation. Membrane waterproofing. Unit: Square feet.

The operation of applying the joint membrane waterproofing material for the wall structure construction joints is calculated in the following manner for the total required quantity of membrane waterproofing. The operation, calculated for each secondary component, is summarized to the primary unit.

Required: Asphaltic tar (0.10 gal per sf)
          Membrane fabric (6 mil)
          Asphaltic tar (0.15 gal per sf)

The quantity calculation for the first asphaltic tar application is performed by multiplying the surface coverage area of width by the height or length for each required waterproofing area per joint, then summarizing each joint location for square feet and gallons required.

$$2 \text{ each} * (2.0' * 19.0') = 76.0 \text{ sf tar}$$

$$76 \text{ sf} * 0.10 = 7.6 \text{ gal}$$

The quantity calculation for the fiber membrane application is performed by multiplying the surface coverage area of width by the height or length for each required waterproofing area per joint, then summarizing each joint location.

$$2 \text{ each} * (2.0' * 19.0') = 76.0 \text{ sf membrane}$$

The quantity calculation for the second asphaltic tar application is performed by multiplying the surface coverage area of width by the height or length for each required waterproofing area per joint, then summarizing each joint location for square feet and gallons required.

$$2 \text{ each} * (2.0' * 19.0') = 76.0 \text{ sf tar}$$

$$76 \text{ sf} * 0.15 = 11.4 \text{ gal}$$

Unit quantity summary: Waterproofing component quantity summary

Retaining wall	
Waterstop, 6"	= 36.0 lf
Asphaltic material (0.10 gal per sf)	= 76.0 sf and 7.6 gal
Fiber membrane (6 mil)	= 76.0 sf
Asphaltic material (0.15 gal per sf)	= 76.0 sf and 11.4 gal

From the quantity summary list, the time duration and production factoring are compiled. The quantity summary list is formed by grouping the operations of similar components.

Example: Production factoring, time duration

Operation. 6" waterstop device.

Crew members	= 2.0 each
Shift duration	= 8.0 hours
Production unit	= lineal feet
Project quantity requirement	= 36.0
Unit cycle time	= 10 min/lf
Productivity factor	= 50 minutes
Cycles per hour: 50/10	= 5.0 cycles
Production per hour	= 5.00 lf/hour
Production per shift: 5.0 * 8	= 40.00 lf/shift
Required shifts: 36.0 lf/40.0 lf	= 0.90 shift
Total craft-hours: 0.90 * 8 * 2	= 14.40 craft-hours
Operation production factor: 14.40/36.0	= 0.400
Operation factor	= 0.400 craft-hour/lf

Operation. Apply asphaltic material, 0.10 gallon per sf.

Crew members	= 2.0 each
Shift duration	= 8.0 hours
Production unit	= square feet
Project quantity requirement	= 76.0
Unit cycle time	= 2.0 min/sf
Productivity factor	= 50 minutes
Cycles per hour: 50/2.0	= 25.0 cycles
Production per hour	= 25.00 sf/hour
Production per shift: 25.0 * 8	= 200.00 sf/shift
Required shifts: 76.0 sf/200.0 sf	= 0.380 shifts
Total craft-hours: 0.38 * 8 * 2	= 6.080 craft-hours
Operation production factor: 6.08/76.0	= 0.080
Operation factor	= 0.080 craft-hour/sf

Example: Material allocation

The material costs associated with each operation are the direct permanent and indirect or incidental materials and supplies required per specification.

The material listed is summarized by the operation and allocated to the direct pay item, which is waterproofing, membrane material, for the square feet required, including the preformed neoprene waterstop, which is considered incidental to the item in this condition.

The material allocation is performed for each operation, for each type of material required, permanent direct and indirect incidental supplies. The material cost is disbursed against the control operation item quantity(*), not the work item quantity.

Operation. Waterproof construction joint, 76.0 sf.*

Direct material: asphaltic material	
0.10 gallon per square foot application rate, 1st coating	
0.15 gallon per square foot application rate, 2nd coating	
Overrun and waste 10.0%	
(0.10 * 76.0) * 1.10	= 8.36 gallons
(0.15 * 76.0) * 1.10	= 12.54 gallons
	= 20.90 gallons
Indirect material:	
Overrun and waste 10.0%	
Fiber membrane fabric (76.0 * 1.10)	= 83.60 sf
Incidental material:	
6" neoprene waterstop	= 36.00 lf

This material list is comprised of the direct permanent and indirect and incidental material required to perform the operations within the controlling item. The direct and indirect material is outlined and required by specification, and the incidental supplies are required by the operation to be performed. All materials are priced with all applicable sales and use taxes included.

The ratio of joint lineal feet to the control waterproofing square feet gives a prospective for a relationship of cost, both for the current job and for future reference. This is taken from the total joint footage of the operation against the neat line waterproofing square feet for the operation. The same can be done with the cost ratio.

Joint ratio, sf per lf

$$76.0 \text{ sf}/18.0 \text{ lf} = 4.22 \text{ square feet per lineal foot}$$

Takeoff Quantification Section Summary

The structure waterproofing components are specified in different types and for different structure components; therefore, a thorough and detailed quantity takeoff must be completed to perform the cost summary accurately and to fully understand the scope of the operation.

COMPONENT TAKEOFF

Structure Waterproofing

The following contains the normal work operation accounts for the performance and cost coding of structure waterproofing construction components.

1. **Membrane waterproofing**

    All work related to the installation of the specified type of membrane waterproofing to the proper line, grade, and dimension as detailed. Quantity should be reported in square feet of membrane installed along with the quantified accounting of all incidental materials consumed, including waste and lapping splices.

2. **Waterstops**

    All work related to the installation of the specified type of waterstop to their proper line, grade, and dimension as detailed. The quantity should be reported in lineal feet, by size, as specified, including overrun and waste.

3. **Joint fillers**

    All work related to the installation of the specified joint filler to their proper line, grade, and dimension as detailed. Quantity should be reported by type and size in either square feet or lineal feet installed, including overrun and waste.

4. **Joint mortar**

    All work related to the installation of the specified joint mortar along with all incidental consumed materials. Quantity should be reported in either square feet or lineal feet installed along with overrun and waste.

5. **Gaskets**

    All work related to the installation of the specified type of gasket to the defined item of work. Quantity should be reported in each of the type and size of gasket specified according to the primary work item.

6. **Waterproofing and stops, other**

    This account is used for other operation accounts not specified here. Quantity should be reported in a representative format relating to the work performed and the appropriate unit of measure.

Chapter 21

Exampled Component Cost Allocation

COST ESTIMATE ANALYSIS

The given examples, taken from Part 4, detail the method of direct cost evaluation for a given component. This is the application of labor and equipment crew determination, the size or capacity evaluation, and cost association directly related thereto. Along with this format are the material tabulation and cost application associated with both the direct and indirect material components.

These methods of cost evaluation are derived from the methods detailed in Chap. 2.

The takeoff evaluation has determined and verified the following information and criteria. It is recommended that this method be performed for each component operation within the scope of the outlined work.

Example 1. Pile Driving
(Chapter 12, Foundation Piling: Takeoff Quantification)

The following example defines a conventional pile takeoff with a determination being resolved for a quantified summary and productivity factor.

There are two pile-driving locations, a north and south abutment. Each given foundation site requires a cluster of 15 steel H piles, having a section of 12″ × 12″, and a weight of 53 pounds per lineal foot. The anticipated design length is 44 lineal feet. The spacing is symmetrical at three rows of five piles, 5 feet on centers. The front row of piles at each location is required to be battered at a 0.5′:6.0′ ratio; in other words, for every 6 vertical feet of pile, the batter is ½ foot. The top of the pile is designated to extend into the footing of the structure 1.0 lineal foot.

636 Takeoff and Cost Analysis Techniques

Washington Baum Bridge, US 64/264 over Roanoke Sound between Manteo and Nags Head, N.C.; structure type: fixed span; owner: North Carolina Department of Transportation. (*Courtesy of Ralph Whitehead and Associates.*)

Detail of the 5543′ long Washington Baum Bridge. The 4-lane structure consists of 73 prestressed concrete girder spans supported by cast-in-place concrete bents. (*Courtesy of Ralph Whitehead and Associates.*)

The pile tips are fitted with a rock point shoe for specified seating at the bearing point. The cutoff or trim waste is calculated at 1.5 lineal feet per pile, or 30 * 1.5 = 45 additional feet required for the project. This requires an order length per piece of 45.5 lineal feet, or a total requirement of 1365.0 lineal feet of 12″ × 53# steel H pile.

Example: Operation analysis

Operation. Furnish and drive pile.

Type	= steel H 12″×53#
Spacing	= 3 rows @ 5′ centers
Design length	= 44 lf
Quantity	= 30.0 each
Template requirement	= simple staking
Points	= 30.0 each
Splices	= not required
Battered pieces	= 10.0 each
Batter	= 0.5 on 6
Batter factor	= 1.025
Driving locations	= 2.0 each
Trim and cutoff waste	= 1.5 lf each

Soil boring data:

Material: Fragmented shale and limestone
Split spoon: 2″ @ 50 lb
Hammer stroke: 30″ (2.5′)
Increment: 1.0′

Design resistance:

Elevation	Blow count/lf
−10′	= 15
−18′	= 30
−25′	= 40
−32′	= 70
−37′	= 55
−43′	= 75
Average blow count	= 47.5

From the operation analysis evaluation, the crew production and associated time duration are determined. A crew of six is utilized for this operation. The crew consists of the following craft determinations and personnel.

Crew function. Pile drive, steel H pile.

Pile supervisor	1.0 each
Crane operator (60 ton)	1.0 each
Pile driver (journeyman)	3.0 each
Pile driver (welder)	1.0 each

Example: Production factoring, time duration

Operation. Drive 12″ × 53# steel H pile.

Crew members	= 6.0
Production unit	= lineal feet
Productivity factor	= 50-minute hour
Shift duration	= 8 hours
Relocate time	= 4 hours

With the given information, the pile factoring formula can be implemented using the prescribed format.

1. Type and size	= 12″ × 54# steel H
2. Total pieces required	= 30.0 ea
3. Total lineal feet	= 1320.0 lf
4. Average length per piece	= 44.0 lf
5. Number of driving locations	= 2.0 ea
6. Average pieces per location	= 15.0 ea
7. Craft-hours consumed per day	= 48.0 craft-hours
8. Work hours per shift	= 8.0 hours
9. Number of workers per crew	= 6.0 ea
10. Moving time to each location	= 4.0 hours
11. Productivity time per hour	= 50.0 minutes
12. Number of straight pieces	= 20.0 ea
13. Number of battered pieces	= 10.0 ea
14. Ratio percentage of battered	= 0.33%
a. Batter	= 0.5
b. Multiplier	= 1.025
15. Longest pile	= 44.0 lf
16. Required splices	= none
a. Percent of pieces spliced	= 0.0%
17. Soil analysis	
a. Average blow count	= 47.5 avg blow count
b. Increment per blow	= 1.0 ft
c. Hammer stroke	= 2.5 ft
d. Soil factor (47.5/1)/2.5	= 19.00 factor
e. Theoretical minutes per lf	= 0.38 minute
f. Theoretical driving time/pile	= 16.72 minutes
g. Batter multiplier	= 1.34
18. Cycle time duration (in minutes)	
a. Set template	= 0.00
b. Rig pile	= 2.00
c. Hoist pile	= 1.50
d. Set hammer	= 2.50
e. Set batter, 2.50 minutes	= 0.83
f. Drive pile	= 22.43
g. Unhook pile	= 2.00
h. Splice time	= 0.00
i. Additional piece (splices)	= 0.00
j. Auger prebore shaft	= 0.00
k. Other operation	= 0.00
l. Total duration per pile	= 31.27
19. Driving time per pile	= 31.27 minutes
20. Productivity factor per hour	= 50 minutes

21. Required crew hours per pile	= 0.63 crew hour
22. Average lineal feet per crew hour	= 70.36 lf
23. Average lineal feet per shift	= 562.91 lf
24. Number of production shifts required	= 2.34 shifts
25. Location moving hours	= 8.0 hours
26. Required relocate shifts	= 1.0 shift day
27. Required operation shifts	= 3.35 work days
28. Net production per day (lf)	= 394.63 lf shift
29. Production factor	= 0.1242 craft-hour/lf
Unit cycle time: 31.27 min/44.0 lf	= 0.711 minute per lf
Cycles per hour: (50/31.27)	= 1.599 cycles per hour
Production per hour: 1.599 * 44.0'	= 70.36 lf per hour
Production per shift: 8 * 70.36 lf	= 562.85 lf per shift
Required shifts: 1320.0'/562.85	= 2.345 shifts
Required relocate time	= 1.0 shift day
Total required operation shifts	= 3.35 shifts (rounded)
Total craft-hour 3.35 * 8 * 6 workers	= 160.8 craft-hours
System check: 1320 lf * 0.1242	= 163.9 craft-hours check
Operation production factor: 163.9/1320	= 0.1242 check

Once these procedures have been completed, the task of applying a cost from the production operation analysis to the previously compiled labor and equipment crew can be done.

For calculating purposes, the shift duration for the pile-driving operation is rounded to 3.35 required shifts.

Example: Crew costing and formulating

The formula detailed under this heading shows the distribution and application of the production factoring method. The cost breakdown summarizes the unit labor and equipment, along with the total required labor and equipment cost, independently and total for the operation.

The application of cost is derived from the production factors calculated by an average wage per craft-hour and the equipment crew cost per operation. The labor crew and equipment crew cost, per shift, are the total cost expended, derived from the labor and equipment formulations.

Labor crew cost determination

The following example shows the necessary calculations for the required labor crew average wage determination based on a shift duration of 8 hours.

Example. Labor crew function, drive steel H pile.

Trade type	Number of personnel	Shift duration	Total hours
PDS	1.0	8.5	8.5
OP (60 ton)	1.0	8.0	8.0
PD (JM)	3.0	8.0	24.0
PD (WLD)	1.0	8.0	8.0
Total	6.0		48.5
Average hours per shift		8.083	

640   Takeoff and Cost Analysis Techniques

The exampled labor crew consists of one crew supervisor, skilled as a pile driver, an operator for the crane, three pile drivers to handle and drive the pile, and one welder to assist with the required welding and cutting operations. This labor crew was formed by visualizing the required operation or structuring it based on past experience.

The production is based on and assumes the crew will be productive for the full shift. The crew supervisor is allotted an additional ½ hour per shift for paperwork, tool gathering, site check and preparation, planning and safety evaluations, etc.

Once the labor crew is structured, the associated cost is applied to arrive at an average wage per craft-hour relative to the specific labor crew. This is calculated using the methods to determine the total wage concern, discussed in Chap. 2, Labor Costs.

The required cost allocations are the base wage for each craft, the overtime compensation method, the applicable fringe benefits, and the appropriate payroll tax and insurance factors. For this scenario, the following criteria are used:

The overtime compensation is based on 8 hours per shift and 40 hours per week. The work is performed in a region that does not have labor agreements or fringe benefit allocations. The payroll tax and insurance are as follows:

Social security tax	@	7.65%
State unemployment tax	@	4.50% (with earning limits)
Federal unemployment tax	@	0.80% (with earning limits)
Worker compensation	@	15.00%
Contractor public liability	@	3.00%
Contractor property liability	@	2.00%
Total		32.95%

The base wage determination, as specified by regional wage information, is as follows:

PDS = crew supervisor (piling)	@	$15.00
OP = equipment operators (60 ton)	@	$12.00
PDJ = pile driver (journeyman)	@	$11.00
PDW = welder	@	$12.00

The average wage per hour is determined next by calculating the standard and premium time wages, by the appropriate hours, to arrive at an average wage per hour.

	Average wage/hour
PDS @ (($15.00 * 8) + ($15.00 * 1.5 * 0.5))/8.5	= $15.44
OP @ ($12.00 * 8)/8	= $12.00
PDJ @ ($11.00 * 8)/8	= $11.00
PDW @ ($12.00 * 8)/8	= $12.00

Exampled Component Cost Allocation

From the given information, the labor crew can be cost allocated for the wage concern appropriate for the project.

Trade type	Number of personnel	Shift duration	Total hours	Average wage/hour	Crew cost wages
PDS	1.0	8.5	8.5	$15.44	$131.25
OP	1.0	8.0	8.0	$12.00	$ 96.00
PDJ	3.0	8.0	24.0	$11.00	$264.00
PDW	1.0	8.0	8.0	$12.00	$ 96.00
Total	6.0		48.5		$587.25

From this calculation, the payroll tax and insurance factor are allocated. This is applied by multiplying the burden factor by the total base wages, then summarizing the two amounts to arrive at the budget wage concern.

$587.25 * 0.3295 = $193.50
$587.25 + $193.50 = $780.75 wage concern
Average hours per shift: 48.5/6 workers = 8.083
Average wage per crew hour: $780.75/48.5 = $16.10

The average cost per crew hour for the determined labor crew will become the multiplier for the craft-hour production factor that determines the unit labor component cost.

The labor crew format must be given consideration and realistic production factors applied. Many labor crews may be structured in a similar manner but utilize different trade allocations and appropriations.

Equipment crew cost determination

Example: Equipment crew function, drive steel H pile.

Equipment type	Number of units	Shift duration	Total hours
½-ton pickup (supervisor)	1.0	8.5	8.5
60-ton crane	1.0	8.0	8.0
Diesel pile hammer #520	1.0	8.0	8.0
80′ pile leads	1.0	8.0	8.0
Tools, piling	1.0	8.0	8.0
Total	5.0		40.5

The exampled crew consists of the necessary and required equipment to perform the operation. This equipment crew was formed by visualizing the required operation or structuring it based on past experience and requirements.

The production is based on and assumes the equipment crew will be productive for the full shift. The pickup truck for the crew supervisor is allotted an additional ½ hour per shift, as with the supervisor. The usage and performance are factored on a full shift with an equal proportionate rate applied to each hour worked.

Once the crew is structured, the associated cost is applied and calculated to arrive at a total cost for each designated shift, relative to the specific equipment crew. This is calculated using the methods to determine the total cost concern for equipment ownership and operating, discussed in Chap. 2, Equipment Ownership.

The required cost allocations are the owning and operating allocation for each unit. The equipment cost concern is based on the defined operating hours or hours worked with no additional compensation for overtime. From the equipment ownership factoring, each unit is allocated an ownership rate based on an 8-hour shift. Any variation above or below this should be prorated accordingly.

An example will be structured using assumed owning and operations costs for each unit described, based on an 8-hour shift.

Equipment type	Units	Owning and operating cost	Cost/hour
½-ton pickup	1.0	$ 24.00	$ 3.00
60-ton crane	1.0	$480.00	$60.00
Diesel pile hammer #520	1.0	$320.00	$40.00
80′ pile leads	1.0	$ 60.00	$ 7.50
Tools	1.0	$ 62.00	$ 7.75

From the given information, the specific equipment crew can be cost allocated for the cost concern appropriate for the operation.

Equipment type	Number of units	Shift duration	Total hours	Average cost/hour	Crew cost concern
Pickup	1.0	8.5	8.5	$ 3.00	$ 25.50
60-ton crane	1.0	8.0	8.0	$60.00	$480.00
Diesel hammer	1.0	8.0	8.0	$40.00	$320.00
Pile leads	1.0	8.0	8.0	$ 7.50	$ 60.00
Tools	1.0	8.0	8.0	$ 7.75	$ 62.00
Total	5.0		40.5		$947.50 shift

To determine the component operation budget, the total operation cost concern for the equipment requirement can be calculated by multiplying the shift cost by the required shifts for each operation.

The total cost, for both labor and equipment, is calculated by the wage and cost concern per shift times the number of required shifts. This format produces the total labor and equipment cost required to perform the given operation.

To arrive at the total labor cost, multiply the unit craft-hour production factor by the average wage per hour by the required operation quantity. To arrive at the unit labor cost, multiply the unit craft-hour production factor by the average wage per hour. For the total equipment cost, multiply the cost per shift by the required shifts. For the unit equipment cost, divide the total equipment cost by the required quantity for the operation. To determine the average wage per craft-hour, multiply the labor cost per shift by the required shifts divided by the required craft-hours for each operation.

Operation. Drive 12″ × 53# steel H pile.

Required quantity	1320.00 lf
Labor cost per shift	$780.75
Equipment cost per shift	$947.50
Total per shift	$1728.25
Required shifts	3.35
Required craft-hours	160.8
Average wage per craft-hour	$16.10
Labor production factor	0.1242
Production per shift (revised)	= 394.03 lf
Unit labor 0.1242 * $16.10	= $2.00 lf
Unit equipment $947.50/394.03	= $2.405 lf
Unit production cost	= $4.405 lf
Labor 0.1242 * $16.10 * 1320.0	= $2639.50
Equipment 3.35 * $947.50	= $3174.13
Total cost, labor and equipment	= $5813.63
Check: $5813.63/1320.0 lf	= $4.404 lf

The labor and equipment summary has summarized the direct costs associated with each work operation. Each operation must have the full direct, field related costs required to perform the listed operation, including any delay or incidental moving within the site. This will be all costs required to perform the tasks, directly associated to the work operation.

NOTE: The operations of trimming and pile layout have not been shown or calculated in this example.

Material allocation

After the production has been determined and the labor and equipment crews have been cost allocated, the direct and incidental consumed materials for the operation must be summarized and cost allocated.

Operation. Furnish 12″ × 53# steel H pile.

Direct material:	
Required pieces	= 30.0 each
Design length	= 44.0 lf
Overrun and waste	= 1.5′ each piece
Order length	= 45.5 lf each
Total feet required	= 1365.0 lf
Splices required	= none
Pile points required	= 30.0 each
Indirect material:	
Welding rod ⅜″	= 60.0 pounds
Cushion material (Micarta)	= 6.0 pieces
Oxygen and acetylene	= 1.0 tank ea

The material list for this operation is comprised of direct materials that are accounted for and detailed by the project specification requirements as permanent embedded items. The indirect materials are those which are needed to perform the operation, regardless of the specific direct materials.

The project required 30 pieces of 12″ × 53# steel H pile to be 44.0 lineal feet long. This length is a permitted length to be transported to the project without special variances; thus the pieces are ordered in the full length required including overrun. This eliminates the need of splicing.

The need for trimming exists owing to any variance in actual driving conditions, the possibility of slight damage to the top of the pile from driving, and the specified top or cutoff elevation required in the footing foundation. With regard to these parameters, the determination to allow or add 1.5 lineal feet to the order length has been allocated. This makes the total order length the contractor must allow for 45.5 lineal feet per piece, which is an indirect cost or nonreimbursable cost relating to the additional footage from that specified.

The specifications detail a pile point at each pile location; thus the quantity of points are one per piece or 30 points required.

The indirect material costs related to the pile-driving operation are the cushion material required within the pile hammer. The assumptions are made that each piece of cushion material will last for five pile drives, equating to the requirement of 6 pieces of cushion material to be ordered.

The pile points are required by specification to be welded to the tip of the pile. The calculation is made that each point will require 2 pounds of ⅜″ welding rod to complete the operation per specification.

The cutoff or trimming is somewhat of a variable, but the assumption is made that 90 percent of the pile will need to be trimmed to the proper elevation after driving. This will require that one tank each of oxygen and acetylene be present at the site.

This material listing must be allocated for cost for the operation. The cost of all associated materials and applicable sales taxes, shipping charges, etc., is directed toward the prime operation item and project required unit and measurement:

$$12″ \times 53\# \text{ H pile, 30 pieces @ } 44.0' = 1320.0 \text{ lf}$$

Material cost

Payment item and unit	= 1320.0 lf pile
Direct materials:	
Required pile: 1365.0′ @ $11.00/lf	= $15,015.00
Required points: 30.0 @ $25.00/ea	= $ 750.00
Subtotal cost	= $15,765.00
Unit cost	= $ 11.94 lf
Indirect materials:	
Welding rod: 60.0 lb @ $2.50	= $ 150.00
Cushion materials: 6.0 ea @ $20.00	= $ 120.00
Oxygen and acetylene: 2.0 ea @ $30.00	= $ 60.00
Subtotal cost	= $ 330.00
Unit cost	= $ 0.25 lf
Total consumed material cost	= $16,095.00
Unit cost 1320.0 lf	= $ 12.19 lf

The consumed material cost associated with the direct and indirect materials should be detailed as listed with the example. This method provides a clear and detailed list for review and distribution upon becoming the successful bidder.

The consumed material list associated with the direct and indirect materials should be detailed as required for the operation. This method provides a clear and detailed tabulation for review and distribution upon becoming the successful bidder.

Summarized direct field cost

1320.0 lineal feet 12" × 53# steel H pile	
Drive labor 1320' @ $1.999	= $ 2,639.50
Drive equipment 1320' @ $2.404	= $ 3,174.13
Permanent material 1320' @ $11.940	= $15,765.00
Indirect material 1320' @ $0.25	= $ 330.00
Total	= $21,908.63
Unit cost	= $ 16.60 lf

This is a summary of direct field expenses without contingency costs, project and corporate overhead, and net profit. From these direct field related costs of labor, equipment, materials, and supplies, the related project and corporate overhead, contingency factor, and desired profit margin are allocated.

This example of operation cost analysis is only for comparison and review to detail how one should structure an operation component cost budget analysis. Each situation is different and will have distinct controlling features specific to itself; therefore, an exact and detailed takeoff along with an individual cost analysis must be performed.

Example 2. Abutment Wall Construction
(Chapter 13, Substructure Takeoff Quantification Unit)

Example: Operation analysis

The following lists of required component dimensions per project plans.

Abutment unit (north and south symmetrical)	
Abutment footer	3' thick 10' wide 40' long
Abutment wall	3' wide 5' high 36' long
Abutment backwall	1.5' thick 4.5' high 36' long
Beam type	Precast concrete type IV AASHTO
End area	789 square inches
East wing wall	1' thick 9.5' high 12' long
West wing wall	1' thick 9.5' high 14' long
Beam seats	1.5' × 2' ea

Abutment wall component quantity summary			
		North side	South side
Formwork place	sf	360.0	360.0
Formwork removal	sf	360.0	360.0
Bulkheads place	sf	30.0	30.0
Bulkheads remove	sf	30.0	30.0
Hardware and embeds	ea	20.0	20.0
Concrete placement	cy	21.6	21.6
Rub and patch	sf	210.0	210.0

Takeoff and Cost Analysis Techniques

From the operation analysis evaluation, the crew production and associated time durations are determined. The crew sizing determined and utilized for this operation is structured for each component. This operation, consisting of multiple components, requires different crews consisting of the following craft determinations and personnel.

Crew function. Abutment wall construction.

Abutment walls	
formwork placement:	
Carpenter supervisor	1.0 each
Crane operator (45-ton)	1.0 each
Carpenter (journeyman)	2.0 each
Carpenter (apprentice)	1.0 each
Laborer (skilled)	1.0 each
Teamster (5-ton truck)	1.0 each
Abutment walls	
formwork removal:	
Carpenter supervisor	1.0 each
Crane operator (45-ton)	1.0 each
Carpenter (journeyman)	1.0 each
Carpenter (apprentice)	1.0 each
Bulkheads	
formwork placement:	
Carpenter (journeyman)	1.0 each
Carpenter (apprentice)	1.0 each
Laborer (skilled)	1.0 each
Bulkheads	
formwork removal:	
Carpenter (journeyman)	1.0 each
Carpenter (apprentice)	1.0 each
Embed installation and placement:	
Carpenter (apprentice)	1.0 each
Abutment walls	
concrete placement:	
Concrete finisher supervisor	1.0 each
Crane operator (45-ton)	1.0 each
Concrete finisher	3.0 each
Laborer (skilled)	2.0 each
Teamster (5-ton truck)	1.0 each
Abutment walls	
rub and patch:	
Cement mason	1.0 each

Example: Production factoring, time duration

With the given information, each component craft-hour production factor and crew time duration can be implemented using the prescribed format. This method uses the given component informational facts along with the factoring methods described in Chap. 2.

Exampled Component Cost Allocation

Operation. Formwork, place, abutment wall.

Abutment wall component	
Production unit	= square foot, surface contact area
Project requirements	= 720.0 sq ft
Crew members	= 7.0 ea
Formwork type	= wood
Forming condition	= average, 5
Shift duration	= 8.0 hours
Unit cycle time	= 0.85 minute/sf
Productivity factor	= 50-minute hour
Cycles per hour: (50/0.85)	= 58.82 cycles/hour
Production per hour: 58.82 * 1	= 58.82 sf/hour
Production per shift: 8 * 58.82	= 470.56 sf/shift
Required shifts: 720.0/470.56	= 1.53 shifts
Total craft-hours: 1.53 * 8 * 7	= 85.68 craft-hours
Operation production factor: 85.68/720	= 0.119 craft-hour/sf
Place formwork	= 0.119 craft-hour/sf

Operation. Formwork, remove, abutment wall.

Production unit	= square foot, surface contact area
Project requirements	= 720.0 sq ft
Crew members	= 4.0 ea
Shift duration	= 8.0 hours
Unit cycle time	= 0.40 minute/sf
Productivity factor	= 55-minute hour
Cycles per hour: (55/0.40)	= 137.5 cycles/hour
Production per hour: 137.5 * 1	= 137.5 sf/hour
Production per shift: 8 * 137.5	= 1100.0 sf/shift
Required shifts: 720.0/1100.0	= 0.66 shift
Total craft-hours: 0.66 * 8 * 4	= 21.12 craft-hours
Operation production factor: 21.12/720	= 0.029 craft-hour/sf
Remove formwork	= 0.029 craft-hour/sf

Operation. Formwork, place, end bulkhead, abutment wall.

Production unit	= square foot, surface contact area
Project requirements	= 60.0 sq ft
Crew members	= 3.0 ea
Shift duration	= 8.0 hours
Unit cycle time	= 3.75 minutes/sf
Productivity factor	= 50-minute hour
Cycles per hour: (50/3.75)	= 13.33 cycles/hour
Production per hour: 13.33 * 1	= 13.33 sf/hour
Production per shift: 8 * 13.33	= 106.64 sf/shift
Required shifts: 60.0/106.64	= 0.56 shift
Total craft-hours: 0.56 * 8 * 3	= 13.44 craft-hours
Operation production factor: 13.44/60	= 0.224 craft-hour/sf
Place formwork, end bulkhead	= 0.224 craft-hour/sf

Operation. Formwork, remove, end bulkhead, abutment wall.

Production unit	= square foot, surface contact area
Project requirements	= 60.0 sq ft
Crew members	= 2.0 ea
Shift duration	= 8.0 hours
Unit cycle time	= 1.25 minutes/sf
Productivity factor	= 55-minute hour
Cycles per hour: (55/1.25)	= 44.0 cycles/hour
Production per hour: 44.0 * 1	= 44.0 sf/hour
Production per shift: 8 * 44.0	= 352.0 sf/shift
Required shifts: 60.0/352.0	= 0.17 shift
Total craft-hours: 0.17 * 8 * 2	= 2.72 craft-hours
Operation production factor: 2.72/60	= 0.045 craft-hour/sf
Remove formwork, end bulkhead	= 0.045 craft-hour/sf

Operation. Place embeds, abutment wall.

Production unit	= each
Project requirements	= 40.0 ea
Crew members	= 1.0 ea
Shift duration	= 8.0 hours
Unit cycle time	= 10 minutes/ea
Productivity factor	= 55-minute hour
Cycles per hour: (55/10)	= 5.5 cycles/hour
Production per hour: 5.5 * 1	= 5.5 ea/hour
Production per shift: 8 * 5.5	= 44.0 ea/shift
Required shifts: 40.0/44.0	= 0.91 shift
Total craft-hours: 0.91 * 8 * 1	= 7.28 craft-hours
Operation production factor: 7.28/40	= 0.182 craft-hour/ea
Place embeds	= 0.182 craft-hour/ea

Operation. Place concrete, abutment wall.

Production unit	= cubic yard
Project requirements	= 43.2 cy
Crew members	= 8.0 ea
Shift duration	= 8.0 hours
Unit cycle time	= 10 minutes/cy
Productivity factor	= 50-minute hour
Cycles per hour: (50/10)	= 5.0 cycles/hour
Production per hour: 5.0 * 1	= 5.0 cy/hour
Production per shift: 8 * 5.0	= 40.0 cy/shift
Required shifts: 43.2/40.0	= 1.08 shifts
Total craft-hours: 1.08 * 8 * 8	= 69.12 craft-hours
Operation production factor: 69.12/43.2	= 1.60 craft-hours/cy
Place concrete	= 1.60 craft-hours/cy

Exampled Component Cost Allocation

Operation. Rub and patch concrete, abutment wall.

Production unit	= square foot, surface area
Project requirements	= 420.0 sf
Crew members	= 1.0 ea
Shift duration	= 8.0 hours
Unit cycle time	= 3 minutes/sf
Productivity factor	= 50-minute hour
Cycles per hour: (50/3)	= 16.67 cycles/hour
Production per hour: 16.67 * 1	= 16.67 sf/hour
Production per shift: 8 * 16.67	= 133.36 sf/shift
Required shifts: 420.0/133.36	= 3.15 shifts
Total craft-hours: 3.15 * 8 * 1	= 25.2 craft-hours
Operation production factor: 25.2/420	= 0.060 craft-hour/sf
Rub and patch concrete surface	= 0.060 craft-hour/sf

Once these procedures have been completed, the task of applying a cost from the production operation analysis to the previously compiled labor and equipment crews can be done.

For calculating purposes, the shift durations for each operation are rounded to two decimal places for each required shift.

Example: Crew costing and formulating

The formula detailed under this heading shows the distribution and application of the production factoring method. The cost breakdown summarizes the unit labor and equipment, along with the total required labor and equipment cost, independently and total for the operation.

The application of cost is derived from the production factors calculated by an average wage per craft-hour, and the equipment crew cost per operation. The labor crew and equipment crew cost, per shift, are the total cost expended, derived from the labor and equipment formulations.

Labor crew cost determination

The following example shows the necessary calculations for the required labor crew average wage determination based on a shift duration of 8 hours.

Crew example. Labor crew function, formwork placement, abutment wall.

Trade type	Number of personnel	Shift duration	Total hours
CSC	1.0	8.5	8.5
OP (45 ton)	1.0	8.0	8.0
CP (JM) journeyman	2.0	8.0	16.0
CP (APP) apprentice	1.0	8.0	8.0
LB (SK) skilled	1.0	8.0	8.0
TM (5 ton)	1.0	8.0	8.0
Total	7.0		56.5
Average hours per shift		8.071	

The exampled labor crew on page 649 consists of one crew supervisor, skilled as a carpenter, an operator for the crane, three personnel skilled as carpenters to construct the formwork, one laborer to assist with the required operations, and one teamster to move and gather the form equipment.

Crew example. Labor crew function, formwork removal, abutment wall.

Trade type	Number of personnel	Shift duration	Total hours
CSC	1.0	8.5	8.5
OP (45 ton)	1.0	8.0	8.0
CP (JM) journeyman	1.0	8.0	8.0
CP (APP) apprentice	1.0	8.0	8.0
Total	4.0		32.5
Average hours per shift		8.125	

The exampled labor crew consists of one crew supervisor, skilled as a carpenter, an operator for the crane, and two personnel skilled as carpenters to remove, clean, and stack the formwork.

Crew example. Labor crew function, formwork placement, bulkhead.

Trade type	Number of personnel	Shift duration	Total hours
CP (JM) journeyman	1.0	8.0	8.0
CP (APP) apprentice	1.0	8.0	8.0
LB (SK) skilled	1.0	8.0	8.0
Total	3.0		24.0
Average hours per shift		8.00	

The exampled labor crew consists of two personnel skilled as carpenters to construct the formwork and one laborer to assist the carpenters.

Crew example. Labor crew function, formwork removal, bulkhead.

Trade type	Number of personnel	Shift duration	Total hours
CP (JM) journeyman	1.0	8.0	8.0
CP (APP) apprentice	1.0	8.0	8.0
Total	2.0		16.0
Average hours per shift		8.00	

The exampled labor crew consists of two personnel skilled as carpenters to remove, clean, and stack the formwork.

Crew example. Labor crew function, hardware and embed installation.

Trade type	Number of personnel	Shift duration	Total hours
CP (APP) apprentice	1.0	8.0	8.0
Total	1.0		8.0
Average hours per shift		8.00	

The exampled labor crew consists of one person skilled as a carpenter to install the required hardware and embedded items required for the formwork.

Crew example. Labor crew function, concrete placement, abutment wall.

Trade type	Number of personnel	Shift duration	Total hours
CSF	1.0	8.5	8.5
OP (45 ton)	1.0	8.0	8.0
CF	3.0	8.0	24.0
LB (SK) skilled	2.0	8.0	16.0
TM (5 ton)	1.0	8.0	8.0
Total	8.0		64.5
Average hours per shift		8.063	

The exampled labor crew consists of one crew supervisor, skilled as a concrete finisher, an operator for the crane, three personnel skilled as finishers to place and finish the concrete, two laborers to assist with the required operations, and one teamster to move and gather the placing and finishing equipment.

Crew example. Labor crew function, rub and patch concrete.

Trade type	Number of personnel	Shift duration	Total hours
CM	1.0	8.0	8.0
Total	1.0		8.0
Average hours per shift		8.00	

The exampled labor crew consists of one person skilled as a cement mason to rub and patch the required concrete surfaces of specification requirements.

These labor crews were formed by visualizing the required operations and functions or structuring them based on past experience.

The production is based on and assumes the crews will be productive for the full shift. The crew supervisor is allotted an additional ½ hour per shift for paperwork, tool gathering, site check and preparation, planning and safety evaluations, etc.

Once the labor crews are structured, the associated costs are applied to arrive at an average wage per craft-hour relative to the specific labor crew.

This is calculated using the methods to determine the total wage concern, discussed in Chap. 2, Labor Costs.

The required cost allocations are the base wage for each craft, the overtime compensation method, the applicable fringe benefits, and the appropriate payroll tax and insurance factors.

For this scenario, the following criteria are used:

The overtime compensation is based on 8 hours per shift and 40 hours per week. The work is performed in a region that does not have labor agreements or fringe benefit allocations. The payroll tax and insurance are as follows:

Social security tax	@	7.65%
State unemployment tax	@	4.50% (with earning limits)
Federal unemployment tax	@	0.80% (with earning limits)
Workers' compensation	@	15.00%
Contractor public liability	@	3.00%
Contractor property liability	@	2.00%
Total		32.95%

The base wage determination, as specified by regional wage information, is as follows:

CSC	= crew supervisor (carpenter)	@	$15.00
CSF	= crew supervisor (finisher)	@	$14.00
OP	= equipment operators	@	$12.00
CPJ	= carpenter (journeyman)	@	$11.00
CPA	= carpenter (apprentice)	@	$ 9.00
CF	= concrete finisher	@	$11.00
CM	= cement mason	@	$11.50
LB	= labor (skilled)	@	$ 8.50
TM	= teamster	@	$ 9.50

The average wage per hour is determined next by calculating the standard and premium time wages by the appropriate hours, to arrive at an average wage per hour.

	Average wage/hour
CSC @ (($15.00 * 8) + ($15.00 * 1.5 * 0.5))/8.5	= $15.44
CSF @ (($14.00 * 8) + ($14.00 * 1.5 * 0.5))/8.5	= $14.41
OP @ ($12.00 * 8)/8	= $12.00
CPJ @ ($11.00 * 8)/8	= $11.00
CPA @ ($9.00 * 8)/8	= $ 9.00
CF @ ($11.00 * 8)/8	= $11.00
CM @ ($11.50 * 8)/8	= $11.50
LB @ ($8.50 * 8)/8	= $ 8.50
TM @ ($9.50 * 8)/8	= $ 9.50

Exampled Component Cost Allocation

From the given information, each labor crew can be cost allocated for the wage concern appropriate for each operation component of the project.

Labor crew function. Formwork placement, abutment wall.

Trade type	Number of personnel	Shift duration	Total hours	Average wage/hour	Crew cost wages
CSF	1.0	8.5	8.5	$15.44	$131.24
OP	1.0	8.0	8.0	$12.00	$ 96.00
CPJ	2.0	8.0	16.0	$11.00	$176.00
CPA	1.0	8.0	8.0	$ 9.00	$ 72.00
LB	1.0	8.0	8.0	$ 8.50	$ 68.00
TM	1.0	8.0	8.0	$ 9.50	$ 76.00
Total	7.0		56.5		$619.24

From this calculation, the payroll tax and insurance factor are allocated. This is applied by multiplying the burden factor by the total base wages, then summarizing the two amounts to arrive at the budget wage concern.

$619.24 * 0.3295   = $204.04
$619.24 + $204.04   = $823.28 wage concern
Average hours per shift:
                   56.5/7 = 8.071
Average wage per crew hour:
              $823.28/56.5 = $14.57

Labor crew function. Formwork removal, abutment wall.

Trade type	Number of personnel	Shift duration	Total hours	Average wage/hour	Crew cost wages
CSC	1.0	8.5	8.5	$15.44	$131.24
OP	1.0	8.0	8.0	$12.00	$ 96.00
CPJ	1.0	8.0	8.0	$11.00	$ 88.00
CPA	1.0	8.0	8.0	$ 9.00	$ 72.00
Total	4.0		32.5		$387.24

From this calculation, the payroll tax and insurance factor are allocated. This is applied by multiplying the burden factor by the total base wages, then summarizing the two amounts to arrive at the budget wage concern.

$387.24 * 0.3295   = $127.60
$387.24 + $127.60   = $514.84 wage concern
Average hours per shift:
                   32.5/4 = 8.125
Average wage per crew hour:
              $514.84/32.5 = $15.84

Labor crew function. Formwork placement, bulkhead.

Trade type	Number of personnel	Shift duration	Total hours	Average wage/hour	Crew cost wages
CPJ	1.0	8.0	8.0	$11.00	$ 88.00
CPA	1.0	8.0	8.0	$ 9.00	$ 72.00
LB	1.0	8.0	8.0	$ 8.50	$ 68.00
Total	3.0		24.0		$228.00

From this calculation, the payroll tax and insurance factor are allocated. This is applied by multiplying the burden factor by the total base wages, then summarizing the two amounts to arrive at the budget wage concern.

$228.00 * 0.3295 = $75.13
$228.00 + $75.13 = $303.13 wage concern
Average hours per shift: 24.0/3 = 8.00
Average wage per crew hour: $303.13/24.0 = $12.63

Labor crew function. Formwork removal, bulkhead.

Trade type	Number of personnel	Shift duration	Total hours	Average wage/hour	Crew cost wages
CPJ	1.0	8.0	8.0	$11.00	$ 88.00
CPA	1.0	8.0	8.0	$ 9.00	$ 72.00
Total	2.0		16.0		$160.00

From this calculation, the payroll tax and insurance factor are allocated. This is applied by multiplying the burden factor by the total base wages, then summarizing the two amounts to arrive at the budget wage concern.

$160.00 * 0.3295 = $52.72
$160.00 + $52.72 = $212.72 wage concern
Average hours per shift: 16.0/2 = 8.00
Average wage per crew hour: $212.72/16.0 = $13.30

Labor crew function. Hardware and embed installation.

Trade type	Number of personnel	Shift duration	Total hours	Average wage/hour	Crew cost wages
CPA	1.0	8.0	8.0	$9.00	$72.00
Total	1.0		8.0		$72.00

From this calculation, the payroll tax and insurance factor are allocated. This is applied by multiplying the burden factor by the total base wages, then summarizing the two amounts to arrive at the budget wage concern.

$72.00 * 0.3295 = $23.72
$72.00 + $23.72 = $95.72 wage concern
Average hours per shift: 8.0/1 = 8.00
Average wage per crew hour: $95.72/8.0 = $11.97

Labor crew function. Concrete placement, abutment wall.

Trade type	Number of personnel	Shift duration	Total hours	Average wage/hour	Crew cost wages
CF-SP	1.0	8.5	8.5	$14.41	$122.49
OP	1.0	8.0	8.0	$12.00	$ 96.00
CF	3.0	8.0	24.0	$11.00	$264.00
LB	2.0	8.0	16.0	$ 8.50	$136.00
TM	1.0	8.0	8.0	$ 9.50	$ 76.00
Total	8.0		64.5		$694.49

From this calculation, the payroll tax and insurance factor are allocated. This is applied by multiplying the burden factor by the total base wages, then summarizing the two amounts to arrive at the budget wage concern.

$694.49 * 0.3295 = $228.83
$694.49 + $228.83 = $923.32 wage concern
Average hours per shift: 64.5/8 = 8.063
Average wage per crew hour: $923.32/64.5 = $14.32

Labor crew function. Rub and patch, concrete surface.

Trade type	Number of personnel	Shift duration	Total hours	Average wage/hour	Crew cost wages
CM	1.0	8.0	8.0	$11.50	$92.00
Total	1.0		8.0		$92.00

From this calculation, the payroll tax and insurance factor are allocated. This is applied by multiplying the burden factor by the total base wages, then summarizing the two amounts to arrive at the budget wage concern.

$92.00 * 0.3295 = $30.31
$92.00 + $30.31 = $122.31 wage concern
Average hours per shift: 8.0/1 = 8.00
Average wage per crew hour: $122.31/8.0 = $15.29

The average cost per crew hour for the determined labor crew will become the multiplier for the craft-hour production factor that determines the unit labor component cost.

The labor crew format must be given consideration and realistic production factors applied. Many labor crews may be structured in a similar manner but utilize different trade allocations and appropriations.

Equipment crew cost determination

Example. Equipment crew function, formwork placement and removal, abutment wall.

Equipment type	Number of units	Shift duration	Total hours
½-ton pickup (supervisor)	1.0	8.5	8.5
5-ton flatbed truck	1.0	8.0	8.0
45-ton crane (conventional)	1.0	8.0	8.0
Tools, formwork	<u>1.0</u>	8.0	<u>8.0</u>
Total	4.0		32.5

Example. Equipment crew function, formwork placement and removal, bulkhead.

Equipment type	Number of units	Shift duration	Total hours
Tools, formwork	<u>1.0</u>	8.0	<u>8.0</u>
Total	1.0		8.0

The exampled crews consist of the necessary and required equipment to perform the operations of both formwork and removal. Unlike the labor crews for these operations, the equipment crews remain the same for both functions of each operation.

Example. Equipment crew function, hardware and embed installation.

Equipment type	Number of units	Shift duration	Total hours
Tools, formwork	<u>1.0</u>	8.0	<u>8.0</u>
Total	1.0		8.0

Example. Equipment crew function, concrete placement, abutment wall.

Equipment type	Number of units	Shift duration	Total hours
½-ton pickup (supervisor)	1.0	8.5	8.5
5-ton flatbed truck	1.0	8.0	8.0
45-ton crane (conventional)	1.0	8.0	8.0
1½ cy concrete bucket	1.0	8.0	8.0
Tools, concrete	<u>1.0</u>	8.0	<u>8.0</u>
Total	5.0		40.5

Example. Equipment crew function, rub and patch, concrete surface.

Equipment type	Number of units	Shift duration	Total hours
Mortar mixer ¼ cy	1.0	8.0	8.0
Tools, concrete	1.0	8.0	8.0
Total	2.0		16.0

The exampled crews consist of the necessary and required equipment to perform each operation. These equipment crews were formed by visualizing the required operation or structuring it based on past experience and requirements.

The production is based on and assumes the equipment crew will be productive for the full shift. The pickup truck for the crew supervisor is allotted an additional ½ hour per shift, as with the supervisor. The usage and performance are factored on a full shift with an equal proportionate rate applied to each hour worked.

Once the equipment crew is structured, the associated cost is applied and calculated to arrive at a total cost for each designated shift, relative to the specific crew. This is calculated using the methods to determine the total cost concern for equipment ownership and operating, discussed in Chap. 2, Equipment Ownership.

The required cost allocations are the owning and operating allocation for each unit. The equipment cost concern is based on the defined operating hours, or hours worked, with no additional compensation for overtime. From the equipment ownership factoring, each unit is allocated an ownership rate based on an 8-hour shift. Any variation above or below this should be prorated accordingly.

Each example is structured using assumed owning and operating costs for each unit described, based on an 8-hour shift.

Equipment type	Units	Owning and operating cost	Cost/hour
½-ton pickup	1.0	$ 24.00	$ 3.00
5-ton flatbed truck	1.0	$ 36.00	$ 4.50
45-ton crane (conventional)	1.0	$420.00	$52.50
1½ cy concrete bucket	1.0	$ 10.00	$ 1.25
Mortar mixer ¼ cy	1.0	$ 18.00	$ 2.25
Tools, formwork	1.0	$ 62.00	$ 7.75
Tools, concrete	1.0	$ 56.00	$ 7.00

From the given information, the specific equipment crew can be cost allocated for the cost concern appropriate for the operation.

658 Takeoff and Cost Analysis Techniques

Example. Equipment crew function, formwork placement and removal, abutment wall.

Equipment type	Number of units	Shift duration	Total hours	Average cost/hour	Crew cost concern
Pickup	1.0	8.5	8.5	$ 3.00	$ 25.50
5-ton flatbed	1.0	8.0	8.0	$ 4.50	$ 36.00
45-ton crane	1.0	8.0	8.0	$52.50	$420.00
Miscellaneous tools, formwork	1.0	8.0	8.0	$ 7.75	$ 62.00
Total	4.0		32.5		$543.50 shift

Example. Equipment crew function, formwork placement and removal, bulkhead.

Equipment type	Number of units	Shift duration	Total hours	Average cost/hour	Crew cost concern
Tools, formwork	1.0	8.0	8.0	$7.75	$62.00
Total	1.0		8.0		$62.00 shift

Example. Equipment crew function, hardware and embed installation.

Equipment type	Number of units	Shift duration	Total hours	Average cost/hour	Crew cost concern
Tools, formwork	1.0	8.0	8.0	$7.75	$62.00 shift
Total	1.0		8.0		$62.00 shift

Example. Equipment crew function, concrete placement, abutment wall.

Equipment type	Number of units	Shift duration	Total hours	Average cost/hour	Crew cost concern
Pickup	1.0	8.5	8.5	$ 3.00	$ 25.50
5-ton	1.0	8.0	8.0	$ 4.50	$ 36.00
45-ton crane	1.0	8.0	8.0	$52.50	$420.00
1½ cy bucket	1.0	8.0	8.0	$ 1.25	$ 10.00
Tools, concrete	1.0	8.0	8.0	$ 7.00	$ 56.00
Total	5.0		40.5		$547.50 shift

Example. Equipment crew function, rub and patch, concrete surface.

Equipment type	Number of units	Shift duration	Total hours	Average cost/hour	Crew cost concern
¼ cy mixer	1.0	8.0	8.0	$2.25	$18.00
Tools, concrete	1.0	8.0	8.0	$7.00	$56.00
Total	2.0		16.0		$74.00 shift

Example:
Crew costing and formulating, abutment wall

The formula detailed under this heading shows the distribution and application of the production factoring method.

The cost breakdown summarizes the unit labor and equipment, along with the total required labor and equipment cost, independently and by total, for the operation.

The application of cost is derived from the production factors calculated by an average wage per craft-hour and the equipment crew cost per operation. The labor crew and equipment crew cost per shift are the total cost expended, derived from the labor and equipment formulations.

To determine the component operation budget, the total operation cost concern for the equipment requirement can be calculated by multiplying the shift cost by the required shifts for each operation.

The total cost for both labor and equipment is calculated by the wage and cost concern per shift times the number of required shifts. This format produces the total labor and equipment cost required to perform the given operation.

To arrive at the total labor cost, multiply the unit craft-hour production factor by the average wage per hour by the required operation quantity. To arrive at the unit labor cost, multiply the unit craft-hour production factor by the average wage per hour.

For the total equipment cost, multiply the cost per shift by the required shifts. For the unit equipment cost, divide the total equipment cost by the required quantity for the operation. To determine the average wage per craft-hour, multiply the labor cost per shift by the required shifts divided by the required craft-hours for each operation.

Operation. Formwork, place, abutment wall.

Required quantity	720.00 sf
Labor cost per shift	$823.30
Equipment cost per shift	$543.50
Total per shift	$1366.80
Required shifts	1.53
Required craft-hours	85.68
Average wage per craft-hour	$14.57
Labor production factor	0.119
Production per shift	470.56 sf
Unit labor 0.119 * $14.57	= $1.734 sf
Unit equipment $543.50/470.56	= $1.115 sf
Unit production cost	= $2.889 sf
Labor 0.119 * $14.57 * 720.0	= $1248.36
Equipment 1.53 * $543.50	= $831.56
Total cost, labor and equipment	= $2079.92

Operation. Formwork, remove, abutment wall.

Required quantity	720.00 sf
Labor cost per shift	$514.85
Equipment cost per shift	$543.50
Total per shift	$1058.35
Required shifts	0.66
Required craft-hours	21.12
Average wage per craft-hour	$15.84
Labor production factor	0.029
Production per shift	1100.00 sf
Unit labor 0.029 * $15.84	= $0.459 sf
Unit equipment $543.50/1100.0	= $0.494 sf
Unit production cost	= $0.953 sf
Labor 0.029 * $15.84 * 720.0	= $330.74
Equipment 0.66 * $543.50	= $358.71
Total cost, labor and equipment	= $689.45

Operation. Formwork, place, end bulkhead, abutment wall.

Required quantity	60.00 sf
Labor cost per shift	$303.13
Equipment cost per shift	$62.00
Total per shift	$365.13
Required shifts	0.56
Required craft-hours	13.44
Average wage per craft-hour	$12.63
Labor production factor	0.226
Production per shift	106.64 sf
Unit labor 0.226 * $12.63	= $2.854 sf
Unit equipment $62.00/106.64	= $0.581 sf
Unit production cost	= $3.435 sf
Labor 0.226 * $12.63 * 60.0	= $171.26
Equipment 0.56 * $62.00	= $34.72
Total cost, labor and equipment	= $205.98

Operation. Formwork, remove, end bulkhead, abutment wall.

Required quantity	60.00 sf
Labor cost per shift	$212.72
Equipment cost per shift	$62.00
Total per shift	$274.72
Required shifts	0.17
Required craft-hours	2.72
Average wage per craft-hour	$13.30
Labor production factor	0.045
Production per shift	352.00 sf
Unit labor 0.045 * $13.30	= $0.599 sf
Unit equipment $62.00/352.0	= $0.176 sf
Unit production cost	= $0.775 sf
Labor 0.045 * $13.30 * 60.0	= $35.91
Equipment 0.17 * $62.00	= $10.54
Total cost, labor and equipment	= $46.45

Exampled Component Cost Allocation

Operation. Place embeds, abutment wall.

Required quantity	40.00 ea
Labor cost per shift	$95.72
Equipment cost per shift	$62.00
Total per shift	$157.72
Required shifts	0.91
Required craft-hours	7.28
Average wage per craft-hour	$11.97
Labor production factor	0.182
Production per shift	44.00 ea
Unit labor 0.182 * $11.97	= $2.179 ea
Unit equipment $62.00/44.0	= $1.409 ea
Unit production cost	= $3.588 ea
Labor 0.182 * $11.97 * 40.0	= $87.14
Equipment 0.91 * $62.00	= $56.42
Total cost, labor and equipment	= $143.56

Operation. Place concrete, abutment wall.

Required quantity	43.20 cy
Labor cost per shift	$923.33
Equipment cost per shift	$547.50
Total per shift	$1470.83
Required shifts	1.08
Required craft-hours	69.12
Average wage per craft-hour	$14.32
Labor production factor	1.600
Production per shift	40.00 cy
Unit labor 1.60 * $14.32	= $22.912 cy
Unit equipment $547.50/40.0	= $13.688 cy
Unit production cost	= $36.600 cy
Labor 1.60 * $14.32 * 43.2	= $989.80
Equipment 1.08 * $547.50	= $591.30
Total cost, labor and equipment	= $1581.10

Operation. Rub and patch concrete, abutment wall.

Required quantity	420.00 sf
Labor cost per shift	$122.31
Equipment cost per shift	$74.00
Total per shift	$196.31
Required shifts	3.15
Required craft-hours	25.20
Average wage per craft-hour	$15.29
Labor production factor	0.060
Production per shift	133.36 sf
Unit labor 0.060 * $15.29	= $0.917 sf
Unit equipment $74.00/133.36	= $0.555 sf
Unit production cost	= $1.472 sf
Labor 0.060 * $15.29 * 420.0	= $385.31
Equipment 3.15 * $74.00	= $233.10
Total cost, labor and equipment	= $618.41

(Owing to the rounding in the mathematical calculations, the unit prices may vary from the computed component total.)

The labor and equipment summary has summarized the direct costs associated with each work operation. Each operation must be the full direct, field related costs to perform the listed operation, including any delay or incidental moving within the site for the operation. This is all costs directed to the work operation required to perform the task.

Material allocation: Abutment wall

After the production has been determined and the labor and equipment crews have been allocated, the direct and incidental consumed materials for the operation must be summarized and cost allocated.

The material costs associated with each operation are the direct permanent and indirect incidental materials and supplies required per specification. The cost of the formwork itself is allocated as formwork, which is treated as an asset to the company. The formwork is treated as equipment and has an ownership rate applied to it for its use. This allocation, detailed in the formwork section (Chap. 3), deals with the cost associated with the formwork purchase or construction and reuse disbursing.

The consumable formwork required, that will not be reused, is considered direct to the job as an incidental supply.

If there is a requirement for formwork that is not in inventory or is classified as specialty, the allocation of price or rental should be factored in the operation over any direct allocation.

The material listed is summarized by operation and allocated to the direct pay item(*), which is the volume of neat line cubic yards of concrete.

Operation.   Construct, abutment wall.

Direct material:	
*Concrete, 3500 psi (gross)	= 43.2 cubic yards
Indirect material:	
Embeds, form hardware	= 40.0 each
Form oil (2 oz per sf)	
((720 + 60) * 2)/128	= 12.19 gal
Formwork:	
Walls: 2×6 wood panel forms	= 720.0 sf
End bulkhead 2×4 wood panel forms	= 60.0 sf
Consumed lumber	= 40.0 lf
Incidental hardware, snap ties 1 each per 5 sf = 780 sf/5	= 156.0 ea
Nails: 2.5 lb per 100 sf = (780/100) * 2.5	= 19.5 lb

The material list is comprised of direct and indirect materials required to perform the operations. The direct material outlined and required by specification and the indirect material are required by the operation to be performed. All materials are priced with all applicable sales and use taxes included.

The formwork is detailed separately owing to the different structural size of the formwork and the intended reuse, or life, of the panels.

The consumed material list associated with the direct and indirect materials should be detailed as required for the operation. This method provides a clear and detailed tabulation for review and distribution upon becoming the successful bidder.

Material cost. Abutment wall.

Payment item and unit		= 40.0 cubic yards (plan)
Direct material:		
Required concrete = 43.2 cy @ $55.00 cy		= $2376.00
Subtotal		= $2376.00
Unit cost		= $ 59.40/cy
Indirect material:		
None required		
Subtotal		= $ 0.00
Unit cost		= $ 0.00/cy
Incidental material:		
Embed anchors	= 40.0 ea @ $6.00	= $ 240.00
Form oil	= 12.19 gal @ $4.50	= $ 54.86
Consumed lumber	= 40.0 lf @ $3.25	= $ 130.00
3′ snap ties	= 156.0 ea @ $2.25	= $ 351.00
Nails	= 19.5 lb @ $0.75	= $ 14.63
Subtotal		= $ 790.49
Unit cost		= $ 19.76/cy
Total consumed material cost		= $3166.49
Unit cost 40.0 cy		= $ 79.16 cy
Formwork (ownership):		
2×6 wood panels	= 720.0 sf @ $1.50	= $1080.00
2×4 wood panels	= 60.0 sf @ $2.25	= $ 135.00
Rental and purchase	= none required	= $ 0.00
Subtotal		= $1215.00
Unit cost		= $ 30.38/cy
Total formwork material cost		= $1215.00
Unit cost 40.0 cy (neat line)		= $ 30.38 cy

Once the operation items have been priced, the costs associated with each operation must be summarized. The labor and equipment cost per shift must be calculated by the number of required shifts to arrive at direct unit labor and equipment cost for the controlling pay item. The materials are then calculated in the same manner, giving a direct cost associated with the given pay item.

Summarized direct field cost

Pay item: 40.0 cubic yards concrete

Labor 40.0 cy	@ $81.213	= $ 3248.52
Equipment 40.0 cy	@ $52.909	= $ 2116.35
Permanent material 40.0 cy	@ $59.40	= $ 2376.00
Indirect material		= $ 0.00
Incidental material 40.0 cy	@ $19.76	= $ 790.49
Formwork 40.0 cy	@ $30.375	= $ 1215.00
Total cost		= $9,746.36
Unit cost 40.0 cy		= $ 243.66 cy

The ratio of surface contact area to the control cubic yards gives a prospective for a relationship of cost, both for the current job and for future reference. This is taken from the total form area of the operation against the neat line cubic yards for the operation. The same system can be utilizeddone with the cost ratio.

Surface ratio,
square foot per cubic yard

$$780.0 \text{ sf}/40.0 \text{ cy} = 19.5 \text{ square feet per cubic yard}$$

Cost ratio,
dollars per square foot

$$\$9746.36/780.0 \text{ sf} = \$12.50 \text{ per square foot}$$

The operation summary cost is derived by an accumulation of the different category subtotals throughout the worksheet detail and listing, in a format as shown.

These various subtotals are then divided by the controlling bid item quantity to arrive at a unit cost for each category or operation.

This is a summary of direct field expenses without contingency costs, project and corporate overhead, and net profit. From these direct field related costs of labor, equipment, materials, and supplies, the related project and corporate overhead, contingency factor, and the desired profit margin are allocated.

This example of operation cost analysis is only for comparison and review to detail how one should structure an operation component cost budget analysis.

Each situation is different and will have distinct controlling features specific to itself; therefore, an exact and detailed takeoff along with an individual cost analysis must be performed.

Part

5

Putting It Together

Sunshine Skyway Bridge, Tampa Bay, Fla. Structure type, cable stay, concrete segmental. Owner, Florida Department of Transportation. (*Courtesy of Figg Engineering Group.*)

Chapter 22 moves on to material specifications. Every project has specific requirements for materials used on and within that project. It shows the estimator the responsibilities of detailing specific components, both major and minor, and describes the differences and uses of each. This section illustrates how to distinguish the permanent or direct materials from the incidental materials and supplies.

In following this detailed format and procedures, Chap. 22 defines bid preparation. This forms an accurate accounting and summarization of the project takeoff quantities. It also encompasses the direct relationship with areas of work to be performed to the translated method of the owner's item summary; in other words, categorizing specific work task codes to the owner's pay items or, in some cases, a lump-sum bid. The estimator will be applying production factors to the quantities to arrive at craft-hours; determining subcontractors, suppliers, and other specialties needed; setting up the final worksheets and spreadsheets needed for the pricing stage; and establishing final prebid status and notifications to the subcontractors and suppliers for timely quotations of services. Again, this defines and illustrates the importance of obtaining a system of procedure with good detail and organization.

This section includes the final part of this book, and hard dollar cost. In this, it shows the areas of summarizing cost, detailing labor, equipment, and materials, and compiling direct costs related to a project. At this point, an estimator will be finalizing projected or "plug" prices of suppliers, finalizing subcontractors, and attaching contingencies, overhead costs, and profits. This section details the overhead areas of a construction company from the field level to the corporate level, what is involved in each of these areas, the general conditions of a project, bond costs, and special contingencies related.

This chapter demonstrates how to strategically apply the overheads needed, both direct job and company related, along with profit structures and review techniques and compiles a final checklist for the job. It also discusses how an estimator should professionally negotiate prices with subcontractors and suppliers and obtain a good but firm relationship.

Chapter 23, "Low Bid and Award," first deals with fundamentals and practices needed once you are a successful low bidder. It shows, again in sequential order, methods and procedures one should follow, giving detail to *the bid file preparation for the project team, owner negotiations, finalizing suppliers and subcontractors, and building a reputation to detailing specifics. Also detailed is the system needed for future data input from current projects and the maintaining of past bid files.*

Within Chap. 23, the vital importance of tracking data from projects and the accuracy required for a competent bid proposal are reviewed. As discussed in Part 1, this ultimately becomes the historical data from which you compile your costing and production factors for estimating. At present, it represents the direct job cost. This affirms the importance of job costing and establishing tracking codes. This method also maintains communication between the estimating team and the project management personnel for information such as: (1) was the project being built as it was bid? and (2) is there any adjustment that has to be made in the relationship thereto? This section reviews the performance required of the estimator and the procedures needed to thoroughly track and finalize a project.

Highlighted are the skills the estimator needs to acquire and thoroughly understand the challenge at hand, and habits and traits required to be successful.

Chapter 22

Material Specifications and Bid Preparation

Material Specifications

The material specifications for each project are outlined and defined in the standard specifications and guidelines for the project, or within the special project provisions.

The material requirements define the specific component composition details, fabrication requirements, handling and storage, installation specifications, and specific supplier and vendor requirements if applicable.

The designated grade, classification, and construction details pertaining to the actual material needs of the specific project must be analyzed during the quantity takeoff procedure. The many different types and grades of certain materials can greatly alter a bid proposal, with both specific composition and strength characteristics.

Completing a thorough and detailed material summary assures that competitive price quotations are obtained, the required delivery schedules and durations are identified, and an accurate listing of the required and specified material requirements for the project are detailed.

Bid Preparation and Closing

The preparation of the bid proposal consists of summarizing all the information, facts, decisions, and analysis compiled to this point.

Along with the quantity takeoff, labor and equipment crew determinations, schedule resources, and material identification, the specific project evaluation and conditions are recognized.

The project scope with regard to the type of the project and work component characteristics; the financial requirements and commitments along with the bonding and prequalification obligations; the time duration, including multiple mobilizations; interim phasing requirements; and total schedule responsibility must be reevaluated and confirmed.

From the component quantity takeoff, crew and equipment cost analysis, and material summary, a preliminary budget for the estimate can be formulated. Added to this will be the subcontractor requirement for the component items anticipated to be contracted out by others. This would include all specialty or other items not performed by your own forces.

For the preliminary estimate, material and subcontract prices not yet quoted firmly should be "plugged," that is, given approximate value of the specified material component or item, an example being redi-mix concrete that normally is quoted delivered to a project in the range of $50.00 to $55.00, which may be plugged for $52.50 per cubic yard until the firm and binding quote has been established. The plug number values must be systematically organized and summarized for reevaluation and correction at bid closing time.

At this stage of the preliminary estimate, a direct project cost has been established. This will be the basis for the overhead cost distribution operations. From this stage of the procedure, a second reevaluation of the project will is done, giving a more realistic view of the project, in both time duration and budget analysis.

The overhead cost distribution procedure is performed next. This includes the direct job overhead and the corporate overhead allocation. When applying associated costs to a project, three elements of cost and cost allocations exist.

The first element is direct project cost association. This element is the specific direct cost for each bid item or component item, that is, each cost that can be identified and associated with a particular item; if the item does not exist, neither does the cost. These costs are referred to as direct project costs.

The second element of cost is that which is directly related to the project and not specifically to a particular item within the project. Project costs are identified as the project mobilization and setup, project office and field staff, operating costs associated with the project, the bond and financing related costs for the project, special insurance or use tax requirements, and any other specific costs that may be required to operate and function the project. This cost element is referred to as project overhead and general condition requirements.

The third element for cost allocation is the corporate cost structure. This element of cost is present regardless of any ongoing projects. It is the cost of having and doing business. It is understood that as the field operations increase, so do the corporate overhead costs, but for general purposes, this cost element is considered as associated directly with the home office and not directly related or associated with any one particular project. These cost components include the company's real estate, property, and plant ownership; corporate wages, utilities, taxes, and insurance; travel expenses; the cost of bidding projects; financing and capital costs; and other specific costs associated directly with the general company operation and not any one specific project. This group of costs is identified as the corporate overhead costs. This element of cost is normally distributed and apportioned as a percentage applied to the project, possibly based on anticipated gross yearly revenue to cost allocation or prorated budget distribution based on project time duration.

At this time, the profit margin amount is determined and allocated. This is the anticipated net income revenue from the total project cost excluding all direct expenses, project, and corporate overhead allocations.

From this complete summary of associated costs, a complete estimated cost for the project is forming. A review at this stage identifies the project schedule, labor and equipment allocations, direct material and supply requirements, project and corporate overhead distributions, and the anticipated profit margin.

Once the estimator, owner, and managers are comfortable and satisfied with the direct bid, schedule, equipment allocation, and profit margin percentage, the final bid closing and hard dollar determination is prepared. At this point of the bid, the only main open item for closing should be the "plug," or estimated material and subcontract numbers. From the acceptance of quotations and proposals, of material vendors and subcontractors, the final agreed and responsible prices are entered in the bid, replacing the plug prices.

These prices are derived from the formal acceptance of prices from responsible negotiations between the prime contractor and the associated suppliers of materials and service subcontractors. This procedure is done in a professional and aboveboard manner so as to earn respect in reputation and business dealings for future projects. It also assures you of obtaining competitive and honest pricing for future projects.

The final stage of the bid proposal closing, preparation and review, along with the mathematical calculation checks, is to once again reassure yourself that all component costs are justified and allocated. This procedural format aids in the preparation, justification, and confidence of an accurate and competitive proposal to perform the requested work.

Chapter 23

Low Bid and Award

Low Bid

The reward of an accurate and competitive bid proposal is being determined the successful low bidder. This justifies the commitments demanding dedication to the reliability and factual input of project records, meticulous quantity takeoff procedures, and honest negotiations with suppliers and subcontractors.

Once the bidding stage has been completed, the analyzing of the data, quantity verification, and supplier determination begin. These methods must be performed within a reasonable period of time. Normally, from the time of the bid closing to the project award and notice to proceed, only a short time may be allotted, in some cases 30 days.

The estimator must maintain accurate and precise notes and records during the bidding process. Creating spreadsheets and tables of adjustments can help tremendously. Knowing the lowest and most responsible subcontractor and supplier during the bidding stage not only aids in becoming the low bidder initially but eases the task of analyzing the quotations afterward.

Commitment to the suppliers is an important and timely procedure. They committed to you for your bid proposal and deserve the same commitment if they were attributable in your obtaining the contract award. This will also expedite and maintain the project schedule as they can prepare for their work operations. Negotiating lower prices or price shopping after the fact is not a reputable method for the prime contractor to engage in. This may result in not obtaining the most competitive price quotation with future projects.

Establishing a procedure list is a good practice for low bid preparation. First and foremost, the operation production requirements should be established from the bid documents. These are the specific work operation components from which the budget and labor performance data will be produced.

This information will structure not only the project for daily performance, schedule, and project budget but also the component data for the historical record tracking for future reference.

The next procedure will be to analyze and determine the lowest, most responsible, and qualified subcontractors and suppliers. Obtain their formal and confirmed quotation, verify the contract terms, and commit the contract in the form of a *subcontract agreement* and/or *purchase order*.

From this a formal production schedule must be confirmed and established. This will orchestrate the project operation and daily production schedule, and redefine interim critical dates and milestones for the management team. It will also detail the equipment requirements and demands.

Along with this, and preferably from the onset, the management team must be formed. The project staff will start to initiate their input to third- and fourth-tier management and organization. A preconstruction meeting between the owner and prime and between the prime and subcontractor and supplier groups should be scheduled. Next would be the scheduling of an operation planning meeting to organize the specific details of the project.

The project site must be addressed next. This will establish the project office and yard location, mobilization requirements, access commitments, and issues specifically associated with the project site. A second visit and evaluation from the initial prebid visit would be recommended.

A well-organized and procedural bid proposal, project records, and detail to specific requirements will aid in the expedient and efficient project startup procedure.

Award

This book was structured to produce an effective method for quantity takeoff and bidding productively and competitively. The importance of accurate analysis, procedural steps and measures, and common sense is the basis of this book. The commitment to detail and to established fundamentals will prove to be an effective benefit to one's knowledge and development of estimating skills.

The requirement of accurate project tracking and record keeping will produce a tool of great importance for both current project budget control and future project performance estimating.

The secret to good estimating is experience, good judgment, attention to details, good historical and current job factors and cost data, and sustaining thorough knowledge of your marketplace in economical conditions and of your labor force and equipment availability.

A bridge becomes not only a link between two points but a link to world transportation, communication, and infrastructure.

The pride and accomplishment of the completed structure is a result of a committed and long-term endeavor of the estimating and construction team, "the building of a bridge," their signature, a permanent and lasting fixture. A bridge becomes the bold statement of a technological construction achievement.

Supporting Cause for Publishing

The signing of the Surface Transportation Act of 1992, and the Surface Transportation Bill of 1994 (Federal Highway Budget) has had a dramatic and favorable impact on the construction industry of this country. Through the six years of this bill, approximately 151 billion dollars will be spent on the nation's deteriorating highway and transportation system. The highway, railway, and bridge infrastructure is the backbone of this country and must be constantly maintained and upgraded for survival and growth to continue.

Some 270,000 of the nearly 600,000 bridges, a specific part of the highway system, are deteriorating rapidly. Much concern with the state department of transportation agencies is with the thousands of outdated, dilapidated bridges around the nation. The federal funding allocations will have a direct impact on this market arena. Highway departments are dealing not only with aging of the structures but also with the congestion areas of the fast-growing metropolises and their needs for future growth. Highway departments must face the future growth potentials of a demographic area along with rebuilding and restructuring a 30- to 40-year-old, 43,000 mile interstate highway system that has grown obsolete.

Rebuild America will be the motto of the civil oriented construction projects in the years to come. Bridges and structures will be a high-demand arena. It will take the combination and team effort of the current professionals along with the structuring and training of America's youth to overcome this challenge.

Nehansic Station Bridge, Somerset County, N.J. Structure type, lenticular truss bridge. (*Courtesy of Federal Highway Administration.*)

Glossary

AASHTO American Association of State Highway and Transportation Officials.

Absolute refusal The bearing value of a pile of which 200 percent of the required bearing is achieved determined by the dynamic formula method, or refusal to which damage will occur at continued driving.

Abutment The part of a bridge that supports the end of the span that meets the embankment. It also supports the embankment to keep it from sliding under the bridge.

ACI American Concrete Institute.

Admixture Material added to a concrete mixture to increase workability, strength, imperviousness, and freeze-thaw capability and to accelerate or retard the set.

Air content The volume of air voids in cement paste, mortar, or concrete, exclusive of space between aggregates. It is expressed in percent.

Air hammer A pile hammer that is operated by compressed air.

Air lift An air-injected pump used for raising material from a submerged area through an open discharged pipe. A suction method of cleaning out a foundation area or pile component, normally used in underwater conditions using compressed air and casing pipe.

Approach slab The roadway section that lies directly before or after the end span of a bridge, adjoining the deck section.

ASCE American Society of Civil Engineers.

ASTM American Society for Testing and Materials.

Augered piles Piles that are augered or drilled to a specified depth or tip rather than driven.

AWS American Welding Society.

Backwall The portion of the abutment substructure unit that is adjacent to, or within, the beam structure unit. It rests on the abutment and also retains the backfill material within the embankment.

Baffles Obstructions set in a surface, or a surface used for deflecting or dissipating material. They are usually in the form of a plate or wall.

Barrier railing (*See* Parapet.)

Bascule A moving span that rotates in a vertical plane about an axis that may be either fixed or movable. An apparatus in which one end is counterbalanced by the other.

Batter pile A pile that is angled from the vertical position to widen the area of support and to resist thrust and horizontal load.

Beam hanger A wire, sheetmetal strap, or other hardware device that supports formwork from the structural members.

Bearing capacity The maximum unit pressure which soil or a structural member will withstand without failure or settlement to an amount detrimental to the integrity of function of a structure.

Bearing pile A pile which carries a vertical load. It may be a friction pile or an end bearing or a tip pile.

Bearing seat The portion of a cap structure that supports the beam member, beam structure, or superstructure. An area onto which a bearing pad or assembly is placed.

Blow count The count within a defined increment of measure to which a pile is driven, counted as strikes by the pile hammer. Normally stated in blows per lineal foot.

Bond breaker A material used to prevent adhesion of a concrete to another surface.

Box beam A beam or self-supporting structural member in the shape of a box.

Bridge A single or multiple span structure, including supports, erected over a depression or an obstruction, such as water, a highway, or a railway.

Bridge deck The load-bearing floor of a bridge which carries and distributes the load to the beam structure.

Bridge length The greater dimension of a structure measured along the centerline between backs of abutments or ends of the floor.

Bridge width The clear width of a structure measured at right angles to the centerline of roadway, between parapets.

Bulb-Tee beam A concrete beam in which the top flange is larger in section than the bottom flange.

Bulkhead A horizontal or vertical partition in formwork. A closure form or end form at a construction joint.

Caissons A foundation system in which vertical holes are augered to the bearing strata and then filled with concrete.

Camber An upward curvature of a beam or structural member to compensate for anticipated or theoretical deflection when a load is applied. A rise or crown of the center of a bridge.

Chamfer A beveled corner, or depression, of a concrete structure, which is formed in the formwork.

Chill ring A metal ring insert conforming to the shape of a steel pile used to aid in welding splices. It is placed between the butt ends of the pile to be spliced.

Cofferdam A watertight enclosure structure, usually built of piles or earth, for which excavation and foundation work can be done.

Composite section A structural section of components interconnected so that the elements act together as a single flexural unit.

Compressive strength The measured maximum resistance of concrete to axial loading, expressed as force per unit cross-sectional area.

Concrete mix design (*See* Mix design.)

Continuous span A bridge superstructure span which extends over three or more given supports in a given direction.

Counterforts Vertical members, normally cast monolithically with the unit, used to stiffen walls and give support.

Cross frame An intermediate horizontal structural member which connects framing or main structural members laterally to make a continuous unit. A cross piece that transmits and diverts load and adds stability.

Cross section The section of a structure perpendicular to a given axis; the vertical end section view.

CRSI Concrete Reinforcing Steel Institute.

Dead load An inert, inactive load in structures caused by the weight of the members, the supported structure and permanent attachments of the deck unit.

Deflection A deviation from a straight line; the bending of a beam or structure under an applied load.

Design load The weight or other force for which a structure is designed to withstand; the worst possible combination of loads.

Design strength The load-bearing capacity of a member for the allowable stresses in design. The assumed values for the strength of concrete. The yield stress of a member for theoretical ultimate section strength.

Diaphragm A partition between chambers; a stiffener and distribution member between bridge girders.

Differing site conditions Site conditions and/or soils which change and vary from that which was specified.

Edge fascia An edge form used to shape the outside vertical parallel edge of a bridge superstructure and limit the horizontal spread of concrete.

Elephant trunk Slang term for a tremie. (*See* Tremie.)

Embedded item Components or "embeds" permanently placed within the cast-in-place concrete structure, other than form hardware such as mechanical and electrical supports, contingent or subsequent form system supports, or hangers.

End bent The pile supported end substructure unit of a bridge.

Equipment crew A composite equipment fleet designated for a specific component.

Expansion joint A device within a separation of two adjoining segments of a concrete structure that is provided to allow relative and independent movement caused by temperature changes.

Fabricated steel A structural steel member or girder which is independently fabricated for a specific supportive function.

Falsework A temporary structure, scaffolding, or shoring, erected to provide interim support during construction.

Fascia The outer edge or overhang portion of the deck section.

Fascia jacks A scaffold support suspended from the beam or superstructure used to form the overhang deck section.

FHWA Federal Highway Administration.

Fillet A narrow flat structural member which adds strength by avoiding sharp angles in connecting two members or parts of them.

Friction pile A bearing pile with supportive capacity produced by friction between the pile skin and the soil.

Fringe benefits The additional benefits and compensation provided for an employee by the company, e.g., health insurance or vacation.

Foot pound The unit of measurement of energy of a pile hammer delivered per blow of the ram of the hammer.

Foundation piling Structural piling supporting a foundation of a structure—wall, pier, abutment, footing, column, etc.

General conditions The part of contract documents, or bidding requirements of a company, which set forth recurring common and similar conditions or items within work.

Grade beam A horizontal load-bearing foundation member or end support which forms a foundation for a superstructure wall.

H-pile A structural steel pile member with an "H" cross section.

Hair pin A hairpin shaped steel unit used to drive pile.

Hammerhead pier cap A cap section of a pier unit horizontally cantilevered from a single column, forming a shape of a "T."

Haunch A poured concrete section that extends beyond a beam, both horizontally and vertically, to support the bearing deck. It also provides the variable adjustment area between the deck section and camber of the beam member. A thickened area near the support bearing area.

Jetting A method of utilizing high pressure water or air to sink piles when a pile hammer is not practical or functional. Used to assist a pile hammer or to move or remove obstructions or components.

Laitance A layer of weak and nondurable material containing cement and fines from aggregate; an improper or overworked surface finish.

Lateral support The method of supporting a structure from horizontal movement by columns, pilasters, battered piling, or cross walls.

Leads A vertically hanging or supported steel member or rack that contains the hammer and guides the pile to its proper alignment and location in a pile driving operation. Pronounced leeds.

Ledger A horizontal formwork member or stringer, attached to a beam side, that supports joists and other supportive members.

Live load The nonpermanent load to which a structure is subjected to in addition to its own weight . It includes people, equipment, and free-standing materials.

Mandrel In pile driving, a cylindrical inner member pushed through an outer skin to create a hole for a pile shell.

Median The center portion of a divided highway separating opposing traffic.

Median barrier A double-faced divider in the median of two adjacent highways.

Mix design A proportioned selection of ingredients for concrete. The proper and most economical mixture of cementitious material, fine and coarse aggregate, water, and admixtures to produce 27 cubic feet (1 cubic yard), to the desired and ultimate composite design strength and properties.

Modulus Unit of measure used in describing the strength of materials, by section.

Monolith A body of plain or reinforced concrete cast or erected as a single integral mass or structure.

Monolithic Concrete Concrete which is cast with no joints other than construction joints.

Multispan bridge A bridge consisting of more than one span.

Neat line A line defining the proposed or specified limits of an excavation or structure component.

Noncomposite section A structural section of nonconnected components in which the elements act independently from one another in flexural movement.

Operating costs The cost associated with conducting a function or operation to a specific component.

Parapet A structural railing component along the longitudinal edge of a superstructure unit, parallel with the roadway, extending vertically above the deck, used for containment of traffic or backfill.

Pay item A designated item of work within a contract which both parties agree will be paid for.

Payroll burden The payroll taxes and insurances (workers' compensation, liability, etc.) which are in addition to the base wages and benefits of an employee's wage. Add-ons for which the company is responsible.

PCI Precast/Prestressed Concrete Institute.

Pier cap The top part of a bridge pier unit which distributes the concentrated loads of the superstructure uniformly over the pier.

Pier column The vertical component of a bridge pier unit which supports the cap and distributes the concentrated loads of the pier cap uniformly to the pier footer.

Pier footer The bottom part of a bridge pier unit which distributes the concentrated loads from the pier unit uniformly over the foundation area or piling.

Pier strut A horizontal structural component between two columns providing lateral support from resisting pressures.

Pilaster A right-angled columnar projection with a capital and a base from a pier or wall; a square engaged pillar.

Pile A long slender structural shaped component of timber, steel, or concrete. It is either driven, jetted, or embedded into the ground to form a supportive element for the load bearing and load distribution of a structure.

Pile bent Two or more piles driven in a row transverse to the long dimension of a structure and fastened together with pile capping or bracing.

Pile cap A horizontal structural component placed on a pile or pile bent to uniformly transfer load to the piling.

Pile point A metal shoe or point attached to the tip of a pile to aid in soil or rock penetration and prevent damage to the pile tip.

Pile refusal The depth or point past which a pile cannot be driven because of geological structure.

Pile tip The bottom of a pile member.

Pipe pile A steel cylinder driven to firm bearing often filled with concrete or sand to form a structural bearing foundation component.

Posttension A method of prestressing reinforced concrete in which steel tendons are tensioned after the concrete has hardened.

Pozzolan A siliceous material, which in itself possesses little or no cementitious value but will, in the presence of moisture, chemically react to form compounds possessing cementitious properties.

Practical refusal The bearing value of a pile of which 150 percent of the required bearing is achieved determined by the dynamic formula method.

Precast A concrete member that is cast and cured in other than its final position and then transported and erected at the permanent site.

Prestressed concrete Concrete members in which internal stresses and distribution are introduced so that tensile stresses resulting from service loads are counteracted to a specified and desired degree; usually accomplished by steel tendons.

Proctor A method for determining the density-moisture relationship in soil for construction requirements and compaction.

Profile grade The trace of a vertical plane, in elevation or gradient, intersecting a particular surface of the roadway.

Ram weight The weight of the moving driving component of a pile-driving hammer.

Retaining wall A structural component used to retain earth.

Rolled steel A structural steel "I" beam made by hot rolling molten steel through a mold to form its cross-sectional shape.

Safety factor The ratio of the ultimate breaking strength of a member or material component to the actual working stress or safe load when in use or applied.

Scuppers Outlets within walls or deck for the overflow of water. A steel catch basin located at the low point of a bridge deck.

Set The penetration, usually in inches, of a pile into the ground per blow of the hammer.

Shear connectors In composite construction, a set of keys or members which are used to provide shear strength at the junction of dissimilar materials. Members which are capable of resisting deformation when loaded by in-plane shear forces.

Sheet pile A structural sheet type of pile used to form a continuous, interlocking vertical line or row. Steel or concrete that provides a tight wall to resist the lateral pressure of adjacent earth, water, or other structures.

Simple span A single superstructure span that offers no resistance to rotation at the supports.

Skin friction The resistance of soil surrounding a pile or caisson to its movement. It is proportional to the area of pile in contact with the soil and increases with penetration depth.

Slipform A method in which a concrete form is moved during the placement procedure that molds the concrete to its final shape; opposed to fixed formwork.

Slope pavement The component, normally of rock or concrete, used for erosion protection and stability of a slope.

Slump A measure of consistency of freshly mixed concrete.

Soil boring log The complete drilling record of soil borings pertaining to a particular project; used for foundation analysis.

Soldier pile A vertical pile member for supporting timbers against an excavation. Vertical wales for strengthening or alignment.

Specification A description of materials, workmanship, methods, and quality, required in a structure for specific components.

Specific gravity The ratio of the mass of a unit volume of a material at a stated temperature to the mass of the same volume of a gas-free distilled water at a stated temperature.

Spoon blows The blows of a 140-pound hammer falling 30 inches onto the sampler rods.

Stirrup A reinforcing device to resist shear and diagonal tension stresses in a beam unit, perpendicular to or at an angle to the longitudinal reinforcing.

Strand pattern The location of the prestressing tendons within the structural member. It is determined by design property and deflection.

Stringer A horizontal member supporting joists, resting on vertical supports or wales.

Stroke The distance, up or down, traveled by the ram of a pile hammer.

Structures Bridges, culverts, drainage components, retaining walls, cribbing, falsework units, and supportive units.

Strut A supporting piece. An inside brace resisting pressure along its length.

Subaquifer (Subaqueous) A water bearing bed or stratum of porous earth, gravel, or stone. Being under or beneath the surface of water. Formed or occurring in or under water.

Substructure The lower portion of a bridge structure unit, that which is below the bearings of simple and continuous spans, forming the section between the foundation and the superstructure or beam structure.

Superstructure The portion of a bridge structure that is above the bearing seats of simple and continuous spans, spring line of arches, or beam structure, which is carried by the supporting under structure components.

Test pile A foundation pile that is installed for the purpose of performing a load test to determine the load-bearing capability of the soil and pile, and the number and size of required piles.

Timber pile A foundation pile of specified wood of designated size and length, rather than of steel or concrete.

Tip elevation The elevation to which a pile is driven or predetermined to be driven.

Top slab elevation The finished grade of a bridge deck slab.

Tremie A pipe, articulated tube, or chute through which concrete may be deposited and placed under water or over long vertical distances. (Also referred to as an elephant trunk.)

Tremie seal Concrete placed under water, through a tremie and within a cofferdam or caisson, to form a seal for dewatering.

Vertical piles Piles which are driven in vertical position.

Viaduct An elevated roadway, with narrow arches of masonry or concrete supporting piers, over a ravine, waterway, or gorge.

Wale A horizontal member for supporting formwork.

Waler A horizontal or vertical brace used to support formwork.

Water-cement ratio The ratio of the amount of water, exclusive of that absorbed by aggregates, to the amount of cementitious material in a concrete mixture. It is stated as a decimal by weight by gallons of water (in pounds) per sack (94 lb) of cement.

Waterstops Thin sheets of rubber, plastic, or other nonpermeable material inserted within a construction joint to obstruct the seepage of water through the joint.

Wave equation A method used for determining the bearing value, driving and hammer properties, and resistance and reaction to the pile during the driving operation.

Web walls The cross wall connecting two adjacent walls or structural components.

Wingwall A secondary supporting wall extending from an abutment or other component structure to support or retain earth.

Working load The load imposed by workers, materials, and equipment.

X-dimension The variable vertical dimension between the top of a beam or girder and the bottom of the proposed deck (normally in the haunch area). The difference between the theoretical design elevation and the actual field elevation, which is determined from the actual camber in the beam minus designed deflection.

Yield The difference in the volume of material required to fill a known area to the amount actually used for the same area, including overrun and waste.

Index

Mississippi River Bridge, Cairo, Ill. Structure type, proportionate continuous truss arch. (Courtesy of Federal Highway Administration.)

AASHTO Gradation for Coarse
    Aggregates, 108
AASHTO Specifications for
    Concrete, 111
AASHTO Specifications for
    Structural Steel, 360
absolute refusal, 258, 679
absolute volume, 106, 108, 115,
    116
  analysis, 106
  method, 115
abutment, 274, 679
  backwall, 274, 288
  beam seat, 284, 355
  footer, 278, 286
  formwork, 28–32, 54, 83–85,
    280
  quantity takeoff, 285, 333
  unit, 285
  wall, 274, 287
  wing, 275
  wing wall, 275, 293
accessories:
  formwork, 90, 91, 93, 94, 100,
    102
  reinforcing steel, 581, 582
activity production factor, 25
add-on, 203
admixture, 67, 109, 679
  air entraining, 110
  calcium chloride, 110
  chemical, 106
  corrosion inhibiting, 110
  fiber reinforced, 110
  fly ash, 110
  latex, 110
  microsilica, 110
  nonchloride, 110
  pozzolan, 106
  retarding, 110
  superplasticizing, 110
aggregate:
  coarse, 106–108
  fine, 106, 111, 115, 116, 121, 122
air content, 107, 109, 114, 115,
    126, 679
air entraining admixtures, 110
air hammer, 229
air/steam hammer, 140, 141, 211,
    221, 225, 226, 229
air tests, 106
airlift, 165, 187, 188, 198, 679
  excavation, 165, 198
ambient air, 114
anchor assemblies, 101
anchor bolt, 278, 387
anchor device, 463
anchor set, 437
anchorage block, 431
anchorage device, 433
anvil, 227, 243, 259
applications for conventional
    formwork systems, 84
approach unit (slab), 519, 679
  concrete placement, 528
  curing, concrete placement, 529

approach unit (slab) (*cont.*):
  embedded hardware, 525
  finish machine, setup, dry run,
    527
  formwork, 524, 525
    interior bulkhead, doweled,
      525
    interior bulkhead, plain, 524
    perimeter bulkhead, 524
  foundation preparation, 523
  grade preparation, 523
  parapet and median:
    bulkhead, 531
    concrete placement, 531
    embedded hardware, 531
    formwork, 530
    rub and patch, 531
    slipform, 532
  placement preparation, 528
  quantity takeoff, 521
  reinforcing steel, 522, 527
  required placement location,
    525
  screed rail, 526
appurtenance, 585–591
arch bridge, xiv, xx, 27, 161,
    688
asphalt plank waterproofing, 620,
    627
assembly criteria, 69
asset cost, 98, 102
augered pile, 241, 243, 247, 679
average wage per hour, 17

backer rod, 621
backfill, 166–168, 174–176, 586,
    588, 595
backwall, 274, 288, 679
  concrete placement, 292
  dependent, 290
  formwork, 288–293
  independent, 288, 289
  quantity takeoff, 288–293
  substructure, 274
baffle, 431, 679
Bar Coding and Classification
    System, 552
bar list, 559
bar ratio, 558
bar ratio multiplier, 576
bar support, 562
  devices, 554
barrel pin, 362, 363
barrier (parapet, railing), 306,
    424, 425, 680
barrier railing (parapet), 424
bascule bridge, 4, 680
basic factoring (reinforcing), 575
battered pile (batter), 223, 224,
    244, 255–257, 680
bay span, 32, 442
beam, 339–347
  camber, 414, 474
  cast-in-place, 351, 352
  deflection, 414
  hanger, 442, 450–452, 680

beam, hanger (*cont.*):
  full, 422, 450, 451
  half, 450, 451
  length, 348–351
  monolithic, 356
  properties, 340–347
  seat, 274, 284, 302, 355
    abutment, 284
    pier cap, 302
  stirrups, 351, 356
  structure, 339
beams/girders:
  quantity takeoff, 366, 368, 379,
    408
  types:
    box, 342
    bulb-tee, 343
    channel, 346
    concrete, 340–348
    double stem, 347
    fabricated, 356
    I-beam, 341
    rolled, 356
    single stem, 345
    span lengths, 348–350
    steel, 355–358
    void slab, 344
  (*see also* girders)
bearing:
  area, 353
  capacity, 220, 680
  elastomeric, 360, 370, 389
  expansion, 360, 371, 389
  fixed, 360, 371, 389
  pad, 355, 360
    bituminous, 458, 459, 463
  pile, 199–201, 680
  plate, 355
  pot, 360
  rocker, 360, 389
  structural steel, 389
  treatment, 360
  value, 200
bearing and tip elevation, 219
  (*see also* tip elevation)
bedding material, 197, 584, 587
beds, 350, 433, 435
bent bar, 551
bid preparation, 159, 671, 675
bid preparation and closing, 671
bituminous bearing pad, 458, 459,
    463
blockout, 284, 424, 472
blow count, 200, 221, 236, 244,
    245, 680
blows per foot, 219, 227, 244, 245,
    253
board feet, 92
bolster, 553
bond breaker, 126, 680
bonding (requirements), 671
bonnet, 227
boot, 202
bottom flange, 374, 383, 463
bottom spotter, 225, 227
box beam, 340, 342, 680

Index

box girder, 446–449, 483–493
   section, 447, 484
   unit, 446
bracing, 67, 91, 95, 96, 98, 101
bridge:
   abutment, 274, 284–295
   approach slab, 519
   backwall, 274, 288
   beam, 339–347
   cap, 301
   continuous span, 382
   footer, 286, 295
   length, 153, 680
   pier, 277
   piling, 199–202, 206–211
   simple span, 379
   span, 348–352
   strut, 298
   substructure, 274, 285
   superstructure, 411
   width, 153, 680
   wing, 275
bridge deck, 32, 72, 417, 424, 480, 680
   waterproofing, 619
bridge types, 3
   cable stay, concrete segmental, vi, 667
   composite concrete arch, xiv
   concrete girder (prestressed), viii, 636
   continuous span, viii
   fixed span, 636
   lenticular truss, 678
   segmental (concrete), vi
   spandrel concrete arch, 27
   steel arch, xx, 161
   structural steel girder, viii
   suspension, 145
   tied deck arch, 161
   truss arch, 688
   wooden covered, 1
buildup, 215, 247, 249, 458, 513, 622, 629
bulb-tee beam, 340, 343, 680
bulk specific gravity, 115, 117
bulkhead, 284, 302, 305, 442, 524, 531, 680
   construction joint, 278, 419, 430
   doweled, 442, 471
   edge, 419
   end, 300
   expansion joint, 420, 471
   formwork, 284
   interior, 442, 470
   perimeter, 442, 470
   plain, 442, 470
   quantity takeoff, 294
   transverse, 419
burden factor, 20
butt weld, 210, 216, 384

cable stay bridge, 667
CAD welding, 555
caisson, 224, 241, 243, 247, 278, 680
calcium chloride, 110

calibrated wrench, 363
camber, 414, 474, 680
cantilevered deck (overhang section), 463, 467, 477
cap, 278, 301
cap component, 301
capblock, 227, 259
cast-in-place, 207
   beam structure, 373
   box girder:
      anchor block, 492
      bottom, 486
      diaphragm component, 490
      end wall, 489
      wall, 488
   concrete pile, 207
   concrete slope paving protection, 598, 601
   superstructure, 427, 491
casting bed, 433–435
categorizing components, 282
cell, 165, 187–189, 192
cement content, 106, 107, 109, 114, 115, 121
cementitious, 108–111
chair, 553
chamfer, 71, 83
chamfered corner, 280
channel beam, 346
chemical admixture, 106
chill ring, 210, 213, 216, 680
chinked, 599
chute, 118
clamshell, 165, 186, 187
cluster, 206, 244–246
CMP (corrugated metal pipe), 584
coarse aggregate, 106–108
cofferdam, 165, 186, 680
   airlift, 165, 679
   cell, 165, 187, 188, 192, 193
   dewatering, 165, 192
   excavation, 165
   quantity takeoff, 168
   sheet pile, 232, 233
cofferdam cell dewatering, 165, 192
cold joint, 120
cold poured silicone, 623
collection pipe, appurtenances, deck drain fixture, 591
colored finish, 124
column, 278, 296
   component, 297
   form system, 570
combination form, 280
combined multiplier (factor), 577–579
company structure, 4
component contact area, 32, 103
composite concrete arch bridge, xiv
composite deck section, 412, 413, 681
compression joint seal, 422
compressive strength, 105, 106, 108–110, 125, 127
   test, 106

concrete:
   AASHTO Specifications, 111
   admixture, 106
   aggregate:
      coarse, 106
      fine, 106
   air content, 107, 109, 114, 115, 126, 679
   bucket, 137, 138
   cement content, 106, 107, 109, 114, 115, 121
   composition, 108
   compressive strength, 105, 106, 108–110, 125, 127
   creep, 120
   curing, 124–126,
   diaphragm, 351, 372
   fill, 250, 266, 269
   finishing, 118, 134
   flowability, 105
   minimum required strength (28 day), 281, 427
   mix design (criteria), 105–107, 681
   pile, 203–207
   placement, 118, 120, 134
      deck unit, 476
   plastic state, 105
   portland cement, 106
   producing, 113
   properties, 115
   quantity takeoff, 282, 293, 301, 303, 306, 439–443, 521–528, 545
   retempering, 111
   sampling and testing, 126
   seal, 166, 208
   slump, 106, 108, 109, 111, 114, 115, 126
   specific gravity, 109
   strength, 106, 108–110
      minimum required (28 day), 281, 427
   temperature, 114
   testing requirement, 106
   time limitation, 112
   type, 106, 109
   unit weight, 107, 109, 115, 126, 127
   volume calculation, 114
   water-cement ratio, 107, 108, 113–115
   yield, 108, 109, 112, 117
concrete approach slab, 519
concrete beam charts:
   beam sections, 340
   box beam properties, 342
   bulb-tee beam properties, 343
   channel beam properties, 346
   double-stem beam properties, 347
   I-beam properties, 341
   single-stem beam properties, 345
   void beam properties, 344
concrete beam types:
   box beam, 342, 348

Index

concrete beam types (*cont.*):
  bulb tee, 343, 349
  channel beam, 346
  double-stem beam, 347, 349
  I-beam, 341, 348
  precast, prestressed concrete girder, 350
  single-stem, 345
  voided prestressed slab beam, 344, 349
concrete deck unit: rub and patch, 480
concrete pile lifting points, 205
concrete placement, deck unit:
  cantilevered overhang section, 477
  haunch section, 476
  required locations, 472
  substructure, 280, 438
  superstructure, 443
  uniform deck section, 478
concrete placing and finishing equipment:
  concrete bucket, 137, 138
  concrete pump, 134, 136
  deck finish machinery, 137
  feeder conveyor, 135
concrete precast deck formwork, 418
conduit, 351, 424, 431, 435, 438
conical point, 209
construction joint, 278, 419, 430
construction requirement (concrete), 149
contact area, 32, 103
continuous span, 382, 383, 681
  bridge, viii
contract, 675, 676
control joint, 278, 279
controlling component, 443, 449
conventional deck, 415, 452
  formwork, 416
  overhang deck formwork, 419
conventional form system, 449
  beam hanger assemblies, 450
  haunch section, 449
  interior deck formwork, 451
copper sheet, 621
copper waterstop, 624
corrosion, 110, 555, 557
  inhibiting admixture, 110
corrugated metal permanent steel formwork, 72
cost distribution, 15, 98, 99
counterfort, 308, 681
coupler, 555
covered bridge, 1
craft hour, 25
crane mat, 130, 187
cranes, 129
  types:
    conventional, 129, 132, 133
    crawler, 129
    hydraulic, 130, 132–135
    lattice, 130, 132, 133
    lift, 130

cranes, types (*cont.*):
  rough terrain, 133
  truck, 132
crash wall, 277
creep, 120, 430, 435
crew costing and formulating, 97
crew cycle time, 577–579
crew manpower, 24
crib, 165
cribbing, 163, 167, 429
cross frame, 359, 385, 386, 681
cross section, 452, 681
cubic yard, 107, 109, 110, 169
  calculation, 169
  concrete, 114
  earthwork, 169
curb and gutter section, 524, 532
curing compound, 111, 124–126
curing concrete, 125, 480
  concrete placement, 480
cushion material, 140, 221, 227
cut off, 203, 260
cycle time, 24
cycles per hour, 24, 25
cylinders, 114, 127

dampproofing, 620
datum line, 164
Dayton Superior (charts), 32–66, 86, 87, 416
dead load, 32, 426, 681
deck, 25, 30, 32, 70–72, 83, 88, 89, 91
  blockout, 475
  bulkhead, 420
    transverse, 419
  camber, 414
  cantilever, 463, 477
  composite, 413
  concrete placement, 438, 443, 472, 476
  conventional, 415, 416, 419, 452
  cross section, 447, 452
  curing, 480
  deflection, 414
  design, 29–32, 68–71, 85, 414
  drain:
    fixture, 475, 589
    outlets, 594
    scuppers, 585, 594
  edge fascia, 419
  embedded hardware, 471
  finish machine, 137, 139, 442
  formwork, 415, 425
  haunch, 413, 449, 476
  interior, 415, 416
  load, 426
  noncomposite, 414
  overhang, 417
  pan (deck), 456–458
  precast concrete panel, 417
  preparation, 443, 474
  quantity takeoff, 439, 512
  rub and patch, 480

deck (*cont.*):
  uniform section, 478
  unit, 412
deflection, 67–69, 414, 681
deformation, 550
deformed round steel bar, 550
deformed steel reinforcing bar, 558
  bar ratio, 558
  production base, 560
  tie method, 559
  tie ratio, 559
  (*see also* reinforcing steel)
density, 164, 166, 167
dentil, 179, 181
dependent backwall, 288, 290
depreciation, 11–13, 15, 99
design load, 32, 681
design strength, 681
detailing, 551, 563, 564
detension, 433, 436, 518
dewatering, 185, 198
  cofferdam cell, 165, 192
  normal, 189
  and subaquifer preparation, 185
diaphragm, 351, 359, 372, 681
  concrete, 351, 372, 401
    placement, 351, 355
  formwork, 352
  quantity takeoff, 372
  structural steel, 359, 384
  types, 373
diesel hammer, 141, 142, 211, 220, 225–227, 230
differing site conditions, 167, 681
dimension and camber, 414
dimensional tolerances, 438
direct, 9, 15, 23
displacement, 207
dolphin, 201
dosage, 110, 112
double-acting hammer, 141
  formulas, 238
double handling, 176
double ledger, 449, 451–453, 501
double-stem tee beam, 347
dowel, 564, 567
doweled, 442, 471
downspout piping, 595
drain:
  castings, 424
  fixture, 589
  system, 584
    subsurface, 583
    superstructure, 584
drainage pipe installation, 587
drainage trench:
  backfill, 588
  excavation, 586
drainage weep holes, 431
drape pattern, 350, 351, 433
drift pin, 362
drilled caisson, 241, 243, 247, 278
driving, 199–203
  condition, 201, 203
  head, 200, 209, 221, 227, 243
  method, 222, 251

driving (cont.):
  resistance (pile), 244, 268
drop hammer formulas, 236
dry excavation, 164
dry run, 123, 474
dry weight, 109, 116, 117
ductwork, 431
dummy joint, 598
dunnage, 271
dynamic pile formula, 258

earthwork and excavation takeoff, 169
edge and bulkhead fascia formwork, 420
edge fascia, 419, 681
edge form, 481
elastic compression, 435
elastomeric bearing pad, 360, 370, 389
elastomeric expansion, 424
elephant trunk, 681
elongation, 433
embanking, 168
embedded, 71
  hardware, 471
  item, 355, 681
encasing pile, 270
end bent, 274, 681
end bulkhead, 300
end treatment, 359
energy, 140–143
entry port, 621
epoxy, 550, 552, 553
  bonding, 404, 406
  coating, 552
equipment:
  concrete placing, 118
  crew, 21–23, 681
  formulation, 17, 18, 21
  lifting and hoisting, 129
  major and primary, 129
  minor, 143
  ownership, 10
  pile driving, 140
  placing and finishing, 118, 134–138
  secondary and support, 143
  specialty, 144
equipment crew:
  cost determination, 641
  determination, 21
  function, 22
estimating bearing capacity, 220
excavation, 163
  airlift, 165, 187, 188, 198
  clamshell, 165, 186, 187
  cofferdam, 165, 196
  cycle time, 173
  drainage trench, 586
  dry excavation, 164, 196
  foundation, 163–168, 179
  incidental, 167, 168, 171, 196
  pipe, 586
  quantity takeoff, 168
  rock: drilling, 164, 197

excavation, rock (cont.):
  excavation, 164, 197
  structure, 163–168
  unclassified, 164, 196
  undercut, 179–181, 197
  unsuitable, 164, 197
  wet, 164, 196
expansion end, 356
expansion joint, 420, 681
  device, 471
  bulkhead formwork, 471
expansive, 110
exterior overhang form system:
  beam hanger assemblies, 464
  deck formwork, 467
  haunch section, 464
  support bracket, 466

fabricated beam, 356
fabricated steel, 356, 681
factoring formula, 255
false decking, 70
falsework, 28–32, 67, 69–71, 83, 84, 391, 682
falsework and foundation criteria, 69
falsework systems for beam structures, 70
fascia:
  edge, 419, 681
  formwork, edge and bulkhead, 420
fascia section, 417
feeder conveyor, 135
felt membrane, 417
fiber reinforcing admixture, 110
filled joint, 422
fillet, 71, 83, 682
filter blanket, 584
filter fabric, 584
final grading, 184
fine aggregate, 106, 111, 115, 116, 121, 122
finish machine, 123, 137, 139, 442
  screed rail, 123, 473
  setup, dry run, 474
finishing exposed surfaces, 120
fixed bearing, 360
fixed end, 355, 371
fixed lead, 235
fixed span bridge, 636
flashing, 621
flat bearing, 360
flexible watertight gasket, 624
flowability (concrete), 105
fly ash, 106, 108–111, 116, 127
fogging (concrete), 123, 125
foot pound, 682
footer, 278, 286, 295, 303
form asset cost, 98, 102
form bolt requirement, 101
form liner, 280
form reuse, 89, 91
form setting, 546
form system, 29
  cantilevered, 30, 32, 70, 71, 88, 89

form system (cont.):
  component, 448
  conventional, 449–451
  edge form, 481
  Dayton Superior formwork charts, 32–66, 86, 87, 416
  Efco (Economy Forms Corp.) steel form system charts, 77–82
    column, round, 81
    column, square, 80
    handset, 82
    lite, 79
    plate girder, 78
    radius form, 77
  formwork, 29–103
  gang, 85, 88, 91
  hand set, 88, 91
  horizontal, 32
    bay span, 32
    dead load, 32, 441
    design load, 32, 441
    impact load, 32, 441
    live load, 32, 441
    overhang dimension, 32
    rated load, 32, 441
    safe working load, 32, 441
    safety factor, 32, 441
    ultimate load, 32, 441
  job built, 85
  metal, 77, 91, 99–101
  overhang, 442
  placement, 83
  prebuilt, 85
  removal, 83–85
  stay-in-place metal, 72
  timber criteria, 67
  vertical, 31, 54, 85
  void, 76, 77, 431
  wood, 90–94
forming, 90, 91, 93, 94, 100, 102
forming condition, 24
formwork:
  abutment, 30, 85, 88, 89
  accessories, 90, 91, 93, 94, 100, 102
  applications, 84
  asset cost, 98, 102
  bracing, 67, 91, 95, 96, 98, 101–103
  cantilevered, 30, 32, 70, 71, 88, 89
  cap, 70
  column, 67, 68, 70
  condition, 24, 68, 71, 77, 94, 100, 103
  contact area, 32, 103, 132
  conventional, 30, 84, 85, 94
  cost distribution, 99
  counterfort, 308, 309
  deck, 30, 32, 70–72, 83, 88, 89, 91, 425
  deflection, 67–69
  depreciation, 99
  design, 32–66
  diaphragm, 30
  embeds, 71, 83, 90, 91

Index

formwork (*cont.*):
  fascia section, 417
  footer, 69, 95, 101
  form asset cost, 98, 102
  gang, 30, 85, 88, 91
  girder, 69, 70, 72, 83, 84,
  horizontal (charts), 32–53, 67, 85
  interior, 30, 70, 83, 84
  loading, 32, 67, 426
  metal, 77
  minimum design wind presure, 31
  ownership cost, 96
  overhang, 417
  precast concrete panel, 417
  perimeter bulkhead, 442
  placement, 83
  quantity takeoff, 96, 102
  removable, 415
  setting, 129, 546
  stay-in-place, 72, 76, 415
  stringer, 70, 485, 486
  substructure, 30, 280
  superstructure, 30, 54
  type, 24
  vertical (charts), 31, 54–66, 85
  void, 76, 77, 431
  wale, 72, 85, 88, 89, 94, 95
  wood system, 71, 90–94
  yoke, 101
formwork removal guide, 84
foundation, 163–168
  detailing, 179
  grading, 546
  preparation, 166, 197
foundation piling:
  add-on, 203
  augered, 241, 243, 247
  battered, 223, 224, 256–257
  bearing, 199–203
  cluster, 206, 244–246
  concrete, 203–211
  driving, 140, 142, 143, 199–203
  friction, 199–201
  H-pile, 212
  load test, 220–222, 265
  pipe, 209
  point, 199, 217
  quantity takeoff, 195
  refusal, 200, 201, 203, 217, 219, 244, 258
  shell, 200, 208–211
  steel, 212
  test, 265
  timber, 200–202
  tip, 199–201
  tube, 209
  vertical, 223
freeboard, 192
friction, 199–201
friction pile, 247, 682
fringe benefit, 9, 15–17, 19, 20, 682

gang form system, 88
gasket, 624, 629
general condition, 672, 682

girders, 355–360
  bolts, 362
  cast-in-place, 351, 352
  cross frame, 359, 385, 386
    bolt, 385
    weld, 386
  diaphragm, 359, 384
    bolt, 384
    weld, 385
  handling, 378
  quantity takeoff, 365
  splice, 359, 383
    bolt, 384
    butt weld, 384
  stringer beam:
    bolt, 387
    weld, 387
  weight, 363
  (*see also* beams/girders)
grade beam, 278
grade preparation, 523
grid, 549, 553
grillage, 433
gross, 13, 31, 68
gross area, 442, 481–483, 485
ground displacement, 207, 209
grout (grouting), 437
  mixture, 623
  pads, 355, 417, 458–460, 463, 514
grouted riprap slope protection, 604
grouted rock, 597
grouted stone riprap, 599

H pile, 212
hair pin, 682
half beam hanger, 451
hammerhead, 278, 682
hammerhead pier cap, 682
hand chip and dentil work, 181
hand-set, 88, 91, 280
  panel, 88
hanger assemblies, 450–452
haunch section, 413, 442, 449, 476, 682
header, 599, 601–603
historical data, 5
historical production factors, 25
hoisting, 129, 133, 134
hold-downs, 350
horizontal formwork design charts, 32–53
  form system, 32, 67, 85
hot poured, 623
hydraulic boom, 132–134

impact load, 32
independent backwall, 288, 289
indirect costs, 2, 9, 15, 16
indirect material, 263, 333, 406, 510, 544, 579, 594, 613, 633
injection equipment, 622
insert, 372, 388, 389, 591
insurance factor, 17, 20
interior:
  bay, 442

interior (*cont.*):
  component, 448
  deck, 25
    bulkhead, 419, 442, 444, 470–475
      plain, formwork, 470
      doweled, formwork, 471
    concrete placement, 442–446, 448, 476
    deck preparation, 443–445
    finishing, 442, 443
    formwork systems, 441, 451
    haunch area, 442
    interior span, 442
    joint, 419, 442, 444
    loads, 441
    overhang, 417, 442
    surface, 454
  diaphragm, 374, 403, 431
  section forming, 415
  stiffener, 359
intermediate strut, 298
I beam, 341

jacking:
  beam, 222, 265
  force, 248, 265
  platform, 265
jetting (method), 203, 223, 682
job built (formwork), 85
joints:
  bulkhead, 284, 442
  cold, 114, 120, 124
  compression, 422
  construction, 278, 419, 421, 424, 430
  contraction, 414
  control, 278–279
  doweled, 442
  drawing, 279
  dummy, 598
  elastomeric, 424
  expansion, 420
  filled, 422
  fillers, 623, 628
    cold poured silicone, 623
    foam, 623
    hot poured, 623
    poured, 623
    preformed, 623
  gasket, 629
  horizontal, 618, 619
  keyed, 420
  keyway, 278
  mortar, 623, 628
  open, 420
  optional, 278
  silicone, 623
  steel, 422
  waterstop, 627

labor costs, 9, 15
  direct, 9

694 Index

labor costs (*cont.*)
  indirect, 9
labor crew, 9, 97
  cost determination, 639, 649
  determination, 18
  formulation, 17, 18, 21
  function, 639, 649
laitance, 90, 122, 123, 682
lap splice, 553
lateral eccentricity of prestress, 434
lateral stability, 31, 201, 219, 224, 235, 296
lateral support, 207, 264, 277, 386, 682
latex (admixture), 110
lattice boom, 130, 132, 133
laydown yard, 213
leads, 140, 143, 224, 225, 235, 243, 682
  fixed, 224, 235, 243
  spotter, 225, 235, 243
  swinging, 224, 235, 243
ledger, 449, 451–453, 501, 682
lenticular truss bridge, 678
lift, 85, 88, 89, 306
lift pour (retaining wall), 306
lifting cable, 351
lifting crane, 241, 248
live load, 32, 85, 88, 89, 426, 682
load:
  bearing, 351
  capacity, 69, 70
  dead, 32, 426
  deck, 441
  live, 32, 85, 88, 89, 426, 441
  pile, 235
  safe working, 32, 441
  test, 220–222, 265
  transfer, 30, 433
  ultimate, 32, 433, 441
loading, 30, 32, 67, 69, 70, 91
locks, 234
loose rubble, 604
low bid, 675

machine riprap, 599
magnetic particle, 555
major equipment, 129
mandrel, 208, 209, 259, 683
masonry, 603
  precast concrete stone slope protection, 603
mass, 106, 115, 119
mat, 553
material allocation:
  approach unit, 543
  beam structure, 406
  piling, 263
  reinforcing steel, 579
  slope protection, 612
  structure drainage, 593
  substructure, 332
  superstructure, 509
  waterproofing and joint filler, 633
material specification, 671

matting, 429
mechanical coupling, 249
mechanical splice, 216
median, 683
  barrier, 443, 683
  rail, 443–446
  (*see also* parapet and median)
membrane:
  curing compound, 126
  filter fabric, 584
  waterproofing, 618, 626
    with asphalt plank protection, 620
    with mortar protection, 620
metal:
  angle support, 461
  deck, 421, 455–457
  formwork system, 77
    stay-in-place beam attachment device, 454
    stay-in-place deck pan, 457
    stay-in-place support angle, 456
  taper ties, 101
microsilica admixture, 110
minimum design wind presure, formwork, 31
minimum required percentage 28-day strength:
  substructure, 281
  superstructure, 427
minor equipment, 143
mix design, 105–109, 683
  criteria, 107
modulus, 208, 683
monolithic, 428
  beam, 428
mortar, 624, 628
multispan, 683
  bridge, 683

natural units, 601
neat line, 167, 683
negative movement, 412
net area, 442, 453, 468
netting, 515
nonchloride admixture, 110
noncomposite deck, 414
noncomposite section, 414, 683
normal dewatering, 189
normal time limitations, concrete delivery, 112

oakum, 624
open joint, 420
operating cost, 12, 683
operation time analysis, 172, 173
optional joint, 278
ordinary surface finish, 121
outlet fixture, 595
overhang dimension, 32
overhang (fascia section), 417, 442, 480
overhang gang system, 419, 514
ownership cost:
  equipment, 10

ownership cost (*cont.*):
  formwork, 96
  operating cost format, 12

painting, 363, 392
  structural steel, 410
parapet, 306, 424, 443, 530, 683
parapet and median:
  bulkhead, 494
  concrete placement, 495
  embedded hardware, 494
  formwork, 494
  rail, 493, 530
  rub and patch, 495
  slipform, 495
pay item, 683
payroll burden, 15, 16, 683
payroll tax, 16, 17, 19, 20
penetration, 200, 201, 203, 207, 208
perimeter bulkhead: formwork, 470
permanent bolt, 363, 384–387, 409
permanent material, 590
permanent pile, 200, 249, 265
pier unit, 6, 277, 294
  beam seat, 284, 302, 355
  cap, 301, 683
  column, 278, 296, 683
  concrete placement, 280
  formwork, 80, 81, 294–303
  footer, 278, 295, 683
  quantity takeoff, 294, 333
  strut, 298, 683
  unit, 6, 277
pilaster, 683
pile:
  absolute refusal, 258
  batter, 223, 224
  bearing, 219
  bent, 213, 249, 684
  buildup, 215
  cap, 243, 684
  cluster, 206, 238, 244–246, 248, 253
  coated, 201, 222
  cutoff, 214
  elevation, 214, 215, 249
  driving, 129, 140, 206, 236, 683
  dynamic formula, 258
  extensions, 216, 236, 248, 270
  factoring formula, 236–238, 255
  hammer, 140–143, 199–203, 225–230, 242–245
    (*see also* pile hammer)
  jetting, 223
  lateral stability, 201, 219, 224, 235
  layout, 264
  leads, 140, 143, 235
  lifting points (concrete), 205
  load test, 201, 213, 220–222, 265, 271
  modulus, 208
  penetration, 201, 203

Index

pile (*cont.*):
  platform, 265, 266
  point and shoes, 217
  practical refusal, 684
  prebore, 223, 251, 255, 257, 261
  production cycles, 209
  properties, 233
  quantity takeoff, 238, 250, 256
  redriving, 215, 249, 250
  refusal, 200, 201, 203, 219, 244, 258, 679, 684
  resistance, 143, 203, 205, 244
  section and sizes (concrete), 204
  set, 143, 221, 236–238, 259
  shoes, 217
  skin friction, 199–201, 203
  splice, 215
  stinger, 249
  stripping, 270
  supported, 69, 199, 274, 277
  template, 226, 234
  tip elevation, 200, 201, 203, 219
  toe, 233, 236, 269
  types (*see* pile types)
pile factoring formula, 236–238, 255
pile hammer:
  air/steam, 140, 141, 221, 225, 226, 229
  anvil, 227, 243, 259
  blow count, 200, 221, 236, 244, 245
  bonnet, 227
  cushion material, 140, 221, 227
  diesel, 140–142, 211, 220, 225–227, 229, 230
  energy, 140–143, 243
  evaluation formula, 259
  foot pound, 682
  hair pin, 682
  ram, 141, 142, 202, 243
  ram weight, 211, 221, 225
  resistance, 143, 203, 205
  striking parts, 141, 142, 220, 221, 226, 229
  stroke, 140–142, 243
  vibratory, 142, 143, 214, 231–233
pile hammer formulas, 143, 236, 237, 238
  double-acting hammer, 238
  drop hammer, 236
  dynamic, 258
  estimating bearing capacity, 220
  evaluation, 259
  single-acting hammer, 237
pile refusal, 203, 244, 679
  absolute, 200, 201, 219, 258
  practical, 217, 684
pile types, 201
  augered, 241, 243, 247
  cast-in-place, 207
  concrete, 203
  drilled, 207
  H pile, 212
  pipe, 209

pile types (*cont.*):
  permanent foundation, 200
  precast, 204, 240
  prestressed concrete, 203, 240
  sheet, 231
  shell, 200, 208–211
  steel (H pile), 212
  test, 265
  timber, 201
pipe:
  backfill, 586, 588
  bedding, 584, 586–588
  CMP, 584
  concrete, 584, 585, 591, 594
  drainage, 583–595
  excavation, 586
  installation, 586–589, 592
  pile, 209–211
  plastic, 586, 686
  PVC, 583
  quantity takeoff, 585–589
  substructure, 583, 586
  superstructure, 583, 584, 586, 589
pipes, conduits, and ducts, 431
placing and finishing (concrete), 118, 125, 134
placement of formwork, 83
plastic (concrete), 105, 120, 124, 126
plastic waterstop, 621, 624
platform, 265
plyform (system), 90–92
plywood, 415, 417, 452, 453, 468, 469
pneumatic, 118
ponding (concrete), 125
portland cement, 106
position dowel, 371
posttension, 351, 436, 684
  cable, 350, 351, 431
  drape pattern, 350, 351, 433
  duct, 431
  elongations, 433
  grouting, 437
  grillage, 433
  jacking, 433
  stressing, 431
  strand, 555
pot bearing, 360
pouring sequence, 426
pozzolan, 106
practical refusal (pile), 684
prebore, 223
prebuilt (formwork), 85
precast clip, 456, 461
precast concrete panel, 417, 458
  formwork system:
    bituminous bearing pad, 458
    deck panel unit, 462
    hanger assemblies and devices, 460
    metal angle support, 461
    nonshrink grout pad, 459
premolded, 422
prestressed (concrete), 339–347

prestressed (concrete) (*cont.*):
  concrete pile, 214
  girder bridge, viii, 636
prestressed precast concrete beam, 350, 366
prestressing, 340, 350, 351
  cables, 340
pretensioning, 434
pretie, 560, 559–561
primary production factors, 24
proctor, 684
production:
  base, 19, 22, 560
  cycle, 331
  unit factoring, 7, 10, 24, 174
productivity factor, 24, 25
profile grade, 684
protection:
  deck unit, 481
  false decking, 481
  netting, 482
protective boots, 202
pumping (concrete), 118, 119
purchase order, 676
PVC, 583, 585, 586

quantity requirement, 24
quantity takeoff:
  abutment, 285, 333
  approach unit (slab), 521
  backwall, 288–293
  beams/girders, 366, 368, 379, 408
  bulkhead, 294
  cofferdam, 168
  concrete, 282, 293, 301, 303, 306, 439–443, 521–528, 545
  deck, 439, 512
  diaphragm, 372
  excavation, 168
  formwork, 96, 102
  foundation piling, 195
  girder, 365
  pier unit, 294, 333
  pile, 238, 250, 256
  pipe, 585–589
  reinforcing steel, 557, 561–563, 580
  retaining wall, 303, 333
  sheet pile, 256
  slope protection, 600, 614
  structural steel, 365, 408
  substructure, 281, 294, 303, 333
  superstructure, 439
  waterproofing, 625, 634
  waterstop, 625, 634
  wing wall, 293

radiography, 555
ram weight, 141, 142, 211, 221, 225, 684
random riprap, 599
rated load, 32
rebar, 557, 558, 560–563
  insert, 355
  mat, 571

rebar (*cont.*):
  standard bending details, 551
  standard lap lengths, 553
  tying, 557, 559
redriving, 215, 249, 250
refusal, 200, 201, 203, 219, 244, 258
reinforcement unloading, storage, rehandling, 562
reinforcing steel, 551
  accessories, 581
  bar list, 559
  bar ratio, 558
  bar supports, 554
  basic factoring, 575
  bending details, 551
  bent, 551, 557
  combined multiplier, 577–579
  epoxy, 550, 570
  fabricated, 550
  fabrication ratio, 577
  inserts, 552
  lap, 553
  lap lengths, 553
  layout, 562
  mat, 553
  plain, 549
  pretie, 560
  production base, 560
  properties, 550
  quantity takeoff, 557, 561–563, 580
  splice, 553
  stirrup, 351, 356, 365, 567
  straight, 551
  support, 554, 562
  tie, 557
  tie method, 559, 576
  tie ratio, 559
  weight calculation, 568, 569
  wire fabric, 582
  wire rope, 550
  (*see also* rebar)
removable form, 415
removal of formwork, 83, 281
required placement locations (concrete), 472, 525, 609
retaining wall: bulkhead, 305
  concrete placement, 306, 309
  counterfort, 308
  footer, 303
  formwork, 308, 310
  lift wall, 306
  panel, 305
  parapet, 306
  quantity takeoff, 303, 333
  unit, 303
  wall, 305–309, 684
retarding admixture, 106, 110
retempering (concrete), 111
reuse factor for formwork, 91
ring gasket, 624
ringers, 193, 233
riser pad, 278, 335, 337, 338
roadway seat, 290–292
roadway surface finish, 122

rocker, 360, 389
rocker bearing, 360
rock drilling, 164, 197
rock excavation, 164, 197
rolled beam, 356, 365
rolled steel, 684
rough terrain crane, 133
rubbed and patched, 281, 426
rubbed finish, 121
rubber composite, 370, 390
rubber waterstop, 621
rubble stone and riprap slope protection, 599, 604

safety factor, 32, 684
safety netting, 515
safe working load, 32
sandblasted finish, 122
scaffolding, 85, 88, 89
screed rail, 473
scribing, 290, 291, 375, 376
scrubbed finish, 122
scupper, 584, 585, 591, 592, 594, 684
seal concrete, 166, 208
section and sizes of concrete pile, 204
section modulus, 70, 201, 485
segmental concrete bridge, vi, 667
segregation, 105, 110–112, 118, 119, 124, 211
selected removal, 183
shear connector, 360, 390, 684
sheet pile (permanent and temporary), 231–234, 684
  bracing, 233, 234
  cantilevered, 233
  cofferdam, 165, 186
  designs, 232
  locks, 234
  properties, 233
  quantity takeoff, 256
  ringer, 233
  strut, 233
  wales, 233
shell pile, 224
shift duration, 24
shim plate, 370, 390
shoring, 31, 69, 84, 89, 163, 165
shrink, 171, 174, 175, 179
shrinkage, 110, 120
sidewalk finish, 124
sieve, 108
silicone, 623
single-acting hammer, 141, 237
  formulas, 237
single-span bridge, 277, 379, 685
site and pad preparation:
  concrete beam, 368
  structural steel girder, 380
skin friction, 199–201, 203, 685
slab deck, 516
slipform, 443, 495, 685
slope protection:
  concrete placement, 601–603

slope protection (*cont*):
  components, 597
    cast-in-place concrete slope paving, 598
    grouted riprap, 599
    header 598, 600
    loose rubble, 604
    machined riprap, 599
    masonry, 603
    random riprap, 599
    rubble stone, 599, 604
    toe header, 598
  formwork, 598, 601–603
  foundation preparation, 601
  pavement, 597, 685
  quantity takeoff, 600, 614
slump, 106, 109, 685
  test, 106, 109
snap ties, 91, 95
soil boring log, 246, 254, 685
soldier pile, 685
solid deck, 428
span: length, 348–350
  single, 68, 277, 379, 685
spandrel concrete arch bridge, 27
specific gravity, 106, 109, 115–117, 685
splice detail, 359
splices:
  pile, 215
  girder, 383
  rebar, 553, 555, 558
spoil, 166, 247
spoon, 251, 253, 260
spoon blow, 685
spotter, 225, 227, 235, 243
spray finish, 124
sprinkling (concrete), 114, 125
square feet, 25, 90, 92, 93, 96, 98, 99, 102, 103
standard bending details (rebar), 551
standard lap lengths (rebar), 553
stay-in-place form, 72, 76, 415
  design criteria, 72, 73, 75, 76
steel:
  beams, 355–357
  cross frame, 359, 385, 386
  diaphragm, 359, 384
  fabricated, 356
  pile, 199–226
  rolled, 356
  sheet pile, 232, 233
  stringer, 386
  structural, 356
  unit weights, 363
  (*see also* reinforcing steel; structural steel)
Steel arch bridge, xx, 161
steel bar reinforcing, 549
steel criteria, 68
steel girder, 356–358, 379, 381
  bridge, viii
  sections, 358
  splices, 383
steel H pile details, 212

Index

steel joint, 422
steel pile:
  details, 232
  properties, 233
steel shell pile, 207, 208
steel wire rope, 549
stinger, 249
stirrup, 351, 356, 685
stockpiling and rehandling, 378, 392
straight bar, 551, 560, 563
strand pattern, 685
strapping, 217
stressing, 428, 431–437
  bar steel, 582
  cable, 428, 431
  strand, 555
striking part, 237
stringer, 386, 685
stringer beam, 386
stripping, 85, 90, 91
stroke, 140–142, 243, 685
structural bonding, 365, 621
structural steel:
  beams, 355–357
  bearing assemblies, 389
  bolts, 362, 384–390
  cross frame, 359, 385, 386
  diaphragm, 359, 384
  fabricated, 356
  girder, viii, 356, 379
  interior stiffeners, 359
  painting, 363, 392
  quantity takeoff, 365, 408
  rolled, 356, 365
  splice detail, 359
  stringer beam, 386
structure collection pipe, 588
strut, 70
  foundation prep, 193
  piling, 233
  substructure, 298
  superstructure, 484
subaquifer, 185, 186, 685
subcontract agreement, 676
substrata, 201
substructure:
  abutment, 274, 285, 287
  backwall, 274, 288–293
    dependent, 289
    independent, 288
  beam seat, 284, 302, 355
  bulkhead, 300
  cap, 301
  column, 278, 296
  counterforts, 308, 309
  drain system, 584
  footer, 278, 286, 295, 303
  formwork, 280–308
  joints, 278, 279
  pier unit, 277, 294
  quantity takeoff, 281, 294, 303, 333
  retaining wall:
    footer, 303
    panels, 305

substructure, retaining wall (*cont.*):
  parapet, 306
  unit, 303
  strut, 298
  unit, 285
  wing wall, 275, 293
subsurface drain system, 583, 584
subsurface ground unit, 586
sump, 195
superelevated, 278
superplasticizing admixtures, 110
superstructure:
  barrier, 424
  box girder, 446
  cantilever, 477
  cast-in-place, 427
  concrete placement, 443
  deck section, 412, 449
  drainage, 584
  edge form, 481
  formwork, 32–53
  haunch, 413, 476
  joints, 419
  overhang, 417, 478
  parapet, 424, 443
  pile supported, 274, 277
  quantity takeoff, 439
  reinforcing steel, 561
  unit, 411, 512
support angle, 76, 454–457
support brackets, 442, 465, 466, 468
support components, 89, 93, 94, 96, 101, 102
support equipment, 143
surface finish (concrete), 120
surface preparation, 601
suspension bridge, 145
swell, 171, 174, 175, 179
swinging leads, 235, 243

takeoff, 24, 25, 96, 102
taper tie, 101
tare, 172
target slump, 109, 115
temperature (concrete), 114
template, 234, 235
temporary bolt, 362
temporary bracing, 378, 391
temporary falsework, 391
  scaffolding, 483
  deck section, 485
  requirement, 447
tendon, 436–439
tensile rod, 365
tensile strength, 274, 362, 433–435, 557
test cylinder, 114
test pile, 221, 265, 685
testing requirement (concrete), 106
threaded rebar insert, 355
through bolt, 355
tie method, 559, 576
tie ratio, 559
  multiplier, 576

tie-rod assembly, 377, 400, 402
tied deck arch bridge, 161
timber criteria, 67
timber pile, 201, 686
time depreciation method, 99
time limitations (curing concrete), 112
tined texture, 124
tip elevation, 200, 201, 203, 219, 686
toe, 165–167, 192, 233, 236, 269
toe header, 598
tooled finish, 122
top flange, 366, 383
top slab elevation, 686
transverse bulkhead, 419
traveler, 415
tremie (placement), 118, 119, 686
tremie concrete, 119
tremie seal, 686
trimming, 234, 249
trough, 118
truck mounted crane, 132
truss arch bridge, 688
tub girder, 360
turn of nut, 363
tying (rebar), 557, 559
types of piling, 201
types of structures, 3

ultimate load, 32, 441
unclassified excavation, 164, 196
undercut, 179–181
underrun, 176
underwater placement (concrete), 119
uniform deck section, 478
uniformity, 106, 107, 127
unit cycle time, 24, 25, 97
unit of measurement use allocation, 99
unit weight, 107, 109, 115, 126, 127
unsuitable excavation, 164, 197
upheaval, 166, 207, 245

vertical formwork charts, 54–66
  form system, 31, 85
  load, 67, 199, 281
vertical pile, 223, 686
viaduct, 686
vibratory hammer, 142, 143, 214, 232, 233
void form, 76, 77, 431
volume calculations:
  concrete, 114
  earthwork, 170
volume proportioning device, 112

wage concern, 17
wale (waler, wales), 72, 85, 88, 89, 94–96, 193, 449–453, 686
walers, double ledger members, stringers, 452
water-cement ratio, 107, 108, 113–115, 686

waterproofing:
  asphalt plank, 620, 627
  bridge deck waterproofing, 619
  dampproofing, 620
  membrane waterproofing, 618, 626
    with asphalt plank protection, 620
    with mortar protection, 620
  quantity takeoff, 625, 634
waterstop, 422, 620, 686
  copper, 624
  copper sheet, 621
  flashing, 621
  gasket, 629
  joint, 623, 627
  oakum, 624

waterstop (cont.):
  plastic, 621, 624
  rubber, 621, 624
  quantity takeoff, 625, 634
water jetting, 223
wave equation, 219, 686
web, 68, 70, 72, 409
web wall, 277, 686
weep hole, 431
  piping, 431
weight calculation, 568, 569
welded wire fabric, 555
wet excavation, 164, 196
wing wall, 275, 293, 686
  concrete placement, 280, 283, 287
  formwork, 280–308
  quantity takeoff, 293

wire brush, 122
wire rope, 555
wooden covered bridge, 1
wooden joist waler, 449
wood form, 85–88, 90–98, 280
wood stringer, 485, 486
workability (concrete), 106, 107, 110, 115, 123
workers compensation, 640, 652
working load, 32, 441, 686

x-dimension, 414, 415, 474, 476, 686

yield, 68, 108, 109, 112, 117, 686
yoke, 101

ABOUT THE AUTHOR

David Nardon has more than 23 years of experience in the construction industry ranging from buildings to segmental bridges. He estimates and manages highway, bridge, dam, airport, and marine projects for the KIEWIT Construction Group, a leading global civil based construction firm. Throughout his career, he has been responsible for and managed the largest reconstruction projects in both Michigan and Tennessee, and the longest four lane bridge constructed in North Carolina. Three of his projects have set records of utilizing the longest precast concrete beams for a project in various states. Since 1984, Mr. Nardon has been responsible for the estimating and construction of some 200 bridge structures, and has estimated over 470 projects in 26 states totaling over 2.2 billion dollars.